contact info: 2013 26@r

Science 9

Authors

Unit 1
Donald Plumb

Unit 2
Bob Ritter

Unit 3
Edward James

Unit 4
Alan J. Hirsch

Program Consultant
Marietta (Mars) Bloch

Contributing Authors

Skills Handbook
Nancy Dalgarno Alldred
Stephen Haberer

Nelson
Thomson Learning™

Australia • Canada • Denmark • Japan • Mexico • New Zealand • Philippines
Puerto Rico • Singapore • South Africa • Spain • United Kingdom • United States

Contributing Writers
Ron Thorpe

Jim Dawson

Program Assessment Consultants
Damian Cooper

Nanci Wakeman-Jones

Expert Review
Wojciech Fundamenski
Institute of Aerospace Studies, University of Toronto

David Logan
York University, Faculty of Pure and Applied Science

John Percy
University of Toronto, Department of Astronomy

Safety Review
Peter Au

Materials Review
Paul Hannan

Nelson Science 9 **Project Team**
Publisher: Susan Green

Project Manager: Beverley Buxton

Marketing Manager: Kevin Martindale

Developmental Editors: Colin Bisset, Ruta Demery, Julia Lee, Michael Redhill

Copy Editors: Geraldine Kukuta, Rosemary Tanner, Wendy Thomas

Proofreader: Shirley Tessier

Photo Research: Maureen de Sousa, Peggy Ferguson, Karen Taylor

Designers: Peggy Rhodes, Monica Kompter

Senior Composition Analyst: Marnie Benedict

Managing Editor: Theresa Thomas

Production Coordinator: Renate McCloy

Illustrators
Cynthia Watada

Myra Rudakewich

Deborah Crowle

Bart Vallecoccia

Jane Whitney

Frank Netter

Nelson
Thomson Learning

1120 Birchmount Road
Scarborough, Ontario M1K 5G4
www.nelson.com
www.thomson.com

Copyright © 1999 Nelson, a division of Thomson Learning. Thomson Learning is a trademark used herein under license.

Printed and bound in Canada
4 5 6 7 8 9 0 /ML/ 8 7 6 5 4

Canadian Cataloguing in Publication Data

Main entry under title:
Nelson Science 9
Includes index.
ISBN 0-17-612032-7

1. Science – Juvenile literature. 2. Science – Experiments – Juvenile literature. I. Plumb, Don, -. II. Title: Science 9.

| Q161.2.N45 1999 | 500 | C99-931363-0 |

Reviewers

Table of Contents

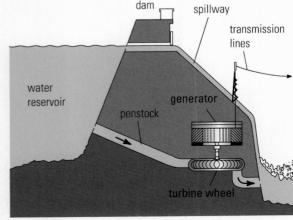

dam

spillway

transmission lines

water reservoir

penstock

generator

turbine wheel

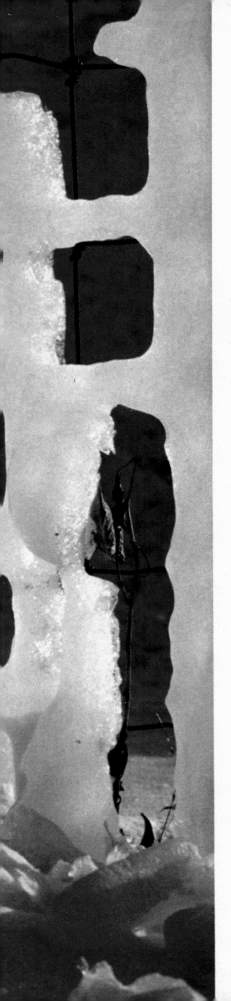

Matter

Unit 1 Overview

What do stars and dust have in common? What do they have in common with the thousands of substances we encounter every day? All of these things are matter. Learning about matter helps us to understand why ice melts, rocket fuel burns, and cars rust. Understanding matter helps us make decisions about manufacturing and using products.

1. Matter and Change

Matter has characteristic physical and chemical properties. Understanding these properties can enable us to make useful products.

In this chapter, you will be able to:
- conduct experiments to identify physical and chemical properties of matter

- observe and classify changes that can occur in substances

- handle chemicals safely

- describe how properties of matter can lead to useful technology

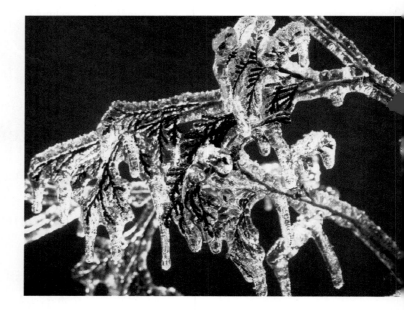

2. Elements and Compounds

Everything is composed of pure substances, which can be classified as elements or compounds. We classify matter to organize the vast amount of information about elements and the thousands of compounds in our world.

In this chapter, you will be able to:
- describe different classifications of matter

- interpret models, symbols, formulas, and names to describe pure substances

- assess some economic and environmental aspects of the production of these substances in Canada

3. Models for Atoms

The ultimate building block of matter is the atom. Throughout history scientists have tried to understand the behaviour of matter by developing models of the atom.

In this chapter, you will be able to:

- describe the nuclear atom and compare it to other models of the atom

- conduct experimental tests to identify atoms based on their atomic structure

- describe examples of technology that have developed from our knowledge of the atom

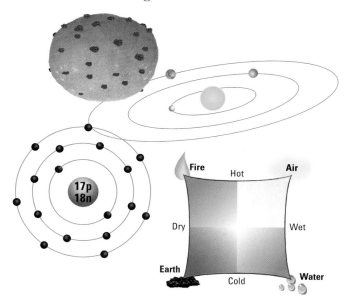

4. The Periodic Table

Properties of substances can be explained by organizing elements into a table. The positions of the elements in the table can be used to predict the properties of elements and the types of compounds they will form.

In this chapter, you will be able to:

- relate properties of elements to their atomic structure and to their position in the periodic table

- investigate ways of organizing elements into families or groups

- describe and identify properties and applications of elements that make them valuable substances

Challenge

In this unit, you will be able to... demonstrate your learning by completing a Challenge.

Building from the Past to the Present

As you learn about elements and compounds and ways of explaining the behaviour of matter, think about how you would accomplish these challenges.

1 **Models for Matter**

Build a display to represent the evolving models of matter.

2 **Marketing Matter**

Create a proposal that promotes a new material to a manufacturing company.

3 **A "Time Machine" Simulation**

Prepare a presentation for Dmitri Mendeleev that explains the evolution of the Periodic table since 1869.

To start your own Challenge see page 128.

Record your ideas for the Challenge when you see

Challenge

H																	He
Li	Be											B	C	N	O	F	Ne
Na	Mg											Al	Si	P	S	Cl	Ar
K	Ca	Sc	Ti	V	Cr	Mn	Fe	Co	Ni	Cu	Zn	Ga	Ge	As	Se	Br	Kr
Rb	Sr	Y	Zr	Nb	Mo	Tc	Ru	Rh	Pd	Ag	Cd	In	Sn	Sb	Te	I	Xe
Cs	Ba	La	Hf	Ta	W	Re	Os	Ir	Pt	Au	Hg	Tl	Pb	Bi	Po	At	Rn
Fr	Ra	Lr															

Matter and Change

1 What is matter? Matter is the leather in a soccer ball and the air that is used to inflate it. Matter is anything that has mass and takes up space. Inquiring about the nature of the visible world often starts with observations of matter and leads to attempts to organize those observations. What makes the particles different from each other? What are some different ways in which matter could be classified? Each kind of matter has different properties. Make a list of adjectives that describe some of these properties.

2 How can we use the various properties of matter? In March 1999, Bertrand Piccard and Brian Jones were very glad of the special properties of their balloon. The two European adventurers became the first people to fly a balloon all the way around the world, non-stop. Their balloon was kept aloft by a lighter-than-air combination of helium and propane-heated air. The balloon's silver-coloured envelope was made of carbon fibre kevlar: very light and strong. Cocooned inside a pressurized capsule, they took 19 d, 21 h, and 55 min to fly eastward around the globe. Why did they use a combination of gases to provide the "lift"? Why were carbon and kevlar chosen for the envelope? Why was the cabin pressurized? An understanding of the properties of matter can help you answer these questions.

3 Imagine you are sitting around a campfire with some friends. You put a pot of water near the fire. When the water begins to bubble and steam, you add some powdered hot chocolate mix and stir until the powder dissolves. Meanwhile, your friends are toasting hotdogs and marshmallows. Suddenly, one marshmallow catches fire, burning brightly for an instant. Your friend blows out the flame and looks at the black crispy chunk left on the stick. Several changes took place around this imaginary campfire. You made a mixture of several substances with water. Some water changed state from a liquid to a gas. But the marshmallow did not change state—it was still a solid at the end. However, the marshmallow seemed to become something else: a black substance. There was a change in the chemical makeup of the marshmallow.

In a small group, brainstorm a list of as many changes in matter as you can. Use a different action word for each change (for example, concrete *hardens*, wood *rots*, snow *melts*, paper *yellows*, fireworks *explode*). In which changes do you think a new substance is formed? Which changes do you think add materials to the air? Indicate these with a check mark and an asterisk, respectively.

Reflecting

Think about the questions in **1,2,3**. What ideas do you already have? What other questions do you have about matter and how it changes? Think about your answers and questions as you read the chapter.

Try This   Fill a Balloon

Examine some samples of baking soda and vinegar. Add one spoonful of baking soda to a glass holding two spoonfuls of vinegar. Describe the change that occurs. Use what you observe to design an apparatus that can blow up a balloon without using your breath. Produce a labelled diagram of your apparatus. As you do this activity, consider the following questions:

1. Describe each substance you started with.

2. What happened when the vinegar and baking soda were mixed?

3. How would you describe the products of the change?

4. If you were evaluating another group's balloon-inflating apparatus, what characteristics would you be looking for in the group's product?

Chemicals and Safety

Matter includes both helpful and harmful solids, liquids, and gases. For example, oxygen is a gas that all animals must take in to survive, but nitrogen dioxide gas from car exhaust is poisonous. Among liquids, water is essential for human survival, but sulphuric acid can cause serious burns if it contacts skin. We put salt crystals (sodium chloride) on our food, but crystals of sodium cyanide are a lethal poison.

How do we know whether or not a given substance is safe to use? How do we work safely with any chemical in the laboratory, at home, or at work? One source of information about hazardous substances is the warning symbols that are placed on containers of potentially dangerous materials. **Tables 1** and **2** show these symbols. In this activity, you will learn about these warning symbols, and you will identify hazards in your own laboratory. You will also learn how to deal with hazards and set some rules for working safely.

Materials

- samples of laboratory chemicals with WHMIS labels
- samples of household products with hazardous product labels

Procedure

Part 1: Safety Scavenger Hunt

1. Copy **Table 3** into your notebook.

2. Look around you and find each of the safety devices listed in **Table 3**. Record the location of each safety device in the table along with any information you found with the device.

3. Examine the samples of chemicals and household products provided. For each sample, record the name of the product, the active ingredient (if listed on the label), and a description of the warning symbol.

| Table 1 | Hazardous Household Product Symbols |

The warning symbols on household products were developed to indicate exactly why and to what degree a product is dangerous.

poisonous   flammable   explosive   corrosive

danger

warning

caution

| Table 2 | WHMIS Symbols |

The Workplace Hazardous Materials Information System (WHMIS) symbols were developed to standardize the labelling of dangerous materials used in all workplaces, including schools. Pay careful attention to any warning symbols on the products or materials that you handle.

compressed gas

dangerously reactive material

flammable and combustible material

biohazardous infectious material

oxidizing material

poisonous and infectious material causing immediate and serious toxic effects

corrosive material

poisonous and infectious material causing other toxic effects

Part 2: Setting Safety Rules

4. In groups, or as a class, record answers to the following questions.
(1B)

    (a) Must safety goggles be worn for all activities? Explain.

    (b) Are normal glasses ever an acceptable substitute for safety goggles?

    (c) Should you inform your teacher if you receive a cut during an activity? Explain.

    (d) What should you do if you have an allergy, medical condition, or physical problem?

    (e) Is it acceptable to bring food or drink into the science classroom?

    (f) When can you touch, taste, or smell chemicals in a science activity?

    (g) What should you do if you accidentally touch chemicals during normal activity?

    (h) What procedure should you follow if chemicals accidentally splash into your eyes?

    (i) How should you dispose of chemicals, broken glass, or sharp scalpels?

    (j) What should you do if any of your equipment is damaged or defective?

5. In your notebook, draw a map of the route your class should follow when the fire alarm sounds.

Table 3   Safety in Science—Precautions and Devices

Hazard	Some precautions	Safety device	Location of device	Device information
Fire	• tie back loose clothing and long hair • avoid contact with the flame or the hot part of the burner • avoid sudden movements	• fire extinguisher • fire blanket • fire alarm switch	?	?
Splashes	• wear safety goggles • wear an apron • avoid sudden movements	• safety goggles • aprons • eye wash station	?	?
Broken glass	• keep glass containers away from edge of counters • avoid sudden movements	• beaker clamp • disposal container for broken glass	?	?

Understanding Concepts

1. Why is it important to standardize safety symbols?

2. Briefly describe the procedure to follow if exposed skin comes in contact with any chemical substance.

3. Do you think it is always safe to pour waste chemicals and solutions down the sink with lots of water? Explain.

4. Think about safety precautions that are taken in different industries.

    (a) Why do people wear face masks when spraying a car?

    (b) Why do hairdressers wear gloves when using chemicals to straighten hair?

    (c) Why do firefighters wear breathing apparatus when entering a burning building?

Exploring

5. Investigate the containers of products in your home for hazardous product symbols.

    (a) Make a table to summarize your findings, with headings: Brand Name of Product, Type of Product, Type of Container, Hazard Symbol.

    (b) Do you see any similarities in the types of containers that are used for hazardous materials? Explain.

    (c) Do you see any similarities in the types of hazards that seem to be associated with particular groups of products? Explain.

Challenge

What types of hazards or safety issues should you consider when marketing a new substance?

Properties of Matter

When you choose your clothes, your lunch, or your shampoo, you are making choices based on the properties of matter. Considering how important these properties are to our daily lives, it's not surprising that people have always been curious about matter and how it changes. Through observation, scientists have found it useful to categorize properties as physical or chemical.

Physical Properties

When you observe matter—whether you see it, touch it, hear it, smell it, or taste it—you are observing its characteristics, called physical properties. A **physical property** is a characteristic or description of a substance that may help to identify it. Unlike a chemical property, a physical property does not involve a substance becoming a new substance. For instance, colour is a physical property. A substance simply has a certain colour: its colour has no relationship to the substance's ability to change into new substances.

Some physical properties that can be observed by using your senses are summarized in **Table 1**. Pick one of the materials shown in **Figure 1** and describe it, mentioning all of the properties listed in **Table 1**.

There are other physical properties you might choose to describe. Simple tests and measurements can aid your senses in observing these properties.

Table 1

Physical Properties Observed with the Senses

Property	Describing the Property
colour	Is it black, white, colourless, red, blue, greenish-yellow...?
texture	Is it fine, coarse, smooth, gritty...?
odour	Is it odourless, spicy, sharp, burnt...?
lustre	Is it shiny, dull...?
clarity	Is it clear, cloudy, opaque...?
taste	Is it sweet, sour, salty, bitter...?

Figure 1

The States of Matter

One of the physical properties of matter is its state—whether it is solid, liquid, or gas at room temperature.

	Solid	**Liquid**	**Gas**
Example			
Shape	Definite: has a fixed (unchanging) shape.	Indefinite: always takes the shape of its container.	Indefinite: always takes the shape of its container.
Volume	Definite: has a fixed volume.	Definite: has a fixed volume.	Indefinite: always fills the entire container.

Hardness

Because they are harder than glass, diamonds are used to cut glass. **Hardness** is a measure of the resistance of a solid to being scratched or dented. A harder material will scratch or dent a softer one. For instance, a diamond stylus is used to cut a large sheet of glass into different sizes. Rank the following substances by hardness: steel nails, chalk, glass, diamond.

Malleability

Gold can be hammered into thin sheets, so it is said to be **malleable**. If a solid is malleable, it can be hammered or bent into different shapes. Aluminum foil is malleable, which makes it useful for wrapping food as it cooks. Many materials, glass for example, are not malleable. Instead of flattening out when hammered, they shatter. **Brittle** objects shatter easily.

Ductility

One of the reasons copper is used for electrical wiring is that it can be drawn out into long, thin wires. If a solid is **ductile**, it can be pulled into wires. What other materials can you think of, besides copper, that are ductile?

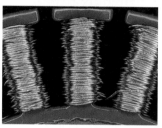

Melting and Boiling Points

The temperatures at which substances change state are characteristic physical properties. For example, under controlled conditions, water always changes from solid ice to liquid water at 0°C—its melting point is 0°C. Similarly, the boiling point of water, when it changes from liquid to vapour, is 100°C.

Crystal Form

Solids can exist in different forms. **Crystals** are the solid forms of many minerals in which you can see a definite structure of cubes or blocks with a regular pattern. For example, when you look closely at salt crystals, you can see that they are tiny cubes.

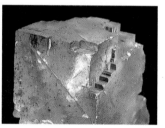

Solubility

When salt and pepper are added to water, the salt dissolves but the pepper does not. **Solubility** is the ability of a substance to dissolve in a solvent such as water. Salt is described as soluble and pepper as insoluble. Drink mixes, for example, contain powdered substances that are soluble in water.

Viscosity

Maple syrup is "thicker" than water—it flows more slowly than water when you pour it. **Viscosity** refers to how easily a liquid flows: the thicker the liquid, the more viscous it is.

Density

When people describe lead as "heavier" than feathers, what they really mean is that lead is more dense than feathers. **Density** is the amount of matter per unit volume of that matter. This is usually expressed in kilograms per cubic metre (kg/m^3) or grams per cubic centimetre (g/cm^3). For example, the density of water is $1.0 \ g/cm^3$. (It can also be expressed as g/mL and g/L.)

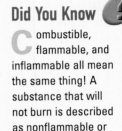

Chemical Properties

In nature, substances often combine or react with each other. When one substance can interact with another, that characteristic behaviour can be called a chemical property. For example, dynamite explodes when exposed to a flame because the dynamite combines with oxygen in the air. This reaction produces new substances. A **chemical property** describes the behaviour of a substance as it becomes a new substance.

Did You Know ❓

Combustible, flammable, and inflammable all mean the same thing! A substance that will not burn is described as nonflammable or noncombustible.

Combustibility

Combustibility is a property that describes the ability of a substance to react with oxygen to produce carbon dioxide, water, and energy. When a flame is brought close to a mixture of gasoline and air, the gasoline ignites and burns. However, water not only does not burn, but can be used to put out some fires. If a substance is **combustible** or **flammable**, it will burn when exposed to a flame. A substance that will not burn is described as nonflammable. What other materials can you think of that are combustible?

Try This   Just How Thick—Comparing Viscosity

You will need an apron, safety goggles, rubber gloves, a stopwatch, a marble, and a graduated cylinder to compare the viscosity of water, several cooking oils, syrups, and liquid detergents.

As a measure of viscosity, you can time how long it takes the marble to fall from the top of each cylinder to the bottom when the cylinder is filled with each of the liquids. (If you use water in your first trial, that will give you a standard for comparing the other liquids.)

1. Before you get started, which liquid do you think has the highest viscosity? Comment on the accuracy of your prediction: which liquid is the "thickest"?
2. Could you identify a liquid based on your data? Try it out: ask a friend to bring you an unidentified liquid, and then use the marble test to identify the liquid.

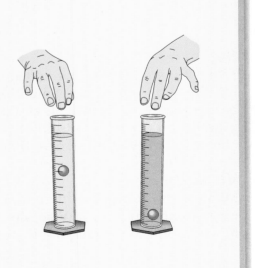

Reaction with Acid

When magnesium metal is added to acid, it produces bubbles of gas and the metal quickly disappears. However, when gold is added to acid, no visible change occurs. The ability of a substance to react with acid is a chemical property. For example, geologists use acid to test samples of rock. A chemical property of limestone is that it reacts with acid to produce bubbles of gas.

Using the Properties of Matter

Matter can be grouped as metals and nonmetals. Metals are suitable for different uses because of their special properties.

Metals have been used by people for thousands of years: first copper, then bronze, iron, and steel. Now, many different mixtures of metals, called **alloys**, are used. Whatever the purpose, whether for airplane parts, the bottoms of cooking pots, or braces for teeth, the metal chosen has properties useful for the job.

The metals used in the braces in **Figure 2**, for instance, must have specific chemical properties: they must not react with saliva or chemicals in food. They must also have specific physical properties. Some of these are shown in **Table 2**.

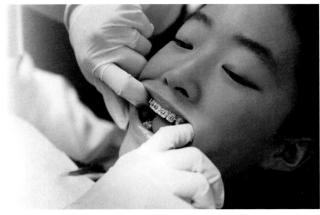

Figure 2
Braces should not be made from toxic metals!

Table 2	Physical Properties of Some Metals Used in Braces				
Metal	Stiffness	Springiness	How easily does it bend?	How easy is it to join?
stainless steel	high	good	fair	fair
gold alloy	medium	fair	fair	easy
nickel/titanium alloy	low	excellent	poor	difficult

Understanding Concepts

1. What property is described by each of the following statements?
   (a) Copper metal can be bent into different shapes.
   (b) A steel blade can scratch glass.
   (c) Alcohol boils at 60°C.
   (d) Under a magnifying glass, sugar appears to be made of tiny cubes.
   (e) A nickel coin is shiny.

2. Make a chart listing physical properties that you can observe qualitatively by using your senses or by doing some simple tests.

   Qualitative Observations

Using Senses	Doing Tests
?	?
?	?

3. Distinguish between a physical property and a chemical property.

Exploring

4. Think of one use of metal.
   (3A) Research the suitability of two different metals for that use, considering the advantages and possible risks. Explain.

Reflecting

5. What are some other properties of matter that were not discussed in this section? For example, do any substances change when they are exposed to air? Can any substances carry electricity?

6. Look at the list of adjectives that you made for Getting Started ❶, on page 12. Do any of these adjectives represent the properties in this section? What other properties are suggested in your list?

Challenge

List the materials used in some everyday products, and identify their useful physical and chemical properties. How would you display this information?

Identifying Substances Using Properties

Have you ever confused the salt and sugar in your home? Because they look the same, you must use other properties to distinguish between them. A mechanic working on a car uses many different solutions. Gasoline, oil, transmission fluid, antifreeze, and brake fluid are just a few of these solutions. Different colours are added to these solutions to make it easier to identify leaks. For example, antifreeze is often green and transmission fluid is red. Physical and chemical properties are the key to identifying substances.

In this investigation, you will determine the identities of five unknown substances, using your laboratory skills. The substances are all white solids, but have other different properties that are described in **Table 1**.

Materials
- safety goggles
- apron
- original containers for salt, baking soda, chalk, sodium nitrate, and sodium thiosulphate
- numbered samples of five unknown solids
- toothpicks
- hand lens
- spot plate or microtray
- medicine dropper
- small beaker with distilled water
- small beaker with dilute hydrochloric acid (3–5%)

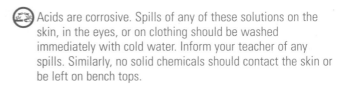 Acids are corrosive. Spills of any of these solutions on the skin, in the eyes, or on clothing should be washed immediately with cold water. Inform your teacher of any spills. Similarly, no solid chemicals should contact the skin or be left on bench tops.

Never use your sense of taste to identify substances—they may be poisonous.

Table 1 **Data About Solids**

Property	Salt	Baking Soda	Chalk	Sodium Nitrate	Sodium Thiosulphate
state (at room temperature)	solid	solid	solid	solid	solid
colour	white	white	white	white	white
clarity	clear	opaque	opaque	clear	clear
crystal shape	small cubes	powder	powder	granular	hexagons
behaviour in water	soluble	soluble	insoluble	soluble	soluble
behaviour in acid	dissolves	fizzes and dissolves quickly	fizzes and some dissolves	dissolves	turns cloudy yellow

Question
How can physical and chemical properties be used to identify substances?

Hypothesis
An unknown substance can be identified by testing its properties and comparing them with those of known substances.

Procedure

1 Observe the WHMIS labels on the materials that you will be using in this investigation.

(a) Record the information for each substance.

2 Make a table in your notebook similar to **Table 2** to record your observations.

3 Put on your apron and safety goggles.

Table 2 **Data Table for Identifying Substances**

Property	Unknown Substance				
	1	**2**	**3**	**4**	**5**
state	?	?	?	?	?
colour	?	?	?	?	?
clarity	?	?	?	?	?
crystal shape	?	?	?	?	?
behaviour in water	?	?	?	?	?
behaviour in acid	?	?	?	?	?
identity of solid	?	?	?	?	?

4 Obtain a small sample of each of the five unknown solids on separate numbered scraps of paper.

5 Look at the samples using a hand lens. Describe their appearance.

✎ (a) Record your description of state, colour, clarity, and crystal shape in your table.

(b) Without specific directions from your teacher, which of your five senses is the only one you can use for your observations? Why?

6 Using the hand lens and a toothpick, count out roughly 20 crystals of whichever solid appears to be salt. Measure out roughly equal amounts of each of the other solids. Place each sample in a different well of the spot plate or microtray. Make sure you have numbered the samples 1 to 5.

7 Using a medicine dropper, add two drops of water to each sample. Observe what happens to the solids.

Step 7

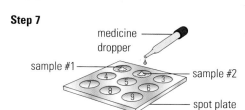

medicine dropper
sample #1
sample #2
spot plate

✎ (a) Record your observations in the table.

(b) What kinds of solids dissolved or mixed with water faster?

8 Rinse and dry the spot plate. Repeat steps 6 and 7, using dilute hydrochloric acid in the medicine dropper instead of water.

✎ (a) Record your observations in the table.

9 Dispose of the contents of your spot plate and put away your materials as directed by your teacher. Clean up your work station. Wash your hands.

Analysis and Communication

10 Analyze your observations by answering the following questions:

(a) What was the identity of each of the five solids? Record their names in your data table. Do your results support your hypothesis?

(b) For each of the five solids, explain how you decided on its identity.

(c) Which physical properties did you examine in this activity?

(d) Which chemical properties did you examine?

(e) Which samples were the most difficult to identify? Explain.

11 Write your investigation as a lab report. (8A)

Understanding Concepts

1. Describe three everyday situations in which it would be useful to identify unknown substances. Explain how you would identify the substances.

Exploring

2. At home, collect five small samples of white powder. Choose from flour, cornstarch, icing sugar, baking powder, cream of tartar, citric acid, powdered milk, and coffee whitener. Put each sample in a small bottle and label the bottles A to E. Keep a list of your samples, labelled with the correct letters. Trade samples with a friend and design an experiment to identify each other's samples.

⊤ Do not use your sense of taste when you identify these substances.

In Search of Safer Paint

What does an artist have in common with a house painter? Both work with paints—mixtures of substances that have been carefully chosen for their special physical and chemical properties. These properties include not just colour but also the ability to flow and stick to the canvas, walls, or other surfaces. People who work with these chemicals also have to consider safety issues.

You may have noticed a strong smell in a freshly painted room. This odour means that particles of solvent—the liquid part of the paint—have evaporated into the air and entered your nose. The solvent is just one of many different substances that have been mixed together to make the paint.

As the solvent evaporates into the air, the solid components of paint are left behind to coat the wall or other surface. Unfortunately, when some solvents evaporate, their fumes are more than just strong or unpleasant—they are dangerous. These fumes may burn, or even explode if concentrated in a poorly ventilated room. Many solvents are also toxic and can poison a person who inhales large amounts.

(a) What are two problems associated with solvents in paint?

Inside a Can of Paint

Paint is a mixture of pigments, resins, and solvent. Each substance in the mixture is chosen for its physical and chemical properties. The components of paint are summarized in **Table 1**.

(b) What are the three components of paint and the purpose of each?

Latex Paints

Water is the major solvent in latex paint, so the fumes from latex paint are harmless. However, water is not very good at dissolving resins, the part of paint that provides strength and durability. As a result, latex paint contains less resin than alkyd paint, and is less strong and durable. Latex paints are not as sticky and so cannot be used on all surfaces. The water in latex paints damages some surfaces. On the other hand, latex paints dry very quickly and you can use water to clean the brushes or rollers.

(c) What are two advantages and two disadvantages of latex paint?

Alkyd Paints

Alkyd paint does not contain water as a solvent. Instead it uses mineral spirits and/or turpentine. These solvents are able to dissolve the resins that make this type of paint useful. Alkyd paint is strong and durable. These properties make it useful on surfaces that are often washed, such as kitchen cabinets. It sticks to metal and other surfaces. But the solvent fumes are hazardous, so very good

Table 1	Inside a Can of Paint		
What Is in Paint	**Comes from**	**Properties**	**Possible Hazards**
pigment	soil, metals, other coloured substances	long-lasting colour	some components harmful if consumed
resin	plant or insect secretions, plastics (alkyds, acrylics, urethanes)	forms hard film that sticks	some components harmful if consumed
solvent	water, mineral spirits, turpentine (from plants)	dissolves other parts of paint	fumes may be toxic; fumes may burn easily (combustible)

ventilation is necessary when painting. Cleaning up requires the same solvents, adding more fumes to the air.

(d) What are two advantages and two disadvantages of alkyd paint?

Choosing a Paint

Think about the decisions you would have to make if you had to do some painting. Painters need to choose products that have the properties they require. Examine the label for "Colour Depot Enviro-safe Paint," shown in **Figure 1**. The label makes claims about the paint that you may or may not agree with.

(e) In a group, list the advantages of Enviro-safe Paint that are claimed in the advertisement.

(f) Discuss these claims. Consider the following questions:

Figure 1

COLOUR DEPOT

NEW

INTERIOR LATEX

ENVIRO-SAFE PAINT
Satin Tint
Scrubbable
Odourless
Stain Resistant
Solvent Free (No V.O.C.'s)
The Healthy Choice for
Your Home Enviro...

FANTASTIC VALUE

Enviro-safe Paint

Painters prefer Enviro-safe!
• Environmentally Safe; Better for Your Health
• Completely Solvent-Free Paint
• Enviro-safe releases no pollutants into the air you breathe!
• Odour-free
• Available in over 100 pastel shades.
• No V.O.C.'s (volatile organic compounds)
• Easy Cleanup
• Help keep the environment clean with Enviro-safe Paint. Each can covers 25% more wall surface, reducing packaging waste by 20% over other paints.

Interior Latex
(Not suitable for bare wood, metal, or unfinished walls.)

• What evidence is presented to support each claim?
• What evidence would you need to be convinced the claim is accurate?
• Are there any confusing or incorrect statements in the advertisement? If so, why do you think this has happened?

(g) Rank the claims as very important, less important, or not important to health and safety.

(h) Rank the claims as very important, less important, or not important to the function of the paint (colouring and protecting a wall).

Understanding Concepts

1. What are three types of physical properties that are important for paint?

2. When is paint most hazardous to human health: while it is in the can, while it is being applied, or after it is dry? Explain your reasons.

Making Connections

3. Which paint would you buy for painting each of the following areas? Explain your choice.
 (a) the walls of a hospital
 (b) the walls of a school hallway
 (c) the window sills in a home

4. What WHMIS or Hazardous Household Product Symbols would you expect to see on a can of (a) latex paint? (b) alkyd paint?

5. Would you prefer to use Enviro-safe Paint or regular latex paint if you were painting your room? Why? If you are not sure, what other issues are important in making this decision?

Exploring

6. Research some of the changes that scientists have made to make paints safer. For example, paint used to contain lead. Why? Now it does not. Why? What replaces the lead and why is it better? Can you think of any ways to make paint safer still?

Challenge

What are the health and environmental issues related to the substance you are marketing?

1.5 Investigation

SKILLS MENU
- Questioning
- Hypothesizing
- Planning
- Conducting
- Recording
- Analyzing
- Communicating

Identifying Substances Using Density

Which is heavier: a kilogram of feathers or a kilogram of lead? Once you think about it, the answer is obvious. They have the same mass but very different volumes and therefore different densities. As you have learned earlier, density is a physical property of matter (**Figure 1**). Each substance has its own characteristic density. Look at **Table 1** to see the densities of some common solids, liquids, and gases.

Density is the amount of matter per unit volume of that matter. Density can be expressed as a formula:

$$\text{Density } (D) = \frac{\text{Mass } (m)}{\text{Volume } (V)}$$

If you know the value of any two of the three variables (D, m, or V) in this formula, you can solve for the third. For example, if a metal has a mass of 30 g and occupies a volume of 6 cm^3, its density can be calculated as

$$D = \frac{m}{V} = \frac{30 \text{ g}}{6 \text{ cm}^3} = 5.0 \text{ g/cm}^3$$

In this investigation, you will use density calculations to identify unknown liquids.

Table 1

Approximate Densities of Some Common Materials

Substance	Density	
	kg/m^3	g/cm^3
gold	19 300	19.3
silver	10 500	10.5
aluminum	2700	2.7
ice	920	0.92
wood (birch)	660	0.66
wood (cedar)	370	0.37
glycerol	1260	1.26
distilled water	1000	1.0
vegetable oil	920	0.92
isopropanol	790	0.79

Materials
- samples of glycerol, vegetable oil, and isopropanol, labelled Unknown A, Unknown B, and Unknown C
- 3 beakers, each 100 mL
- balance
- graduated cylinder

Isopropanol (rubbing alcohol) is flammable and toxic.

Question

1 The labels have fallen off three bottles of liquid. The liquids are glycerol, vegetable oil, and isopropanol. Write a testable question for determining the identity of the unknown liquids.

2A

Hypothesis

2 The densities of the unknown liquids can be compared with the known densities of…
The unknown densities can be calculated if experimental measurements are made of…

(a) Copy these statements into your notebook and complete the sentences.

Experimental Design

3 You will be given samples of three liquids, identified as Unknown A, Unknown B, and Unknown C.

4 Design an experiment to determine the identity of the three unknown liquids, using the materials in the list above.

(a) Write out the design as a series of numbered sentences.

5 Include with your design a fully labelled data table to record your observations.

(a) Make a ruled table for observations.

6 Note that your design must include suggestions on appropriate safety procedures.

(1B) (a) Make a list of safety precautions that you will follow.

7 Show your procedure, data table, and safety suggestions to your teacher.

Procedure

8 When your design has been approved by your teacher, obtain the necessary materials and perform your experiment.

✎ (a) Record observations in your data table.

9 Calculate the densities of the unknown liquids.

(a) Show calculations, including formula, substitution, and units.

✎ (b) Record your results on a class data sheet, if your teacher suggests you do so.

Analysis and Communication

10 Analyze your observations by answering the following questions:

(a) What were the identities of Unknowns A, B, and C?

(b) What other physical properties might have helped you decide what the unknown liquids were?

(c) List any difficulties that you experienced when making measurements.

(d) How would you modify your experimental design to improve it?

(e) Determine the average of all the density values obtained by the groups in your class. How does your class average compare with the expected values for glycerol, vegetable oil, and isopropanol?

11 Write your investigation as a lab report. (8A)

Making Connections

1. You are designing a new transparent bottle for an oil and vinegar salad dressing. Vinegar is mostly water. Some spices dissolve into the oil, some into the vinegar, and some spices remain separate. Look for the densities of oil and water in **Table 1**. Where in the bottle would you expect to find the oil layer? How would this affect the design of your bottle? Explain, using a sketch of your bottle design and label.

2. Calculate the densities of the following substances, given mass and volume information:

   (a) mass = 200 g and volume = 40 cm^3

   (b) mass = 4 g and volume = 3.20 cm^3

   (c) mass = 36 g and volume = 54 cm^3

3. The density formula can be used to calculate mass or volume if density is given. Use the densities of substances given in **Table 1** to calculate:

   (a) the mass of 100 cm^3 of silver

   (b) the volume of 270 g of aluminum

   (c) the mass of a 20 cm^3 block of birch wood

4. An unknown metal has a volume of 20 cm^3 and a mass of 54 g. Use **Table 1** and calculations to guess the likely identity of the unknown metal.

5. Which do you think is more dense: an (2A) unpeeled orange or a peeled orange? You may be surprised to find that one floats in water and the other does not. Design and perform an experiment, using "home apparatus," to determine their densities.

Challenge

Identify two instances in which an understanding of density is important in our lives. Consider how to include these examples in your display.

Figure 1

Although these three substances all have the same volume, their masses are very different.

 1 cm^3 gold 1 cm^3 birch   1 cm^3 vegetable oil

SKILLS MENU
○ Questioning   ● Conducting   ● Analyzing
○ Hypothesizing   ● Recording   ● Communicating
○ Planning

Chemical Magic

A chemical property, such as combustibility, describes the ability of a substance to interact with another substance. These interactions result in change. Change can be subtle—a leaf slowly changes colour from green to yellow in the fall. Change can be dramatic—gasoline explodes in a fireball. Experimenting with different substances and recording observations have led scientists to form new hypotheses to classify some of these changes. This investigation will allow you to identify some changes in chemical properties as new substances are formed. It may also lead you to hypothesize about what kinds of changes you observe.

Materials
- safety goggles
- apron
- 2 small test tubes
- test-tube rack
- 4 labelled medicine droppers
- 2 mL distilled water
- indicator solution (phenolphthalein) in a dropper bottle
- 2 mL of Solution A (0.5% sodium hydroxide)
- 2 mL of Solution B (2.0% sulfuric acid)
- 2 mL of Solution C (2.0% calcium chloride)
- 2 cm² of aluminum foil
- 10-mL graduated cylinder
- 2 mL of Solution D (2.0% copper (II) chloride)

Solutions A–D are corrosive. Specifically, Solutions A, B, and C are drain cleaner, car battery acid, and de-icing salt, respectively. Spills of any of these solutions on the skin, in the eyes, or on clothing should be washed immediately with cold water. Inform your teacher of any spills.

Phenolphthalein solution is flammable and harmful if ingested. Inform your teacher of any spills.

Question
How do various substances interact with each other?

Hypothesis
When some substances are mixed, they will form new substances with new properties.

Procedure

1 Put on your apron and safety goggles.

2 Obtain a test tube and put it in a test-tube rack. Use one of the medicine droppers to add 5 drops of water to the test tube. Then add 2 drops of indicator solution to the water.

Step 2

(a) Record your observations.

(b) Is there any evidence that a new substance was produced? Explain.

3 Use a second dropper to add 5 drops of Solution A to the water/indicator solution.

(a) Record your observations.

(b) Is there any evidence that a new substance was produced? Explain.

4 Use a third dropper to add 5 drops of Solution B to the solution.

(a) Record your observations.

(b) Is there any evidence that a new substance was produced? Explain.

5 Use the fourth dropper to add 5 drops of Solution C to the solution.

(a) Record your observations.

SKILLS HANDBOOK: (6A) Obtaining Qualitative Data

(b) Is there any evidence that a new substance was produced? Explain.

6 Crumple a small piece of aluminum foil and place it in a second test tube. Place the test tube in a rack. Using a graduated cylinder, measure 2 mL of Solution D and add it to the test tube.

(a) Describe the initial colour of Solution D.

(b) Describe the colour of the solution after 3 min.

(c) Describe the change in the aluminum foil.

(d) Other than any colour changes, what evidence do you have that a chemical change has occurred?

7 Dispose of the contents of your test tubes and put away your materials as directed by your teacher. Clean up your work station. Wash your hands.

Analysis and Communication

8 Analyze your observations by answering the following questions:

(a) How could you tell when a new substance was produced?

(b) Describe two particular physical properties of substances that changed during this activity.

9 Write a summary paragraph explaining to others how you can tell when a new substance is produced.

Challenge

How would you display the information in your summary paragraph so that it makes sense to a younger audience?

Understanding Concepts

1. List all the changes in matter you can think of that might occur in a kitchen. Do any of the products of these changes have new properties? Explain.

Making Connections

2. Imagine that you are a magician and that you want to design a new magic trick to amaze your audience. You go to a magicians' supply store and discover three new products, illustrated in **Figure 1**. Write a script describing how you could use any or all of these products to create a magic trick.

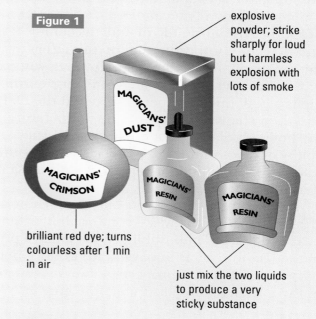

Figure 1

explosive powder; strike sharply for loud but harmless explosion with lots of smoke

MAGICIANS' DUST

MAGICIANS' CRIMSON

MAGICIANS' RESIN

MAGICIANS' RESIN

brilliant red dye; turns colourless after 1 min in air

just mix the two liquids to produce a very sticky substance

Exploring

3. In the investigation you just completed, what would happen if you changed the order in which water, indicator, and Solutions A, B, and C were mixed? Design an experiment to test this idea, and predict the observations you would expect. Check your design and predictions with your teacher before mixing any substances. Compare your results with your predictions.

Reflecting

4. After observing different kinds of changes, can you suggest a way of categorizing changes in matter?

Physical and Chemical Changes

Some of the most useful and powerful properties of matter are those related to how and why matter changes. There are countless changes in matter that affect us every day: for example, applying heat to an egg, burning gasoline, freezing water, and mixing oil and vinegar, to name a few. Understanding and categorizing kinds of change are an important first step to making use of change.

You can discover a great deal about matter simply by observing a candle. The physical properties of the candle include its colour, texture, and density—properties that do not affect the ability of the wax to change in any way. However, some other physical properties do change wax. For example, the wax melts at a definite temperature called the melting point and then changes to vapour at a temperature called the boiling point. These physical properties affect the ability of wax to undergo physical change—the wax, whether solid, liquid, or vapour, is still the same substance.

As the candle burns, you can observe another kind of property that affects change in the wax—combustibility. Candle wax burns, producing heat and light. Combustibility is a chemical property: it describes the ability of the wax to react with oxygen to produce new substances. Unlike physical properties, chemical properties always involve change in a substance. Some of these changes are illustrated in **Figure 1**.

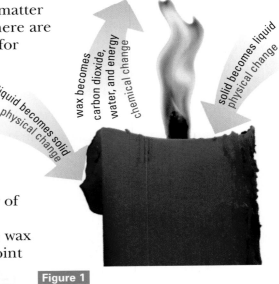

liquid becomes solid physical change

wax becomes carbon dioxide, water, and energy chemical change

solid becomes liquid physical change

Figure 1

The wax of this candle is undergoing both physical and chemical changes. Which change results in new substances? Which does not?

Physical Change

In a **physical change**, the substance involved remains the same substance, even though it may change state or form. When the candle wax has melted or vaporized, it is still wax.

Changes of state—melting, boiling, freezing, condensation, sublimation—are physical changes (**Figure 2**). You can see a physical change when you pour melted chocolate over ice cream. Liquid chocolate forms a thin, even coating over the ice cream. The chocolate becomes solid as the ice cream cools it, but once it's in your mouth, it tastes the same in both states because its particles have not changed.

Dissolving is also a physical change. When you dissolve sugar in water, the sugar particles spread out, but they are still there, as sugar particles. You can reverse the process by evaporating the water and collecting the sugar. Most physical changes are easy to reverse.

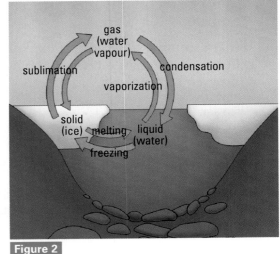

gas (water vapour)

sublimation   condensation

vaporization

solid (ice)   melting   liquid (water)

freezing

Figure 2

Changes of state

Chemical Change

In a **chemical change**, the original substance is changed into one or more different substances that have different properties.

As the wax in a candle melts and vaporizes, some of the wax particles join up with oxygen from the air. The result of this chemical change is the production of water vapour, carbon dioxide gas, heat, and light. The wax particles that seem to disappear are actually changing into something else.

Chemical changes always involve the production of new substances. Most chemical changes are difficult to reverse. Burning, cooking, and rusting are all examples of chemical changes, as is the change shown in **Figure 3**.

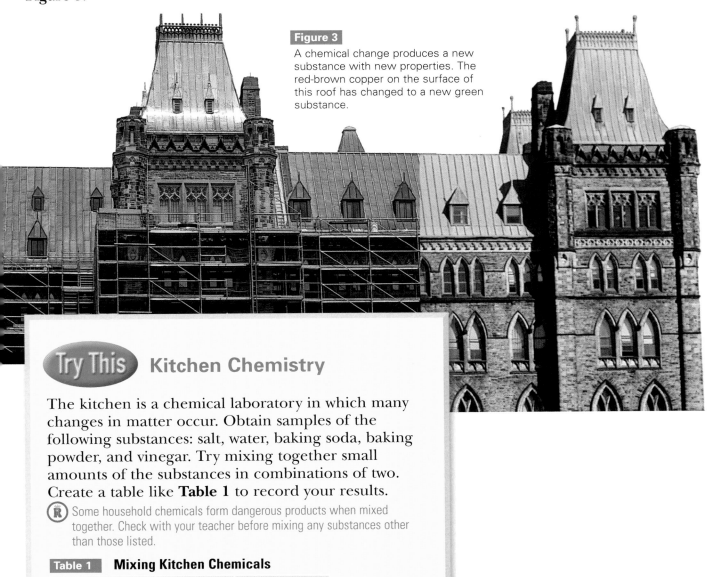

Figure 3
A chemical change produces a new substance with new properties. The red-brown copper on the surface of this roof has changed to a new green substance.

Try This Kitchen Chemistry

The kitchen is a chemical laboratory in which many changes in matter occur. Obtain samples of the following substances: salt, water, baking soda, baking powder, and vinegar. Try mixing together small amounts of the substances in combinations of two. Create a table like **Table 1** to record your results.

ℝ Some household chemicals form dangerous products when mixed together. Check with your teacher before mixing any substances other than those listed.

Table 1 **Mixing Kitchen Chemicals**

Substances Mixed	Observations	Evidence of a new substance?
salt and water	?	?
?	?	?

Chemical or Physical?

You can't see the chemical change in the wax just by looking at a burning candle. So how can you tell if a chemical change has occurred? The clues listed in **Table 2** can help you decide. But do not come to a conclusion too quickly. All of these clues *suggest* that a new substance has been produced, but any one of them could also accompany a physical change. You must consider several clues in order to determine what type of change has taken place.

Table 2 **Clues That a Chemical Change Has Happened**

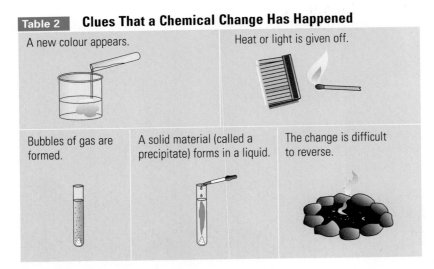

| A new colour appears. | Heat or light is given off. |
| Bubbles of gas are formed. | A solid material (called a precipitate) forms in a liquid. | The change is difficult to reverse. |

Figure 4

Operating a car involves many changes.

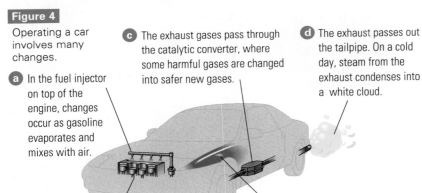

a In the fuel injector on top of the engine, changes occur as gasoline evaporates and mixes with air.

b Inside the engine cylinders, the explosion of the gasoline-air mixture produces hot exhaust gases, including water vapour, carbon dioxide, and nitrogen oxides.

c The exhaust gases pass through the catalytic converter, where some harmful gases are changed into safer new gases.

d The exhaust passes out the tailpipe. On a cold day, steam from the exhaust condenses into a white cloud.

e As the steel of the car is exposed to air and water, a crumbly reddish-brown substance forms: the steel has changed into rust.

Challenge

Identify physical and chemical changes that are useful to us. How would you display these examples?

In what ways can the substance you are marketing change physically or chemically? How does this make it useful?

Understanding Concepts

1. Explain how a physical change differs from a chemical change.

2. Classify each of the following as a physical or a chemical change. Explain why.

    (a) garbage rotting

    (b) cutting up carrots

    (c) a silver spoon turning black

    (d) making tea from tea leaves

    (e) bleaching a stain

    (f) boiling an egg

3. During a power failure, Blair lit four identical candles. He placed three candles very close together on a table, and one on a different table. When the power came back on an hour later, Blair was surprised to see that the candles in the group were much shorter than the one by itself. There was also more melted wax around the base of each of the three candles. Account for Blair's observation. What kind of candles should you keep on hand for emergencies?

Making Connections

4. Which of the changes described in **Figure 4** involve chemical changes? Which involve physical changes? For each identified change, determine the impact it could have on the environment. Which changes must car designers pay attention to in order to minimize damage to the environment?

Reflecting

5. Go back to the list you brainstormed for Getting Started **3** on page 13. Write beside each change on your list whether it is chemical or physical. If you are not sure of any, just leave a blank space. You can return to your list as you work through this chapter.

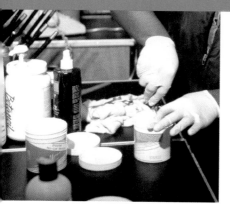

Hair Colourist

If you have the right personality and an interest in the aesthetic side of chemistry, then Helder Sousa recommends the job of colour technician. After studying for 10 months in a school for hair stylists, he soon gravitated to the technician side of the work.

While his high school chemistry helped give him an understanding of chemical processes, Sousa did most of his learning as an apprentice at a well-known Toronto salon. He started as an assistant, doing the shampooing and sweeping while he learned about colour theory. Gradually, by working with an experienced colourist, he learned how to achieve the effects the clients wanted. In many ways, learning how to colour hair is more complex than cutting. "Now, I have no nervousness about it. With the years comes more knowledge, like any job."

The business begins and ends with the clients. "You may have an idea of what auburn means, but the client may mean something else—and you have to figure out just exactly what colour they want."

People tend to confide in their colourists, so psychological skills are in order too. "Good listening skills are a must. This is not the industry for someone who doesn't enjoy working with people." But if you do like working with people, and you're good at matching the right colours to the client, you can follow a range of careers—in film, television, or magazines. Sousa's advice: "Study long and hard and work in a good salon."

> You have to love what you do— it's hard work to become good at it, and people are very picky about their hair.

Exploring (3A)

1. There are two basic types of hair colour: temporary and permanent. Research how they differ chemically and how they affect the hair.

2. Find out if there are any courses for hair stylists in your area. What background is required? How long is the training?

3. What are the advantages of apprenticing, compared with college courses?

4. Create a pamphlet to inform others who may be interested in a career as a colour technician.

Observing Changes

Using their knowledge of physical and chemical properties and current technology, chemists cause many useful changes, including transforming crude oil into plastics, and changing minerals from the ground into copper and iron.

In this investigation, you will learn more about physical and chemical changes. Remember that a change is probably physical unless there is almost certain evidence that a new substance has been produced.

Materials

- safety goggles
- apron
- 4 test tubes
- test-tube rack
- distilled water
- 2-mL measuring spoon
- copper(II) sulfate powder
- test-tube stopper
- iron (a piece of steel wool about 1 cm × 1 cm × 2 cm)
- stirring rod
- dilute sodium carbonate (5% solution)
- dilute hydrochloric acid (3% solution)
- magnesium ribbon (2-cm strip)
- tongs

(T) Copper(II) sulfate is poisonous. Report any spills to your teacher.

Hydrochloric acid is corrosive. Any spills on the skin, in the eyes, or on clothing should be washed immediately with cold water. Report any spills to your teacher.

Question

How can we recognize physical and chemical changes?

Hypothesis

1 Write a statement to answer the question.

(4A)

Procedure

Part 1: Copper(II) Sulfate and Water

2 Put on your apron and safety goggles.

3 Make a table similar to **Table 1** to record all your observations and inferences.

Table 1	**Physical and Chemical Changes**				
Part	Starting Substances		Observations after Mixing	Inference Physical? Chemical?	Evidence
	Name	**Properties**			
1	water	?	?	?	?
	copper(II) sulfate	?	?	?	?
2	?	?	?	?	?

4 Obtain a small amount of copper(II) sulfate in a test tube. Put the test tube in the test-tube rack. Obtain some distilled water.

(a) In your table, describe the water and the copper(II) sulfate.

5 Pour distilled water into the test tube containing the copper(II) sulfate, to a depth of about 4 cm. Put a stopper in the test tube to seal it. Take the tube out of the rack and mix the contents by turning the tube upside down several times. Return the test tube to the rack.

Step 5

water
copper(II) sulfate

(a) Was there a change? Record your observations.

(b) Make an inference based on your observations: if there was a change, was it physical or chemical? How do you know? Record your inference and the evidence to support it.

Part 2: Copper(II) Sulfate and Iron

6 Into another clean, dry test tube in the rack, pour some of your mixture of copper(II) sulfate and water, to a depth of about 2 cm. (Save the remainder of your copper(II) sulfate mixture to use in Part 3.) Obtain a piece of steel wool (iron).

(a) Describe the steel wool and the solution before you continue.

7 Using a stirring rod, push the steel wool into the copper(II) sulfate mixture.

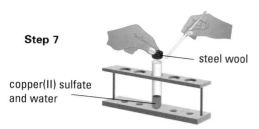

Step 7

copper(II) sulfate and water — steel wool

✎ (a) Record your observations.

(b) Was there a physical or a chemical change? What is the evidence?

Part 3: Copper(II) Sulfate and Sodium Carbonate

8 Into another clean, dry test tube, pour sodium carbonate solution to a depth of about 1 cm.

✎ (a) Describe the sodium carbonate solution and the remainder of your copper(II) sulfate mixture.

9 Pour one solution into the other.

Step 9

copper (II) sulphate and water — sodium carbonate solution

✎ (a) Record your observations.

(b) Was there a physical or chemical change? What is the evidence?

10 Dispose of the mixtures in the test tubes as instructed by your teacher.

Part 4: Hydrochloric Acid and Magnesium

11 Into another clean, dry test tube, pour dilute hydrochloric acid to a depth of about 2 cm. Obtain a small piece of magnesium ribbon.

✎ (a) Describe the dilute hydrochloric acid and the magnesium.

12 Using tongs, carefully add the magnesium ribbon to the test tube without splashing. As any change occurs, feel the bottom of the tube to check for any temperature change.

Step 12

dilute hydrochloric acid — magnesium

✎ (a) Record your observations.

(b) Was there a physical or chemical change? What is the evidence?

13 Dispose of the mixtures as instructed by your teacher. Wash your hands.

Analysis and Communication

14 Analyze your observations using your completed table by answering the following questions:

(a) What kind of change took place when you mixed the substances in each part of the investigation? What evidence do you have? Does this support the hypothesis?

(b) In each part of this investigation, identify what physical properties changed?

(c) Look at **Table 2** on page 30. Which of those clues did you observe?

(d) Which of the changes you observed might be reversed? Explain how.

Understanding Concepts

1. What are some examples of physical and chemical changes in the home? Give reasons for your classification.

Reflecting

2. If you wanted to test more properties of a new substance formed in Part 2, how could you separate it from other materials in the test tube?

Corrosion

Have you ever wondered why metal car bodies rust but plastic bumpers do not? As you know, different substances have different physical properties, such as colour and hardness, and different chemical properties, such as combustibility and reactivity with acid. One chemical property that has great economic importance is the tendency of a substance to undergo **corrosion**—the slow chemical change that occurs when a metal reacts with oxygen from the air to form a new substance called an oxide.

Kinds of Corrosion

The most dramatic example of corrosion is rusting—the reaction of iron with oxygen to form iron oxide. Iron is usually found in the form of steel, a mixture of iron, carbon, and other substances. This alloy or mixture of metals is much harder and tougher than the original iron.

Figure 1

Rust damages many steel car bodies.

Rusting is a chemical change that involves iron, oxygen from the air, and water, as well as the salts or other minerals dissolved in the water. Every year, millions of dollars of damage are caused to building structures, vehicles (like the car in **Figure 1**), and other iron and steel products due to this process. Rust is particularly damaging because of one of its physical properties: rust is porous, absorbing water almost like a sponge. As a result, it dissolves or flakes off, leaving another layer of fresh metal underneath to be attacked by oxygen. The process continues until the rust has eaten its way through the metal.

By contrast, aluminum has a similar chemical property in that it also reacts with oxygen, but the aluminum oxide that forms is strong and unaffected by water. The corrosion stops and the oxide acts as a protective coating. If you have aluminum cooking pans at home, you will be familiar with the grayish, dull coating of aluminum oxide.

Even silver develops a surface coating of tarnish if it comes into contact with sulfur-containing foods such as eggs or mustard. The black coating seen in **Figure 2** is silver sulfide. Silver tarnishes slowly if left out in the air; the more sulfur-containing pollutants in the air, the more quickly it tarnishes. The black layer can be removed by polishing the silver.

Figure 2

Silver slowly corrodes in air.

Preventing Corrosion

There are several ways to prevent corrosion. One method is to paint the surface of the metal (**Figure 3**). As long as the painted surface is not broken or cracked, oxygen in the air cannot get at the metal. For the same reason, cars are often sprayed with oil to coat the bottom and inner surfaces of the car body. Iron can also be protected by coating it with other metals.

Some metals have the chemical property of being easier to corrode than iron. This property is used to protect the hulls of ships and boat motors. For example, as seen in **Figure 4**, a plate of zinc is attached to steel boat motors. The steel motor parts remain untouched as the zinc is slowly used up. The zinc is replaced when it has completely corroded.

A special alloy of steel, made by mixing iron with nickel and copper, is now used in some building structures. The metals corrode quickly but the nickel and copper oxides form a protective layer that prevents further rusting.

Another way to prevent corrosion is to use materials that have different chemical properties. Plastics are being used increasingly in car bumpers and panels that get frequent bumps and scratches. Steel loses its strength if air and water penetrate through a scratch in the paint, but plastics never corrode and remain strong and flexible.

Figure 4

Another metal attached to a boat motor can protect it from corrosion in a process called cathodic protection.

Figure 3

Some bridges are so large that painters take years to finish the whole structure. Then they have to begin again!

Understanding Concepts

1. What is "corrosion"?
2. How is an oxide formed?
3. Describe two processes that form two different oxides.
4. Make a poster describing three ways to protect a metal from corrosion.

Making Connections

5. Corrosion of automobiles causes millions of dollars of damage every year. Which parts of the automobile corrode the most? Why? Describe how car owners and manufacturers can help to reduce the effect of corrosion.
6. Make a list of the products that you have in your home that can corrode. What decisions or steps can you take to protect these products from corrosion?

Reflecting

7. Engineers design pipelines to carry oil or natural gas over hundreds of kilometres. These pipelines are made of steel, but do not corrode. The engineers attach other metals to the pipelines every kilometre or so. How does this protect the steel?

SKILLS MENU
- Questioning
- Conducting
- Analyzing
- Hypothesizing
- Recording
- Communicating
- Planning

Preventing Corrosion

What kinds of decisions do designers and engineers make when they design products for people to buy? And what kinds of research do they do to help them with those decisions? You have learned that corrosion is a kind of chemical change that affects many everyday products. Being able to design and conduct experiments on corrosion is part of research and development for many companies. In this activity you will take on the role of someone in one of these industries and design an experiment to try to improve a product.

Part 1: Designing Engineering Team (2A)

You are part of the design engineering team for PDQ Automobile Corporation that has just designed a revolutionary commuter car (**Figure 1**). It is very lightweight and, therefore, fuel-efficient because it uses metals such as aluminum and magnesium instead of steel in the frame and body. However, critics have suggested that these metals are more likely to corrode, especially in Canadian winters when salt is used on the roads.

Materials
- sample strips of aluminum, magnesium, and steel
- a sheet of emery paper (to polish the metal)
- 6 beakers
- labels and marking pens
- salt
- water

Figure 1

Question

1 What question does your design team need to answer to support its choice of materials?

Hypothesis

2 Write a hypothesis for this experiment, based on the claims of PDQ's critics.

Experimental Design

3 Design an experiment, using the materials provided, to test your hypothesis.

Procedure

4 When you have your teacher's approval for your design, make a data table that you can use to record observations.

5 Assemble the necessary materials and conduct your experiment.

Analysis and Communication

6 Analyze your observations by answering the following questions:

(a) Describe the appearance of the materials before you started.

(b) Describe the appearance of the materials at various times during the testing period.

(c) How did you design your experiment to make sure that the type of metal was the only variable?

7 Present your data and conclusions to the
8C Board of Directors at PDQ. You may want to design a logo, motto, or mission statement for PDQ Corporation as part of your presentation.

SKILLS HANDBOOK: (2A) Controlled Experiments (8C) Science Projects

Part 2: Corrosion Laboratory Team (2A)

You are part of a small corrosion laboratory which has a contract with Ace Rust-Prevent Corporation. Your research team has been told to investigate the effectiveness of various methods of rust prevention, especially in Canadian winter conditions.

Materials

- steel nails
- a sheet of emery paper (to polish the metal)
- 6 beakers
- salt
- water
- rust-proofing chemicals (if available)
- motor oil
- paint
- paint brush

Question

1 What question does your laboratory need to answer for Ace Rust-Prevent Corp?

Hypothesis

2 Write a hypothesis for the experiment your laboratory will perform.

Experimental Design

3 Design an experiment, using the materials provided, to test your hypothesis.

(1B) (🛑) Include safety precautions in your design.

Procedure

4 When you have your teacher's approval for your design, make a data table that you can use to record observations.

5 Assemble the necessary materials and conduct your experiment.

Analysis and Communication

6 Analyze your observations by answering the following questions:

(a) Describe the appearance of the materials before you started.

(b) Describe the appearance of the materials at various times over the testing period.

(c) How did you design your experiment to make sure that the different types of protection were tested fairly?

(d) Compare the effectiveness of the different methods of protection.

7 Discuss your data with your team and write (8C) a proposal for the most effective method. You may want to design a name and logo for your research laboratory as part of your presentation.

Reflecting

1. Review the scientific inquiry process outlined in section 2A of the Skills Handbook.

(a) Which steps did you follow in this investigation?

(b) Which steps were easiest, and which were most challenging. Explain why.

2. How did you organize your team? How successful were you?

3. What improvements would you make in your design or investigation if you were to do it again?

Challenge

What factors made the proposals effective? How can you use them to market your substance?

Combustion

What chemical reaction occurs in the gas furnace that heats your home? What kind of chemical reaction occurs when you light a match? What caused the forest fire in **Figure 1**? What makes a car engine work? These, and the fires shown in **Figures 2** and **3**, are examples of an important type of chemical reaction called combustion. In **combustion**, a substance reacts rapidly with oxygen and releases energy. The energy is observed as heat and light. Many substances, such as wood, kerosene, and diesel oil, burn readily in air, which is only about 20% oxygen. This makes them useful as fuels.

The three necessary components of combustion are illustrated in **Figure 4**, called the fire triangle.

Fossil Fuels and Combustion

Coal, oil, natural gas, and gasoline are all fuels. They are called **fossil fuels** because they were formed from plants, animals, and microorganisms that lived millions of years ago. When these organisms died, they did not decompose completely. Instead, they were buried by sediments and the energy in their cells remained "locked up."

Human technology, developed over the centuries, depends on these long-buried organisms. Their stored energy powers homes, industries, and various means of transportation.

When any fossil fuel burns, the main products of the reaction are carbon dioxide and water vapour. The particles that make up fossil fuels are called **hydrocarbons**. To represent the combustion of a fossil fuel simply, the following word equation can be used:

hydrocarbon + oxygen → carbon dioxide + water

In a **word equation**, the substances you start with are written on the left and are called the **reactants**. The resulting substances, written on the right, are called **products**.

Figure 1
Some combustion reactions are destructive. Forest fires consume thousands of hectares of trees every year in Canada.

Figure 2
The quick reaction of magnesium with oxygen is combustion. Magnesium is often used as a component of emergency flares, which produce a bright light even in rain or snow.

Figure 3
Fires can rage for months when oil wells burn out of control. The fires can be extinguished using explosives and other methods that seal the leaking oil. Which component of the fire triangle is removed to stop the fire?

Figure 4
The fire triangle is a convenient way to remember the three components of any combustion reaction. Removing any one of these makes the triangle incomplete and puts out the fire.

Combustion and Air Pollution

Under ideal conditions, the combustion of hydrocarbons produces only carbon dioxide and water. But ideal conditions rarely exist. Fossil fuels are not pure hydrocarbons but rather are mixtures of many different substances. Also, the chemical action of combustion can be less efficient if there is not enough oxygen or heat. When not enough oxygen is available, two other products may be produced: carbon monoxide and carbon. Carbon monoxide is a poisonous gas that you will learn more about in Section 2.8.

When gasoline burns in an automobile engine, the carbon dioxide that is produced increases the greenhouse effect, which may be causing global warming (**Figure 5**). Other products include carbon monoxide, smaller hydrocarbons, sulphur dioxide, and nitrogen oxides, all of which can harm people and the environment. In fact, combustion is the major source of air pollution in the environment.

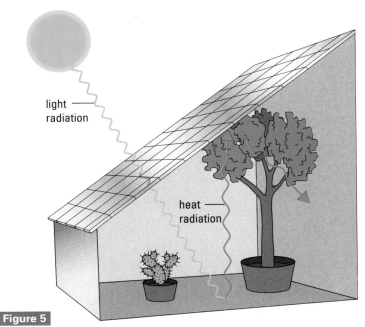

light radiation

heat radiation

Figure 5
Carbon dioxide gas produced by combustion in industry and automobile engines increases the so-called greenhouse effect. The glass panes of a greenhouse allow sunlight to pass through but prevent heat from escaping. A similar situation occurs in the atmosphere: carbon dioxide in the atmosphere acts like the glass in a greenhouse, trapping heat close to Earth's surface. Many scientists believe this is causing a gradual increase in Earth's temperature.

Understanding Concepts

1. What is "combustion"?

2. **(a)** When fossil fuels burn, what are the reactants?

   (b) What are the two main products of this combustion?

   (c) Why are other products also formed?

3. Illustrate which part of the fire triangle is removed when each of the following methods is used to stop combustion.

   (a) closing the valve on a propane tank that supplies propane to a barbecue

   (b) dropping and rolling if your clothing catches fire

   (c) pouring water on a campfire

   (d) pouring baking soda on a grease fire

   (e) blowing on a flaming marshmallow

Making Connections

4. Why should you never operate a gas or charcoal barbecue inside a building?

5. Explain why building codes require an external source of air for fireplaces in new homes.

Exploring

6. Fossil fuels may be obtained from
   (3A) the black oil that is pumped from oil wells. Use the Internet or CD-ROM database to find out how these substances are found and separated. Some key words you might want to use in your search are: petroleum, fractionation, fractional distillation, and oil refining.

Reflecting

7. Wood is combustible but chalk is not. Is combustibility a physical or chemical property of substances? Explain.

Chapter 1 Review

Key Expectations

Throughout this chapter, you have had opportunities to do the following things:

- Recognize physical and chemical properties of everyday substances, such as solubility and combustibility. (1.2, 1.3, 1.9, 1.11)

- Distinguish between physical and chemical changes. (1.7, 1.8)

- Distinguish between metals and nonmetals. (1.2, 1.9)

- Describe chemical changes using indicators such as change in colour, or the production of a gas, precipitate, heat, or light. (1.8, 1.9, 1.11)

- Use density calculations to identify unknown substances. (1.5)

- Investigate physical and chemical properties, and organize, record, analyze, and communicate results. (1.3, 1.5)

- Investigate physical and chemical change, and organize, record, analyze, and communicate results. (1.6, 1.8, 1.10)

- Formulate and research questions related to the properties of matter and communicate results. (1.1, 1.4, 1.11)

- Compare the physical and chemical properties of substances and assess their potential uses and associated risks. (1.2, 1.4, 1.9, 1.11)

- Explore careers requiring an understanding of the properties of matter. (Career Profile)

KEY TERMS

alloy	hardness
brittle	hydrocarbon
chemical change	malleable
chemical property	physical change
combustible (or flammable)	physical property
combustion	precipitate
corrosion	products
crystal	reactants
density	solubility
ductile	viscosity
fossil fuel	word equation

Reflecting

- "Matter has characteristic physical and chemical properties. Understanding these properties enables us to make useful products." Reflect on this idea. How does it connect with what you've done in this chapter? (To review, check the sections indicated above.)
- Revise your answers to the questions raised in Getting Started. How has your thinking changed?
- What new questions do you have? How will you answer them?

Understanding Concepts

1. Make a concept map to summarize the material that you have studied in this chapter. Start with the word "matter."

2. Look again at the Getting Started activity at the beginning of this chapter. Classify the changes as physical or chemical.

3. (a) What is the difference between a physical property and a chemical property?
   (b) Give an example of one physical property and one chemical property for each of the following: wood, gasoline, and baking soda.

4. For each of the following, replace the description with one or two words:
   (a) the starting substances in a reaction
   (b) the substances formed in a reaction
   (c) a change in which a new substance is produced
   (d) a change in which no new substance is produced
   (e) able to dissolve in a solvent
   (f) breaks when hit against a hard surface

5. The sentences below contain errors or are incomplete. Write complete, correct versions.
   (a) A physical change produces a new substance.

(b) The formation of frost is a chemical change.

(c) A chemical change may produce a new substance called a predominate.

(d) A new colour indicates a physical change.

(e) Ability to react with acid is an example of a physical property.

(f) Some substances are safe to taste in the lab.

(g) Malleability is a chemical property.

(h) A chemical change is a change of state or form.

(i) Corrosion is the reaction of a metal with nitrogen in the air.

(j) Goggles may be taken off if a student has finished his or her experiment.

6. Suggest five clues you could consider before deciding whether a change is physical or chemical.

7. Indicate whether each of the following is a physical or a chemical change. Give a reason for each.

(a) water freezing on a pond

(b) soap removing grease from hands

(c) an electric bulb glowing

(d) a cake baking

(e) wood burning

(f) kitchen scraps composting

(g) a paper clip bending

(h) dynamite exploding

8. Copy the properties in Column A (**Table 1**) into your notebook. Match each set of properties with the appropriate substance in Column B.

Table 1

Column A	Column B
colourless, low viscosity, liquid at 20°C, noncombustible	salt
white, crystalline, solid at 20°C, soluble in water	sulfur
yellow, powdery, solid at 20°C, insoluble in water, burns in air with a blue flame	alcohol
colourless, low viscosity, liquid at 20°C, combustible	water

9. Which of the properties described in the previous question were (a) physical properties? (b) chemical properties?

Applying Skills

10. Name four materials or pieces of equipment that you used in your investigations to ensure lab safety. Explain the function of each.

11. Find the missing quantity in each of the following density problems, given any two of mass, volume, and density. Show the appropriate formula and units.

(a) mass = 40 g; volume = 2 mL

(b) mass = 8 kg; density = 2 kg/L

(c) density = 5 kg/m^3; mass = 400 kg

12. A yellow solid is heated and is observed to change to a brown liquid. Analyze whether the change is physical or chemical.

13. A white solid is heated and is observed to change to a liquid at 65°C. When the solid is cooled, it becomes a white solid again at 65°C. Analyze whether the change is physical or chemical. Explain.

Making Connections

14. Carbon dioxide ejected from a fire extinguisher is so cold that it changes to snow.

(a) Is this a physical or chemical change?

(b) The carbon dioxide snow, when applied to a burning object, is said to smother the flame. What kind of chemical change is the carbon dioxide snow preventing? How does the carbon dioxide stop the fire?

(c) Are there any potential risks of using carbon dioxide in this way?

15. A welder uses heat from the combustion of acetylene (a fossil fuel) to weld steel plates together.

(a) Name all the physical changes that occur during and after this process.

(b) Describe the chemical change that produces the heat the welder needs.

(c) What safety precautions must the welder take?

16. Corrosion is the reaction of metals with oxygen in a chemical change. What kinds of changes occur in other substances over time? For example, what happens to plastic products left out in sub-zero temperatures or in intense sunlight for long periods of time? In what instances could the use of plastics practically replace the use of metals?

2 Elements and Compounds

Getting Started

1 Douglas Cardinal is an architect who often creates models of structures to give shape to a concept. Architects use models not only to help them uncover design problems and solve them, but also to make the concepts much easier for other people to understand and visualize. Have you ever built a model? What did it teach you? Brainstorm a list of the possible uses of models. ➤

2 In science, models are not just physical constructions. A model can be a diagram, a classification system, a mental picture, a theory. In fact, a model is any means of representing a thing or process. Models are made to help us visualize things we cannot see or things we find difficult to understand. What scientific models are you already familiar with?

Choose one and ask yourself what the model seeks to explain, and how complete that explanation is. ➤

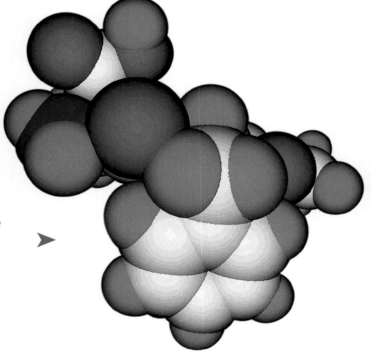

3 A Montreal woman is helped from her home during the 1998 ice storm. Emergency generators used to provide electricity spread carbon monoxide fumes through her apartment complex. Carbon monoxide is a poisonous gas that kills people every year.

How can the particle theory help us understand substances like carbon monoxide? The particle theory suggests that matters are made of basic building blocks, called elements. An element is composed of only one kind of atom. Atoms can be assembled in different ways. For example, carbon and oxygen atoms can be combined to form the gas you exhale with every breath (carbon dioxide) or the dangerous gas that affected this woman (carbon monoxide). What other combinations of atoms are you aware of, and how can they affect our daily lives?

Reflecting

Think about the questions in **1**,**2**,**3**. What ideas do you already have? What other questions do you have about scientific models and elements and compounds? Think about your answers and questions as you read the chapter.

Try This Finding Pure Substances

1. Pure substances are samples of matter that contain only one type of particle. In a small group, make a list of at least 20 pure substances that you know about. Include everyday substances such as water, copper, and gold. Also be sure to include substances you have used in the laboratory. Remember that pure substances can be solids, liquids, or gases, and their names can have one word or more than one word in them.

2. Scientists have found that chemical changes can split the particles in some pure substances into smaller particles. Other pure substances cannot be broken down. In your group, discuss your list of pure substances. Which ones do you think can be split?

Models of Matter: The Particle Theory

Observing and experimenting with matter has led scientists to theorize about what matter really is. Their working definition of matter is anything that takes up space and has mass. When we observe matter, we see that it behaves in countless ways and can be put to countless uses. But is matter one, definable thing? Over the centuries, scientists have created many models to explain what matter is, to explain what lies beyond what we can physically observe. One of the most enduring models of matter is the particle theory.

Building Blocks of Matter: The Particle Theory

More than 2000 years ago in Greece, a philosopher named Democritus suggested that matter was made up of tiny particles too small to be seen. He thought that if you kept cutting a substance into smaller and smaller pieces, you would eventually come to the smallest possible particles— the building blocks of matter. The basic principles of the particle theory are reviewed in **Table 1**.

Pure Substances and Mixtures

Using the particle theory, we can understand two categories of substances: pure substances and mixtures. A **pure substance** contains only one kind of particle. For example, a piece of aluminum foil contains only aluminum particles. Sugar is a pure substance. It contains only sugar particles. A scoop of sugar made from Canadian sugar beets contains exactly the same kind of particles as a scoop of sugar made from Australian sugar cane.

A **mixture** contains at least two different pure substances, or two different types of particles. When you drink a glass of milk or eat a cookie, you are consuming mixtures of different substances. Most common substances are mixtures. **Figure 1** shows the relationship between pure substances and mixtures.

Table 1	The Particle Theory of Matter
Principle	**Illustration**
1. All matter is made up of tiny particles.	
2. All particles of one substance are the same. Different substances are made of different particles.	substance A substance B
3. The particles are always moving. The more energy the particles have, the faster they move.	hot cold
4. There are attractive forces between the particles. These forces are stronger when the particles are closer together.	particles far apart—force weak particles close together—force strong

Classifying Mixtures

Imagine you have a sample to identify that you think is a mixture. What type of mixture might it be? Perhaps your mixture is a solution, like sugar in water. If you add a small amount of sugar to water and stir, the sugar disappears as it dissolves. A **solution** may be made up of liquids, solids, or gases. Air is a solution of gases. Perfumes are solutions of alcohol and fragrances. Many alloys are solutions of solid metals.

The sample of matter may be a **heterogeneous mixture**, like pizza. When you make a pizza, tomato sauce, cheese, mushrooms, and pepperoni are mixed together or scattered on top of the crust. Each part can be easily seen. Garden soil, oil and vinegar, salad dressing, and garbage are all heterogeneous mixtures.

Elements and Compounds

Many properties of matter can be explained by using the particle theory. But what are these particles? Two hundred years ago, scientists like the one in **Figure 2** already knew of thousands of pure substances and were constantly discovering more. Scientists hoped that by breaking down these substances, they would discover the building blocks of matter—the particles that Democritus suggested couldn't be broken down any further. Once they knew all the building blocks, they believed they would be able to predict the properties of a pure substance from the properties of the "blocks" or "particles" of which it was made.

Figure 1

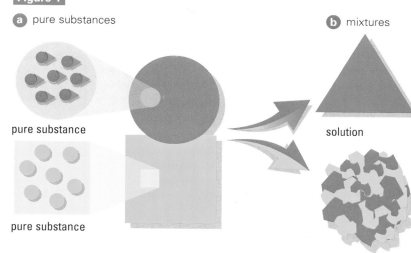

a pure substances
b mixtures

pure substance

pure substance

solution

heterogeneous mixture

a Pure substances: Pure substances contain only one kind of particle.

b Mixtures: When two pure substances are mixed together, sometimes they mix smoothly and sometimes they mix unevenly. If the particles mix very well with one another—so well that you can see only one phase or visible part—the mixture is called a solution. You say that one substance dissolves in another. If the particles don't mix well with one another, you will see more than one phase. This type of mixture is called a heterogeneous mixture.

Figure 2

Using equipment like this, scientists of several centuries ago conducted experiments on matter.

As a result of these efforts, scientists now know of over 100 of these building blocks, which they call elements. **Elements** are pure substances that cannot be broken down into simpler substances. Water, which is formed from two of these elements (hydrogen and oxygen), is a compound. **Compounds** are pure substances that contain two or more different elements in a fixed proportion. They are formed when elements combine together in chemical reactions. For example, in water, there are always twice as many particles of hydrogen as particles of oxygen. **Table 2** shows how the classification of matter can be expanded to include elements and compounds.

Table 2	**Pure Substances and Mixtures**	
Elements in substance	**Type of substance**	**Description**
element A alone	pure substance (element)	Only element A is visible.
elements A and B	pure substance (compound)	The compound may have completely different properties from each of the elements; the elements are chemically combined.
	mixture (solution)	The solution may look very similar to one of the elements, as particles of the other element are hidden within it.
	mixture (heterogeneous)	Particles of each element can be seen separately.

Atoms and Elements

How can we use the particle theory to understand elements and compounds? Once again, water provides a clue. At one time, people thought that water was made up of particles that could not be broken down further. We now know, however, that water can be broken down into hydrogen and oxygen and that, when we measure the volumes of the gases, we always get twice as much hydrogen as oxygen. One water particle is formed from the particles of two elements: two particles of hydrogen and one particle of oxygen as shown in **Figure 3**.

Scientists now call the particles in the particle theory **atoms**. Each element is made of only one kind of atom. Since there are more than 100 elements, there are more than 100 kinds of atoms.

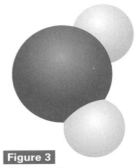
Figure 3
A model of a water molecule

Molecules

Atoms join together in combinations. When two or more atoms join together, a molecule is formed. **Molecules** can contain two atoms or many thousands of atoms. The atoms in a molecule can be all the same kind of atom. For example, in the element oxygen there are two oxygen atoms in each molecule. Molecules can also be made of two or more different kinds of atoms. For example, a compound called butane (a fuel) contains four carbon atoms and ten hydrogen atoms in each molecule.

Figure 4

a Atoms of the element oxygen join together in pairs, forming oxygen molecules.

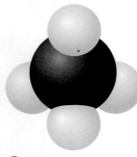

b In the compound methane (the main component of natural gas), each molecule contains one carbon atom and four hydrogen atoms.

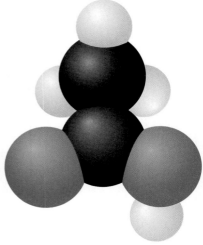

c In acetic acid (vinegar), each molecule has eight atoms: two carbon atoms, two oxygen atoms, and four hydrogen atoms.

Models of Molecules

Examine the models of molecules shown in **Figure 4**. The first shows a molecule that contains two atoms of oxygen. This is a molecule of the element oxygen. The other two show molecules of compounds.

Different Molecules from the Same Elements

In compounds, atoms of one element join together in a fixed ratio with atoms of other elements. For example, when hydrogen and oxygen combine in the ratio of 2:1, the compound they form is always water. What happens if hydrogen and oxygen combine in a different ratio? The result is a different compound with different properties. In nature, there is only one other compound that contains only hydrogen and oxygen: hydrogen peroxide. In hydrogen peroxide, hydrogen and oxygen atoms are in a 2:2 ratio, as shown in **Figure 5**.

The properties of water and hydrogen peroxide are very different. Water is everywhere on Earth, part of all living things. Hydrogen peroxide is a much more reactive compound. In **Figure 6** it is shown interacting with a drop of blood.

Other elements can also be combined in different ratios to produce different compounds. For example, acetic acid (vinegar) contains atoms of hydrogen, oxygen, and carbon. These same elements also combine to form many other compounds, including sugars and fats.

Figure 5
A hydrogen peroxide molecule has two hydrogen atoms and two oxygen atoms. Compare this with the model of water shown in Figure 3.

Figure 6
A hydrogen peroxide solution bubbles as blood is added. Hydrogen peroxide is used in antiseptics. It bubbles when it touches dirt and blood in cuts or scrapes. Water in a similar situation does not react.

Understanding Concepts

1. Use diagrams to explain the difference between

   (a) a pure substance and a mixture

   (b) a solution and a heterogeneous mixture

2. Create a summary chart for the concepts in this section: element, compound, atom, mixture, and molecule. Include the following headings: term, definition, example.

3. Give examples of two molecules that are made from the same types of atoms.

4. State whether each of the following pure substances is an element or a compound. Explain your reasoning.

   (a) A clear, colourless liquid that can be split into two gases with different properties.

   (b) A yellow solid that always has the same properties and cannot be broken down.

   (c) A colourless gas that burns to produce carbon dioxide and water.

Reflecting

5. Look back to the Try This activity in Getting Started. Are there any changes you would make in your answers?

Challenge

The particle theory is a model used to help explain matter. How would you use this model in the challenge you have chosen?

Classifying Elements

When you go into a drugstore, you notice that products with similar properties are grouped together. Soaps and skin care products are in one section, vitamins and herbal products in another, and so on. In the same way, hardware stores often group nails, screws, and other fasteners together. You have already investigated the physical properties of pure substances, including some elements. Can we use these properties to classify elements—the basic building blocks of matter? In this investigation you will examine a number of elements and classify them on the basis of lustre, malleability, and other physical properties.

Materials
- safety goggles
- apron
- small pieces of paper or baking cups
- samples of some of the following elements: aluminum, chromium, iron, nickel, tin, cobalt, lead, copper, magnesium, silicon, zinc, carbon, sulfur, iodine
- magnet
- conductivity apparatus (flashlight bulb and holder, leads with alligator clamps, 1.5 V-battery)

Question
How may a selection of elements be grouped or classified?

Hypothesis

1 Look at the list of elements in the Materials list. Which elements would you group together, based on previous knowledge? Explain your grouping.

Procedure

2 Draw a data table in your notebook as shown in **Table 1**.

3 Put on your apron and safety goggles.

4 Print the name of each element on a small piece of paper or baking cup. Obtain a sample of each element on the paper or in the cup labelled with its name. Throughout this activity, make sure not to mix the samples.

☠ Your teacher will demonstrate properties of silicon and iodine. Students should not handle iodine, which is poisonous.

5 Examine each of the elements and note its colour and lustre (shininess). Your teacher may have scraped the surface of some samples to expose the element underneath.

✏ (a) Record your observations.

6 Try to bend or break the samples. (Your teacher will demonstrate with some elements, including silicon and iodine.) You will recall that substances that can be bent or hammered into a sheet are described as malleable. Substances that shatter or crumble may be described as brittle.

✏ (a) Record your observations.

7 Which elements seem heavy or light for their size, compared with the other elements? (That is, how would you rate their densities?)

✏ (a) Record your observations.

Table 1

| Element | Properties | | | | | |
	Colour	Lustre	Malleability	Density	Magnetism	Electrical Conductivity
?	?	?	?	?	?	?
?	?	?	?	?	?	?

8 Use a magnet to determine which elements are magnetic and which are nonmagnetic.

🖉 (a) Record your observations.

9 Assemble the conductivity apparatus. Touch the electrical leads or alligator clamps to opposite ends of each sample of element. If the lamp glows, the element is a conductor. If the lamp does not glow, the element is an insulator.

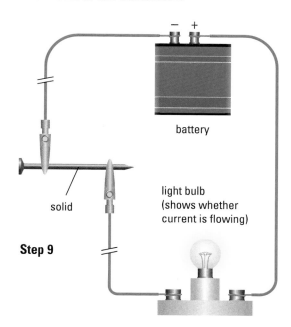

Step 9

battery

solid

light bulb
(shows whether
current is flowing)

🖉 (a) Record your observations.

10 Return the samples of the elements to your teacher for recycling and further use. Put away any materials as directed by your teacher. Clean up your work station. Wash your hands.

Analysis and Communication

11 Based on your observations, can you classify elements according to different properties?

12 Write a summary supporting your view, providing evidence from your investigation.
⑺ᴰ

Making Connections

1. How do the properties of the following elements determine their use?

(a) Copper and aluminum are used in electrical wiring.

(b) Carbon rods are used in some batteries.

(c) Steel (iron) cans can be separated easily from aluminum cans.

2. You may have noticed that when you put a metal object against your skin, it feels cold. Why does this happen? What physical property of metals explains this observation? What uses does this property have?

Exploring

3. Research other ways of classifying elements. Use these new strategies to classify the elements you studied in this activity. Compare your new groupings with the previous ones.

Reflecting

4. Think of situations where it might be useful to classify elements.

Putting Metals to Work

You have investigated the classification of elements as metals and nonmetals based on their physical properties (**Figures 1** and **2**). Similar characteristics can be observed in their chemical behaviour. The physical and chemical properties of **metals** make them very useful substances in many applications. Because copper and aluminum conduct electricity well, they are used in wiring and electrical circuits. The lack of reactivity of gold and silver make them valuable in jewellery. The hardness and strength of metals make them useful in buildings and automobiles. Their ability to be shaped easily and conduct heat have made metals useful for centuries in cooking pots and utensils.

Figure 1
Metals are generally shiny and malleable solids that are good conductors of heat and electricity.

Metals and Industry

People have used metals for tools and jewellery for thousands of years. More recently, many other uses have been found for different metals and mixtures of metals. For instance, platinum is a very unreactive metal that can be used in precision instruments. Uranium is a very dense metal that is used in nuclear power plants.

Many less dense, or "light" metals, also have useful applications. Sodium can be used in chemical reactions. Magnesium is a light, strong metal that can be used in automobile wheels and luggage, especially in cases designed for photographic equipment, which requires extra protection. Aluminum pots and pans are sometimes used in kitchens because of their light weight and good heat conductivity.

One group of metals is shown in **Table 1** and **Figure 3**. These elements are called **heavy metals** because they show the typical shininess and other physical properties of metals, as well as having very high density. Many of these elements are more than four times as dense as water.

Compounds of many heavy metals are essential in small amounts in healthy plants and animals. However, in larger quantities they can cause damage to living things. Heavy metal compounds may be absorbed directly from the air or water through leaves, roots, skin, lungs, or gills. Animals can also absorb them from the organisms they consume. In a

Figure 2
Nonmetals are generally dull, brittle, nonconducting solids or gases at room temperature.

Figure 3
Heavy metals, such as gold, silver, nickel, and copper are used in coinage.

Table 1	Heavy Metal Elements and Their Uses	
Metal	**Selected Properties**	**Typical Uses**
tungsten	very high melting point	light-bulb filaments
chromium	resists corrosion	chrome plating
iron	forms strong alloys	structural steel
copper	good conductor	electrical wire
nickel	resists corrosion	coinage
lead	resistant to acid, soft	batteries
zinc	forms protective coating	galvanized containers
tin	resists corrosion	coating for steel cans
mercury	conductor	home thermostat switches

normal food chain, a predator eats many smaller animals over time and may accumulate poisonous levels of metal compounds in the liver or other organs. Generally, concentrations increase up a food chain from plants to herbivores to carnivores.

Lead is a heavy metal that can have damaging environmental effects. Lead poisoning can harm the nervous system and the brain. Until the 1960s, lead was often used in drain pipes, in paints, and even in gasoline. Unfortunately, young children sometimes ate flakes of paint and people of all ages were harmed by inhaling lead-containing dusts from automobile exhausts. Today, most drain pipes are made from plastic, and alternative paints and lead-free gasolines have been developed. Mercury, shown in **Figure 4**, is another heavy metal that can be dangerous.

Heavy metals in the environment today can come from many sources. Industries, mining operations, and power plants must take care not to carelessly release substances into the air or water.

Figure 4
Mercury metal is unusual because it is a liquid at room temperature. Discharge of mercury compounds into the environment from pulp mills and industrial plants is dangerous because mercury damages the nervous systems of fish and humans.

Did You Know

Mercury was used in felt-making for top hats, so many hatters went mad with mercury poisoning, hence the expression: "Mad as a hatter."

Understanding Concepts

1. Make a chart listing the properties of metals and nonmetals.

2. **(a)** Why are some elements called heavy metals?

   (b) Give examples of three heavy metals and list their uses.

Making Connections

3. Elements are chosen for various uses. Research and write a summary paragraph explaining the following choices of elements:

   (a) Copper is better than aluminum in most electrical wiring.

   (b) Lead is no longer used in the manufacture of food containers.

   (c) Mercury is no longer used in most thermometers.

   (d) Titanium is used in aircraft wings.

Exploring

4. Record the substances described (7B) as minerals that are listed on the label of a multiple vitamin from your medicine cabinet or a local pharmacy. Make a graph and identify which elements are provided in the largest amounts.

5. Plumbers and electricians work (3A) with copper metal as well as solder, a mixture of metals used to join metals together. Find out how they choose particular metals to use for specific purposes.

Challenge

What examples would you display to illustrate how the properties of metals determine their use in everyday products? Which metals were not yet discovered in the 1800s?

2.4 Investigation

SKILLS MENU
- Questioning
- Conducting
- Analyzing
- Hypothesizing
- Recording
- Communicating
- Planning

Breaking Compounds into Elements

When you eat your favourite candy bar, you probably don't give the ingredients too much thought. Maybe one day, though, you'll start thinking about the ingredients that make it taste so good: nuts, caramel, crispy wafers, chocolate, creamy fillings. You'll notice that different ingredients account for different flavours, textures, and smells. You may go further and ask yourself, for instance, what it is in chocolate that makes it sweet or allows it to coat the candy bar. Eventually you'll start asking questions that you can't answer by observation alone. You may even start imagining different models that could explain what the smaller and smaller particles of matter are in your candy bar.

The underlying notion of the models of matter we have considered is that all matter is made up of the same basic ingredients—elements—and that these can be combined in millions of different ways, producing substances with very different properties. For instance, hydrogen peroxide and water are both compounds made up of the elements hydrogen and oxygen, but they are very different liquids used for different purposes—one as an antiseptic on wounds and the other for drinking.

People were aware of water long before they were aware that it is made up of hydrogen and oxygen. In fact, early scientists assumed that water was an element because they had not discovered a way of breaking it down further. One way of breaking water down into hydrogen and oxygen is by **electrolysis**—the use of electricity to cause chemical changes in solutions. Water is a pure substance made up of only one kind of molecule, but this molecule can be broken down, releasing atoms of hydrogen and oxygen (**Figure 1**).

You may recall that chemical changes can be represented by a word equation—a short way of representing a chemical reaction that tells you simply what substances (reactants) are used up and what substances (products) are produced. For the electrolysis of water, the word equation is:

$$\text{water} \xrightarrow{\text{electrical energy}} \text{hydrogen} + \text{oxygen}$$

Materials
- safety goggles
- electrolysis apparatus
- water
- 5 g sodium sulfate
- power supply (6V DC)
- electrical leads
- wooden splints

⚡ Electrical devices can be hazardous.

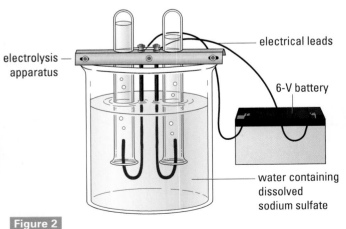

electrolysis apparatus

electrical leads

6-V battery

water containing dissolved sodium sulfate

Figure 1
Hydrogen and oxygen molecules are made of two atoms each. Water molecules contain three atoms, two of hydrogen and one of oxygen.

Figure 2
Apparatus used to show electrolysis of water

Question

(2A) **1** Write a question that is being investigated.

Hypothesis

2 Write a statement that compares the volumes of hydrogen and oxygen that will be produced.

Procedure

3 Put on your safety goggles.

4 Assemble the apparatus as shown in **Figure 2**. (Sodium sulfate or an equivalent solute is used to allow electric current to pass through the water.)

5 Turn on the power.

(a) Note what happens as soon as the power is turned on.

✎ (b) Record your observations.

6 Turn off the power as soon as one of the test tubes is full of gas.

(a) What was the effect of turning off the power?

(b) Compare the relative amounts of gas in the tubes.

(c) Describe the physical properties of the gases without removing the tubes from the water.

7 Examine the tube containing a smaller volume of gas. Place your thumb over the mouth of the tube, remove it from the solution, and hold it mouth up. Light a wooden splint. Test the gas by uncovering the tube and quickly bringing the flaming splint close to the mouth of the tube.

✎ (a) Record your observations.

8 Repeat the same process for the tube full of gas.

✎ (a) Record your observations.

⊛ This test should be attempted on only a small amount of gas in an open-mouthed, shatter-proof container. It should be done only under teacher supervision. Safety goggles should always be worn.

9 Dispose of the contents of your apparatus and put away materials as directed by your teacher. Clean up your work station. Wash your hands.

Analysis and Communication

10 Analyze your observations by answering the following questions:

(a) What was the identity of the gas in the full tube? Explain.

(b) What must have been the identity of the gas in the partially filled tube? Explain.

(c) Write a word equation to describe each of the following: the electrolysis reaction; the reaction that occurred at the mouth of the full tube.

(d) What safety precautions did you take during this activity?

(e) What were the relative volumes of hydrogen and oxygen gases produced? Can you think of a possible explanation? (Hint: Consider the models of the molecules at the beginning of this section.)

11 Summarize the results of this investigation in a paragraph.

Exploring

1. Research how electrolysis is used in industry to produce hydrogen and other elements. Try an Internet search using key words "electrolytic + cells" and "electrolysis + water." Design a poster to present your research.

2. Hydrogen peroxide antiseptic slowly splits apart into oxygen gas and water. But this process can be made to happen much more quickly. Some contact lens cleaning systems use a disc that is coated with a substance that makes this happen. Get a disc from your local drug store and try the reaction.

Case Study

Testing for Elements and Compounds

Imagine that you are a technician in a chemical lab. While cleaning up after a flood in the laboratory, you discover that the labels of three gas cylinders have been damaged. You suspect that the cylinders contain carbon dioxide, oxygen, and hydrogen. How could you find out which cylinder contains which gas and whether the cylinders have been contaminated by the flood water? Do they now contain water vapour? Four common tests are used to identify four common gases.

(a) Why would it be difficult to identify the three gases by their physical properties?

Oxygen—The Glowing Splint Test

One common chemical reaction is combustion (burning). Oxygen must be present for combustion to take place. Substances such as wood and oil burn readily in air, which is about 20% oxygen. In pure oxygen, they burn much more intensely. This chemical property of oxygen—supporting combustion—allows you to identify it. You could test for it as follows (**Figure 1**):

- Light a wooden splint.
- Blow out the flame but leave the splint glowing.
- Hold the glowing splint in a small amount of the unknown gas.
- If the splint bursts into flame, the gas is oxygen.

(b) Why should you blow out the flame first?

(c) Why should you only test a small amount of the gas?

(d) Do you know of any other gases that would give a similar result?

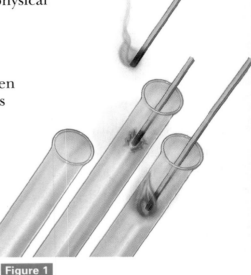

Figure 1
The glowing splint test for oxygen

Hydrogen—The Burning Splint Test

You may have already used the second test (**Figure 2**), based on hydrogen's explosive reaction in air.

Ⓡ This test should be attempted on only a small amount of gas in an open-mouthed, shatter-proof container. It should be done only under teacher supervision. Safety goggles should always be worn.

- Light a wooden splint.
- Hold the burning splint in a small amount of the unknown gas.
- If you hear a loud "pop," the gas is hydrogen.

(e) What property of hydrogen are you testing for?

(f) Why should you only test a small amount of the gas?

(g) Which should you test for first: oxygen or hydrogen? Why?

(h) If the flame goes out, what can you say about the identity of the gas?

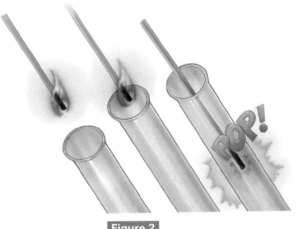

Figure 2
The flaming splint test for hydrogen

Carbon Dioxide—The Limewater Test

Carbon dioxide does not burn and does not allow other materials to burn. If you put a burning splint into carbon dioxide, the flame will go out.

The chemical test for carbon dioxide uses a liquid called limewater, a clear, colourless solution of calcium hydroxide in water. Carbon dioxide reacts with the dissolved calcium hydroxide, producing a precipitate (**Figure 3**). A **precipitate** is a solid, insoluble material that forms in a liquid solution. The precipitate causes the limewater to appear cloudy or milky. If you suspect that one of the cylinders contains carbon dioxide, carry out this test on a small amount of the gas in a test tube.

- Bubble the unknown gas through limewater solution, or
- Add a few drops of limewater to the gas and swirl it around.
- If the limewater turns cloudy or looks milky, the gas is carbon dioxide.

(i) Do you think oxygen or hydrogen would also turn limewater cloudy? Why?

(j) In what order would you carry out the three tests?

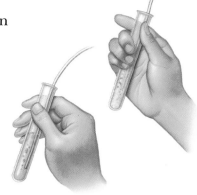

Figure 3

The limewater test for carbon dioxide

Did You Know

Natural gas is a common heating fuel. It is safe and efficient when used properly, but it is odourless. Gas companies add tiny amounts of a strong-smelling substance to the gas before it is piped to consumers. This enables you to detect escaping gas, well before the amounts in the air become dangerous.

Understanding Concepts

1. How would you test for the gas produced in each of the following, and what observations would you expect to make?

   (a) A can of pop fizzes.

   (b) A nail added to a strong acid produces a combustible gas.

   (c) When potassium chlorate is heated, a gas that supports burning is produced.

2. (a) If you placed a glowing splint in a test tube full of a clear, colourless gas, and the glowing stopped, which of the gases discussed here is most likely present in the test tube?

   (b) How could you confirm the identity of this gas?

Exploring

3. Gases such as those mentioned in this case study can be a safety hazard, especially if people are not aware of their properties. Research how labelling, colour-coding, and other systems are used so that gases can be handled safely. Create a poster to communicate this safety information to others.

4. The gases hydrogen and helium are both less dense than air, so either gas may be used in lighter-than-air ships. Use references to research what properties hydrogen and helium have in common, and in what ways they differ. Explain why modern blimps are filled with helium gas instead of hydrogen.

Water Vapour—The Cobalt Chloride Test

Water is a liquid at room temperature, but many chemical reactions produce water vapour as a product. When water vapour touches a cold surface, it condenses to liquid water. To test for the presence of water, use the following test (**Figure 4**):

- Hold a cold surface near the suspected water vapour.
- Touch a piece of blue cobalt chloride paper to any liquid that condenses.
- If the paper changes from blue to pink, water is present.

(k) Is it likely that there would be water vapour in the cylinders? Why?

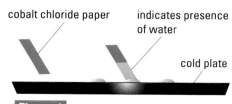

cobalt chloride paper   indicates presence of water

cold plate

Figure 4

The cobalt chloride test for water

2.6 Investigation

SKILLS MENU
○ Questioning   ● Conducting   ● Analyzing
● Hypothesizing   ● Recording   ● Communicating
○ Planning

Identifying Mystery Gases

You have learned how to use chemical tests to detect oxygen, hydrogen, and carbon dioxide gases. In this investigation, you will observe some chemical reactions that produce gases and use the tests to infer which gas is produced.

Hydrochloric acid is corrosive. Any spills on the skin, in the eyes, or on clothing should be washed immediately with cold water. Report any spills to your teacher.

Hydrogen peroxide is poisonous and a strong irritant. Manganese dioxide is also toxic. Report any spills to your teacher.

Materials

- safety goggles
- apron
- 4 test tubes
- test-tube rack
- hydrogen peroxide (3% solution)
- manganese dioxide powder
- toothpick
- 3 wooden splints
- lighter for splints
- hydrochloric acid (5% solution)
- magnesium ribbon (3 to 5 cm long)
- tongs
- limewater solution
- sodium bicarbonate (baking soda)
- test-tube stopper

Question

Each of the following reactions produces a colourless, odourless gas. Which gas is produced in each reaction?

Hypothesis

1 After reading the Procedure, write a
(4A) hypothesis to predict the gas that is produced in each reaction.

Procedure
Part 1: Hydrogen Peroxide and Manganese Dioxide

2 Put on your apron and safety goggles.

(6D) **3** Make a table to record your observations.

4 Put a clean, dry test tube in the test-tube rack. Pour about 4 mL of hydrogen peroxide solution into the test tube. Obtain a tiny amount of manganese

dioxide powder on the blunt end of a toothpick. Observe the two reactants.

(a) Record your observations of the two substances in your table.

5 Add the manganese dioxide to the hydrogen peroxide. Allow the reaction to proceed for 15 s, noting any changes. Bring a burning splint close to the mouth of the test tube. If no reaction occurs, blow out the flame and insert the glowing splint halfway into the test tube.

(a) Record your observations of the reaction.

(b) Record the results of the splint tests.

(c) What gas was produced in the reaction?

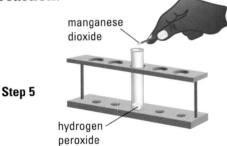

manganese dioxide

Step 5

hydrogen peroxide

Part 2: Hydrochloric Acid and Magnesium

6 Put another clean, dry test tube in the rack. Carefully pour about 4 cm depth of hydrochloric acid solution into the test tube. Obtain about 3 to 5 cm of magnesium ribbon.

(a) Record your observations of the reactants.

7 Roll the magnesium into a ball and use tongs to add it carefully to the acid. Avoid splashing. Allow the reaction to proceed for 15 s, noting any changes. Bring a burning splint close to the mouth of the test tube. If no reaction occurs, blow out the flame, and insert the glowing splint halfway into the test tube.

SKILLS HANDBOOK: (4A) Asking Questions and Hypothesizing   (6D) Creating Data Tables

✎ (a) Record your observations of the reaction.

Step 7

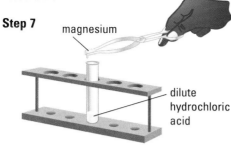
magnesium
dilute hydrochloric acid

✎ (b) Record the results of the splint tests.

(c) What gas was produced in the reaction?

Part 3: Hydrochloric Acid and Sodium Bicarbonate

8 Put two clean, dry test tubes into the rack. Pour about 4 mL of fresh limewater into the first test tube. Pour about 4 mL of hydrochloric acid into the second tube. On a piece of paper, obtain a small amount (about enough to cover a penny) of sodium bicarbonate.

✎ (a) Record your observations of the reactants.

9 Slowly add the sodium bicarbonate to the test tube containing the hydrochloric acid. After about 5 s, put a burning splint close to the mouth of the test tube. If there is no reaction, blow out the splint and insert the glowing end into the tube.

✎ (a) Record your observations.

✎ (b) Record the result of the splint tests.

10 If the splint went out, carefully pour the product gas from the reaction tube into the limewater tube.

⊘ Be careful—do not allow any of the liquid to pour from the reaction tube.

Step 10

limewater
hydrochloric acid and sodium bicarbonate

11 Put a stopper into the limewater test tube to seal it. Mix the limewater and gas by turning the tube upside down several times.

✎ (a) Record your observations.

(b) What gas was produced in this reaction? Do you know for certain? Explain.

12 Dispose of the mixtures in your test tubes as directed by your teacher. Clean up your work station. Wash your hands.

Analysis and Communication

13 Analyze your observations by answering the following questions:

(a) Why did you record your observations of the reactants before proceeding with each chemical reaction?

(b) What gas(es) were you testing for with the burning splint? the glowing splint?

(c) What gas were you testing for with limewater?

(d) What other indication did you have that this gas might be present?

(e) What kinds of changes occurred in each part of the investigation?

(f) What evidence do you have of each change?

14 Make a table listing the physical and chemical properties of the gases produced in this investigation.

Making Connections

1. Which gas seemed to be the most hazardous in this activity? Create a safety sheet outlining the steps you would take to handle this gas safely.

Reflecting

2. List any problems you encountered when following the procedure and suggest ways the procedure could be improved to eliminate the problems.

Chemical Symbols and Formulas

A Canadian chemical company that hopes to sell its products in Romania may need to hire an interpreter to communicate with potential customers. But the interpreter does not need to translate the formulas of any chemicals. That is because all countries use the same chemical symbols to represent elements and compounds, even when the name is different. For example, the same symbol, Fe, represents the element that people call *iron* in Canada, *fer* in France, and *fier* in Romania. A chemist in any country can identify the contents of the bottle in **Figure 1**.

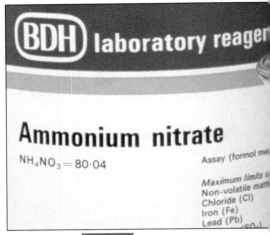

(BDH) laboratory reagen

Ammonium nitrate

$NH_4NO_3 = 80 \cdot 04$

Assay (formol me

Maximum limits
Non-volatile mat
Chloride (Cl)
Iron (Fe)
Lead (Pb)

Figure 1

Chemists rely on symbols and formulas to help them keep track of chemicals.

Chemical Symbols

The alchemists in the Middle Ages were among the first to recognize that it would be convenient to represent chemical substances using symbols. A circular chart of symbols, developed in China in the thirteenth century, linked the elements to a cycle of space and time. Another system of symbols of elements and compounds was devised by the English chemist, John Dalton, in 1808. He represented elements as circles with letters or designs inside them, but scientists found that Dalton's system was difficult to use. These symbols are compared in **Figure 2**.

Today, a common set of symbols for elements is accepted around the world, even though some names may vary. A **chemical symbol** is an abbreviation of the name of an element. Some symbols are shown in **Table 1**. The names and symbols for elements come from many sources.

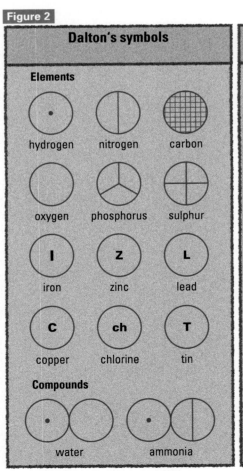

Figure 2

Dalton's symbols			Alchemists' symbols
Elements			
hydrogen	nitrogen	carbon	iron
oxygen	phosphorus	sulphur	tin
iron (I)	zinc (Z)	lead (L)	mercury
copper (C)	chlorine (ch)	tin (T)	lead
Compounds			silver
water		ammonia	sulphur

Table 1

Some Modern Symbols for Elements

aluminum	Al
bromine	Br
calcium	Ca
carbon	C
chlorine	Cl
copper	Cu
fluorine	F
gold	Au
hydrogen	H
iron	Fe
lithium	Li
magnesium	Mg
neon	Ne
nickel	Ni
nitrogen	N
oxygen	O
phosphorus	P
potassium	K
silicon	Si
silver	Ag
sodium	Na
sulfur	S

Hydrogen comes from the Greek word for "water-former." Mercury was named after a Roman god, but its symbol, Hg, comes from the Latin word *hydrargyrum* for "liquid silver." Sodium was named for sodanum, a headache remedy, and its symbol, Na, came from the Latin word *natrium*. Notice that a single-letter symbol is always capitalized, and that the first letter of a two-letter symbol is capitalized while the second letter is not.

Chemical Formulas

Just as single symbols are used to represent elements, combinations of these symbols are used to represent compounds. A **chemical formula** is the combination of symbols that represents a particular compound (**Table 2**). The chemical formula indicates which elements are present in the compound and in what proportion, as shown in **Figure 3**.

Table 2	Some Examples of Chemical Formulas
Name of substance	**Formula**
sodium bicarbonate (baking soda)	$NaHCO_3$
calcium carbonate (chalk)	$CaCO_3$
sodium nitrate (fertilizer)	$NaNO_3$
calcium phosphate (fertilizer)	$Ca_3(PO_4)_2$
sodium chloride (salt)	$NaCl$
acetylsalicylic acid (ASA or aspirin)	$C_9H_8O_4$
acetic acid (vinegar)	$C_2H_4O_2$

Each chemical symbol in a formula represents an element. If only one atom of an element is present in a compound, no number is included. If there is more than one atom of that element in the compound, the symbol is followed by a number written below the line. This number (called a *subscript*) tells how many atoms of that element are present in one molecule. For example, the formula for water—H_2O—tells you that the elements are present in the ratio of two atoms of hydrogen to one atom of oxygen. The formula for sodium bicarbonate—$NaHCO_3$—tells you that the elements are present in a ratio of one atom of sodium to one atom of hydrogen to one atom of carbon to three atoms of oxygen.

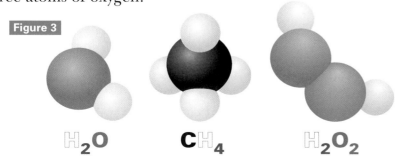

Figure 3

H_2O CH_4 H_2O_2

Understanding Concepts

1. Why are symbols useful in describing chemicals?

2. What are the symbols for the following elements?
   (a) calcium   **(b)** iron
   (c) chlorine   **(d)** phosphorus
   (e) copper

3. Write a chemical formula for the following:
   (a) a molecule of hydrogen gas that is made up of two atoms of hydrogen
   (b) a molecule of propane gas that is made up of three atoms of carbon and eight atoms of hydrogen

4. For each of the compounds in **Table 2**, state the number of different elements present and the number of atoms of each element. What is the total number of atoms in each molecule?

5. Molecules of nitrous oxide, used by dentists as an anaesthetic, contain two atoms of nitrogen and one atom of oxygen. Write the chemical formula for nitrous oxide.

Making Connections

6. Research a common use for each (3A) of the following pure substances:
   (a) helium gas   **(b)** sucrose
   (c) acetone   **(d)** tartaric acid
   (e) propane

7. Prepare a chart describing each substance above, where it is used, chemical formula, and the number of atoms in each type of element in the compound.

Challenge

What chemical symbols and formulas will you need in the challenge you have chosen?

Atoms, Molecules, and the Atmosphere

Slowly breathe in a lungful of air. As you do, think about the billions of molecules that you are inhaling. Air is mostly nitrogen and oxygen, but you are breathing in a mixture that contains other gases as well. In this section, you will learn about some of the gases found in air (**Figure 1**).

Nitrogen (N_2)

Two atoms of the element nitrogen combine to form a molecule of the gas nitrogen (**Figure 2a**). Nitrogen makes up approximately 80% of the atmosphere. It is not very reactive, which means we can breathe it safely without causing chemical changes in our lungs. However, under certain conditions, such as in a car engine, nitrogen gas reacts with oxygen to produce nitrogen dioxide (NO_2), a very toxic red-brown gas. Nitrogen dioxide in low concentrations causes the yellow haze of air pollution you may have seen in some cities.

Argon (Ar)

Argon atoms do not combine with other atoms to form molecules. As a result, argon gas is composed of single atoms of argon (**Figure 2b**). Almost all of the argon in the atmosphere has leaked out from inside the Earth. This gas is completely harmless and quite useful, especially for filling electric light bulbs and fluorescent tubes.

Oxygen (O_2 and O_3)

Atoms of the element oxygen can combine to form two different molecules. The more common of these contains two atoms of oxygen. This is the form that makes up about 21% of the air you breathe and is what we commonly call oxygen gas (**Figure 2c**). Almost all organisms need oxygen to survive, as it is used in cellular respiration.

The less common oxygen molecule, called ozone (O_3), contains three atoms of oxygen (**Figure 2d**). Ozone is formed naturally in the upper layers of the atmosphere. It is very important to life on Earth because it absorbs most of the ultraviolet radiation from the Sun. If all of this radiation reached the surface of Earth, it would harm all living things exposed to it.

Unfortunately, air pollutants such as chlorofluorocarbons (CFCs) have been destroying the ozone layer at an alarming rate. Worldwide measures to stop this pollution have begun, but it will be years before the risk to the ozone layer is past. Because of damage to the ozone layer, more ultraviolet light is now reaching Earth's surface. Ultraviolet light damages skin. As a result, scientists now encourage people to protect their skin with clothing or sunscreen when they go out in the sunshine.

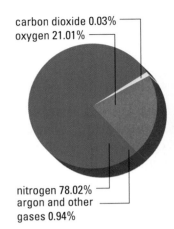

carbon dioxide 0.03%
oxygen 21.01%
nitrogen 78.02%
argon and other gases 0.94%

Figure 1
Gases in Earth's atmosphere

Figure 2

a nitrogen molecule

b argon atom

c oxygen molecule

d ozone molecule

While the layer of ozone several kilometres up in the upper atmosphere is necessary for life, ozone at ground level is hazardous to living things. It can damage plants and causes respiratory problems in people and other animals because it reacts with lung tissue. It is produced when certain gases, produced mainly by vehicles, react with each other and with the more common O_2 molecule.

Carbon Dioxide (CO_2) and Carbon Monoxide (CO)

Two atmospheric gases contain only atoms of carbon and oxygen (**Figure 3**). One, carbon dioxide, is necessary for life on Earth. The other, carbon monoxide, is extremely poisonous to vertebrate animals.

When fossil fuels burn, the two main products are carbon dioxide and water. However, if there is a shortage of oxygen during combustion, carbon monoxide is also produced. How can the supply of oxygen be limited? If you burn propane indoors, for instance in a gas barbecue or heater, you might use up most of the oxygen from the air in the room. The same might happen if you run an automobile engine in a closed garage.

The carbon monoxide molecule (CO) is similar to the oxygen molecule (O_2). This similarity makes carbon monoxide poisonous. When carbon monoxide molecules enter the lungs, the body's red blood cells treat CO molecules as if they were O_2 molecules. Instead of oxygen, the cells carry CO through the body. The cells of the body are starved of the oxygen they need. Death can result.

In order to prevent accidental fatalities due to CO poisoning, many municipalities are passing bylaws that require CO detectors in every home.

Did You Know

A sign of carbon monoxide poisoning is that the skin turns redder than normal.

Figure 3

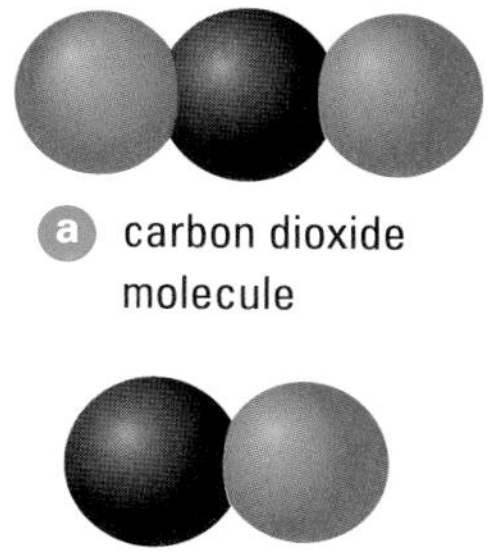

(a) carbon dioxide molecule

(b) carbon monoxide molecule

Try This What Is Air?

Is air an element or a mixture of several substances? Moisten a piece of steel wool. Drop it into an empty jar and then invert the jar over a pan of water. Mark the liquid level in the jar. Leave the jar for 24 h. Mark the liquid level again. What has happened to the air in the jar? Test the jar for oxygen using a glowing splint. What do your observations suggest?

Understanding Concepts

1. Identify (as elements or compounds) each of the following molecules:

   (a) carbon dioxide

   (b) carbon monoxide

   (c) oxygen

   (d) ozone

   (e) nitrogen

2. Carbon dioxide and carbon monoxide contain only carbon and oxygen. Which of these molecules is dangerous to breathe? Why is it dangerous?

3. **(a)** How is ozone formed at ground level?

   (b) What effect does this ozone have on living things?

4. If humans and all other animals are constantly producing carbon dioxide, and it is a fairly stable gas, why does it make up only a tiny percentage of Earth's atmosphere?

Exploring

5. People with some diseases, such as emphysema, are given air with an increased percentage of oxygen. Find out what proportion of oxygen is in this air. What safety precautions would you advise for people near the patient? Create an information brochure.

6. Visit an automotive service centre.

   (a) What device is used to prevent any buildup of carbon monoxide inside the garage?

   (b) How is fresh air brought into the garage?

   (c) How would fresh air reduce the amount of carbon monoxide produced by cars?

Reflecting

7. Would you make any changes to the list of substances that you produced at the beginning of the chapter?

Building Models of Molecules

What do a toy car and the particle theory of matter have in common? Both are models. **Models** can be mental pictures, diagrams, or three-dimensional constructions. The particle, model helps us to understand how matter behaves, just as a model car helps a child to understand how an automobile works.

For example, engineers test models of new jet aircraft in wind tunnels to find out how well the design operates at supersonic speeds. Architects and designers make models of floor plans to design efficient buildings. Such three-dimensional models may be held together with nails or glue, or drawn on a computer screen.

In the same way, scientists make models of molecules in order to understand them and to predict how the molecules will behave (**Figures 1, 2a, 4a**). In these models, the atoms are held together by connections called **bonds**. The connections represent electrons that "glue" or bond the atoms together. The molecules can also be represented by drawings on paper called **structural diagrams**. In these diagrams (**Figures 2b, 3, 4b**), each atom is represented by its chemical symbol and each bond is represented by a straight line drawn between the symbols.

Each kind of atom generally forms a set number of bonds. For example, each hydrogen atom forms only one bond. Each oxygen atom forms two bonds: either two single bonds with two other atoms, or one double bond with one other atom. In this activity, you will build models of some common molecules and see how bonds can link atoms in molecules.

Materials

- molecular model kit, made up of "atoms" and connectors (the atoms may be represented by balls, soaked chickpeas, or small coloured marshmallows; connectors may be sticks, springs, or toothpicks)
- coloured pens

Figure 1
The structure of the DNA (deoxyribonucleic acid) molecule, which carries our genetic characteristics, was discovered by Francis Crick and James Watson. They figured out the structure by working with molecular models similar to yours.

Figure 2
Rubbing alcohol is actually named isopropanol. It can be represented by **a** building a ball-and-stick model or by **b** drawing a structural diagram.

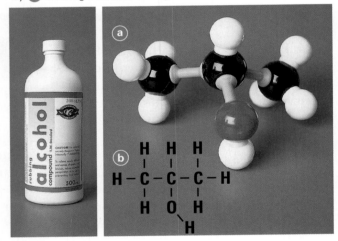

Figure 3 **a** $O=O$ **b**
Structural diagrams of oxygen and water

Figure 4
The molecule acetylene, C_2H_2, is used to melt and join metals together. Note that each carbon atom still makes a total of four bonds, but one bond is a triple bond. Each hydrogen bond makes a single bond.

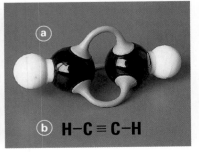

b $H-C \equiv C-H$

Procedure

1 Examine the model kit that you have been provided. If you have a ball-and-stick kit, you will notice that different atoms have different numbers of holes to represent the number of connections that they must make with other atoms. The number of connections for each atom is summarized in **Table 1**.

Table 1	
Atom	**Number of Connections per Atom**
hydrogen	one
oxygen	two
nitrogen	three
carbon	four

2 Make a model of hydrogen gas by connecting two atoms of hydrogen with a connector.

(a) Draw a structural diagram of the molecule. Write the name and formula, H_2, beside your diagram.

3 Make a model of oxygen gas by joining two atoms of oxygen with two connectors, to represent the two connections that each oxygen atom usually makes.

(a) Draw a structural diagram of the model. Write the name and formula, O_2, beside your diagram.

4 Make models of the following molecules: nitrogen (N_2), ammonia (NH_3), methane (CH_4), water (H_2O), ethene (C_2H_4), and carbon dioxide (CO_2). Make sure that each atom makes the correct number of connections.

(a) Draw structural diagrams of the models. Write the formula beside each diagram.

5 If you have time, obtain three carbons and eight hydrogens. See how many different molecules you can make with some or all of the atoms. Remember how many connections each atom can have.

(a) Draw structural diagrams of the models. Write the formula beside each diagram.

Understanding Concepts

1. Why do chemists find making models of molecules useful?

2. Explain, with an example for each, how oxygen can make two single bonds with other atoms or one double bond.

3. Compare the advantages and disadvantages of representing molecules using space-filled models or structural diagrams.

4. Usually, more bonds between two atoms make a stronger connection. Which of all the molecules you made probably has the strongest bond?

5. Sulfur dioxide (SO_2) is an air pollutant that is one of the causes of acid rain. In this molecule, sulfur makes four bonds. Draw a structural diagram of the molecule.

6. Try to make molecules and draw structural diagrams of the following molecules: formaldehyde (H_2CO, used in preserving biological specimens), methanol (H_3COH, used as a de-icer), dimethyl ether (H_3COCH_3, used in refrigeration).

Exploring

7. Use a molecular computer program to build models of the molecules in this activity. Which type of model do you think is more useful? Explain your answer.

Reflecting

8. All simple models have limitations. Have you encountered any substances in this unit for which you cannot construct models using the number of bonds described above?

Challenge

What are some other examples of models? Why do people find models useful? How can you use models in the challenge you have chosen?

Names and Formulas for Compounds

You have learned that chemical formulas can be written to represent compounds. But how do we know the proportions of each element? For example, we may know that table salt, sodium chloride, contains the elements sodium (Na) and chlorine (Cl), but is the compound NaCl or $NaCl_2$ or Na_2Cl or some other formula? The key is in knowing how atoms combine. Some of the basic principles are summarized in **Table 1**.

Table 1	How Elements Combine
Rule 1: Metals combine with nonmetals in many compounds.	
Rule 2: Write the name of the metal first and the nonmetal second.	
Rule 3: Change the ending of the nonmetal to "ide."	
Rule 4: Each atom has its own combining capacity.	
Rule 5: Atoms combine so that each can fill its combining capacity.	

Combining Capacity

After performing many experiments, scientists discovered patterns in the ability of different elements to combine to form compounds. They analyzed how many atoms of each element were present in a molecule of each of the compounds. This ability to combine with other elements is called the **combining capacity**. Combining capacity is similar to the number of connections that an atom can make.

Scientists have given a numerical value to the combining capacity of each metal or nonmetal to explain the compounds that they form. For example, both sodium and chlorine were assigned a combining capacity of 1. Sodium chloride has the formula NaCl: it contains one atom of sodium for each atom of chlorine. Calcium chloride has been found experimentally to have the formula $CaCl_2$. Each of the chlorine atoms has a combining capacity of 1, so the combining capacity of calcium must be 2.

You can better understand this idea by thinking of the work you have done with models. Aluminum has a combining capacity of 3. That is, it needs to make three connections. Each chlorine atom only needs to make one connection. Thus, when aluminum combines with chlorine, the resulting compound is aluminum chloride ($AlCl_3$), a compound used as an antiperspirant.

Tables 2 and **3** list the combining capacities of metals and nonmetals. You can use these numbers to predict the chemical formulas of the compounds that such elements form. Sodium bromide (**Figure 1a**), used in photography, is made up of sodium and bromine. Each of these elements has a combining capacity of 1. Thus, the chemical formula is NaBr.

The compound calcium oxide (**Figure 1b**), or lime, is made up of the elements calcium and oxygen. For both of these elements, the

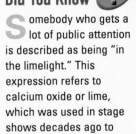

Did You Know

Somebody who gets a lot of public attention is described as being "in the limelight." This expression refers to calcium oxide or lime, which was used in stage shows decades ago to produce a brilliant white light for footlights.

Table 2

Combining Capacities of Some Metals

Element	Symbol	Combining capacity
aluminum	Al	3
barium	Ba	2
calcium	Ca	2
magnesium	Mg	2
potassium	K	1
silver	Ag	1
sodium	Na	1
zinc	Zn	2

Table 3	Combining Capacities of Some Nonmetals		
Element	**Symbol**	**Combining capacity**	**Combined name**
bromine	Br	1	bromide
chlorine	Cl	1	chloride
fluorine	F	1	fluoride
iodine	I	1	iodide
oxygen	O	2	oxide
sulfur	S	2	sulfide

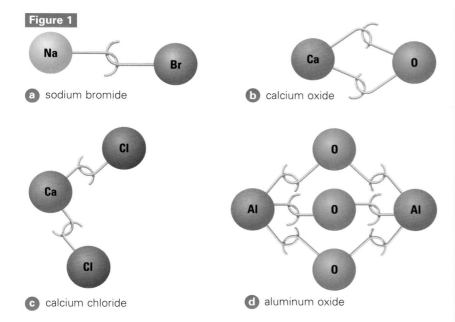

Figure 1

(a) sodium bromide

(b) calcium oxide

(c) calcium chloride

(d) aluminum oxide

combining capacity is 2. Thus, one atom of calcium can combine with one atom of oxygen. The chemical formula of calcium oxide is CaO.

If the combining capacities of the elements are different, then the numbers of atoms are also different. For example, calcium has a combining capacity of 2, and chlorine has a combining capacity of 1. Therefore, in the compound calcium chloride (**Figure 1c**), one atom of calcium combines with two atoms of chlorine. The chemical formula of calcium chloride is $CaCl_2$.

Aluminum has a combining capacity of 3 and oxygen has a combining capacity of 2. Therefore, in the compound aluminum oxide (**Figure 1d**), two aluminum atoms must combine with three oxygen atoms. The chemical formula of aluminum oxide is Al_2O_3.

Metals with More Than One Combining Capacity

Some metals have different combining capacities in different compounds. You can see in **Table 4** that lead, copper, tin, and iron all have more than one possible combining capacity. In naming compounds of these metals, their combining capacity is shown in Roman numerals following the name of the metal. For example, there is a compound named iron(II) oxide ("iron-two-oxide") and another compound named iron(III) oxide ("iron-three-oxide"). Similarly, there are two compounds of copper and oxygen.

Table 4

Some Elements That Have More Than One Combining Capacity

Element	Symbol	Combining capacity
copper	Cu	1, 2
iron	Fe	2, 3
lead	Pb	2, 4
tin	Sn	2, 4

Understanding Concepts

1. What does the term combining capacity mean?

2. Elements can be classified as metals or nonmetals. Which elements change their names when they form compounds? Explain, with an example.

3. What are the names of the following compounds?

   (a) $CaCl_2$, used in bleaching powder and for melting ice

   (b) CaO, used in plaster for construction

   (c) CuCl, used to make red-coloured glass

   (d) KI, added to "iodized" table salt to prevent a condition called goitre

   (e) AgCl, used in photography

4. Use the values for combining capacities shown in the tables to write chemical formulas and draw "hook-and-ball" diagrams for:

   (a) sodium fluoride

   (b) magnesium fluoride

   (c) potassium bromide

   (d) zinc oxide

   (e) silver oxide

   (f) aluminum fluoride

   (g) aluminum sulfide

5. (a) Tin forms two different compounds with chlorine: $SnCl_2$ and $SnCl_4$. Build a simple model of each compound showing tin's combining capacity.

   (b) Give the names and formulas of the two compounds that tin could form with oxygen.

Exploring

6. How do sodium chloride and calcium chloride melt ice? What are the advantages and disadvantages of each compound? Why is urea sometimes preferred to either of these compounds?

Plant Nutrients and Fertilizers

A farmer looks at a vegetable crop about to be harvested (**Figure 1**). The growing season has been a good one with lots of sunlight, warm weather, and rain. But the farmer has made some deliberate and important decisions to increase yield and make the plants as healthy as possible. The plants grow in soil that has been carefully enriched by the addition of chemical substances. These substances added to the soil to help plants to grow are called **fertilizers**. Why have fertilizers been added? What chemicals are in each type of fertilizer? How does the farmer choose which kind of fertilizer to use?

Figure 1

Farmers need to make decisions about how to increase the amount of vegetables or fruit that their fields can produce.

What Plants Need from Soil

Plants take in and process chemical compounds called **nutrients** in order to grow. The plants take in water through the roots, and carbon dioxide from the air through openings in the leaves. Using energy from the Sun, they combine these compounds in a chemical change called **photosynthesis** to produce sugars and oxygen gas. This change can be represented by the word equation:

$$\text{carbon dioxide} + \text{water} \xrightarrow[\text{chlorophyll}]{\text{sunlight}} \text{sugars} + \text{oxygen}$$

Table 1	Nutrient Elements Essential for Plants	
Elements Needed...		
...in large amounts	**...in medium amounts**	**...in tiny amounts**
carbon	calcium	iron
hydrogen	sulfur	manganese
oxygen	magnesium	boron
nitrogen	copper	—
phosphorus	zinc	—
potassium	chlorine	—
—	cobalt	—

However, **Table 1** shows that all living things need elements other than carbon, oxygen, and hydrogen. These additional nutrients are dissolved in the water that the plant takes in through its roots. Some of the nutrients come from minerals in the soil and some from decomposing plant and animal matter. As the plants grow, they use more and more of these nutrients and the soil becomes less rich in the elements that plants need. Nutrients can also be lost as rainwater moves down through the soil to the water table, taking water-soluble nutrients with it.

The farmer must then decide how to replace these elements. One method is to remove just the useful part of the plant, leaving most of the plant to be ploughed back into the soil. Corn is often harvested in this way. Another choice is to add fertilizers to the soil to put back the elements that plants need (**Figure 2**). The fertilizers can be either chemical or organic. Don't be confused by the names: organic fertilizers have some of the same chemicals as chemical fertilizers, but they are produced naturally.

Figure 2

The crop on the left was given chemical fertilizer. The one on the right was not fertilized.

Chemical Fertilizers

The most important nutrient elements, after carbon, hydrogen, and oxygen, are nitrogen (N), phosphorus (P), and potassium (K) (**Figure 3**). But these elements have to be absorbed as compounds, not as elements. In a chemical reaction called nitrogen fixation, a few plants are able to change nitrogen from the air (N_2) into compounds called nitrates (for example, ammonium nitrate or NH_4NO_3). The nitrates can dissolve in water and be absorbed through the roots. Plants absorb phosphorus in the form of compounds called phosphates (for example, sodium phosphate or Na_3PO_4). Potassium forms many compounds that are soluble in water. It can even form compounds with nitrate or phosphate. Potassium phosphate is K_3PO_4 and potassium nitrate is KNO_3. Other sources of nutrients are summarized in **Table 2**.

Chemical industries mix these compounds together in very precise amounts depending on the particular crop or the time of year that the fertilizer will be spread on the fields. These three nutrients are so important that, typically, fertilizers used on the farm or in the home garden are described by three numbers corresponding to the percentages of these three elements in the mixture. For example, 15-10-5 fertilizer contains 15% nitrogen (as nitrate), 10% phosphorus (as phosphate), and 5% potassium.

SKILLS HANDBOOK: (8A) Reporting Your Work

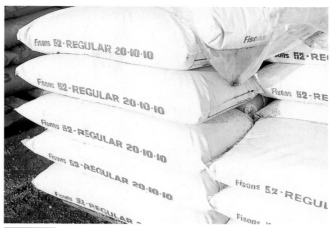

Figure 3

The number rating on a bag of fertilizer indicates, in order, the ratios of nitrogen (as nitrate), phosphorus (as phosphate), and potassium in the mixture.

Table 2

Some Chemical Sources for Major Plant Nutrients

Element	Source
carbon	carbon dioxide (CO_2)
hydrogen	water (H_2O)
oxygen	water (H_2O)
nitrogen	ammonia (NH_3) ammonium nitrate (NH_4NO_3) urea (N_2H_4CO)
phosphorus	calcium dihydrogen phosphate ($Ca(H_2PO_4)_2$)
potassium	potassium chloride (KCl)

Try This Read the Labels (8A)

Visit a garden supply centre, and note the information and instructions on the packaging for (a) at least three kinds of fertilizer that are not labelled "organic," and (b) at least one fertilizer that is labelled "organic." Select a label that has a high percentage of nitrogen and note which plants this fertilizer is recommended for. Repeat this step, looking for a fertilizer high in phosphorus. Note what other nutrients besides nitrogen (nitrate), phosphorus (phosphate), and potassium are mentioned on the labels. Note what the organic fertilizers are made from.

Organic or Chemical Farming?

When the farmer wants to add fertilizers to the soil, there are two choices: natural or chemical. The farmer can choose a natural fertilizer, such as manure or compost. Such fertilizers are often called "organic" fertilizers, because they are natural materials produced by living organisms. Many people favour this kind of fertilizer because it recycles materials that otherwise might be wasted. Such fertilizers can also save energy because the manure or compost may be produced on or near the farm and will probably not require transportation over long distances. A disadvantage of using natural fertilizers, apart from a rather powerful smell, is that farmers cannot know the exact amounts of nutrients that they are adding. Very large farms also may have difficulty obtaining enough natural fertilizer for their needs.

Organic farmers not only use natural fertilizers, but also avoid chemical pesticides and herbicides. Still, most farmers choose chemical fertilizers that are produced by industries. They can include exact amounts of any individual nutrient, so that the farmer can deliver just the right mixture of nutrients to a particular field or crop. These fertilizers are often packaged conveniently, and are so concentrated that a small amount of a chemical fertilizer can deliver a large amount of nutrients. Chemical fertilizers can also be mixed with water and delivered through pipes or sprayed over large areas of fields (**Figure 4**).

Figure 4
Fertilizer spreaders are used by farmers to add nutrient elements to the soil.

Table 3

Percentages of Nutrients Removed by Different Crops

Nutrient Element	N	P	K	Ca	Mg	S
wheat (for bread)	1.7	0.3	0.5	0.05	0.1	0.1
grass (for cattle feed)	1.4	0.3	1.5	0.3	0.1	0.1
potato	0.3	0.04	0.5	0.02	0.02	0.03

Understanding Concepts

1. What elements are needed in large amounts by plants?

2. What are the major natural sources of nutrients for plants?

3. Use **Table 2** to state what types of atoms and how many of each are found in a molecule of

   (a) urea

   (b) calcium dihydrogen phosphate

Making Connections

4. Plants take in nitrogen, phosphorus, and other atoms through their roots and use the atoms to build roots, stems, and leaves. The following questions are based on **Table 3**, which shows the percentage of nutrients removed from the soil in a growing season by different crops:

   (a) Which element is most used up as crops grow?

   (b) Which crop uses potassium (K) at the greatest rate?

   (c) Which crop would require a fertilizer mix with the most calcium (Ca)?

Exploring

5. Find out from a fertilizer supplier which fertilizer mixtures are used at different times of the year and why. For example, one mixture might be used in the fall to promote strong root growth to help the plant survive the winter.

6. Consider the following statement:
   (3B) "Chemical fertilizers are preferable to organic fertilizers." Decide whether you agree or disagree with this statement and write a persuasive essay putting together as many arguments as possible to support your point of view. Use arguments and evidence from this text or from research. Consider examples from large- and small-scale farming, city parks, home gardening, or any other situation where fertilizers are used.

Science Teacher

Anya Martin is a high school science teacher whose aim is not to produce scientists, but to produce science-minded people.

Her parents encouraged her interest in the world around her with frequent nature walks and family outings to the Ontario Science Centre: a great place for children with a love of discovery. She also attended many weekend environmental workshops for young people.

Throughout high school her interest in science and the environment grew, leading Martin to enroll in the University of Toronto's Biology program. She followed this up with a teaching degree in Science and Environmental Science and has taught in Toronto-area high schools ever since.

How does Martin share her enthusiasm with her students? She tries to communicate her excitement and interest in the events— past, present, and future—that shape our view of how the world works. To make science more real and immediate, she relates it to our everyday lives. In chemistry, for example, she focuses on familiar reactions—"kitchen chemistry"—and helps her students realize that they already know a lot about the subject; they just need a little help interpreting what they see all around them.

She is fascinated by the fast-paced changes in so many areas of science but is concerned that we may be ignoring ethical issues in our rush toward achieving what used to be impossible. "Perhaps we should slow down; take control of the changes and developments," she suggests. "Students need to be comfortable with science; to understand it well enough to be able to participate in the debates on where science should go from here."

Martin advises students to seize the opportunity to study science. It feels good to be able to say "I know what that means!" Take that science class and enjoy yourself with it!

> Include science in all aspects of your life. It's there anyway.

Exploring

1. Martin has certain characteristics that make her well-suited to her profession. What are they?

2. Make a chart to illustrate the educational and life choices you would have to make if you wanted to become a science teacher.

3. Martin is concerned about the ethics of some recent science developments. List four aspects of science that have some ethical questions associated with them.

Metal Extraction and Refining in Canada

Think of all the materials around you that are made of metals. Very few of these metals occur naturally as elements. Gold nuggets are still panned from stream beds, and silver and copper are often found as elements. But most metals in nature are found in the form of compounds because they combine very readily with oxygen, sulfur, or other elements. These compounds are called **minerals** (**Table 1**). For example, most iron mines produce the compound iron oxide (Fe_3O_4) rather than the element iron. However, minerals are rarely found as pure substances in the ground. They are mixed with many less useful compounds in rocks called **ore** (**Figure 1**). How are minerals separated from the rest of the ore? How are elemental metals produced from the mineral compounds?

Table 1   **Types of Minerals**

Element	Mineral Name	Mineral Formula	Mineral Sample
silver, gold, platinum	silver, gold, platinum	Ag, Au, Pt	
calcium	limestone	$CaCO_3$	
aluminum	bauxite	Al_2O_3	
lead	galena	PbS	
mercury	cinnabar	HgS	
iron	magnetite	Fe_3O_4	
copper	malachite	$CuCO_3$	

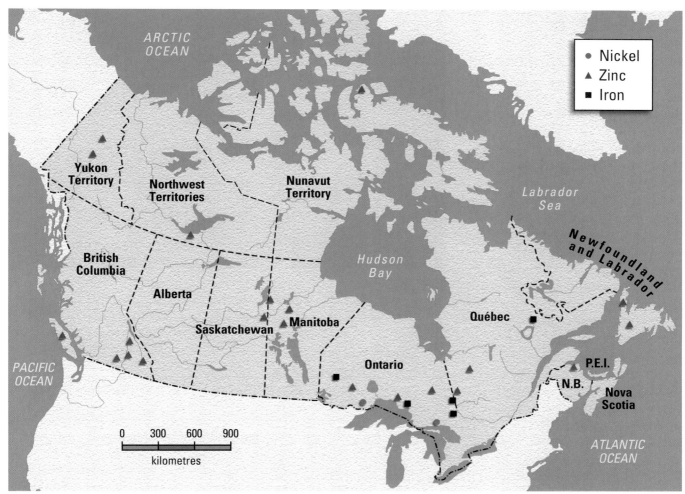

Figure 1

Mining of metals occurs across Canada. Some of the most important metals produced in Ontario are nickel, zinc, and iron.

Metals in History

In prehistoric times, people perhaps learned by accident how to make metals from minerals. Previously, they used stone and bone tools to plant crops, hunt animals, and chop wood. About 3500 B.C., people learned how to smelt copper—that is, to react the copper and draw it out of the other materials in the copper ore in a chemical change. Centuries later, when copper and tin were mixed together in the process, the result was a much harder, more flexible metal alloy called bronze. Over the centuries, other "new" metals were produced, each having new advantages of hardness, durability, or flexibility (**Figure 2**).

Figure 2

This furnace, called a *tatara*, is the type used to make Samurai swords. Making steel was a complex technology developed early in Japan's history. At a time when there was no written language, the process of sword making became part of a ritual, which was passed from generation to generation.

Mining and Metallurgy

Today metal production is one of the most important natural resource industries in Canada. Mining and metal refining employ thousands of people across the country and contribute billions of dollars to the economy (**Figure 3**). Canada is the world's top exporter of minerals such as iron, nickel, copper, aluminum, and zinc.

The technology of separating a metal from its ore is called **metallurgy** (**Figure 4**). It requires a knowledge of both the physical and chemical properties of minerals and metals.

Figure 3

Large ore carriers are used to remove nickel ore from this mine in Sudbury, Ontario, for further processing.

Preparing the ore involves crushing the rocks and separating the desired mineral from waste materials, often clay and other minerals. Magnets are used to separate the magnetic elements, such as iron. Gold and silver are dissolved in mercury to form alloys, which can be separated from the rocks.

Producing the metal from its mineral involves chemical changes. In iron production, the mineral iron oxide reacts with carbon monoxide to produce iron metal. This can be represented by the word equation:

iron oxide + carbon monoxide → iron + carbon dioxide

Metal production has many environmental implications. For example, the extraction of lead from galena involves two stages. The first step is to change lead sulfide into lead oxide. A second easier chemical reaction changes lead sulfide to pure lead. An environmental problem lies in the first reaction, which is shown by the word equation:

lead sulfide + oxygen → lead oxide + sulfur dioxide

Sulfur dioxide is one of the main sources of acid rain. Industries try to remove this gas from smokestacks in a process called scrubbing, instead of releasing it into the atmosphere.

Purifying the metal involves removing any impurities from the metal. The metal can be distilled if it has a low boiling point. (The metal is heated to evaporate it away from the impurities.) Electricity can also be used in a process similar to the electrolysis of water.

Figure 4

Iron metal is produced at temperatures of over 1000°C in a blast furnace.

Using Metals to Make Alloys

Many metals are more useful if they are melted and mixed with other metals to form alloys (**Table 2**). Changing the composition of the mixture, even by tiny fractions, can have dramatic effects on the properties of the alloy. For example, hundreds of different mixtures of elements are used to produce various steels.

Solder is an alloy of lead and tin that melts at a lower temperature than either of the pure metals (**Figure 5**). Electronic technologists, plumbers, and electricians melt it to join metal pieces such as electronic components, pipes, or wires.

Table 2	Examples of Alloys of Steel	
Type of Steel	**Composition**	**Uses**
stainless steel	70% iron, 20% chromium, 10% nickel	cutlery
plain steel	98% iron, 2% carbon	car bodies
high-strength steel	95% iron, 2% manganese, 1% carbon, 1% chromium, 1% other metals	steam turbines

Figure 5

A technologist uses a soldering iron to align protective coatings over circuits in the manufacture of flexible, electronic circuits.

Understanding Concepts

1. Design a concept map on mining. Begin with (9D) the word "element."

2. Using **Table 1**, identify the elements in each of the following minerals: (a) bauxite, (b) cinnabar, and (c) galena.

3. What are the chemical symbols for the metals that can be obtained from the following minerals: (a) limestone, (b) magnetite, and (c) malachite?

4. (a) What are three steps in obtaining a metal from its ore?

   (b) Which step(s) include a (i) physical change? (ii) chemical change?

5. Using the word equations in this section as a guide, complete these word equations for metal production:

   (a) lead oxide + carbon monoxide → lead + ?

   (b) zinc sulfide + oxygen → ? + sulfur dioxide

Making Connections

6. What practical advantages would the discovery of bronze have given people?

7. Why are metals mixed together to form alloys? What are the environmental implications?

Exploring

8. Research the methods used to obtain one of the following elements in Canada: nickel, copper, iron, zinc, gold, silver, lead, or uranium. Use a variety of sources and media, and prepare a presentation.

9. Recycling of metals is an issue that presents many questions. What metals are recycled in Canada? What elements are in short supply and should be recycled? What problems are associated with recycling? How does the cost of recycling compare with the cost of producing a metal "from the ground"? What pollution issues are addressed by recycling? Develop your own research project on some aspect of metal production and recycling.

10. A car exhaust pipe made of stainless steel corrodes much more slowly than one made from normal steel. However, it costs more than twice as much as a normal exhaust pipe and is more difficult to attach to other metal components. How would you decide whether or not to use a stainless steel exhaust pipe? What factors would you consider? If you can, talk to your local muffler repair shop about these questions and report your findings.

Explore an Issue

A Mine in the Community

In many communities, people are faced with making decisions that balance their need for the products provided by industries with their concerns about the environment. In this activity, you will use a graphic organizer called a "P-M-I" to consider the good and bad points of a project to build a mine in a northern Ontario community called Pemigon. "P" stands for "plus," "M" stands for "minus," and "I" stands for "interesting question."

Figure 1

Mining Council Stresses Benefits

Chris Beany, Information Officer with the Pemigon Mining Council, spoke yesterday to the Chamber of Commerce stressing the benefits that copper mining would bring to the community. "Mining has long been one of this province's most important natural resource industries. We need the minerals that mines provide to make the metal products that we use every day. Over 30 000 jobs in the province depend directly on mining today, and a new copper mine would provide more jobs in this area, too. The people who live here would have more money to spend, so local businesses—everything from hardware stores to car dealerships to house builders—would benefit. Mining companies donate to charities, hospitals, and universities; they pay taxes that support the services that governments provide; and the profits that we earn by selling copper to other countries helps Canada grow. Mining companies today are also very careful to protect the environment. Pemigon and Canada both need this new copper mine."

Figure 2

ENVIRONMENTAL GROUP Opposes Mine

Alex Green, spokesperson for P.A.M.P. (People Against Mining in Pemigon), spoke yesterday to a group of citizens about environmental damage that would be caused by a new copper mine. "Mines are always destructive of the environment. Open-pit or strip mining is the worst because huge areas of the land are ripped open and left bare. But even underground mines involve large amounts of land damage, because after the useful minerals are separated from the ore, large amounts of waste rock or "mine tailings" have to be stored above ground. Worse, both the ore and the tailings contain sulfide minerals. When the ore is refined in a process called smelting, sulfur dioxide is produced that combines with water in the clouds to produce sulfuric acid rain that kills trees. The tailings are equally dangerous, because rainwater and snowmelt dissolve the sulfide minerals, producing acids that run into the soil and get into groundwater. Acid mine drainage poisons plants, wildlife, and fish for many kilometres downstream. The Pemigon forests and animals are beautiful—they should be allowed to live in health. Conserve and recycle copper—don't build a mine."

Should the Mine Be Opened? (8B)

1. Imagine that you are a member of a citizens' group in Pemigon. Your cooperative group needs to help decide whether to establish a copper mine in the area. Read the information in the two newspaper articles provided (**Figures** 1 and 2).

2. Draw two vertical lines on a sheet of paper to divide it into thirds. Write P, M, and I in the spaces, as shown in **Table 1**.

Table 1

A New Mine for Our Area

P (+)	M (−)	I (?)
?	?	?

3. Working with your group, brainstorm as many ideas about the mine as possible and write them in point form in the spaces. The newspaper articles will give you some hints, but use your own knowledge and imagination to add your own ideas. Plus (P) comments describe benefits and positive effects of building a mine. Minus (M) comments describe problems and negative effects of the mine. Interesting Questions (I) describe what the group would like to know about the topic or further information the group would like to have to help them make a decision.

4. Be prepared to present your ideas to the class.

5. Now, imagine that you are an individual member of the community, with a particular job, family situation, and age. Prepare a presentation to the Town Council in which you argue your position. Be prepared to answer questions from Councillors, to be played by class members. The following are some possible roles you might consider:
   - a boy, 13, whose parents are unemployed
   - a girl, 17, who works in a local grocery store
   - a man, 25, single, who sells lumber and building materials
   - a boy, 17, who is planning to go to college to become a lab technician
   - a woman, 38, who stays at home with three children
   - a man, 30, active in the aboriginal rights movement
   - a woman, 45, vice-president of a copper-mining company
   - a man, 52, who is a miner
   - a girl, 14, whose father works in the forestry industry
   - a man, 28, who works as a local fishing guide
   - a woman, 62, who is an environmental activist

Challenge

How would a P-M-I help you prepare a persuasive marketing proposal?

Chapter 2 Review

Key Expectations

Throughout the chapter, you have had opportunities to do the following things:

- Describe the uses and properties of some common elements. (2.2, 2.3, 2.5, 2.6, 2.8, 2.12)

- Explain the particle theory of matter. (2.1)

- Classify pure substances as either elements or compounds. (2.1, 2.4)

- Describe compounds and elements in terms of molecules and atoms, and recognize that compounds may be broken down into elements by chemical means. (2.1, 2.4, 2.8, 2.9, 2.12)

- Explain the value of models in explaining the behaviour of matter, and build physical models of some common molecules. (2.1, 2.9)

- Identify and use the symbols for common elements and the formulae for common compounds. (2.7, 2.10)

- Demonstrate knowledge of laboratory safety procedures while conducting investigations. (2.2, 2.4, 2.6)

- Investigate the properties of elements and compounds, and organize, record, analyze, and communicate results. (2.2, 2.4, 2.6)

- Formulate and research questions related to the properties of elements and compounds and communicate results. (2.2, 2.3, 2.4, 2.8, 2.11, 2.12)

- Compare the physical and chemical properties of elements and compounds in substances such as fertilizers and ores, and assess their potential uses and associated risks. (2.3, 2.8, 2.11, 2.12)

- Describe technologies that have depended on understanding atomic and molecular structure. (2.12)

- Describe methods used for mining and processing metals in Canada, and associated environmental and economic issues. (2.12, 2.13)

- Explore careers requiring an understanding of the properties of matter. (Career Profile)

KEY TERMS

atom	metallurgy
bond	mineral
chemical formula	mixture
chemical symbol	model
combining capacity	molecule
compound	nutrient
electrolysis	ore
element	photosynthesis
fertilizer	precipitate
heavy metal	pure substance
heterogeneous mixture	solution
matter	structural diagram
metal	

Reflecting

- "We classify matter to organize the vast amount of information about elements and compounds." Reflect on this idea. How does it connect with what you've done in this chapter? (To review, check the sections indicated above.)

- Revise your answers to the questions raised in Getting Started. How has your thinking changed?

- What new questions do you have? How will you answer them?

Understanding Concepts

1. Make a concept map to summarize the material that you have studied in this chapter. Start with the phrase "pure substance."

2. The sentences in the following list contain errors or are incomplete. In your notebook, write your complete, correct version of each sentence.

   (a) Elements are made up of compounds.

   (b) Nonmetals are shiny and good conductors.

   (c) Electrolysis is the breaking down of water into hydrogen and nitrogen gases.

   (d) Ozone molecules contain three argon atoms.

(e) Three important fertilizer elements are nitrogen, phosphorus, and sodium.

(f) A mineral contains ore and waste rock.

3. Describe two compounds that contain atoms of the same elements, but in different proportions.

4. State the types of atoms and the numbers of each type that are present in the following molecules: copper phosphate (Cu_3PO_4) and sodium nitrate ($NaNO_3$).

Applying Skills

5. Match the description on the left with one term on the right. Use each term only once.

Description	Term
A smallest particle of an element	1 electrolysis
B mixture of rocks and minerals	2 element
C a poisonous gas	3 bond
D solid product produced when solutions mix	4 nutrient
E substance containing only one type of atom	5 carbon monoxide
F separation of water	6 atom
G connection between atoms	7 molecule
H material necessary to plants and animals	8 precipitate
I particle made of two or more atoms	9 ore

6. What safety precautions did you follow in investigations that you performed in this chapter?

7. Copy the tests in Column A into your notebook. Match each test with the appropriate gas in Column B.

Column A	Column B
limewater test	water vapour
cobalt chloride test	hydrogen gas
glowing splint test	oxygen gas
burning splint test	carbon dioxide

8. Examine the molecules in **Figure 1**.

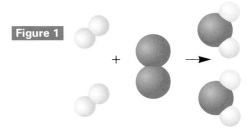

Figure 1

(a) What substances could the drawings represent?

(b) Write a word equation for the reaction.

9. Use a sketch to show the molecules of two different compounds that contain only atoms of carbon and oxygen.

10. Identify the elements in the following compounds, state the relative numbers of atoms of the elements, and name each compound. You will need to look at **Tables 2** and **3** on page 64.

(a) Ag_2S (c) Na_2O (e) CaI_2

(b) $ZnBr_2$ (d) MgS

11. Write the formula, name, and structural diagram for the compound formed by each of the following combinations of elements:

(a) potassium and chlorine

(b) calcium and oxygen

(c) aluminum and sulfur

12. Write word equations to represent the following reactions:

(a) Potassium and water produce potassium hydroxide and a very flammable gas.

(b) Calcium carbonate and hydrochloric acid produce calcium chloride, water, and a gas that turns limewater milky.

(c) Potassium chlorate produces potassium chloride and a gas that causes a glowing splint to burst into flame.

Making Connections

13. Customs officials investigating a crate shipped from Central America wanted to know what it contained before allowing the crate into Canada. Although the labels were in Spanish, the following chemical formulas were printed on the crate: $NaHCO_3$, $NaNO_3$, $Ca_3(PO_4)_2$. Would you recommend that the officials allow the crate to continue or should they call the shipping company to ask for more information? Explain your reasoning.

14. Ozone is needed in the upper atmosphere and is produced as a pollutant at ground level. Why don't we just collect the ozone at ground level and carry it up in aircraft for release in the upper atmosphere?

15. Research and report on different mining techniques used in Ontario. For example, what is landfill mining? Include a description of the substances that are mined, and the reason for using a particular technique.

3 Models for Atoms

1 The better our model of matter, the more we can understand about matter itself, and the more uses we can make of that understanding. Models of matter help us understand the chemical processes that affect every aspect of modern life, from nourishing our bodies to fuelling space shuttles.

You are already familiar with one model, the particle theory, which states that all matter is made of tiny particles. Like all models, this model is a mental picture, a diagram, or a three-dimensional means of representing something. For instance, the particle theory can be used to help understand physical behaviour of substances, such as changes of state. It can also be applied to explain the formation of various molecules from different combinations of atoms. But what makes atoms different from each other? Look at the illustrated model. What is "inside" the particle we call an atom?

2 Over time, it became more and more clear to scientists that the particle theory could not explain all observable behaviours of matter. New evidence required that new models be created. For instance, the particle theory is not useful for understanding static electricity. Why do we sometimes get electric shocks when we touch metal doorknobs? Why is dust more attracted to your television screen than to your television cabinet? How can a different model of matter be used to answer these questions?

3 How do fireworks produce dazzling displays of light, colour, and sound? The answer lies in the nuclear model of the atom. What are some other practical applications of this model?

Reflecting

Think about the questions in **1**, **2**, **3**. What ideas do you already have? What other questions do you have about atoms? Think about your answers and questions as you read the chapter.

Try This   What You Know About Atoms

Make a table similar to **Table 1**. In the first column, write the answers to the following questions:

1. What do you know for certain about the structure of the atom?

2. What do you think you know about the structure of the atom?

3. What do you think you might learn about the structure of the atom? (Include any questions you might have about the structure of the atom.)

In the second column, jot down new ideas, details, examples, or even diagrams

about the structure of the atom that occur to you as you read this chapter. After you have finished the chapter, put a check in the first column beside the ideas that your reading has confirmed, and a question mark beside the ideas left unconfirmed by your reading.

Table 1

Ideas and questions	What I learned
?	?
?	?
?	?
?	?

Making a Logical Model

Imagine that you are standing in front of a pop vending machine (**Figure 1**). You put in a coin, press a button, and a can falls down into the tray at the bottom. How does the machine work? You can't see inside it, so you have to create a model that could explain the workings of the machine.

One possibility is that there is a very small person working inside. When the coin appears, the person checks to see which button was pushed, searches for the right can, and puts it in the tray.

A second possibility might involve a mechanical system with various electronic sensors, levers, motors, and slots that operate when the coin is inserted to release the can.

How could you test these hypotheses? You could pull out the electric plug in the back. If no can appears when the power is off, perhaps the second hypothesis is correct. Or maybe the person inside refuses to work in the dark—the first hypothesis is still a possibility!

The best model you can create is the one that allows you to predict how the vending machine will behave in as many situations as you can imagine. You may come up with a model that works perfectly for all of the evidence that you have. But someone else might try using the vending machine in a completely new situation. If the can doesn't fall and the model can't explain it, then the model needs to be adjusted.

Like a scientist exploring models of matter, all of your testing, thinking, and experimenting is based on the fact that you can't see what's going on inside the vending machine. Atoms can't be seen— we can only see how matter behaves in certain circumstances. The model of matter changes when that model is tested in new circumstances that produce new and unpredictable results. In this investigation, you will follow a similar process, by trying to guess what is inside a sealed box. You will use some basic scientific skills:

- gathering and organizing observations,
- inventing a model to explain these observations, and
- communicating your findings to others.

Figure 1

What model could explain how a vending machine works?

Materials
- a sealed box (e.g., a shoe box) containing an object or objects
- ruler
- magnet

Question
What is inside a sealed box?

Hypothesis
Simple experiments can be designed to make a model of the contents of a sealed box.

Procedure

1 Obtain a sealed box that contains an (6B) object. Measure the outside dimensions of the box.

✎ (a) Record your observations.

2 Without breaking the seal, make all the observations you can by carefully shaking, tilting, or otherwise moving the box.

✎ (a) Record each movement that you chose to use, and the observations you made each time.

(b) Write a description, or model, of what you think the object is. For example, describing it as "a 15-cm-long metal object, branched into four small projections at one end" is better than describing it as "a fork."

3 Examine your observations and invent new movements of the box that will help you determine the size, shape, and other physical properties of the object. In particular, try to make some quantitative observations.

✎ (a) Record each movement that you chose to use, and the observations made each time.

4 With your group, discuss a model for the object in the box.

(a) Make a labelled drawing of your model, including measurements if possible.

(b) Write a short description of the main characteristics of your model.

5 After completing the drawing and description, open the box and look at the object.

(a) Describe the object.

Analysis and Communication

6 Analyze your observations by answering the following questions:

(a) Write a two-paragraph summary of the (7E) similarities and differences between the real object and your model. In the first paragraph, list the characteristics that you were successful in determining. In the second paragraph, list those characteristics that you were unable to determine.

(b) Speculate on why this "black box" experiment is similar to the process that scientists followed when they produced their models for matter.

Making Connections

1. Think about the original example of the vending machine.
   (a) What experiments could you do to find out how the vending machine operates?
   (b) Draw a model of how you think the machine operates.

2. Choose an everyday appliance. Draw and label a model to explain how it works. Suggest ways that your model could be tested.

Exploring

3. Obtain a polystyrene sphere in which your teacher has embedded an object. Take a thin metal probe or knitting needle and carefully insert it into the sphere. Make a systematic series of probings. Record your observations and make a model drawing to describe what is inside the polystyrene "atom."

Reflecting

4. Think about your group's success with determining the identity of the object in the box. Is there any reason why you successfully determined some characteristics but not others?

Developing Models of Matter

Have you ever taken something apart to figure out how it works? Have you ever watched the sand wash away from around your feet as you stand at the edge of the ocean? Like scientists and philosophers through the ages, you are curious about the world around you. You want to know how things will behave in certain circumstances. Perhaps you will develop an idea about something and test it. In just the same way, scientists have observed, questioned, and theorized for centuries about the "stuff" that makes up the world: matter. They looked at the evidence around them and, in an attempt to explain it, developed many models of matter. These have been modified, combined, or rejected as new evidence was discovered.

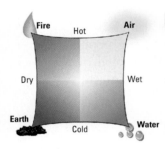

Figure 1
According to the four-element model, each element is a mixture of two properties. For example, fire is a mixture of hotness and dryness.

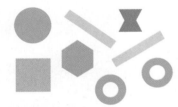

Figure 2
In the atomic model, atoms were conceived to be of different sizes, to have regular geometric shapes, and to be in constant motion.

Figure 3
The Greek philosopher Aristotle also believed that matter was made of four elements: earth, air, fire, and water.

About 450 B.C.

A Greek scholar named Empedocles proposed that matter was composed of four "elements": earth, air, fire, and water (**Figure 1**). These elements mixed together in different proportions to yield different substances. Rust might be one part fire and two parts earth. Volcanic rock might be two parts fire and one part air. Unlike most philosophers of his time, Empedocles checked some of his theories experimentally. He demonstrated that, even though air is invisible, it is not just "nothing." Because it takes up space, it must be a form of matter.

About 400 B.C.

Another Greek, Democritus, suggested that matter was made of tiny particles that could not be broken down further (**Figure 2**). He called the particles atoms, after the Greek word *atomos*, which means "indivisible." Thus, different elements were composed of different kinds of atoms, a revolutionary concept at the time. However, Democritus' ideas were never widely accepted because Socrates, a very influential figure at the time, did not accept them.

About 350 B.C.

The philosopher Aristotle believed in Empedocles' "four element" model despite the more recent "atomic" model (**Figure 3**). Aristotle's influence was so great, and his writings read by so many people, that the "four-element" model was accepted for almost 2000 years.

450 B.C.   400 B.C.   350 B.C.

Figure 4

Antoine Lavoisier is often considered the father of modern chemistry. His involvement in a state organization that collected taxes led to his execution by guillotine in 1791 during the French Revolution.

Figure 5

When Cavendish burned hydrogen in oxygen, he produced water.

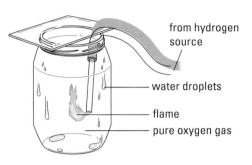

from hydrogen source

water droplets

flame

pure oxygen gas

A.D. 500–1600

Do metals grow like plants, ripening into gold? Many **alchemists** (combination of philosopher, mystic, magician, and chemist) believed that they did. For centuries they performed numerous experiments attempting to make gold from cheap metals such as iron and lead. They devised chemical symbols for substances that we now recognize as elements and compounds. They also invented many laboratory tools that we still use today: beakers, filters, stirring rods, and distillation apparatus. However, despite finding many new substances, they still accepted the four-element model. And no one ever turned lead into gold!

1650

An English scientist, Robert Boyle, did not believe in the four-element model. He devised a new definition for the word *element:* "I mean by element, simple unmitigated bodies." This became the modern definition of an element: a pure substance that cannot be chemically broken down into simpler substances. Boyle also believed that air was not an element, but rather a mixture.

Late 1700s

Joseph Priestley was the first person to isolate oxygen scientifically, but he did not know that oxygen is an element. This fact was soon recognized by Antoine Lavoisier (**Figure 4**). Experimenting with Priestley's oxygen, Lavoisier concluded that air must be a mixture of at least two gases, one of which was oxygen.

Meanwhile, Henry Cavendish experimented by mixing a metal with acid, which resulted in a flammable gas that was lighter than air. He did not know that the gas he had prepared was hydrogen, but discovered that his gas would burn in some of Priestley's oxygen, producing water (**Figure 5**). Until that time, scholars had believed that water was an element.

Figure 7

(a) In Thompson's model, the atom is a positive sphere with embedded electrons.

(b) In Nagaoka's model, the atom is compared with the planet Saturn, where the planet represents the positively charged part of the atom, and the rings represent the negatively charged electrons.

Figure 6

In Dalton's atomic model, an atom is a solid sphere. This model is still useful for explaining chemical charges: atoms combine and molecules come apart as chemical reactions occur.

1808

By this time it was generally accepted that matter was made of elements: the two models had come together. English chemist John Dalton published a theory of why elements differ from each other and from non-elements (**Figure 6**). Dalton's **atomic model** for matter stated that:

- All matter is made of atoms, which are particles too small to see.
- Each element has its own kind of atom, with its own particular mass.
- Compounds are created when atoms of different elements link to form molecules.
- Atoms cannot be created, destroyed, or subdivided in chemical changes.

1800s

However, Dalton's atomic model cannot explain why, on a dry winter day, you get a spark when you touch a metal doorknob. Obviously, matter is able to develop positive and negative **charges**—quantities of electricity that may build up on an object. A new model was developed, introducing tiny negatively charged particles that could be separated from their atoms and moved to other atoms.

In 1831, Michael Faraday found that electric current could cause chemic changes in some compounds in solution. The atoms could gain electric charges and form charged atoms, called **ions**. In this modified version of Dalton's model:

- Matter must contain positive and negative charges.
- Opposite charges attract and like charges repel.
- Atoms combine to form molecules because of electrical attractions between atoms.

1904

J. J. Thomson revised the atomic model further, to explain his discovery of very light negative particles, called electrons. He also did experiments with beams of much heavier positive particles (later identified as protons). The new model became known as the "raisin-bun" model (**Figure 7a**).

- Atoms contain particles called electrons.
- Electrons have a small mass and a negative charge.
- The rest of the atom is a sphere of positive charge.
- The electrons are embedded in this sphere, so that the resulting atoms are neutral or uncharged.

The Japanese scientist H. Nagaoka, working at about the same time, modelled the atom as a large positive sphere surrounded by a ring of negative electrons (**Figure 7b**).

Figure 8
Rutherford's experiment

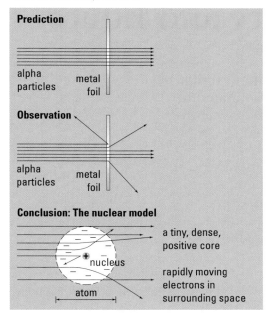

1911

Ernest Rutherford, working at McGill University in Montreal, designed an experiment to test Thomson's and Nagaoka's models. He aimed a type of radiation called alpha particles (positively charged particles smaller than most atoms) at a thin sheet of gold foil. He predicted, based on Thomson's raisin-bun model, that the particles would pass straight through the gold foil, as indeed most of them did. However, a very small number of the alpha particles bounced almost straight back from the gold foil (**Figure 8**). Rutherford was amazed and described the result as being similar to firing bullets at a piece of tissue paper and having one of them bounce back! To explain how the positive alpha particles had been repelled, Rutherford had to come up with another new model—the **nuclear model**:

- An atom has a tiny, dense, positive core called the nucleus (which deflected the alpha particles and contains protons).
- The nucleus is surrounded mostly by empty space, containing rapidly moving negative electrons (through which the alpha particles passed unhindered).

Understanding Concepts

1. How were alchemists similar to and different from modern scientists?

2. Describe the changing definitions of an element.

3. Describe the changing definitions of an atom.

4. What were the four main points in Dalton's theory?

5. Whose work led to a model that suggested that
   (a) the atom contains a dense positive core?
   (b) atoms can form charged particles called ions?
   (c) atoms contain electrons and protons?
   (d) atoms cannot be divided further?
   (e) electrons surround a central positive core?

6. In Rutherford's gold foil experiment,
   (a) what kind of electrical charge did the alpha particles have?
   (b) what might Rutherford have expected to observe, based on Thomson's model?
   (c) what is the relative size of the heaviest part of the atom compared with the whole atom?

Exploring

7. Design an experiment to demonstrate that fire and earth are not elements.

8. Design and build a model to represent one of the early models of the atom.

Reflecting

9. Why do scientists continue to make models of matter if the models keep changing? Write a few sentences to suggest why models might be useful.

Challenge

Draw a series of diagrams to represent the various models of the atom. How would you build and display these? How would you use these in your presentation to Mendeleev?

Biochemistry and Ethics

As a child growing up in India, Shree Mulay was fascinated by biographies of famous scientists. When she was eight years old she read about Marie Curie, the Nobel prize-winner whose work in the chemistry of radioactive materials revolutionized our understanding of the structure of atoms. She was so excited by Curie's work that she decided to pursue a career in chemistry. After obtaining her B.Sc. in Delhi, she came to McGill University to obtain a Master's and then a Doctorate in chemistry.

Today Dr. Mulay is assistant director of the Clinical Laboratory at the Royal Victoria Hospital in Montreal, as well as director of the McGill Centre for Research and Teaching on Women. She teaches endocrinology to students at McGill, and in her laboratory work she examines the roles of hormones as they affect reproduction and pregnancy.

Dr. Mulay believes that scientists have an obligation to take an active part in their society. She has criticized governments and pharmaceutical companies that develop contraceptives for women, for not adequately safeguarding the health of women, especially in Third World countries. "Women were reporting problems and they were not being recorded by the investigators," she says. After discovering in 1999 that some clinical drug trials were not meeting ethical guidelines, Dr. Mulay was featured on David Suzuki's *The Nature of Things*.

Dr. Mulay highly recommends a career in chemistry. A basic knowledge of chemistry, she says, can lead to many, many different areas of medical and scientific research.

> I found it exciting—the possibility of knowing that you can relate the chemical structure of a molecule to its function in the body.

Exploring

1. There are very strict guidelines controlling how pharmaceutical companies can test, and then report on, their products. Research these guidelines and make a brief presentation to your class.

2. Why do you think so many areas of medical and scientific research depend on a knowledge of chemistry?

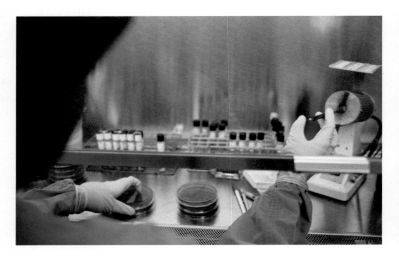

Inside the Atom

What does "splitting the atom" mean? And if the atom is split, what do the pieces look like? Rutherford's experiment was a breakthrough in how people thought about matter. As you have learned, in the nuclear model, most of the atom is empty space, filled with quickly moving electrons. The positive nucleus is so small that it takes up only a tiny fraction of the size of the atom. Yet almost all of the atom's mass is concentrated in this nucleus, which contains protons.

Types of Subatomic Particles

The "pieces" of an atom—the particles of which an atom is composed—are called **subatomic particles**. Electrons and protons are subatomic particles. Experiments conducted by the English scientist James Chadwick, in 1932, led to the discovery of a third subatomic particle with no charge: the neutron. These subatomic particles are described in terms of their mass relative to each other, their electrical charge, and their location.

- **Protons** are positively charged particles with a relative mass of 1, located in the nucleus.
- **Neutrons** are neutral particles with a relative mass of 1, also located in the nucleus.
- **Electrons** are negatively charged particles with a relative mass of approximately 1/2000 of the mass of a proton or neutron, travelling in regions of space around the nucleus.

Protons are especially significant, because the number of protons in an atom determines what the atom is. For example, any atom with one proton is a hydrogen atom (H), and any atom whose nucleus contains 12 protons is magnesium (Mg).

Counting Subatomic Particles

How many electrons, protons, and neutrons are there in an atom? An atom itself has no electric charge, and the negative charge of one electron is as strong as the positive charge of one proton. The number of protons and electrons in an atom is the same, so the charges cancel each other out. Some examples are given in **Table 1**.

Table 1

Element	Number of Protons	Total Positive Charge	Number of Electrons	Total Negative Charge	Net Charge of Atom
hydrogen	1	1+	1	1–	0
oxygen	8	8+	8	8–	0
magnesium	12	12+	12	12–	0
copper	29	29+	29	29–	0
uranium	92	92+	92	92–	0

Did You Know ?

If an atom were the size of a football field, its nucleus would be the size of a grain of sand in the centre.

The number of protons in an atom is called the **atomic number**. If you know the atomic number of an atom, you know how many protons—and how many electrons—that atom contains.

Another significant number is the mass number. The **mass number** represents the sum of protons and neutrons in an atom. (The mass of the electrons, relative to the mass of the protons and neutrons, is insignificant.) Therefore, if you know the atomic number (the number of protons) and the mass number (the sum of protons and neutrons), you can easily calculate the number of neutrons:

number of neutrons = mass number − atomic number

We can represent the numbers of subatomic particles by using **standard atomic notation**, an internationally recognized system that allows anyone to communicate information about any atom. In this notation, we write the chemical symbol of the atom and place the atomic number to the lower left and the mass number to the upper left.

For example, the atomic notation of chlorine is

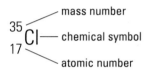

- mass number
$^{35}_{17}$Cl — chemical symbol
- atomic number

Figure 1

Atoms can gain or lose electrons to form charged atoms or ions. The elements are still the same because the number of protons has not changed.

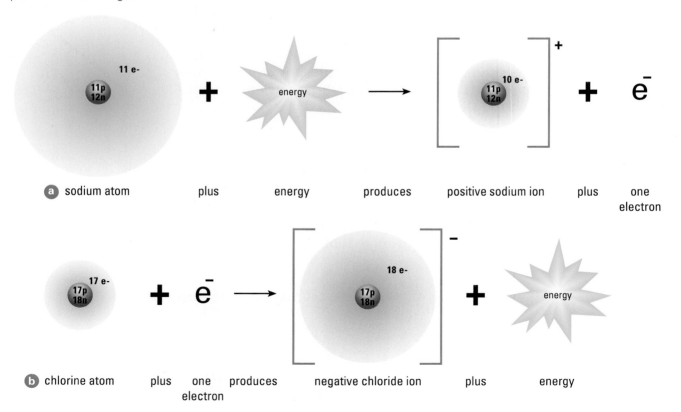

a sodium atom   plus   energy   produces   positive sodium ion   plus   one electron

b chlorine atom   plus   one electron   produces   negative chloride ion   plus   energy

Similarly, the atomic notation for an atom of sodium is

$$^{23}_{11}\text{Na}$$

This notation tells us that sodium has 11 protons and $23 - 11 = 12$ neutrons. Since the atom is neutral, it also tells us that the number of electrons is 11. This sodium atom could also be represented as sodium -23.

Charged Atoms

Why is it dangerous to work with electrical appliances around water? Tap water and rain water can conduct electricity because they are not pure water: they contain charged atoms that can move in the solution and carry electric current. These charged atoms are called ions—they have a charge because the number of electrons is not equal to the number of protons. For example, salt water contains sodium ions, which have 11 protons and 10 electrons—the sodium ions have a charge of +1. Salt water also contains chloride ions, which have 17 protons and 18 electrons—the chloride ions have a charge of −1.

Ions are formed when negatively charged electrons move from one atom to another. If an atom loses an electron, there are more protons in the nucleus than electrons in the space around it. Atoms such as sodium and calcium lose electrons to form such ions with net positive charges (**Figure 1a**). If atoms gain electrons, there are more electrons than protons. Atoms such as chlorine and oxygen gain electrons to form ions with net negative charges (**Figure 1b**). These charged particles, now described as ions rather than atoms, can move in solutions and conduct electricity.

Understanding Concepts

1. Draw and complete a table with three rows and four columns to summarize what you know about the nuclear model of the atom. The column headings should be "particle, proton, neutron, electron." The row headings should be "mass, charge, location in atom."

2. Write standard atomic notation for the following:
   (a) an atom of nitrogen with 7 protons and 8 neutrons
   (b) an atom of bromine with 35 protons and 36 neutrons
   (c) an atom of sulfur with 16 protons and 16 neutrons
   (You may want to look at the table of chemical symbols on page 58.)

3. Assuming that each atom is neutral, copy and complete **Table 2** by filling in the blanks.

Table 2

atomic number	mass number	no. of protons	no. of neutrons	no. of electrons
8	16	?	?	?
11	?	?	12	?
?	?	14	16	?
?	29	?	?	14

4. An ion is an atom that has gained an electrical charge.
   (a) A magnesium atom has 12 protons. How many electrons does a magnesium ion, with a charge of +2, contain?
   (b) A fluorine atom has an atomic number of 9. How many electrons does a fluorine ion, with a charge of −1, have?

Reflecting

5. If the atom is almost entirely empty space, why do atoms not just collapse into each other and into a much smaller volume? (Hint: think about what you know about attraction and repulsion of charges.)

Challenge

If it is useful to build a model of an atom for your challenge, what materials would you use to represent protons, neutrons, and electrons?

A "Planetary" Model of the Atom

Why do some elements and not others combine to form compounds? What part of an atom is involved when chemical reactions occur? Rutherford's nuclear model did not answer these questions. Also, the nuclear model itself had problems. Since electrons are negative and protons in the nucleus are positive, why didn't the atom collapse as the electrons were attracted to the nucleus? To answer these questions, another model was required.

Look at the planetary model in **Figure 1**. How does this model resemble Rutherford's nuclear atom? Further experiments suggested that, as an improvement to Rutherford's model, electrons orbited around the nucleus like planets around the Sun.

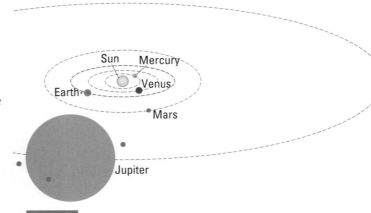

Figure 1
Planetary model of the atom

Atoms and Rainbows

What do atoms and rainbows have in common? Their behaviour can be explained in terms of energy. When white light passes through raindrops or a prism, it is split into a **spectrum**—a rainbow of the many colours that combine to make white light (**Figure 2**). Different colours of light have different energies. For example, blue light has more energy than red light. When elements are heated in a flame, they show a few specific colours (**Figure 3**). These specific colours, seen through a spectroscope, are called a line spectrum. How could these definite energies of light be explained?

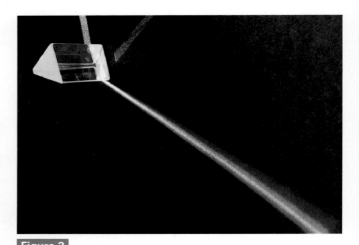

Figure 2
A narrow beam of white light is split by a prism into a continuous rainbow of colours.

Figure 3
When an element is heated in a flame, it produces only certain colours or energies of light. Each element has its own unique line spectrum.

a cesium

b sodium

c lithium

A Danish physicist, Niels Bohr, proposed a "planetary" model of the atom to explain line spectra. Bohr suggested that:

- Electrons move around the nucleus in nearly circular paths called **orbits**, like planets around the Sun (**Figure 4**).
- Each electron in an orbit has a definite amount of energy.
- The farther away the electron is from the nucleus, the greater its energy is.
- Electrons cannot exist between these orbits, but can move up or down from one orbit to another.
- The order of filling of electrons in the first three orbits is 2, 8, and 8.
- Electrons are more stable when they are at lower energy, closer to the nucleus.

Bohr's model explains the spectra of elements in terms of the "jumps" that electrons make from one orbit or **energy level** to another (**Figure 5**). In his model, electrons are arranged in orbits that surround the nucleus like layers of an onion. Space in the orbits is limited, and the electrons are arranged in a definite pattern. Within the orbits, electrons move quickly. When electrons are energized by heat, electricity, or light, they use this extra energy to jump out to a higher orbit (**Figure 6**). We say they are in an **excited state**.

The excited electrons are very unstable and tend to fall back into their normal, more stable orbits. This low-energy state is called the **ground state**. When the electrons drop back to their normal orbits, their extra energy is given off in the form of light. The amount of energy given off is equal to the difference in energy between the higher and lower energy levels. This very specific energy amount corresponds to a very specific colour.

Bohr's model of electrons in energy levels also explains why each element has a different spectrum. For example, hydrogen has only a single proton and a single electron. Sodium has 11 protons and 11 electrons. The orbits in these two atoms are at different distances from the nucleus. Therefore, the energies of the electrons in the two atoms are slightly different. An electron jumping from the third level to the second level in a hydrogen atom may produce a red colour. The colour is evidence of the difference in energy between the two levels. An electron jumping from the third level to the second in a sodium atom may produce a yellow colour, because the orbits are at different distances from the nucleus.

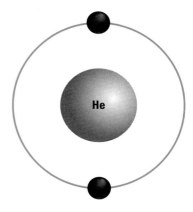

helium

Figure 4

In Bohr's model of the atom, electrons travelled around the nucleus in nearly circular orbits, much like planets around the Sun.

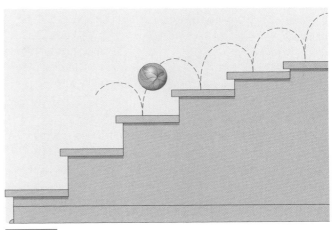

Figure 5

Imagine an electron in Bohr's model as being like a marble on a staircase. The marble can only be at certain definite levels. It can jump up or fall down by only very specific amounts.

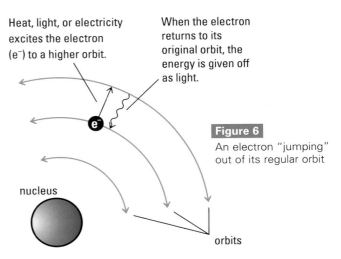

Heat, light, or electricity excites the electron (e⁻) to a higher orbit.

When the electron returns to its original orbit, the energy is given off as light.

Figure 6

An electron "jumping" out of its regular orbit

nucleus

orbits

The Bohr Model of Electron Arrangement

Bohr developed his model of the atom to explain the line spectrum of hydrogen, but it was soon extended to other elements. Scientists drew **Bohr diagrams** to represent the electronic structure of elements. In these diagrams, the symbol of the element is written in the centre to represent the nucleus of the atom. A series of concentric circles is drawn around the nucleus to represent the orbits, and electrons are shown in these orbits.

The element hydrogen, which has an atomic number of 1, has 1 electron in its first orbit. The element nitrogen (symbol N), which has an atomic number of 7, has 7 electrons. Two electrons are in the atom's first orbit, and the remaining 5 electrons are in its second orbit. Phosphorus (symbol P), with an atomic number of 15, has 15 electrons. Two electrons are in the first orbit, 8 are in the second orbit, and 5 are in the third orbit (**Figure 7**).

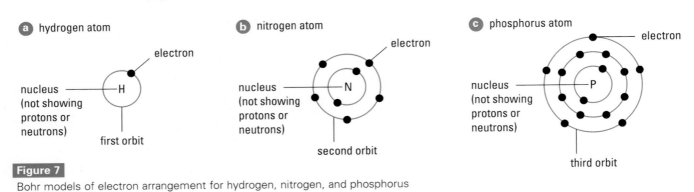

Figure 7

Bohr models of electron arrangement for hydrogen, nitrogen, and phosphorus

Bohr-Rutherford Diagrams

We can combine Rutherford's nuclear model with Bohr's planetary model in diagrams that summarize the numbers and positions of all three subatomic particles in an atom. In these **Bohr-Rutherford diagrams**, a circle is drawn in the centre to represent the nucleus of the atom. The numbers of protons and neutrons are written in this circle. Electrons are again shown in circular orbits about the nucleus. For example, consider a Bohr-Rutherford diagram for magnesium, Mg-24. The atomic number of magnesium is 12 and its mass number is 24.

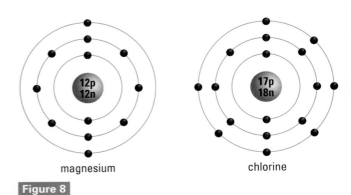

Figure 8

Bohr-Rutherford diagrams for Mg-24 and Cl-35. The electron arrangements can be described as 2, 8, 2 for magnesium and 2, 8, 7 for chlorine.

Therefore, there are 12 protons and 12 neutrons in the nucleus. Mg-24 has 12 electrons, 2 in the first orbit, 8 in the second orbit, and 2 in the third orbit. Similarly, Cl-35 has 17 protons and 18 neutrons in the nucleus, 2 electrons in the first orbit, 8 electrons in the second orbit, and 7 electrons in the third orbit. The Bohr-Rutherford diagrams for these atoms are shown in **Figure 8**.

Electron Arrangements in Ions

Recall that atoms can lose or gain electrons to form charged atoms (ions). But which ions do atoms form? In Bohr diagrams for the first 20 elements from the Periodic Table, there is something special or "stable" about the numbers 2, 8, and 8. Many elements tend to form ions by losing or gaining enough electrons to have 2 or 8 electrons in their orbits. For example, look at the atoms in **Figure 9** and compare them with **Figure 8**. A magnesium atom becomes an ion by losing the two electrons in its outermost orbit. The ion has 2 electrons in the first orbit and 8 in the second. It also has a +2 charge because it has two more protons than electrons. Chlorine forms an ion by gaining one electron. The ion has 2 electrons in the first orbit, 8 in the second, and 8 in the third. It has a –1 charge because it has gained an extra electron.

Did You Know

The Bohr-Rutherford model is useful because it explains how most chemical reactions occur. But, like all models, it can change with further experiments. There is a very complicated model called the quantum-mechanical model that you might learn about in senior chemistry.

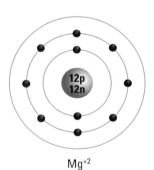

Mg^{+2}

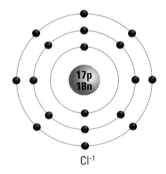

Cl^{-1}

Figure 9
Bohr-Rutherford diagrams for Mg-24 and Cl-35 ions. The atoms gain or lose electrons to have a "stable" arrangement of electrons. The ions are described as Mg^{+2} and Cl^{-1} ions.

Understanding Concepts

1. **(a)** What is meant by the term "spectrum"?

   (b) How is the spectrum seen in a rainbow different from the spectrum of an element?

2. **(a)** In words, describe the structure of the atom using the Bohr-Rutherford model.

   (b) What paths do electrons follow in the Bohr model?

   (c) Why are orbits also called energy levels?

   (d) How do the energies of electrons in different orbits compare?

3. **(a)** How does the Bohr model of the atom explain light given off in line spectra?

   (b) Why do different elements produce different line spectra?

4. Draw Bohr diagrams for

   (a) oxygen (symbol O), atomic number 8

   (b) aluminum (symbol Al), atomic number 13

   (c) calcium (symbol Ca), atomic number 20

5. Draw Bohr-Rutherford diagrams for

   (a) fluorine-20 (symbol F), atomic number 9

   (b) boron-11 (symbol B), atomic number 5

   (c) potassium-40 (symbol K), atomic number 19

6. **(a)** Draw Bohr-Rutherford diagrams for the stable ions formed by gain or loss of electrons in each of the atoms in the previous two questions.

   (b) Write the symbol and charge for the stable ion formed by each atom.

7. For atoms to interact, they must collide with each other. Which subatomic particle do you think has the most important role in chemical change? Explain why.

8. Look at the tables of combining capacities on pages 64–65 and your answers to question 6.

   (a) Which elements are named in both the table and the question?

   (b) Compare the combining capacity and the charge on the ion for each of these elements. What pattern do you notice?

Reflecting

9. Draw a chart to help you organize and remember what you need to know to draw Bohr-Rutherford diagrams.

Using Electrons to Identify Elements

According to the Bohr model of the atom, when the atoms in an element are provided with energy, some of the electrons may "jump" up to higher levels. This energy can be in the form of heat, light, or electricity. The electrons are said to be in an excited state because they are in higher energy orbits than normal.

The electrons then tend to fall back down to their normal, lower energy level or ground state. When this happens, the atoms give out energy in the form of light. Since different elements have slightly different energy levels, different energies or colours of light are given out (**Figure 1**). These colours are like the "fingerprints" of elements, especially metals, even when they are combined with other elements in chemical compounds.

In this investigation, you will use an experimental technique called a **flame test**. You will heat samples of compounds and determine the identity of the metal in each. **Figure 2** shows an alternative method for conducting flame tests.

☠ Some of these solutions are poisonous. Any spills of these solutions on the skin, in the eyes, or on clothing should be washed immediately with cold water. Inform your teacher of any spills.

Materials
- safety goggles
- apron
- Bunsen burner
- flint lighter
- eight 250-mL beakers, each containing one splint for each student group
- each splint soaked in 0.5 mol/L solutions of one of the following: lithium nitrate, sodium nitrate, potassium nitrate, barium nitrate, calcium nitrate, copper nitrate, and two unknown metal nitrates
- beaker containing water for extinguishing splints

Question
What can be observed when various compounds are heated in a flame?

Figure 1

a Sunlight produces a continuous spectrum of light corresponding to many different energies.

b An element like hydrogen produces only certain energies of light that correspond to electrons dropping to lower energy orbits.

Hypothesis

1 Write your own hypothesis about what you
4A might observe.

Procedure

2 Design an observation table in which to record what you see.

3 Review the safety procedures for using a Bunsen burner. Put on your apron and safety goggles.

4 Obtain from your teacher 8 wooden splints that have been soaked for 2–4 h in solutions of 6 known and 2 unknown compounds.

☠ Ⓣ Only the bottom halves of the splints have been soaked. Handle the splints by the "dry" ends only.

5 Use the flint lighter to ignite the Bunsen burner. Adjust the burner to produce the hottest flame possible.

6 Hold the soaked end of the lithium nitrate splint in the flame for a short time. As soon as the flame is no longer strongly coloured extinguish the splint by placing it in the beaker of water.

SKILLS HANDBOOK: **4A** Asking Questions and Hypothesizing

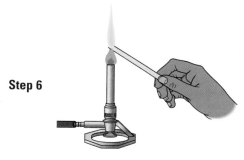

Step 6

✎ (a) Record your observations of the flame colour for lithium in your data table.

🔥 Do not burn the splints.

7 Repeat step 6 for each of the other 5 known solutions.

✎ (a) Record your observations of the flame colours for the other metals in your table.

8 Repeat step 6 for the 2 unknown solutions. What do you think is the identity of the unknown solutions?

✎ (a) Record your observations of the flame colours for the unknown metals in your table.

9 Return the splints as directed by your teacher. Clean up your work station and return your apparatus. Wash your hands.

Analysis and Communication

10 Analyze your observations by answering the following questions:

(a) What was the identity of each of the unknown solutions? How did you decide?

(b) Which metals were easy to identify? Explain.

(c) Which metals were difficult to identify? Explain.

(d) Why were all of the compounds you tested nitrates?

⟨8A⟩ 11 Write a lab report for this investigation.

Figure 2
Another way to do flame tests is to dip a loop of platinum wire into a solution and then into the flame.

Understanding Concepts

1. What is the significance of conducting flame tests?

2. Explain how you might test a sample of an unknown white solid to determine if it was table salt, sodium chloride. Remember that a taste test is never recommended.

Exploring

3. Design an investigation to answer the
⟨2A⟩ question: "What chemicals are used to make the flame colours of burning fire logs?" Conduct the investigation and report on your findings, if your design is approved by your teacher.

4. Your teacher may have cardboard spectroscopes available to view the spectrum of each element. Describe the so-called "line spectrum" that you observe. How would this line spectrum be more useful than a simple flame test?

Isotopes and Radioisotopes

The discovery of isotopes and radioisotopes has changed our lives in many ways: we now have sophisticated medical equipment, nuclear waste, and an ongoing debate about nuclear power.

An **isotope** is any of two or more forms of an element, each having the same number of protons but having a different mass due to a different number of neutrons. For example, chlorine has two common isotopes. Each has 17 protons, but some atoms contain 18 neutrons and others contain 20 neutrons. Thus, one isotope of chlorine is called Cl-35, having an atomic number of 17 and a mass number of 35 (17 protons plus 18 neutrons). The other isotope of chlorine, Cl-37, has an atomic number of 17 and a mass number of 37 (17 plus 20). Hydrogen has three isotopes (**Figure 1**).

Isotopes of the same element have the same physical properties and the same chemical properties—they undergo the same reactions. However, some isotopes are unstable, or **radioactive**, which means that the nucleus has a tendency to break apart and eject very-high-energy particles into its surroundings (**Figure 2**). The huge amount of energy these particles have can be both dangerous and useful. Atoms that have unstable nuclei are called **radioisotopes**.

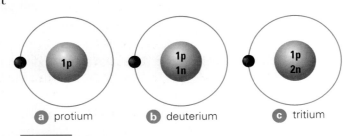

Did You Know?

Marie Sklodowska Curie (1867–1934) won the Nobel Prize in Physics in 1903 for her research into radioactivity, and the Nobel Prize in Chemistry in 1911 for her discovery of the elements radium and polonium. Marie Curie died of leukemia, likely caused by her exposure to radiation during her experiments.

1p
1p
1n
1p
2n
a protium
b deuterium
c tritium

Figure 1
The three isotopes of hydrogen have the same number of protons but different numbers of neutrons. The one with no neutrons is the most common.

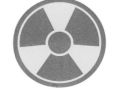

Figure 2
This symbol indicates that a radioactive substance is present and should be handled with care.

Types of Radioactivity

Radioactivity was discovered accidentally in 1896 by a French scientist, Henri Becquerel, when he was studying a sample of uranium. He found that the uranium could produce an image on photographic film even when the film remained sealed inside its package. Some of the unstable uranium nuclei split apart or **decayed**, producing particles that went right through the packaging and reacted with the film. Over the next seven years, three different kinds of radioactivity were identified (**Figure 3**). They were

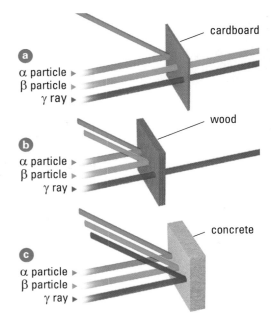

cardboard
a
α particle
β particle
γ ray
wood
b
α particle
β particle
γ ray
concrete
c
α particle
β particle
γ ray

Figure 3
The three types of radioactivity have different penetrating power. Gamma rays are the most dangerous and can only be blocked by thick sheets of concrete or lead.

named alpha (α) and beta (β) particles and gamma (γ) rays. Alpha particles were later found to be helium nuclei, containing two protons and two neutrons. Beta particles are high-energy electrons, and gamma radiation is high-energy electromagnetic radiation with no mass.

Applications of Radioisotopes

Radioisotopes must be treated with caution because they can damage living tissue. They may alter the DNA, which affects how cells divide. This can cause serious diseases, including cancer and birth defects. However, radioisotopes are very useful when used carefully by qualified doctors, scientists, and technicians. A few of the many applications of radioisotopes are shown in **Figures 4** to **6**.

Understanding Concepts

1. What is an isotope?

2. What is meant by the term "radioactivity"? Give an example of a radioisotope.

3. Represent the following radioisotopes using standard atomic notation. How many neutrons does each have?
   (a) technetium-99 (atomic number 43)
   (b) cobalt-60 (atomic number 27)
   (c) carbon-14 (atomic number 6)
   (d) iodine-131 (atomic number 53)
   (e) americium-241 (atomic number 95)
   (f) uranium-235 (atomic number 92)

Making Connections

4. Describe four useful applications of radioisotopes. What are their risks, if any?

Exploring

5. Visit your local medical centre or dentist to find out how X rays are produced. Find out what steps medical professionals take to protect themselves and their patients from radiation.

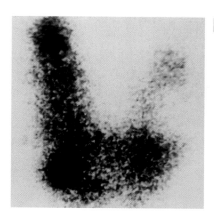

Figure 4

Radioisotopes are often used to diagnose medical problems such as thyroid disease. Doctors inject the radioactive isotope iodine-131 into the body, and the blood carries most of it to the thyroid gland in the neck. Technicians can then take a "photo" of the neck to see the size, shape, and activity of the gland.

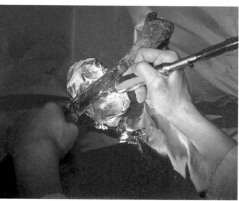

Figure 5

Archaeologists use a technique called "carbon-14 dating" when they want to know the age of ancient humans and their artifacts. Carbon dioxide in the atmosphere naturally contains a measurable amount of carbon-14, which is absorbed by plants in the process of photosynthesis. It is then incorporated into the bodies of animals that eat the plants. This reindeer bone was discovered in a glacier. During the years since the animal's death, the bone's carbon-14 steadily "decayed." When scientists test the bone, they measure the amount of carbon-14 left. They then calculate the length of time since the animal died.

Figure 6

You probably have a radioisotope in your home. Some smoke detectors contain a tiny amount of the radioisotope americium-241. The radioactive particles charge air molecules so that a small electric current flows. When smoke enters the detector, it interferes with this current and the alarm sounds.

Fireworks: Electron Jumps in Action

What creates the colours in fireworks?

A modern firework shell contains black powder that burns to propel the firework up into the air. The shell also contains separate packages of chemicals that produce special effects, such as bursts of colour, flashes, and sound. Some of these materials are described in **Table 1**.

Each firework explosion is a carefully controlled series of chemical changes that occur at just the right times. These chemical changes produce large amounts of heat which make electrons in metal atoms jump up to higher energy levels or orbits. When the electrons drop back to their normal ground state in orbits nearer the nucleus, they give off energy including bursts of coloured light.

Ⓡ Making fireworks is hazardous and should be attempted only by well-trained professionals.

Table 1	Some Chemicals Used for Special Effects	
Material	**Special Effect**
magnesium metal	white flame
sodium oxalate	yellow flame
barium chlorate	green flame
cesium(II) sulfate	blue flame
strontium carbonate	red flame
iron filings and charcoal	gold sparks
potassium benzoate	whistle effect
potassium nitrate and sulfur	white smoke
potassium perchlorate, sulfur, and aluminum	flash and bang

Figure 1

A fireworks shell

mixture of potassium chlorate, strontium carbonate, and paraffin oil

mixture of potassium chlorate, sulfur, and aluminum

mixture of potassium nitrate, sulfur, and charcoal

① The technician lights the first fast-burning fuse (A), which causes an explosion that launches the shell into the air as the black powder explodes.

② Fuse A also lights a slow-burning fuse (B) which ignites a mixture that produces a red burst, when the shell is high in the air.

③ The red explosion lights another slow-burning fuse (C) which ignites a mixture that produces a final flash and loud bang.

Inside a Firework Shell

Suppose a pyrotechnics technician had the job of making a firework shell that would rise 50 m and then produce a red burst of fire followed by a loud bang and a flash. The technician would have to make three different explosive mixtures: one to lift the shell into the air and one for each of the two special effects (**Figure 1**).

The first and most dangerous step is mixing the ingredients. As in other combustion reactions, the ingredients needed in fireworks include a fuel, a source of oxygen (called an oxidizer), and a source of heat to start the reaction (a burning fuse). Typical

Did You Know ❓

The history of fireworks (pyrotechnics) began in China 1000 years ago with the discovery of black powder (gunpowder), a mixture of potassium nitrate (saltpetre), charcoal, and sulfur. At the time, people used saltpetre to preserve meat. They may have found by chance that a mixture of charcoal, iron, and saltpetre produced sparks when sprinkled into a fire. When this mixture burns, it produces large amounts of gases and energy in the form of heat and light.

oxidizers are potassium nitrate, potassium chlorate, potassium perchlorate, and ammonium perchlorate. Each mixture also contains binders like red gum, paraffin oil, or dextrin. The binders act as fuel and hold the mixture together. The technician then wraps each mixture in a cardboard package and links the packages with fuses.

Issue   Should Fireworks Be Banned?

Imagine that your local council has received complaints about fireworks displays in your area. Some people have given the council a proposal.

The council decides to discuss this issue in a private meeting. They invite interested groups of citizens to write position papers, in which they state their opinions and back them up with evidence.

(a) Read the points listed below.

(b) Choose one of the two opinions. (Your teacher may assign these points of view to groups of students.)

(c) Research the issue further, expand upon 3A the points provided, and develop or reflect upon your position.

(d) Write a one-page position paper, clearly 3B stating your opinion and the reasons for it.

THE PROPOSAL
No one should be allowed to buy or use fireworks at any time within city boundaries.

Laurel Bishop

P. Tobian.

Olive Travers

Anthony Petrucelli

Julie F. Fletcher

John Hornstein

Mary Fong

Opinion A

- Fireworks are dangerous mixtures of chemicals. When ignited, they can explode in unpredictable ways. People have been terribly injured through the unsafe use of fireworks.

- Fireworks displays pollute the environment. The reactants involved can produce nitrogen dioxide and sulfur dioxide, both of which are poisonous gases and produce acid rain. Noise pollution is also created.

- Fireworks are very expensive, and they last for only a few seconds. We would be better off using the money to celebrate special occasions in other ways.

Opinion B

- Fireworks are a traditional way of celebrating for some cultures. The move to prevent fireworks could be seen as discriminating against those groups.

- People who want to use fireworks will continue to do so outside the city. In wooded areas or farmlands far from emergency services, the risk of fire or accident would be greater.

- Fireworks displays that mark special events promote tourism and bring economic benefits to the community.

Chapter 3 Review

Key Expectations

Throughout the chapter, you have had opportunities to do the following things:

- Explain the usefulness of scientific models, and describe the evolution of models of the atom. (3.1, 3.2, 3.4)

- Describe the Bohr-Rutherford model of the atom, and draw diagrams for atoms and ions of the first 20 elements. (3.2, 3.4)

- Write standard atomic notation, and state numbers of subatomic particles in a given atom or isotope. (3.3, 3.6)

- Demonstrate knowledge of laboratory safety procedures while conducting investigations. (3.1, 3.5)

- Investigate the relationship between atomic models and properties of substances, and organize, record, analyze, and communicate results. (3.1, 3.5)

- Formulate and research questions related to the properties of elements and compounds and communicate results. (3.2, 3.6, 3.7)

- Describe technologies that have depended on understanding atomic and molecular structure. (3.6, 3.7)

- Explore careers requiring an understanding of the properties of matter. (Career Profile)

KEY TERMS

alchemist	isotope
atomic model	mass number
atomic number	neutron
Bohr diagram	nuclear model
Bohr-Rutherford diagram	orbit
charge	proton
decay	radioactive
electron	radioisotope
energy level	spectrum
excited state	standard atomic
flame test	notation
ground state	subatomic particle
ion	

Reflecting

- "The ultimate building block of matter is the atom. Scientists try to understand the behaviour of matter by developing models of the atom." Reflect on this idea. How does it connect with what you've done in this chapter?
- Revise your answers to the questions raised in Getting Started. How has your thinking changed?
- What new questions do you have? How will you answer them?

Understanding Concepts

1. Make a concept map to summarize the material that you have studied in this chapter. Start with the word "atoms."

2. The sentences in the list below contain errors or are incomplete. In your notebook, write your complete, correct version of each sentence.

   (a) Protons are negative particles in orbits around the nucleus.
   (b) The mass number is the number of neutrons.
   (c) Isotopes have different numbers of protons.
   (d) A Bohr diagram shows protons in orbits.
   (e) When an electron jumps to a higher level, the atom is in ground state.
   (f) A chemical reaction results in new elements.

3. Who
   (a) proposed an atomic model thousands of years ago?
   (b) tried to change lead into gold in the Middle Ages?
   (c) recognized oxygen as an element?
   (d) proposed an atomic model in which compounds were made by combinations of atoms?
   (e) discovered the nucleus?
   (f) proposed that electrons existed in definite orbits?

4. (a) In Thomson's raisin-bun model, what were the electrical charges on the
   (i) raisins? (ii) bun?
   (b) How was Bohr's model different from Thomson's model of the atom?

5. With respect to the Bohr-Rutherford model of the atom,
   (a) where are the protons, neutrons, and electrons found?
   (b) which particles make up most of the mass of the atom?
   (c) which particles take up most of the space in the atom?

6. When any of the first 20 elements form ions, what are the numbers of electrons in the first three energy levels?

7. Describe how you can use the mass number and atomic number to find the numbers of protons, electrons, and neutrons in an atom.

8. (a) What is meant by the term "isotope"?
   (b) How many isotopes of the element hydrogen are there?
   (c) What are the numbers of protons and neutrons in these isotopes?

Applying Skills

9. Copy **Table 1** into your notebook. Fill in the blanks with the missing numbers.

Table 1

Finding the Atomic Number, Mass Number, and Numbers of Subatomic Particles in an Element

Element	Symbol	Atomic No.	Mass No.	No. of Protons	No. of Electrons	No. of Neutrons
helium	He	2	4	?	?	?
oxygen	O	?	16	8	?	?
sodium	Na	11	23	?	?	?
chlorine	Cl	?	37	?	17	?
calcium	Ca	?	?	?	20	22

10. Write standard atomic notation for each of the atoms in the previous question.

11. Draw Bohr-Rutherford diagrams for each of the atoms in the previous question.

12. Match the description on the left with one term on the right. Use each term only once.

Description	Term
A atom with same atomic number but different mass number	1 atomic number
B atom with unstable nucleus	2 mass number
C charged atom	3 proton
D number of protons	4 neutron
E positive subatomic particle	5 isotope
F sum of protons and neutrons	6 ion
G uncharged subatomic particle	7 radioisotope

13. Describe how each of the following atoms gains or loses electrons to have a stable number of electrons in each energy level:
   (a) beryllium, atomic number 4
   (b) nitrogen, atomic number 7
   (c) sulfur, atomic number 16

14. Write the charge for each of the ions in the previous question.

15. Identify the numbers of protons and neutrons in each of the following atoms by interpreting their standard atomic notation:
   (a) $^{40}_{19}K$ (b) $^{28}_{13}Al$ (c) $^{14}_{6}C$

Making Connections

16. You have learned that models are modified as scientists gather new evidence. Has this happened with the atomic model? Explain your answer.

17. For centuries people believed Aristotle's model for matter was true.
   (a) How did Aristotle's model differ from Dalton's model?
   (b) Why did it take so long for the model to evolve?

18. Design and draw a diagram of a firework you think would be suitable for Canada Day. Explain why you chose your design. Include safety and environmental considerations.
   (R) Do not test your design. Fireworks are extremely hazardous.

19. Find out more about radioisotopes. How are they used in medicine to diagnose diseases? How are they used to kill cancerous cells? What radioactive elements are used for various medical purposes?

The Periodic Table

1 There are tens of thousands of different chemical compounds that make up our world. Some of these materials are natural. For example, people have used cotton (a plant) for clothing and flint (a rock) for tools for thousands of years. Other materials are **synthetic**—invented and produced by people. Nylon and steel have similar functions to cotton and flint, but they are synthetic. In fact, the substance "nylon" did not even exist before this century. Which of the objects in the photograph are made of synthetic substances? How are these substances made? ➤

2 All substances, whether natural or synthetic, are made of the same building blocks—atoms. Atoms can combine to form molecules. Molecules that are made of more than one type of atom are called compounds and the number of possible compounds is almost infinite. However, there are only about 100 different kinds of atoms to use as building blocks. For example, each molecule of acetic acid is made up of atoms of carbon, oxygen, and hydrogen. What makes these three atoms different from each other? Do any other atoms have similar properties to these three?

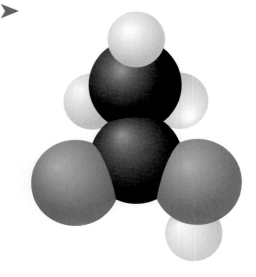

3 How does it help you when things are organized? In the last century, a scientist called Dmitri Mendeleev looked for ways to organize the current knowledge about atoms. He invented a **periodic table**—an organized arrangement of elements that explained and predicted physical and chemical properties. Mendeleev's table was the key to understanding elements and discovering new elements and compounds. How did Mendeleev organize his table? Make a list of some of the things that you organize or are organized for you, in your daily life.

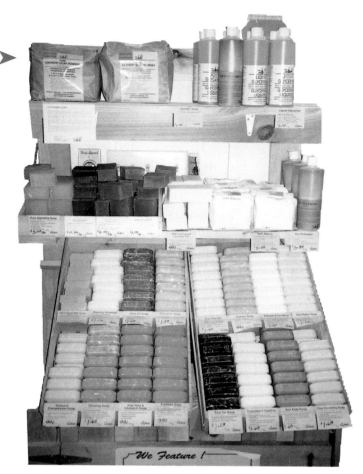

Reflecting

Think about the questions in **1**, **2**, **3**. What ideas do you already have? What other questions do you have about the periodic table? Think about your answers and questions as you read the chapter.

Try This   Organizing Elements

Using graph paper or chart paper, cut a strip of paper 20 cm long by 1 cm wide. Draw vertical lines on the paper to divide it into 20 squares. Write the numbers 1 to 20 consecutively in the squares (**Figure 1**). The numbers represent 20 elements. Cut the tape into pieces by cutting along the lines between (a) elements 1 and 2, (b) elements 2 and 3, (c) elements 10 and 11, and (d) elements 18 and 19. You should now have five pieces of paper.

You are told that elements 3, 11, and 19 have similar physical and chemical properties. Line up these elements in a vertical column, and tape the strips of paper into your notebook. The other columns will also line up.

Look at the table you have made. Which elements should have similar properties to

element 4? What element should have similar properties to element 9?

Assuming that the elements' numbers are also their atomic numbers, draw Bohr-Rutherford diagrams for elements 3, 11, and 19. What do these elements have in common? Where do you think you should put elements 1 and 2 on your table? Tape them onto your table and explain the reason for your choice.

Write numbers in order on a piece of tape as described.

1	2	3
9	10	11
17	18	19

Figure 1

Organizing the Elements

We often organize things to make them more useful. When you bake a cake, for instance, you might think of the ingredients in categories. You need something sweet for flavour, fat for bonding, flour for substance, and something to help it rise. Many different ingredients can fall into each category. You can combine ingredients from these four categories to make different kinds of cakes.

Up to the mid-1800s, scientists were busy discovering elements and recording their properties. Then they tried to organize their experimental observations in a useful way. At first, they listed the elements alphabetically. But every time a new element was discovered, the whole list had to be changed!

They tried other organizing methods. Could elements be grouped by state and colour? No, too many elements look alike. By taste? Definitely not. Too many elements are poisonous. Also, state, colour, and taste cannot be measured. The scientists also used properties that you studied earlier in this unit. For example, properties such as conductivity, malleability, and lustre suggested that elements could be grouped generally into metals and nonmetals (**Figure 1**).

John Dalton and other scientists then found a quantity that could be measured for an element—its atomic mass. The **atomic mass** is the average mass of an atom (i.e., of all occurring isotopes) of an element. Each element has its own unique atomic mass. Several scientists started to arrange the known elements according to their atomic masses (**Figure 2**).

Figure 1

Physical and chemical properties suggest that elements can be organized into metals and nonmetals.

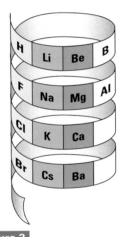

Figure 2

One early scientist organized elements in an arrangement that looked like a coiled spring.

Mendeleev and the First Periodic Table

The best arrangement of elements was produced by a Russian scientist, Dmitri Mendeleev (**Figure 3**). Working with the 64 known elements, he wrote the name of each element on a separate card, along with its atomic mass and other properties such as solubility, density, flammability, and so on. Then he played "chemical solitaire," arranging the cards in different ways to see if he could find any patterns. Perhaps a pattern would explain the behaviour of the known elements. He also hoped this might show a way to discover new elements.

Mendeleev organized his first periodic table by arranging the elements in order of increasing atomic mass. When an element or group of elements seemed to repeat properties he had seen before, he started a new row. Eventually he found that elements with similar properties fit into the same vertical columns. However, often an element seemed to belong in a place based on its mass, but did not fit based on its properties. When that happened, he ignored the mass and moved the element to a column with other elements having similar properties.

Figure 3

Dmitri Mendeleev's organization of the elements into a periodic table made the study of chemistry manageable.

Mendeleev's arrangement showed a regular pattern, which he described as a law. Simply put, a scientific law is a rule that nature appears to follow. It is a generalization of many observations; a statement rather than an explanation. **Mendeleev's periodic law** states:

If the elements are arranged according to their atomic mass, a pattern can be seen in which similar properties occur regularly.

Testing Mendeleev's Periodic Law

To test a scientific law, you use it to predict new observations. When Mendeleev came to a place in his table in which no known element would fit, he used a blank card (**Figure 4**). Mendeleev predicted that elements would eventually be found to fit the spaces. He also predicted the properties of those elements by examining the elements surrounding the blank spaces. **Table 1** shows two of Mendeleev's predictions and the elements that were discovered to fill in the blank spaces. His predictions were accurate! These discoveries helped convince people that the periodic table worked. A century later, it is still used to summarize chemical facts.

Group	I	II	III	IV	V	VI	VII	VIII
Formula of Compounds	R_2O	RO	R_2O_3	RO_2 H_4R	R_2O_5 H_3R	RO_3 H_2R	R_2O_7 HR	RO_4
1	H(1)							
2	Li(7)	Be(9.4)	B(11)	C(12)	N(14)	O(16)	F(19)	
3	Na(23)	Mg(24)	Al(27.3)	Si(28)	P(31)	S(32)	Cl(35.5)	
4	K(39)	Ca(40)	–(44)	Ti(48)	V(51)	Cr(52)	Mn(55)	Fe(56) Co(59) Ni(59) Cu(63)
5	[Cu(63)]	Zn(65)	–(68)	–(72)	As(75)	Se(78)	Br(80)	
6	Rb(85)	Sr(87)	?Yt(88)	Zr(90)	Nb(94)	Mo(96)	–(100)	Ru(104) Rh(104) Pd(105) Ag(108)
7	[Ag(108)]	Cd(112)	In(113)	Sn(118)	Sb(122)	Te(125)	I(127)	

Figure 4

Part of Mendeleev's periodic table, including atomic masses. If he did not know of an element to fit a space, he left it blank and predicted the atomic mass of the element.

Table 1   **Two of Mendeleev's Predicted Elements**

Property	Gallium		Germanium	
	Predicted 1871	Discovered 1875	Predicted 1871	Discovered 1886
atomic mass	68	69.9	72	72.3
density (g/cm^3)	5.9	5.94	5.5	5.47
melting point	low	30°C	high	2830°C
solubility in acids	medium	medium	low	low

Understanding Concepts

1. What are some of the properties that helped scientists organize elements into metals and nonmetals?

2. **(a)** What property of atoms did Mendeleev use to organize elements?

   (b) How did he use this property to organize them?

   (c) When did he ignore this property in building his table?

3. Examine the elements in the first column of Mendeleev's first periodic table.

   (a) All but one are shiny metals. What is the exception?

   (b) The first four elements in this group have atomic numbers 1, 3, 11, and 19, respectively. What do these elements have in common?

4. Examine the first two columns of Mendeleev's periodic table. Refer to **Table 2** on page 64.

   (a) What are the combining capacities of the two elements from the first column?

   (b) What are the combining capacities of the three elements from the second column?

   (c) What does this suggest about the combining capacities of all the elements in Mendeleev's periodic table?

5. Sodium chloride, potassium chloride, and calcium chloride are all used for melting ice in the winter. The chemical formula of sodium chloride is NaCl. Predict

Challenge

Make a list of the key points Mendeleev used in creating his periodic table. Speculate on any strengths or weaknesses based on what you know so far.

Inventing a Periodic Table

Mendeleev invented his periodic table by looking for patterns in the properties of different groups of elements. He looked for regularities that would enable him to put elements into families or groups with similar properties. Assuming that not all elements had been discovered, he deliberately left gaps in his table. He predicted that elements would eventually be discovered that would fill these gaps.

Imagine that you own a hardware store. You have received a shipment of nuts and bolts. Unfortunately, the contents spilled and are all mixed together. You also suspect that some of the nuts or bolts may be missing from your original order. You had planned to arrange the twenty types of nuts and bolts in a logical pattern of four rows and five columns on a display rack.

What would be a logical arrangement of the nuts and bolts? Which nut or bolt is missing? How would you predict its characteristics? In this investigation, you will make a periodic table of nuts and bolts to answer these questions.

Part 1: Hardware Items

Materials
- balance
- ruler
- graph paper
- set of 19 nuts and bolts in a resealable plastic bag

Procedure

1 Obtain a bag of hardware items (nuts and bolts). There should be 19 items in the bag. Count them to make sure.

2 There is one hardware item missing from the shipment.

(a) What nut or bolt do you think is missing? Describe the missing item.

3 Draw a large 4 × 5 grid on a sheet of paper with four rows and five columns. Number the squares from 1 to 20, starting in the top left corner and numbering across the rows. Call these "hardware numbers."

4 In your group, invent a system that you will use to organize your nuts and bolts. Place each item in a square, leaving a space if necessary.

(a) Describe your organizing system.

Step 4

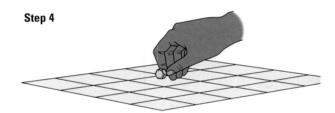

5 Remove the nuts and bolts from the grid. Now, replace them on the grid using the following new system. Place the smallest bolt in the top left corner and the largest nut in the bottom right corner. Put similar items in the same vertical column. Arrange them so that size increases down each column and generally across each row.

(a) What nut or bolt do you now think is missing? Describe the missing item.

6 Use the balance to measure the mass of each item.

(a) Record the masses in the appropriate square on the grid.

7 Use a ruler to measure the longest dimension of each item.

(a) Record the lengths in the appropriate square on the grid.

8 Count the nuts and bolts and return them to the bag, as directed by your teacher.

9 Plot a graph of hardware mass vs.
7B hardware number (1 to 20).

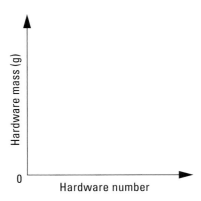

10 Plot a graph of hardware length vs.
hardware number (1 to 20).

11 Examine your graph of mass vs. number
data.

(a) What general trend in mass did you
note
(i) across any row of your table?
(ii) down any column of your table?

12 Examine your graph of length vs. number
data.

(a) What general trend in length did you
note
(i) across any row of your table?
(ii) down any column of your table?

13 Predict the mass and length of the missing
item.

Part 2: Store Manager's Dilemma

14 Imagine that you are the manager at a
store that sells a wide variety of items, such
as a bulk-food store or a stationers'. The
store has many different items for sale,
and the customers have to be able to find
them easily.

(a) Make a list of all the different items
you need to find space for in the store.

15 Devise a classification system that would
make it easy for customers to find the
things they need.

(a) Present your organization system as a
detailed floor plan for the store.

Understanding Concepts

1. (a) How does your hardware periodic table
compare with Mendeleev's periodic table
of elements?

(b) What do you think the mass and length
of the nuts or bolts could represent for
atoms in Mendeleev's periodic table?

2. (a) What groups of elements do you think
nuts represent? Why?

(b) What groups of elements do you think
bolts represent? Why?

(c) Imagine that a nut and bolt are screwed
together. What do you think such a
combination might represent?

3. How does that process compare with
Mendeleev's work?

Reflecting

4. Imagine that you were given a bolt identical
to one of the bolts in your set, except that it
is made of a more dense material. What do
you think this bolt would represent in our
atomic model?

5. Some elements in the periodic table never
combine with any other elements. What kind
of hardware item might you include in this
set to represent such an element?

Challenge

What examples of models have you come
across earlier in this unit? Which are
appropriate for the challenge you have
chosen?

Exploring the Modern Periodic Table

Mendeleev's periodic table was a major breakthrough in the understanding of the elements. However, he found that organizing elements in order of atomic mass did not always work. In some instances he had to decide whether to place an element by its atomic mass or by its properties. He concluded that his calculation of atomic mass was flawed, trusted his instincts, and went with the properties. This decision proved to be a good one when the nuclear atom and subatomic particles were discovered. People realized that the key to the identity of an element was the number of protons in the nucleus—the atomic number—rather than the atomic mass. A new law was born. The **modern periodic law** states:

> *If the elements are arranged according to their atomic number, a pattern can be seen in which similar properties occur regularly.*

These properties include melting and boiling temperatures and the sizes of atoms. The size of a spherical atom is called the **atomic radius**—the distance from the nucleus to the "outer edge" of the atom.

In the "nuts-and-bolts" activity, you organized the hardware items and looked for trends in properties, such as length and mass, both along a row and down a column. What trends in properties can we observe when we organize elements into a periodic table by atomic number? How does the arrangement of elements in the periodic table relate to the arrangement of electrons in the atoms? Work in pairs and refer to the periodic table on the inside back cover to answer these questions.

Procedure

1 Examine the periodic table on the inside back cover. Note whether the elements are solids, liquids, or gases at room temperature.

(a) Which symbols represent elements that are gases at room temperature?

(b) Name two elements that are liquid at room temperature.

2 Look at the symbols and atomic numbers of the elements.

(a) What is the atomic number of helium (symbol He)?

(b) What is the atomic number of gold (symbol Au)?

(c) What is the symbol of the element with atomic number 22?

(d) What is the symbol of the element with atomic number 33?

3 Look at the atomic masses of the elements.

(a) What is the atomic mass of aluminum (symbol Al)?

(b) What is the atomic mass of silver (symbol Ag)?

(c) What is the symbol of the element with atomic mass 40.1?

(d) What is the symbol of the element with atomic mass 83.8?

4 Mendeleev used atomic masses to organize his periodic table.

(a) What two elements are "out of order" according to atomic mass in the fifth row?

5 Look at the densities and melting points of the elements.

(a) Which element has the highest melting temperature?

(b) Which element has the lowest melting temperature?

(c) Which element has the greatest density?

(d) Which element has the lowest density?

6 Elements 1, 3, 11, and 19 are in the first column of the periodic table.

(a) Draw Bohr diagrams for these elements.

(b) How is the electron arrangement in these elements similar?

(c) How many electrons do you think there are in the outer orbit of the elements Rb and Cs?

7 Elements 9 and 17 are in the second last column of the periodic table.

(a) Draw Bohr diagrams for these elements.

(b) How is the electron arrangement in these elements similar?

(c) How many electrons do you think there are in the outer orbit of the elements Br and I?

8 Look at elements 3 to 10.

(a) Draw Bohr diagrams for these elements.

(b) Describe the general pattern that you observe across a row of the periodic table.

9 Elements in the same column tend to form similar compounds. For example, the compounds that hydrogen forms with elements in the first column are LiH, NaH, KH, and RbH. The compounds that hydrogen forms with elements in the second column are BeH_2, MgH_2, CaH_2, and so on. Hydrogen forms the following compounds by combining with other elements: CH_4, NH_3, H_2O, and HF.

(a) What are the formulas of the compounds formed by the combination of hydrogen with the following?
(i) silicon (Si)
(ii) phosphorus (P)
(iii) sulfur (S)

Understanding Concepts

1. In what state are most elements at room temperature?

2. Illustrate how the following properties generally change as you go from left to right in the periodic table.
   (a) atomic number
   (b) melting temperature
   (c) atomic radius

3. Illustrate how the following properties generally change as you go down the columns of the periodic table.
   (a) density
   (b) melting temperature
   (c) atomic radius

4. Look at the portion of the periodic table in **Figure 1**. Facts about calcium have been omitted. Use your understanding of the periodic table to predict the properties of calcium.

11	97.8 883 0.971 180	12	649 1107 1.74 150	**Figure 1**	
Na sodium 23.0		**Mg** magnesium 24.3			
19	63.3 760 0.862 220	**?**		21	1541 2836 2.99 160	22	1660 3287 4.54 140
K potassium 39.1		**Ca** calcium		**Sc** scandium 45.0		**Ti** titanium 47.9
37	38.9 686 1.53 235	38	769 1384 2.6 200	39	1522 3338 4.47 180	40	1852 4377 6.49 155
Rb rubidium 85.5		**Sr** strontium 87.6		**Y** yttrium 88.9		**Zr** zirconium 91.2

5. **(a)** Plot a graph of atomic radius vs. atomic number for the first 20 elements.
   (7B)
   (b) What general trends do you observe in your graph?

Making Connections

6. Elements in the same group have similar properties. Think of two examples in everyday life where similar substances could be substituted for each other. What other factors would you consider before making the substitutes?

Challenge

What were the advantages of working in a group for this activity? What would you do differently in another group situation?

Groups of Elements

When you go to the supermarket, how do you find the foods on your shopping list? If you are looking for yogurt or milk, you look for the aisle marked "Dairy Products." For potatoes or carrots, you look for the "Fruits and Vegetables" aisle. Food in a supermarket is arranged so that people can find things quickly and easily.

In the same way, the periodic table groups elements with similar properties. You can quickly identify an element as a metal if it is on the left or in the centre of the table. Nonmetals are generally found on the right side.

The set of elements in the same column in the table is called a **chemical group**. These elements have similar physical and chemical properties. They form similar kinds of compounds when they combine with other elements. And their behaviour can be explained by the Bohr-Rutherford model of the atom: the elements in a group have the same number of electrons in their outer orbit and tend to form ions by gaining or losing the same number of electrons.

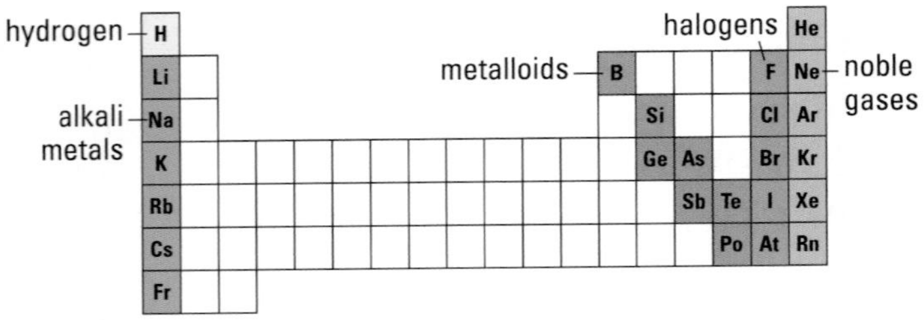

Figure 1
Location of noble gases, alkali metals, halogens, hydrogen, and metalloids on the periodic table. Hydrogen is a unique element.

Figure 2
Noble gases have many applications.

Noble Gases

The elements that occupy the far right column of the periodic table (**Figure 1**) are the **noble gases**. All gases at room temperature, the noble gases are often called inert gases because they are so unreactive, almost never forming chemical compounds with other elements.

You might think that such elements would be rare but argon makes up almost 1% of every breath of air that you inhale. Because it is colourless, odourless, tasteless, and nonreactive, no one notices it.

Their lack of reactivity makes noble gases quite useful (**Figure 2**). Helium gas has very low density and is used in lighter-than-air blimps. Neon is used in colourful light displays. Argon is used to fill light bulbs. If ordinary air were used, the metal filament would catch fire and burn out in seconds.

The lack of reactivity of noble gases is explained by their electronic structure. Chemists describe their outer orbits (shown in **Figure 3**) as

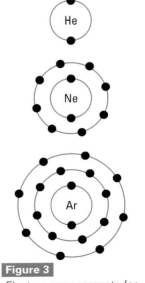

Figure 3
Electron arrangements for three noble gases

"filled." Noble gas atoms do not react with other atoms because they already have a stable arrangement of electrons.

Alkali Metals

The elements that occupy the far left column of the periodic table (**Figure 1**) are called the **alkali metals**. Lithium, sodium, potassium, etc. are all shiny, silvery metals. Unlike the noble gases, they are extremely reactive (**Figure 4**). Because they combine so readily with other elements, they are found in nature only as compounds.

Alkali metal compounds are found everywhere on Earth (**Figure 5**). The most common are sodium compounds, which occur in plants, animals, soil, and sea water. Do you remember the chemical name for salt? Another compound, sodium hydroxide, is used in making soap and paper. Potassium and sodium compounds transmit nerve impulses in your body.

Part of the experimental proof that alkali metals are a group is given by the formulas of the compounds they form: LiH, NaH, and KH with hydrogen; Li_2O, Na_2O, and K_2O with oxygen; and so on. They clearly have a lot in common.

The reactivity of the alkali metals is explained by their electronic structure. The outer orbits (**Figure 6**) have one electron—an unstable arrangement—so alkali metals tend to lose this electron, becoming ions with a charge of +1. These ions readily join with other elements such as oxygen and chlorine.

Halogens

Fluorine, chlorine, bromine, and the other **halogens** that occupy the seventeenth column of the periodic table (next to the noble gases) (**Figure 1**) are the most reactive nonmetals. Because of their reactivity, they almost always appear naturally as compounds, not as elements. The pure elements you may see in your lab are artificially extracted. For example, chlorine is extracted by the electrolysis of sea water. As shown in the periodic table on the inside back cover, the halogens occur in different states. Fluorine and chlorine are gases, bromine is a liquid, and iodine is a solid at room temperature.

The reactivity of the halogens makes them very useful. The most common halogen compounds are chlorine compounds found in living things, ocean water, and rocks. Table salt is mostly

Figure 4

Like all alkali metals, sodium reacts vigorously with water. The reaction releases heat and produces pure hydrogen gas. Why could this reaction be dangerous?

Figure 5

Alkali metals form compounds such as sodium chloride (salt) and sodium bicarbonate (baking soda). If potassium bitartrate is added to baking soda, baking powder is formed.

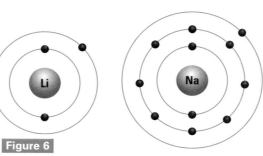

Figure 6

Electron arrangements for three alkali metals

Figure 7

A train derailment involving tank cars containing chlorine gas forced the evacuation of the city of Mississauga, Ontario, in 1979.

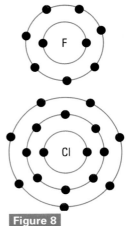

Figure 8

Electron arrangements for two halogens

sodium chloride but is "iodized" by adding a small amount of potassium iodide to prevent disease of the thyroid gland. Chlorine is used to kill bacteria and purify drinking water, but in large amounts it is extremely dangerous (**Figure 7**). Iodine is dissolved in alcohol to make an antiseptic to treat skin cuts. Sodium fluoride is added to toothpaste because the fluorine atoms bond to tooth enamel, making it less likely to develop cavities.

Halogens are so reactive because of their electronic structure. The outer orbits of the fluorine and chlorine atoms (**Figure 8**) have seven electrons. In chemical changes, halogens tend to gain one electron in order to have a stable arrangement of electrons. Halogen ions have a charge of –1.

A Group of One

Hydrogen is a unique element. Its most common isotope has only a single proton and no neutron in its nucleus. Like the alkali metals, it has only one electron in its outer orbit. Losing the electron makes the hydrogen ion positive, so it reacts with other elements, such as the halogens, that need extra electrons to fill their orbits (resulting in negative ions). Hydrogen has little else in common with the alkali metals: it is a colourless, odourless, tasteless, highly flammable gas (**Figure 9**). In other reactions, hydrogen acts like a nonmetal, gaining one electron so it has a complete first orbit. For example,

Figure 9

Hydrogen's low density makes it useful for weather balloons. Why is hydrogen not used in blimps that carry people?

it reacts with the alkali metals to form compounds such as LiH, NaH, KH, and so on.

Almost all of Earth's hydrogen exists in combination with other elements. Its reactivity is too great for it to exist in the atmosphere as a free element. Hydrogen is one of the main elements in all living things, as well as in petroleum, coal, and natural gas.

Metalloids

Metalloids are elements that possess both metallic and nonmetallic properties. Not strictly a group themselves, they are found in different groups on the right side of the periodic table (**Figure 1**), on both sides of the zigzag line that divides the metals from the nonmetals. For example, silicon is a metalloid. It is shiny and silvery, but is not malleable and is only a partial conductor of electricity. Other metalloids are boron, germanium, arsenic, selenium, antimony, tellurium, polonium, and astatine.

The electronics industry uses silicon and germanium, both semi-conductors, to make microcomputer chips. You may have read about arsenic as a poison. Another metalloid, boron, is used in borax water softeners and in the antiseptic, boric acid.

Rows on the Periodic Table

The groups of elements—the columns in the periodic table—have similar physical and chemical properties. These properties, however, vary from element to element in a column. Elements beside each other in the table also show similarities and gradual changes in properties. These horizontal rows of elements are called **periods**. The first period contains two elements: hydrogen and helium. The second period contains eight elements, starting with lithium and ending with neon. As you go from left to right within a row, the atomic number increases and the elements gradually change from metallic (lithium) to nonmetallic (fluorine), and then finally to the noble gases (neon) at the far right.

Challenge

What group of elements do any of the substances in your challenge belong to?

Understanding Concepts

1. Where on the periodic table do you find metals, metalloids, nonmetals, and noble gases?

2. (a) Define the term "chemical group."
   (b) Give three examples of chemical groups.
   (c) Compare the arrangement of electrons in the elements in the same group.

3. (a) Why are noble gases sometimes called inert gases?
   (b) How is the electronic structure of helium different from other noble gases? Why is it still included in the group?

4. (a) List similar properties of the alkali metals.
   (b) What similarities in electron arrangement do the alkali metals show?

5. (a) List similar properties of the halogens.
   (b) What similarities in electron arrangement do the halogens show?

6. (a) What evidence suggests that hydrogen should be in the first column of the periodic table?
   (b) How is hydrogen different from other elements in the first column?

7. What are metalloids?

8. Rubidium (Rb) is an alkali metal. What are the formulas of its compounds with hydrogen and with oxygen?

9. Write the names and formulas of the four compounds that can be formed by combinations of potassium, lithium, chlorine, and/or bromine.

10. Express, as a law, the relationship between position on the periodic table and number of electrons in the outer orbit.

Making Connections

11. Make a chart listing practical applications of alkali metal compounds.

Exploring

12. One meaning for the word "period" is "a portion of time marked by some returning action or phenomenon." Determine the trends that appear across the periods of the periodic table. Is this definition appropriate? Write a paragraph giving evidence to support your view.

SKILLS MENU
- Questioning
- Conducting
- Analyzing
- Hypothesizing
- Recording
- Communicating
- Planning

Groups of Elements and Compounds

How can you use the periodic table to make predictions? You know that the position of an element in the periodic table gives a clue to its physical and chemical behaviour. Elements in the same group (column in the table) tend to react in the same way. In this investigation, you will examine some of the properties of four groups in the periodic table: Group 1 (the alkali metals), Group 2, Group 16, and Group 17 (the halogens). The elements you will examine are shown in **Figure 1**.

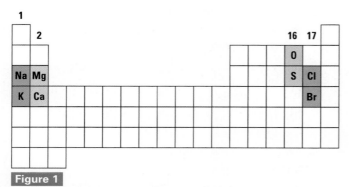

Figure 1

The position of an element in the periodic table can tell something about its properties.

One way to find similarities is to look at the properties of the elements themselves. You will examine Groups 1 and 2 elements directly. Another way is to investigate their compounds. Elements in the same group tend to form compounds with similar properties. In this investigation, you will compare Groups 1 and 2, and Groups 16 and 17, by looking for trends in the behaviour of their compounds. Remember that elements in Groups 16 and 17 change their names slightly when forming compounds: chlorine in a compound becomes chloride, sulfur becomes sulfide, and so on.

Materials
- safety goggles
- apron
- periodic table
- 3 beakers, each 250 mL
- overhead projector
- for teacher demonstration only: sodium metal, potassium metal, calcium metal, phenolphthalein indicator
- microtrays
- toothpicks
- eyedropper
- small samples of the following compounds: copper(II) chloride, copper(II) bromide, copper(II) oxide, copper(II) sulfide, calcium carbonate, magnesium carbonate, potassium carbonate, sodium carbonate

Assume that all powders are poisonous. If you spill any of these powders on the skin, in the eyes, or on clothing, wash the area immediately with plenty of cold water. Inform your teacher of any spills.

Question

1 Write a question that is being investigated. **4A**

Hypothesis

2 Make your own hypotheses at each step, as directed in the procedure.

Procedure

3 Make a full-page data table as shown in **Table 1**.

Table 1

Name of Starting Material	Appearance of Starting Material	Observations
calcium	?	?
sodium	?	?
potassium	?	?
magnesium carbonate	?	?
sodium carbonate	?	?
potassium carbonate	?	?
calcium carbonate	?	?
copper(II) chloride	?	?
copper(II) oxide	?	?
copper(II) bromide	?	?
copper(II) sulfide	?	?

4 Put on your apron and safety goggles.

Part 1: Comparing Groups 1 and 2 Elements

(Teacher demonstration using overhead projector)

5 Place a sheet of plastic wrap or an overhead acetate over the screen of the projector to protect it from any splashes. Turn on the projector. Place 3 beakers on the screen. Fill one-third of each beaker with water. Add 2 drops of phenolphthalein indicator to each beaker.

 Alkali metals are highly reactive. They must be stored under oil to prevent reaction with air. Use of larger amounts may cause an explosion. Do not cover the beaker with a glass plate as dangerous quantities of hydrogen gas may accumulate.

Step 5

6 Examine a small piece of calcium (about enough to cover your little fingernail).

(a) Predict what will happen when the calcium is added to water.

✎ (b) Record the appearance of the element in the appropriate space on your data table.

7 Add the calcium to one of the beakers. After the reaction stops, note the colour of the indicator and whether the resulting solution is clear or cloudy.

✎ (a) Record your observations.

8 Repeat steps 6 and 7, using a tiny piece of sodium (enough to fit on the flat end of a toothpick).

✎ (a) Record your observations.

9 Examine the periodic table. Which groups do calcium, sodium, and potassium belong to? How should potassium behave?

(a) Make a prediction on the appearance and behaviour of potassium.

10 Repeat steps 6 and 7 using a tiny piece of potassium (enough to fit on the flat end of a toothpick).

✎ (a) Record your observations.

Part 2: Comparing Compounds of Groups 1 and 2 Elements

11 Obtain a microtray, a toothpick, and a small amount of magnesium carbonate (enough to fit on the flat end of a toothpick).

(a) Predict what will happen when the magnesium carbonate is added to water.

✎ (b) Record the appearance of the compound.

12 Add the compound to a cell of the microtray that is half-filled with water. Try to dissolve the compound, using a toothpick to stir.

Step 12

✎ (a) Record your observations.

13 Repeat steps 11 and 12 using sodium carbonate.

✎ (a) Record your observations.

14 Examine the periodic table. What groups do sodium, potassium, magnesium, and calcium belong to? How should potassium carbonate and calcium carbonate behave?

(a) Make a prediction on how potassium carbonate and calcium carbonate should behave in water.

15 Repeat steps 11 and 12 using potassium carbonate, and again using calcium carbonate.

✎ (a) Record your observations.

Part 3: Comparing Compounds of Groups 16 and 17 Elements

16 Repeat steps 11 and 12 using copper(II) chloride.

✎ (a) Record the appearance of the compound and your observations.

17 Repeat steps 11 and 12 using copper(II) oxide.

✎ (a) Record your observations.

18 Examine the periodic table. What groups do the elements chlorine, bromine, oxygen, and sulfur belong to? How should copper(II) bromide, and copper(II) sulfide behave?

✎ (a) Record your prediction.

19 Repeat steps 11 and 12, using copper(II) bromide and copper(II) sulfide.

✎ (a) Record your observations.

20 Dispose of the contents of your microtrays and put away your materials as directed by your teacher. Clean up your work station. Wash your hands.

Analysis and Communication

21 Analyze your observations by answering the following questions:

(a) Which metal reacted most vigorously in water?

(b) What evidence did you have that sodium and potassium are alkali metals, but calcium is not?

(c) Compare the appearance and solubilities of the Group 1 and Group 2 carbonates.

(d) Are these results consistent with the location of these elements in the periodic table? Explain.

(e) Compare the appearance and solubilities of the copper(II) chloride, bromide, oxide, and sulfide.

(f) Are these results consistent with the location of these elements in the periodic table? Explain.

22 Write a paragraph summarizing how the periodic table can be used to predict the properties of elements and compounds.

Understanding Concepts

1. Predict what you would see if you added rubidium metal to water.

2. Predict the solubilities in water of

(a) copper(II) iodide

(b) barium carbonate

3. Zinc oxide is a white solid and zinc sulfide is a black solid. Both are insoluble in water.

(a) Are these properties consistent with your experimental observations? Explain.

(b) Are these properties consistent with the periodic table? Explain.

Exploring

4. Obtain a *Handbook of Chemistry and Physics*. Look at the listings of properties of various compounds. Record examples of similarities of compounds of elements in the same chemical groups.

5. Find out what other compounds of Groups 1, 2, 16, and 17 elements are available from your teacher. Plan an investigation, including hypotheses, to investigate the properties of calcium chloride, magnesium chloride, and other compounds in the periodic table. Obtain your teacher's approval before you carry out your investigation.

Challenge

Identify the elements in this investigation that were organized into Mendeleev's periodic table. Compare where he placed them with where they are placed in the modern periodic table.

Elemental Magic

Imagine that you are visiting a sports and camping store. You are thinking of buying skates, and your friend is looking at canoes. How do you decide what to buy? Certainly, price is important! But you also have many choices of materials. The skates could be made of natural leather, a synthetic, or a combination of the two. The canoe could be made of aluminum, fibreglass, Kevlar, ABS plastic, or other substances. Scientists use the "magic" of chemistry to make thousands of materials.

(a) What natural or "traditional" materials can you see around you?

The "magicians" are materials scientists—modern alchemists who make new products by using their knowledge of the periodic table. Understanding the arrangements of elements into groups and periods makes it possible to predict new ways of assembling atoms into molecules (**Figure 1**).

(b) What kinds of elements do materials scientists use?

(c) What are some of the products of this magic?

Over the centuries, people have used four general types of materials to make everything they need: metals, polymers, composites, and ceramics. The relative importance of these substances has changed over time, as you can see in **Table 1**. For example, in 5000 B.C., metals were relatively unimportant. By 1960, they had become by far the most important materials in the world!

(d) What types of metals would have been used in 5000 B.C.?

(e) What types of metals were in wide use by 1960?

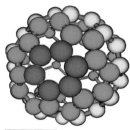

Figure 1

Materials scientists are always looking for new ways of assembling atoms into molecules. "Bucky-balls" are made of carbon atoms arranged in a sphere.

Table 1

Percentage Use of Different Types of Materials in History

Material	5000 B.C.	A.D. 1800	1960	2000
Metals	5	30	80	40
Polymers	40	35	10	25
Composites	15	5	2	15
Ceramics	40	30	8	20
Total	100%	100%	100%	100%

Metals

Metals form the majority of elements in the periodic table (**Figure 2**). In 5000 B.C., almost the only metals used were gold and copper for jewellery and containers. Because the metals were difficult to produce, they were rare and highly prized. As people discovered ways of purifying and mixing metallic elements, metal alloys such as bronze and steel became more important. They could be used for weapons, utensils, building, and many other uses. By the middle of the 20th century, alloys were used in everything from tractor bodies to building structures to tableware (**Figure 3**). Today super alloys can be specially engineered to be, for example, heat-resistant jet engine parts or corrosion-resistant chimney linings for power plants.

Li	Be												Al				
Na	Mg												Al				
K	Ca	Sc	Ti	V	Cr	Mn	Fe	Co	Ni	Cu	Zn	Ga				
Rb	Sr	Y	Zr	Nb	Mo	Tc	Ru	Rh	Pd	Ag	Cd	In	Sn		
Cs	Ba	La	Hf	Ta	W	Re	Os	Ir	Pt	Au	Hg	Tl	Pb	Bi	
Fr	Ra	Lr												

Figure 2
The metallic elements

Figure 3

(f) Why are alloys, rather than elemental metals, used for these highly-specialized applications?

Polymers

Polymers or elastomers are materials made of long molecules, something like miniature lengths of string. The most important elements in these substances are carbon, hydrogen, and oxygen (**Figure 4**). In 5000 B.C., the only elastomers were natural compounds found in animal and plant fibres. Wool, leather, wood, linen, and cotton were used for clothing, shelter, tools, boats, and so on.

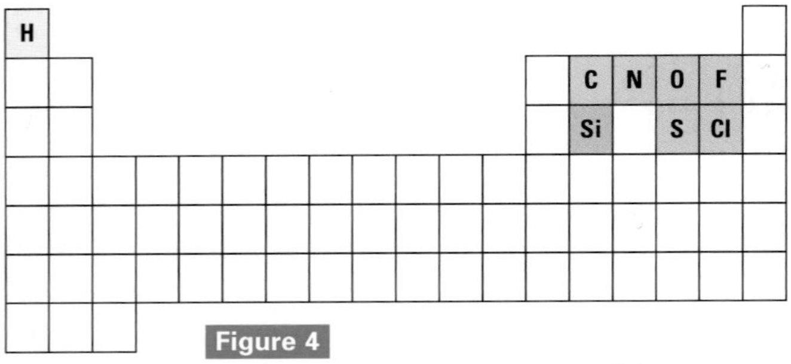

(g) Suggest other natural elastomers, and their uses.

Over the centuries, glues, rubber, and other natural elastomers were discovered. A major breakthrough in polymer chemistry was the invention of plastic polymers (**Figure 5**) in the 20th century, including nylon for clothing, polytetrafluoroethylene for nonstick cookware, and polystyrene for drink cups. Some new polymers are tough, light, and heat-resistant enough to be used in jet aircraft. Try to imagine a world without plastics: no CDs, no disposable food containers, and no plastic toys.

(h) The development of almost indestructible polymers is not entirely good news. What drawbacks do they have?

(i) What is being done to try to overcome these drawbacks?

Ceramics and Glasses

Ceramics and glasses are materials derived from minerals and rocks. The elements that are most important in these compounds include silicon, carbon, and oxygen, but many elements can be used, as you can see in **Figure 6**. In 5000 B.C., the most important ceramics were stone for weapons, flint for tools, and pottery for containers. Glass was invented about 2000 years ago. Today tough new ceramics are being engineered that have very special properties (**Figure 7**).

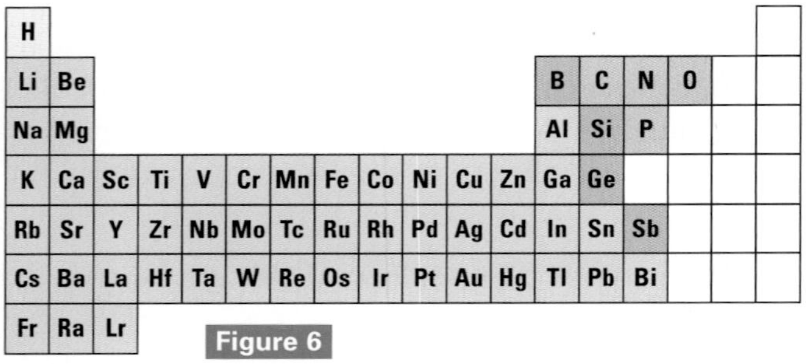

(j) Carbon fibre is a ceramic. What is it used for? Why is it suitable for these uses?

Composites

Composites are materials that are formed by mixing two other materials. The first composite was probably a straw-clay combination used in

Figure 4
The elements in polymeric materials

Figure 5
All plastics are examples of polymers.

Figure 6
The elements used in ceramic materials

Try This Making Gak

You can make your own synthetic polymer by dissolving enough borax cleaning powder to fit on the end 2 cm of a toothpick in half a teaspoon of water. Add about a teaspoon of white glue to the mixture and mix well. As the reaction continues, you will be able to pick up the polymer and knead it in the palm of your hand to squeeze out the water. Instant goo!

Ⓣ Airborne cleaning powder can irritate eyes, skin, and respiratory tract. Do not ingest.

building bricks. When materials scientists make composites, they try to combine the best properties of the polymers, ceramics, or metals that they

put together. For example, fibreglass, a composite of a polymer and tiny glass fibres, can be used to insulate houses or make boat hulls.

(k) Paper is a composite. What substances might it contain?

Materials Science and Canoes

Back in the sports store, your friend is still looking at canoes, and suggests buying one for a canoe trip in northern Ontario. What kind of canoe should you use? What should the canoe be made of? Even for something as simple as a canoe, materials science offers choices in all the categories that you have learned about (**Figure 8**). Depending on whether you want to run rapids or just paddle on a lake, the new alchemists have given you a wide choice of materials.

(l) Research the various materials that are now used for making different kinds of canoes. What are the benefits and drawbacks of each material?

(m) Having decided what kind of paddling you want to do on your imaginary canoe trip, choose the most appropriate canoe to buy.

Figure 8

Canoes for shooting rapids are made of different materials from those used for paddling on quiet lakes.

Understanding Concepts

1. **(a)** Draw a bar graph to summarize the information in **Table 1**.
   (b) Examine your graph or **Table 1**. What types of materials were most important in (i) 5000 B.C., (ii) A.D. 1800, (iii) 1960, and (iv) 2000?

2. Give five examples of each type of material that you would use in a typical day.
   (a) polymers (Classify the polymers you chose as natural or synthetic.)
   (b) ceramics
   (c) metals
   (d) composites

3. Why are materials scientists called "magicians"? Write a paragraph comparing them with stage magicians.

Exploring

4. Why did metals peak in popularity and then decline? Research some aspect of the use of metals in Canada and prepare a written or oral report.

5. Research "bucky-balls" and other fullerenes, both named after the scientist, Buckminster Fuller. How do materials scientists hope to use these new molecules?

Reflecting

6. Compare the four types of materials scientists use with the number of elements used. Do you think this affects manufacturing costs?

Challenge

Identify three products that are valuable to you. What are they made of? Include this information in your display. Which substance would you choose to market?

Ozone: A Global Environmental Hazard

Can elements in the periodic table occur in more than one form? Can an element be helpful in some situations and harmful in others? The answer to these questions is yes, and oxygen is an example of such an element.

You will recall that oxygen exists in two different forms: oxygen gas and ozone. The colourless, odourless oxygen gas that we breathe has the formula O_2. Without this gas, all organisms would die. Ozone is a pale blue gas with the chemical formula O_3. It is formed by the action of sunlight on oxygen, by lightning, and as a side effect of pollutants released from car engines. At ground level, ozone is poisonous but, in the upper atmosphere, it protects us from the Sun's radiation.

Ozone and Sunlight

You will remember that sunlight contains many different energies or colours of light, represented by the visible spectrum. Sunlight also contains invisible ultraviolet (UV) radiation, which has higher energy than any visible colour. UV radiation is believed to cause skin cancer, decrease the body's resistance to diseases, and can blind unprotected eyes. It also harms plant life. Fortunately, less than 10% of the Sun's UV radiation passes through the atmosphere. The reason is the ozone layer: ozone in the upper atmosphere that absorbs the UV radiation, preventing it from reaching the ground.

The Ozone Killer

Chlorofluorocarbons (**CFC**s) are compounds invented by chemists in the 1930s by putting together carbon, chlorine, and fluorine atoms. CFC's seemed very safe because they were stable: they didn't break down and they weren't harmful to living things. CFCs were used to clean, to cool, and to dissolve other substances. The first CFC, Freon, is probably the coolant in your refrigerator at home.

It took several decades for the first CFCs to work their way up through the atmosphere, where an unexpected reaction took place: UV radiation released chlorine atoms from the CFCs. And each chlorine atom broke apart 100 000 molecules of ozone. As the ozone broke down, the UV radiation was able to penetrate farther through the atmosphere, releasing more chlorine from CFCs as it went (**Figure 1**). An ozone killer was on the loose. Could it be stopped?

The Montreal Protocol

Recognizing the need to stop CFCs from reaching the ozone layer, over 100 countries signed the Montreal Protocol in September 1987, agreeing to cut CFC production in half by January 1996. Equipment that already contains CFCs was not banned for three reasons: the expense; the technology to

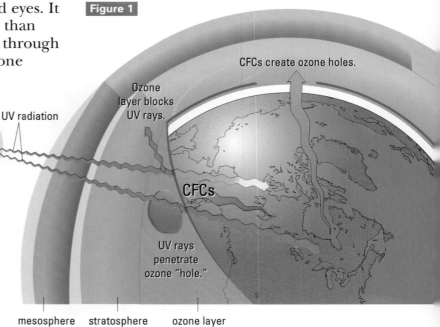

Figure 1

Ozone layer blocks UV rays.

CFCs create ozone holes.

UV radiation

CFCs

UV rays penetrate ozone "hole."

mesosphere   stratosphere   ozone layer

completely replace CFCs is not yet developed; and the technology that depends on CFCs, especially the refrigeration of food and medicine, is too important to just shut it down.

Some countries have still not agreed to the Montreal Protocol. Even in countries that are not producing CFCs, there is evidence that these chemicals are being smuggled in and used illegally. The production and use of CFCs remains a global issue. **Figure 2** shows the use of CFCs by region.

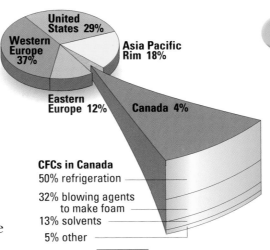

CFCs in Canada
50% refrigeration
32% blowing agents to make foam
13% solvents
5% other

Figure 2
Use of CFCs by region. More than half of all CFCs used each year are released to the air.

Challenge

To prepare a persuasive report, you need to consider all factors related to the points you are trying to make. What factors do you need to consider for the challenge you have chosen?

Issue   Should CFCs Be Banned?

Statement
CFC production should be stopped completely by the end of this year. All refrigerators, air conditioning systems, manufacturing uses, etc., must be converted to allow the use of some other substance.

Benefits of a Ban

Opinion of an atmospheric scientist
If we ban CFCs, damage to the ozone layer may stop getting worse within a few decades. If we don't, the ozone layer will keep getting thinner and thinner.

Opinion of a dermatologist
The sooner CFC use is ended, the less damage will be caused to people's health. The rate of skin cancer may keep on climbing unless we do something now.

Risks of a Ban

Opinion of a consumer group
If there is a total ban, all our fridges and air conditioners will have to be replaced. We can't afford that. Besides, how do we know that the substitutes for CFCs are any safer? We used to think CFCs were safe.

Opinion of a highrise owner
Each of my apartment buildings has a large air conditioner. It would cost $200 000 to replace each one. I'd have to raise my rents.

Opinion of a citizen
It's too late to worry about this. The CFCs are already up there. We can just wear sunglasses and use more sunblock.

What Do You Think?
- Should Canada ban CFCs totally? Should Canada stick to the Montreal
(3A) Protocol? Is there another alternative? Research these questions.

- Decide how you feel about this issue and assemble your thoughts and
(3B) reasons into a position statement. Present your opinions in a letter to your member of parliament or to a local environmental group.

4.8 Investigation

SKILLS MENU
○ Questioning ● Hypothesizing ○ Planning ● Conducting ● Recording ● Analyzing ● Communicating

Linking Atomic Structure and Periodicity

In this chapter, you have learned how Mendeleev and modern chemists arranged the elements in the periodic table. This table helps scientists to understand the elements and to predict their properties. Elements in the same group are similar but show gradual changes in properties down a column in the table. Elements also show changes along a row. By studying the elements in Groups 1, 2, and 13 to 18 (**Figure 1**) you can confirm that these changes occur.

How can we show these changes or **periodic trends** down a column or along a row? One choice is to plot graphs on paper or computer. For example, a plot of atomic radius vs. atomic number shows definite trends. Or we can build three-dimensional models to represent such trends. For example, different lengths of straws can represent different atomic radii of atoms. If these straws are placed in a microtray arranged like a periodic table, you can see trends along rows and down columns. In this investigation, you will use this technique to look for periodic trends in the table.

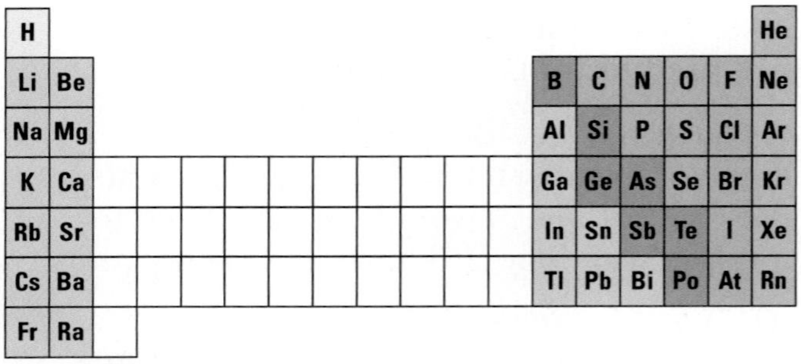

Figure 1

Materials

- periodic table, containing the following properties for each element: atomic radius (in pm), density (in g/cm^3), boiling point (in °C), and melting point (in °C)
- 96-well microtray
- fine-point permanent marker
- straws with diameter the same as the microtray cells
- scissors
- ruler

Question

How can we show trends down a column or along a row?

Hypothesis

1 Write your own hypothesis for the property assigned to you.

Procedure

2 Your group will be assigned a particular periodic property.

3 Using a periodic table, locate the data for your property for each of the elements.

(a) Make a data table similar to **Table 1**, titled with the property you are representing (e.g., atomic radius or boiling point).

Table 1

Element	Element Symbol	Atomic Number	Number Value of Property from Table	Calculated Length of Straw (cm)
hydrogen	H	1	37	1.9
?	?	?	?	?

Periodic properties for elements can be represented with different straw lengths. For example, hydrogen has atomic number 1 and atomic radius 37 pm. At a scale of 1 pm = 0.05 cm, the straw length will be 37 × 0.05 = 1.85 cm, or 1.9 cm to one decimal place.

(b) Copy the names and symbols of the elements and their values into the first and second columns of your data table.

4 You need to find the length of straw in cm that you need for each element. The longest straw should be about 10 cm to 15 cm in length. Use the following calculation for each element:

Length of straw in cm = number value x scale value

The scale value depends on which property you are investigating.

- For atomic radius in picometres, the scale value is 0.05.

- For density in g/cm^3, the scale value is 1.
- For boiling point in °C, to allow for negative values, each straw represents a temperature scale that begins at –300°C and ends at 3600°C. Add 300 to each element's boiling point before multiplying by the scale value, which is 0.003.
- For melting point in °C, again add 300 before multiplying by the scale value of 0.003.

(a) Perform the calculations and complete the table. Record numbers to no more than one decimal place.

5 Obtain a microtray, which will be your "periodic table." Use columns 1 to 8 to represent the eight groups of elements (Groups 1, 2, and 13 to 18). Use rows A to F to represent the five rows of the periodic table.

6 Use scissors to cut lengths of straws to represent the property for each of the elements. As soon as it is cut, place each straw in the appropriate well as shown.

Step 6
An eight-by-five section of a microtray can represent eight columns (Groups 1, 2, and 13 to 18) and five rows of a periodic table.

7 Use the permanent marker to label the outside of your microtray with the property represented.

8 Look at your table and look for general trends in your property (i) across a row and (ii) down a column

(a) Record the trends that you observe.

9 Exchange your microtray with groups that have investigated the other three properties.

(a) Record the trends that you observe for the other properties.

Analysis and Communication

10 Analyze your data by answering the following questions:

(a) What were the general trends in each property across a row?

(b) What were the general trends in each property down a column?

(c) In your opinion, which property showed the most predictable, regular trends in the table? Explain.

(d) In your opinion, which property showed the most unpredictable, irregular trends in the table? Explain.

11 Present your model.

Exploring

1. Each group should remove one straw and hide it. Exchange your microtray with another group. Each group should then try to cut a new straw of a different colour to put into the empty cell. When you have finished, compare the new straw with the hidden straw. Repeat this process with other groups.

2. Use a computer graphing program to plot graphs of
   (a) atomic radius vs. atomic number
   (b) density vs. atomic number
   (c) boiling point vs. atomic number
   (d) melting point vs. atomic number
   Compare the computer graphs with your three-dimensional microtray tables.

Reflecting

3. Discuss, among the members of your group, reasons for the observed trends. Develop hypotheses to account for the trends. How would you suggest testing your hypotheses?
(4A) (4B)

Challenge

What information would you need to add if you were displaying your model rather than presenting it?

Groups of Elements: Profile

In this activity, you will use the combined abilities of a team of students to research, gather, analyze, and produce a profile that represents a group of elements.

Materials
- a container (file folder, large envelope, small pizza box, etc.)
- research materials (library, CD-ROMs, Internet, etc.)
- magazines and newspapers

Procedure

1. You will be a member of a team. Each team should have a variety of different strengths and interests.

2. Choose a group of elements to study: alkali metals, halogens, or any other column in the periodic table. Record your team's choice with your teacher.

3. Research information about the group you have chosen.

4. From the list in **Table 1**, decide on 10 items that will represent your element. Note that
   (a) there must be at least one item from each of the eight categories described;
   (b) you may choose more than one item from a category, but you must have exactly 10 items.

5. Decide which team members will be responsible for each item. Work individually or as a team to produce the items, making sure that they are clearly related to your group of elements.

6. With your team, assemble the items in your team's portfolio. Add to the portfolio a table of contents listing the 10 items that your team chose. Your team members should all sign this table of contents to indicate their participation in producing the portfolio.

Table 1	Choices for the Profile
thinking visually	• decoration for the portfolio • a collage of magazine advertisements or photos • a poster that describes your elements • a picture, cartoon, drawing, or labelled diagram • your own idea, confirmed with your teacher
writing	• a one-page essay describing properties of your elements • a letter written to government or industry • a poem or short story about your elements • your own idea, confirmed with your teacher
thinking logically	• a graph of data related to the group • a mathematical calculation based on data • a flow chart showing how one of your elements is produced • a coloured periodic table containing all the information you have discovered about your elements • your own idea, confirmed with your teacher
using music	• a song, jingle, or rap • a written reflection on a piece of music that reminds you of your elements • your own idea, confirmed with your teacher
reflecting	• a description of two things about your work that you would like the teacher to pay particular attention to in the portfolio • a description of what you found most difficult in this activity • a description of how one of these elements affects your family or life • your own idea, confirmed with your teacher
using your body	• a dance, pantomime, game, or computer activity • a set of hand signals or special handshake • your own idea, confirmed with your teacher
thinking of nature	• a description of how one of your elements affects organisms, positively or negatively • a record or journal of a field trip • your own idea, confirmed with your teacher
working with people	• an evaluation of how well you functioned as a team, including how you could improve • a record of an interview with someone from an environmental group or the workplace involved in one of your elements • your own idea, confirmed with your teacher

Challenge

What profile choices should you include in the challenge you have chosen?

(c) State the modern periodic law.

4. Explain the difference between a period and a group.

5. (a) In the periodic table, where are the metals found?

   (b) Where are the nonmetals found?

6. What kinds of ions, including number and charge, are formed by

   (a) alkali metals?   (b) halogens?

7. Make a chart comparing alkali metals, halogens, and noble gases with respect to

   (a) position in the periodic table

   (b) ability to react with other elements

   (c) number of electrons in the outermost orbit

8. Why is ozone described as hazardous at ground level but helpful in the upper atmosphere?

Applying Skills

9. Match the description on the left with one term on the right. Use each term only once.

Description	Term
A produced by people	1 alkali metal
B very unreactive gas	2 composite
C very reactive metal	3 halogen
D very reactive nonmetal	4 noble gas
E very long molecule	5 polymer
F material formed by mixing two or more other materials	6 synthetic

10. Study the periodic table on the inside back cover.

    (a) How many orbits do the elements in the third row have?

    (b) How many orbits do the elements in the fourth row have?

    (c) What conclusions can you draw about the relationship between the period on the periodic table and the number of orbits that the elements in that period have?

11. (a) What are the symbol and atomic number for the element aluminum?

    (b) Draw a Bohr diagram for this element.

    (c) What noble gas has the closest atomic number?

    (d) How many electrons must aluminum lose or gain to form a stable ion?

    (e) What charge of ion will result?

12. A new element has been discovered, and all you know is that it is an alkali metal. Predict

    (a) its state at room temperature

    (b) the number of electrons in its outer orbit

    (c) its possible atomic number

13. Repeat question 12 for a newly discovered halogen.

14. Repeat question 12 for a newly discovered noble gas.

Making Connections

15. List the four different types of materials that people have used over the centuries and give one ancient and one modern example of each type.

16. Classify each of the following as a metal, polymer, ceramic, or composite and describe why each was chosen for its identified use:

    (a) concrete reinforced with steel bars

    (b) a pottery coffee mug

    (c) a bronze statue

    (d) a polyethylene drink bottle

    (e) a shirt that is 40% cotton and 60% nylon

17. Refer to the section on "Corrosion" on page 34. Using your understanding of the reactivities of elements, suggest a reason why magnesium is sometimes attached to ships' hulls to protect them.

18. Elements are the basic building blocks of all the substances in the world. Think of the structure of an atom.

    (a) Which part of the atom is involved in the chemical reactions that form these substances? Give reasons for your answer.

    (b) Identify one substance that is produced by industry, and describe its potential uses and associated risks.

19. CFCs are damaging to the ozone layer, so scientists are searching for replacement substances. Research what properties the replacements must have and how they are being developed. How do they differ from CFCs? Write a report on your findings.

Challenge

Building from the Past to the Present

Throughout history, scientists have observed the physical and chemical properties of elements and compounds, and have created models of matter to help explain those properties. The collection of ideas that is presently used to explain the nature of matter is called the atomic theory of matter. Each of the challenges below provides an opportunity to explain how our understanding of matter has evolved over the centuries.

1 Models for Matter

You have been commissioned to build a display that explains the evolving models of matter to the general public. Your design must convey the idea that scientific discovery is dynamic and that it affects everyone's life directly. Specifically, your display should clearly demonstrate that understanding matter is a key to understanding every object and substance in our lives.

Design and build a display that includes:
- different atomic models, and an explanation of how those models changed with the discovery of new evidence;
- a representation of how the periodic table evolved and different models of the periodic table;
- instances in which the properties of elements and compounds were documented but not understood;
- examples of how the physical and chemical properties of elements determine how they are used in the design and creation of everyday products.

2 Marketing Matter

Where do the materials for consumer products come from? Most materials that we use come indirectly from the ground. For instance, plastics are manufactured from crude oil and alloys are created from metals obtained in mining and metallurgy. You are the inventor of a substance or material that is commonly used today. Promote your material to a manufacturing company that makes products for which your material is well-suited.

Prepare a proposal that includes:
- a description of the material or substance, identifying its elements and compounds;
- a Bohr model of the main elements, and an illustration of their position on the periodic table;
- a description of the structure and properties of the elements and compounds, and an explanation of how these properties will improve the manufacturer's products;

- any safety considerations related to the processing and manufacturing of the material;
- a flow chart showing the steps that lead from the raw material to possible finished products.

3 A "Time Machine" Simulation

Imagine that a time machine has been programmed to travel back in time to 1869 to bring Dmitri Mendeleev to meet you. Prior to his coming you prepare documentation to present the strengths and weaknesses of his periodic table and to show him how his model evolved as knowledge of the structure of the atom evolved.

Prepare a report that includes:

- comparisons of Mendeleev's periodic table with the modern periodic table;
- a model of the nuclear atom, including subatomic particles;
- an outline of the elements discovered since Mendeleev's time and an analysis of whether their properties could have been explained using Mendeleev's periodic table;
- an illustration of where the elements discovered since Mendeleev's time are located in the modern periodic table.

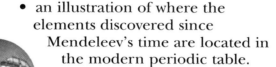

Assessment

Your completed challenge will be assessed according to how well you:

Process
- understand the specific challenge
- develop a plan
- choose and safely use appropriate tools, equipment, and materials when necessary
- conduct the plan applying technical skills and procedures when necessary
- analyze the results

Communication
- prepare an appropriate presentation of the task
- use correct terms, symbols, and SI units
- incorporate information technology

Product
- meet established criteria
- show understanding of concepts, principles, laws, and theories
- show effective use of materials
- address the identified situation/problem

Unit 1 Review

Understanding Concepts

1. In your notebook, write the letters (a) to (p), then indicate the word(s) needed to complete each statement below.

   (a) In a(n) _____?_____ change, a new substance is produced.

   (b) A(n) ___?___ is a mixture of metals.

   (c) A solid produced when two solutions are mixed together is a(n) _____?_____.

   (d) The starting materials in a chemical reaction are called ____?____.

   (e) A(n) ___?___ is a sample of matter containing only one type of atom.

   (f) Water can be split into two elements (hydrogen and oxygen) in a process called _____?_____.

   (g) _____?_____ is added to farmland to help crops grow.

   (h) ___?___ are shiny, malleable, and conduct electricity.

   (i) The _____?_____ is the number of protons in an atom.

   (j) An electrically charged atom is a(n) ___?___.

   (k) An atom with an unstable nucleus is described as _____?_____.

   (l) The _____?_____ is the core of the atom, containing most of its mass.

   (m) ___?___ gases do not form compounds with most other elements.

   (n) The size of an atom is described by its atomic ___?___.

   (o) ___?___ materials are made by humans, rather than naturally.

   (p) Elements can be arranged in a(n) _____?_____ table.

2. Indicate whether each of the statements (a) to (p) is TRUE or FALSE. If you think the statement is FALSE, rewrite it to make it true.

   (a) Combustion is the chemical reaction between a fuel and hydrogen.

   (b) Colour and hardness are examples of physical properties.

   (c) Density is the volume per unit mass of a substance.

   (d) A substance that can be drawn into a wire is called ductile.

   (e) Viscosity describes the flammability of a substance.

   (f) A molecule is a combination of atoms.

   (g) A mineral is a compound of a metal mixed with other materials in rock.

   (h) The chemical symbol for calcium is Cal.

   (i) A flame test can be used to identify an element.

   (j) A neutron is positive and located in the nucleus.

   (k) The mass number is the sum of electrons and protons in the atom.

   (l) A Bohr diagram shows electrons in orbits about the nucleus.

   (m) A row of the periodic table is called a period.

   (n) The sizes of atoms increase down a column of the periodic table.

   (o) Alkali metals include fluorine, chlorine, and iodine.

   (p) The modern periodic table organizes elements by atomic mass.

3. Describe the similarities and/or differences between each pair of terms listed below:

   (a) physical property, chemical property

   (b) combustion, corrosion

   (c) ductile, malleable

   (d) products, reactants

   (e) element, compound

   (f) atom, molecule

   (g) mineral, ore

   (h) metal, nonmetal

   (i) atomic number, mass number

   (j) proton, neutron

   (k) atom, ion

   (l) ground state, excited state

   (m) natural material, synthetic material

   (n) period, group

   (o) metal, metalloid

   (p) noble gas, alkali metal

For questions 4 to 10, choose the best answer and write the full statement in your notebook.

4. Which of the following is an example of a chemical change?
   (a) ice melts
   (b) a car rusts
   (c) a pottery mug shatters
   (d) a snowflake forms
   (e) food is chewed

5. What is the mass of a block of wood if its density is 0.75 g/cm^3 and it has a volume of 10 cm^3?
   (a) 75 g   (b) 7.5 g
   (c) 0.75 g   (d) 13.3 g
   (e) 6.7 g

6. Which of the following is not a property of metals?
   (a) malleability
   (b) ductility
   (c) good conduction of heat
   (d) shiny lustre
   (e) brittleness

7. If potassium (K) has a combining capacity of 1 and oxygen has a combining capacity of 2, what is the formula for potassium oxide?
   (a) K_2O (b) K_2O_2
   (c) KO_2 (d) KO
   (e) K_2O_3

8. Which of the following combinations describes the proton in the modern atomic model?
   (a) negative charge and significant mass
   (b) positive charge and very small mass
   (c) negative charge and located in the nucleus
   (d) positive charge and significant mass
   (e) neutral charge and found in the nucleus

9. Which of the following combinations describes the electron in the modern atomic model?
   (a) negative charge and significant mass
   (b) positive charge and very small mass
   (c) negative charge and located in orbit about the nucleus
   (d) positive charge and significant mass
   (e) neutral charge and found in orbit about the nucleus

10. Which of the following is used to organize the modern periodic table?
    (a) atomic number
    (b) atomic mass
    (c) number of neutrons
    (d) atomic size
    (e) state at room temperature

11. A friend tells you that an antacid tablet bubbling in water is a chemical change, but the water bubbling in a kettle and turning to steam is not. Do you agree? Explain.

12. Choose a familiar substance. Describe as many physical properties of this substance as possible. State a chemical property, if possible.

13. Describe three observations that would help you decide that the burning of a fuel was a chemical change.

14. How is it possible that there are millions of pure substances, even though there are only about 100 different elements?

15. State the types of atoms and the numbers of each type that are present in the following molecules: sodium phosphate (Na_3PO_4) and lead IV sulfate $Pb(SO_4)_2$.

16. Compare the atomic models of (a) Dalton, (b) Rutherford, and (c) Bohr.

17. Describe the charge, mass, and location of the three fundamental particles that make up the modern atomic model of the atom.

18. Describe how the periodic table is organized. Include in your description similarities and differences observed across rows and down columns of the table.

19. Draw a sketch of the periodic table. Using patterns or colours, shade in the areas of the table that represent the locations of the (a) alkali metals, (b) halogens, (c) noble gases, (d) hydrogen, (e) metalloids, (f) metals, in general, and (g) nonmetals, in general.

Applying Skills

20. A substance is white, has a density of 1.2 g/mL, is very hard, has a melting point of 500°C, and fizzes when added to acid.
    (a) Which properties could be described as quantitative? Explain.
    (b) Which properties could be described as chemical? Explain.

21. A student is given the densities of types of wood in **Table 1**.

Wood	Density (g/cm³)
ironwood	1.24
birch	0.66
red cedar	0.37
balsa	0.12

Table 1

(a) Which wood would sink when placed in water? Explain.

(b) Which wood would be the best choice to build a model airplane? Explain.

(c) The student is provided a rectangular block of wood that measures 8 cm × 8 cm × 8 cm. Its mass is 190 g. Use complete calculations to determine the identity of the wood.

22. Write chemical formulas for the following compounds:

(a) magnesium chloride

(b) silver iodide

(c) zinc oxide

23. Suppose someone tells you that a green object contains copper. You are not convinced because you have seen that copper wires and jewellery are reddish-brown.

(a) Is it possible that this green substance really does contain copper? Explain.

(b) With the assistance of a chemist, what test might you carry out to settle this question?

24. Copy **Table 2** into your notebook. Fill in the blanks with the missing numbers.

25. Write standard atomic notation for each of the elements in the previous question.

26. (a) What is the symbol and atomic number for the element, sulfur?

(b) Draw a Bohr diagram for this element.

(c) What noble gas has the closest atomic number?

(d) How many electrons will sulfur lose or gain to form a stable ion?

(e) What charge of ion will result?

27. A new element has been discovered, and all you know is that it is a halogen.

(a) Predict its state at room temperature.

(b) Predict the number of electrons in its outer orbit.

(c) Predict its possible atomic number.

(d) On what do you base your predictions?

28. (a) Magnesium forms a compound with fluorine called magnesium fluoride, MgF_2. What would be the name and formula of the compound formed when magnesium combined with iodine, another halogen atom?

(b) Aluminum forms a compound with chlorine called aluminum chloride, $AlCl_3$. What would be the name and formula of a compound it would make with bromine?

29. A chemist melts a sample of an unidentified mineral and then passes an electric current through the liquid. This produces a solid and a gas. The solid is shiny at first, but turns dull quickly. The gas has a strong, choking odour.

(a) Could the mineral be an element? State reasons for your answer.

(b) Suppose the new solid is an element. To which chemical family might it belong? State reasons for your answer.

(c) Suppose the new gas is an element. To which chemical family might it belong? State reasons for your answer.

Making Connections

30. The four Hazardous Household Product Symbols indicate products that are poisonous, flammable, explosive, and corrosive. Which labels would be on containers of (a) aerosol insect spray, (b) drain cleaner, (c) ant powder, and (d) furniture polish?

Table 2

Element	Symbol	Atomic Number	Mass Number	No. of Protons	No. of Electrons	No. of Neutrons
beryllium	Be	4	9	?	?	?
carbon	C	6	?	?	?	8
silicon	Si	?	?	?	14	14
potassium	K	?	?	19	?	20

31. You have read about two forms of paint. Research another product that comes in two or more comparable forms, with slightly different physical and chemical properties. Examples might include wood or plastic for dock-building, ceramic tile or linoleum for flooring, and wool or synthetic pile for clothing. Compare their properties, methods of production, and use.

32. You and your group are members of a "Think Tank" of concerned business people, environmentalists, and politicians. Your task is to suggest ways to reduce the money lost each year to corrosion of automobiles. These could include new technology, regulations, or changes to people's lifestyles. Write a brief report of your findings, suitable for newspaper publication.

    (a) Brainstorm as many ways as possible to reduce the amount of corrosion that occurs.

    (b) Record all your ideas in a table. Be sure to include benefits, drawbacks, and cost.

33. Problems have been caused in the past when elements and compounds have been released into the environment. Research and report on such problems, using a CD-ROM database or the Internet. Some possible key words and combinations are: groundwater + contamination, chemical + effluent, water + pollution.

34. As advisors to the Minister of the Environment, your group must suggest ways to reduce the amount of air pollution.

    (a) Brainstorm as many ways as possible to reduce the amount of air pollution from vehicles, industries, and power plants.

    (b) Record all your ideas in a table. Be sure to include benefits, drawbacks, and cost.

35. "Radioactivity is more helpful than harmful." Do you agree or disagree with this statement? Give some possible arguments both for and against the statement, and then explain your opinion.

36. Visit a garden supply centre, and note the information and instructions on the packaging for (a) at least three kinds of fertilizer that are not labelled "organic," and (b) at least one fertilizer that is labelled "organic." Select a label that has a high percentage of nitrogen and note what the label says the fertilizer is for. Repeat this step, looking for a fertilizer high in phosphorus. Note what other nutrients besides nitrogen (nitrate), phosphorus (phosphate), and potassium are mentioned on the labels. Note what the organic fertilizers were made from.

37. Think of a group of objects (e.g., leaves, food, animals, children's toys, or drug store products, etc.) and devise a way to categorize them so that, if a new one were to be discovered, it could be included in your system. Create a computer or poster display of your organization system, and explain your decisions.

38. Elements have been named for many reasons. Using your library resources, research the following:

    (a) Germanium, lutetium, and polonium were named to honour the geographic origin of their discoverers. Who were the discoverers and where did they come from?

    (b) Which heavenly bodies were the following named after: mercury, uranium, neptunium, plutonium, tellurium, selenium, palladium, cerium? (Some are very easy, while others are not so obvious.)

    (c) Some elements are named to honour people. Which people were honoured by the following: gadolinium, curium, einsteinium, fermium, mendelevium, lawrencium, nobelium, seaborgium? State their full names, and write a couple of sentences about each person.

    (d) Some elements were named after places. What places are the following named after: europium, hafnium, americium, berkellium, californium?

39. An inventor is trying to sell an idea to protect people from ultraviolet radiation: a headband that releases a cloud of ozone around a person's head. Write a letter to this inventor explaining whether it is a good idea or not.

40. Design a "mini-poster," approximately 20 cm × 20 cm, to represent an element. The design of the poster should reflect the properties of the element, and its use or significance to society. The symbol for the element, its atomic number, and its name should be found somewhere in the box. Your teacher may decide to have a number of students produce these posters in order to build a "Great Wall of Chemistry" on a display board or in the hall at school.

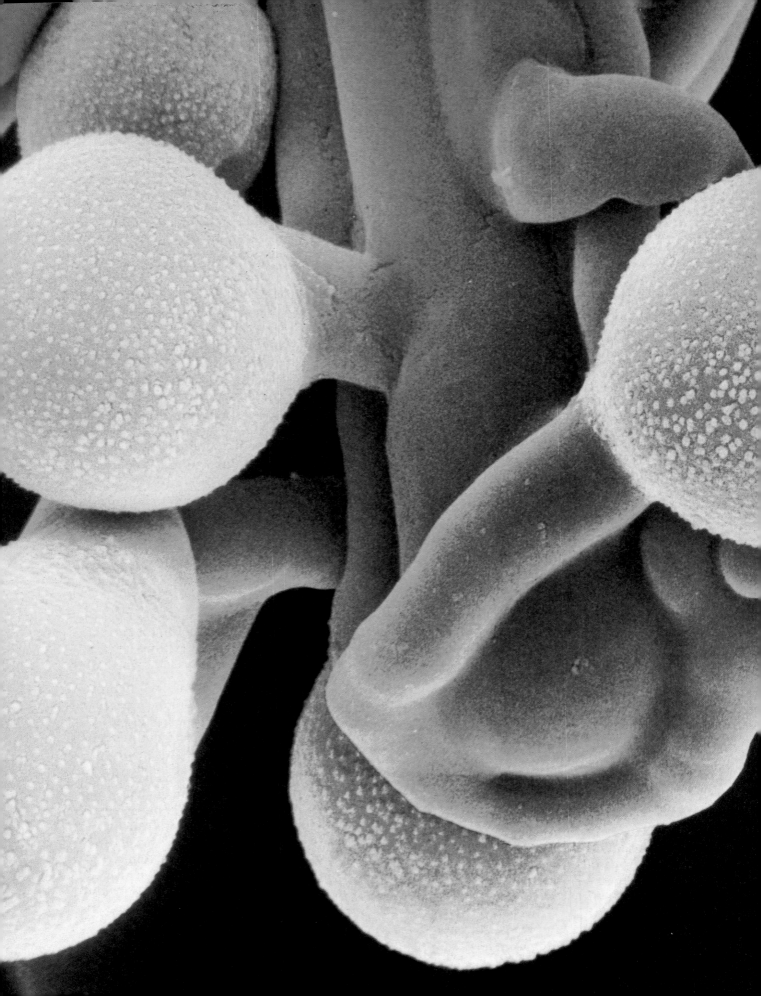

Reproduction

Unit 2 Overview

During the 1800s scientists constructed what has come to be known as the cell theory. The theory is based on three important principles.

- All living things are composed of one or more cells.
- The cell is the functional unit of life.
- All cells come from pre-existing cells.

How do single cells make duplicate cells? How do multicellular organisms produce offspring by combining genetic information from two parents? What are the impacts of scientific research and technological innovations within a social context?

5. Cell Growth and Reproduction

All living things undergo cell division. Cell division is essential for the perpetuation of life.

In this chapter, you will be able to:
- Explain the importance of the cell theory in developing a modern understanding of cell biology.

- Describe the processes and explain the importance of cell division.

- Examine how different organisms use various types of asexual reproduction for propagation.

- Use a microscope and mathematical computations to determine the rates of cell division, and the growth rates of plants and animals.

6. A Closer Look at Cell Division

The way in which a cell functions and divides is determined by genetic information contained in its nucleus.

In this chapter, you will be able to:
- Explain the importance of DNA replication to the survival of an organism.

- Identify factors that can alter the genetic information and explain how changes in the DNA structure can change an organism.

- Describe and carry out experiments on the cloning of plants.

- Explain technological innovations for regeneration and cloning and evaluate social issues related to this technology.

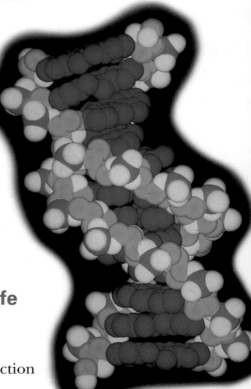

7. Sexual Reproduction and the Diversity of Life

Sexual reproduction creates a diversity of species.

In this chapter, you will be able to:
- Describe the difference between asexual and sexual reproduction and indicate advantages and disadvantages for each strategy.

- Identify and describe adaptive advantages for different strategies of sexual reproduction such as: conjugation, hermaphroditic reproduction, and separate sexes.

- Identify reproductive structures within flowers and humans.

- Describe the process by which sex cells are formed within multicellular organisms and explain why sex cells have half as many chromosomes as other cells.

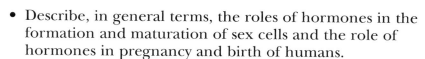

- Describe, in general terms, the roles of hormones in the formation and maturation of sex cells and the role of hormones in pregnancy and birth of humans.

8. Zygotes and Development

Different organisms use very different strategies, and combinations of strategies, to ensure the survival of their offspring.

In this chapter, you will be able to:

- Identify and describe adaptive advantages for different developmental strategies such as: spores, seeds, eggs, and development within the uterus.

- Explore seed formation and germination of plant embryos within a laboratory setting.

- Examine the events of zygote formation, embryo growth, and fetal development of humans within the uterus.

- Examine and evaluate the social implications of various reproductive technologies designed to assist couples who want to have children.

- Explain how materials pass across the placenta between mother and child, and examine the moral implications of drug use by parents during pregnancy.

In this unit you will be able to... demonstrate your learning by completing a Challenge.

Society and Reproductive Technology

As you learn more about reproduction and related technologies, think about how you would accomplish these challenges.

1 **Survey on Reproductive Technology**

Conduct a survey to determine public opinion about various scientific breakthroughs in reproductive technology.

2 **Public Information Display**

Prepare a display for the general public that presents the issues about one type of reproductive technology.

3 **Futuristic Short Story or Play**

Write a futuristic short story or a play that focuses on issues involving reproductive technologies and how they can affect our society.

To start your own Challenge, see page 258.

Record your ideas for the Challenge when you see

Challenge

5 Cell Growth and Reproduction

Getting Started

1 The 100 trillion cells of your body are truly awe-inspiring, when you think that they all started from a single fertilized egg. They stand as proof of the ability of human cells to grow and reproduce. How does one cell grow into a multicellular organism? If all the cells in your body came from the same egg cell, why don't they all look alike? Why are there more of some cell types than others?

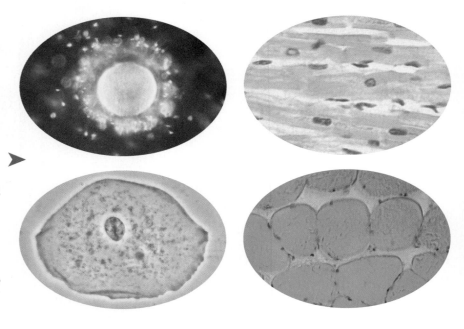

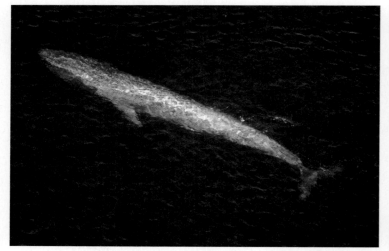

2 The 35-m blue whale is about 18 times longer than the average human. A look at the giants on this planet also reveals the sequoia tree, three times longer than the massive blue whale but also hundreds of years older. Do larger organisms have larger cells? Is cell division in plants similar to that in animals? Do all cells divide at the same rate?

3 The largest known living organism is a quaking aspen plant. In the photograph, all of what appear to be individual aspen trees are actually one organism with a common root system. This single organism covers 43 ha or over 80 football fields! The trees develop from runners: horizontal roots that grow above or below the ground. All the trees have the same genetic information: they are one organism. What are the advantages of reproducing by runners?

Reflecting

Think about the questions in **1**, **2**, **3**. What ideas do you already have? What other questions do you have about how cells grow and reproduce? Think about your answers and questions as you read the chapter.

 **Monitoring Cell Replacement**

Using a permanent marker, place a small drop of ink on the palm and back of your hand. Because the marker ink is not water soluble, the cells that absorb the dye are permanently stained.

Predict in which area the stain will first disappear. Observe the stained areas daily and record your observations. Explain your observations.

The Microscope and Cell Theory

Scientific discovery often depends upon technological innovation. Nowhere is that more evident than in cell biology. Advances in lens grinding led to the development of microscopes, which in turn opened a window to a microscopic world.

Cells were first described in 1665, when the English scientist, Robert Hooke, noticed many repeating honeycomb-shaped structures while viewing a thin slice of cork under his primitive microscope. In his book, *Micrographia*, Hooke used the word "cell" to describe these structures. However, cork, the inner bark from oak trees, has few living cells. What Hooke observed were the rigid cell walls that surrounded the once-living plant cells.

A few years later, Anton van Leeuwenhoek observed living blood cells, bacteria, and even tiny single-cell organisms in a drop of water, using a simple microscope (a microscope with a single lens). As microscopes improved, cells could be observed and described more closely, but it wasn't until 1820 that a scientist, Robert Brown, examined plant cells and described the tiny sphere called the nucleus (plural: nuclei). Nuclei were soon discovered in animal cells as well. In the mid-1800s, a zoologist, Theodor Schwann, and a botanist, Matthias Schleiden, concluded that plant and animal tissues are composed of cells. This discovery provided the foundation for the first part of the cell theory. The first two parts of the cell theory state

- All living things are composed of one or more cells.
- The cell is the functional unit of life.

Table 1	
Magnification Needed to Create a 1-mm Image	
Object	**Magnification**
fish egg	none
human egg	10×
plant cell	20×
animal cell	50×
bacterium	1000×
mitochondrion	1000×
large virus	10 000×
ribosome	40 000×
cell membrane	100 000×
hydrogen atom	10 000 000×

Technological Advances in Microscopy

Microscopes provided scientists with a new window into cells. Greater magnification not only allowed them to discover smaller cells, but it allowed them to gain a better understanding about how cells worked. **Table 1** shows the magnification required to view different objects.

Viewing with the Compound Light Microscope

An important advance in the development of the microscope came when scientists added a second lens to the simple microscope. An image magnified 10× by the first lens and 10× by the second lens could be viewed as if it were 100× larger (**Figure 1**).

Even the most sophisticated techniques limit the light microscope to about 2000× magnification. But in order to see very tiny viruses or the detail within a human cell, greater magnification is required. The electron microscope provides this window.

Figure 1

Algae cells seen through a light microscope

Viewing with the Transmission Electron Microscope

A very crude electron microscope was invented in Germany in 1932. It provided an image of 400× magnification, but the image was grainy. The electron microscope's true value became apparent in 1937 when James Hillier and Albert Prebus unveiled their electron microscope at the University of Toronto. Their instrument was capable of 7000× magnification. Today, transmission electron microscopes are capable of 2 000 000× magnification (**Figure 2**).

Instead of light, the electron microscope uses beams of electrons. Electrons are tiny subatomic particles that travel around the nucleus of an atom. However, electron microscopes have two limitations. First, specimens that contain many layers of cells, such as blood vessels, cannot be examined. A thick specimen would absorb all the electrons and produce a blackened image. Because the electrons pass best through single layers of cells, only thin sections of cells can be used. These thin sections are produced by encasing a specimen in plastic and shaving off thin layers. But mounting cells in plastic kills them, which means that only dead cells can be observed—the second limitation. Although ideal for examining the structures within a cell, the transmission electron microscope does not allow you to examine a living cell as it divides.

Viewing with the Scanning Electron Microscope

The scanning electron microscope provides a new method for investigating thicker specimens by reflecting electrons from their surface. This scanning electron microscope produces a three-dimensional image (**Figure 3**). Electrons are passed through a series of magnetic lenses to a fine point. This fine point of electrons scans the surface of the specimen. The electrons are reflected and magnified onto a TV screen where they produce an image. The scanning electron microscope lacks the magnification and the high resolution of the transmission electron microscope, however, it provides greater depth of field.

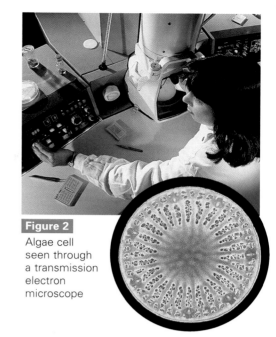

Figure 2
Algae cell seen through a transmission electron microscope

Understanding Concepts

1. Explain how the evolution of the microscope made it possible to develop a cell theory.

2. Give one advantage of using a compound light microscope over a single lens microscope.

3. Give one advantage of using the light microscope over a transmission electron microscope.

Making Connections

4. Which microscope do you think would be best for viewing each of the following? Give reasons for your choice.

   (a) a virus

   (b) a hair mite

   (c) the detailed structure of a cell's nucleus

   (d) a living microorganism

Challenge

As you have seen, all microscopes have limits. When you are creating your display, how could you present the limits on what scientists can know?

Figure 3
Algae cell seen through a scanning electron microscope

Cells: The Basic Unit of Life

Animal-Cell Structures

After many hours spent looking through microscopes, scientists have determined that even though there is no one, common cell, all plant and animal cells have many common factors.

Many of the cell structures shown in **Figure 1** can be seen with a light microscope. You should be able to see the nucleus and possibly some of these other structures.

The entire cell is covered by a **cell membrane**. The membrane acts like a gatekeeper, controlling the movement of materials into and out of the cell.

The **nucleus** of the cell acts as the control centre, directing all of the cell's activities. Genetic (hereditary) information is organized into threadlike structures called **chromosomes**. Each chromosome contains many different genes. **Genes** are units of genetic information that determine the specific characteristics of an individual.

The **cytoplasm** is the area of the cell where the work is done. Nutrients are absorbed, transported, and processed within the cytoplasm.

The cytoplasm contains a number of different organelles that each have a specific form and function. An **organelle** is a specialized structure inside a cell.

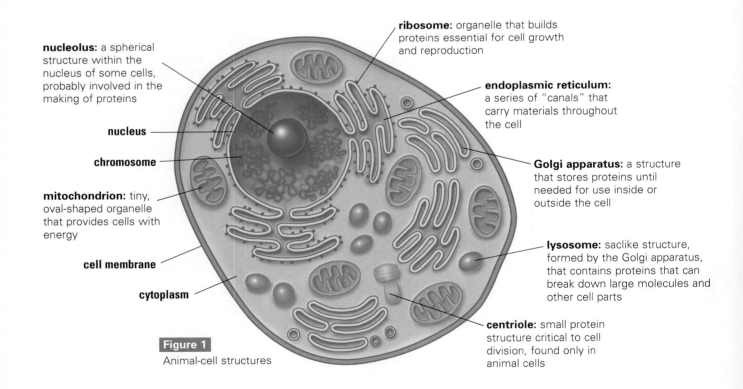

nucleolus: a spherical structure within the nucleus of some cells, probably involved in the making of proteins

nucleus

chromosome

mitochondrion: tiny, oval-shaped organelle that provides cells with energy

cell membrane

cytoplasm

ribosome: organelle that builds proteins essential for cell growth and reproduction

endoplasmic reticulum: a series of "canals" that carry materials throughout the cell

Golgi apparatus: a structure that stores proteins until needed for use inside or outside the cell

lysosome: saclike structure, formed by the Golgi apparatus, that contains proteins that can break down large molecules and other cell parts

centriole: small protein structure critical to cell division, found only in animal cells

Figure 1
Animal-cell structures

Plant-Cell Structures

Plant cells, as shown in **Figure 2**, contain all the organelles found in animal cells plus a few other structures. For example, the cell membrane of a plant cell is surrounded by a **cell wall**. Composed of a rigid material called cellulose, cell walls protect and support plant cells. Gases, water, and some minerals can pass through small pores (openings) in the cell wall. Immediately inside the cell wall is the cell membrane; however, you usually cannot see it when you examine plant cells with a light microscope.

Both animal and plant cells contain a vacuole, but in plant cells they are much larger. Plant cells also contain chloroplasts.

vacuole: fluid-filled space containing water, sugar, minerals, and proteins

chloroplast: organelle containing chlorophyll used in photosynthesis

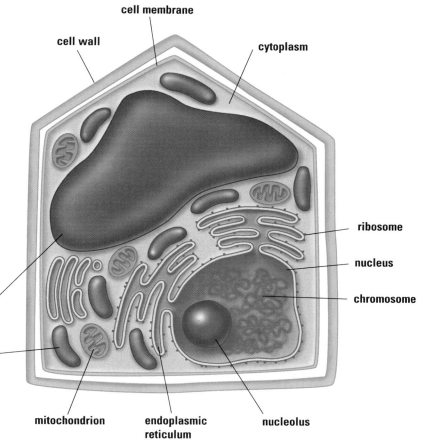

cell membrane

cell wall

cytoplasm

ribosome

nucleus

chromosome

mitochondrion

endoplasmic reticulum

nucleolus

Figure 2
Plant-cell structures

Movement

Outside the cell membrane, some cells have a **flagellum** (plural: flagella), a whiplike tail that helps the cell move. Some cells have many tiny hairs, called **cilia**, that either move the cell or the environment surrounding the cell. The cilia usually work together.

Understanding Concepts

1. What is the function of
   (a) the cell membrane of a cell?
   (b) the cytoplasm?
2. Where in a cell is genetic information found?
3. How does the structure of plant cells differ from that of animal cells?
4. What can a plant cell do that no animal cell can? What plant-cell structure enables it to carry out this function?

Challenge

In the future, new forms of microscopes may be invented. Could a new microscope play a role in your story or play? What would it reveal?

Discovering the Origin of Cells

Which came first, the chicken or the egg? Where did the first cell come from? People have always had theories about the origins of living things. But everyone believed that cells came from non-living things.

Life from the Heavens

Thousands of years ago, scientists (or natural philosophers as they were called then) noticed that when a pond dried up during a long period of drought, no living frogs or fish were found in the mud. When rain finally began to fall, the pond filled with water and was soon teeming with frogs and fish. Some philosophers concluded that the frogs and fish must have fallen to Earth during the rainstorm.

(a) What observations can you make to support or refute the hypothesis that fish and frogs fall to Earth during rainstorms?

Aristotle's Proposal

Aristotle, a great philosopher who lived in Greece in the 4th century B.C., rejected the hypothesis that life came from rain. He proposed that the fish and frogs came from the mud, a non-living thing. Aristotle also believed that flies came from rotting meat, because he had always observed flies on rotting meat. Aristotle's theory, known as spontaneous generation, persisted for nearly 2000 years. Spontaneous generation is a theory that suggests that non-living things can be transformed into living things without any external causes.

(b) What observations can you make to support Aristotle's theory?

(c) What observations can you make to challenge Aristotle's theory?

Testing the Theory of Spontaneous Generation

In 1668, Francesco Redi designed an experiment to test the hypothesis that rotting meat is transformed into flies. Redi placed bits of meat in two jars and sealed one of the jars, as shown in **Figure 1**. The open jar was designated the control, while the closed jar was designated the experimental.

(d) Before you read on, predict what happened in both jars. Provide your reasons.

Apparently, flies were attracted to the meat in the open jar and began laying eggs on this food supply. The eggs hatched into maggots, which then began feeding on the meat. The maggots became flies, and the cycle continued. Redi concluded that flies come from other flies, not from rotting meat!

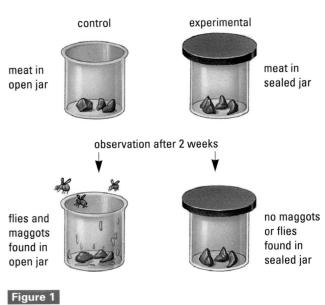

Figure 1
Redi's experiment

Spontaneous Generation and Single-Cell Organisms

Although Francesco Redi helped defeat the theory of spontaneous generation for relatively complex organisms, like flies, many scientists continued to accept the theory for microorganisms (organisms that can only be seen with the aid of a microscope). John Needham (1713–1781) was one of these scientists. Needham noticed that meat broth left unsealed would soon change colour and give off a putrid smell. Microorganisms, sometimes called microbes, were found growing in the broth, but where did they come from?

Needham's Experiment

Needham boiled meat broth in flasks for a few minutes in order to kill the microbes. The broth appeared clear after boiling. The flasks were then tightly sealed and left for a few days, and the murky contents were examined under a microscope. As shown in **Figure 2**, the broth was teeming with microorganisms.

(e) Does this mean that the broth had spontaneously created microorganisms? Give your reasons.

(f) What changes would you make to Needham's experimental procedure before accepting his data?

Needham rushed to retest the experiment. This time he checked for microbes before sealing the flasks completely (**Figure 3**).

When he observed sample drops of broth through his microscope, he found no microbes after boiling. Needham reasoned that the boiling had destroyed the microbes. However, when the flasks were checked a few weeks later, many microbes had reappeared. Needham concluded that microbes came from non-living things in the nutrient broth!

(g) How is it possible to check for microbes immediately after boiling and not find any, but find so many two weeks later?

Figure 3
Needam's second experiment

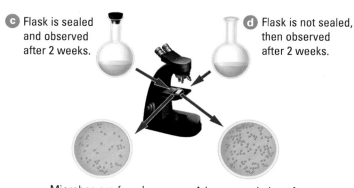

a Flasks containing beef broth are heated and checked for microbes.

b Flasks are checked for microbes.

No microbes are found.   No microbes are found.

c Flask is sealed and observed after 2 weeks.

d Flask is not sealed, then observed after 2 weeks.

Microbes are found.   A large population of microbes is found.

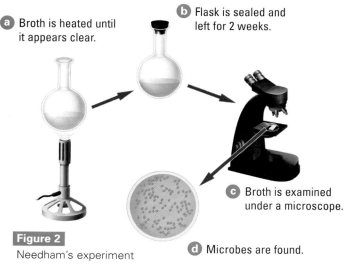

a Broth is heated until it appears clear.

b Flask is sealed and left for 2 weeks.

c Broth is examined under a microscope.

d Microbes are found.

Figure 2
Needham's experiment

Changing or Replacing a Theory

Scientific theories are accepted as long as they can explain observed events. Once evidence is collected that challenges the theory, the scientific theory must either be modified or abandoned.

Needham's conclusions were not challenged until 25 years later. Then, Lazzaro Spallanzani (1729–1799) repeated Needham's experiment, but he boiled the flasks longer and sealed some of them tighter, as shown in **Figure 4**.

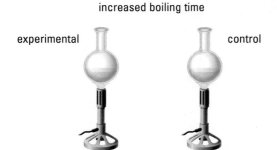

increased boiling time

experimental   control

a Flasks containing beef broth are heated and checked for microbes.

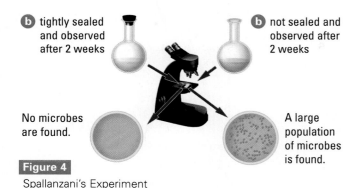

b tightly sealed and observed after 2 weeks

b not sealed and observed after 2 weeks

No microbes are found.

A large population of microbes is found.

Figure 4

Spallanzani's Experiment

(h) How might a longer boiling time affect the experimental results?

(i) How might making the seal on the flasks tighter affect the results?

(j) Predict the results of Spallanzani's experiment and justify your prediction.

Spallanzani found no microorganisms in the tightly sealed flask of broth. The longer heating time with the tight seals employed by Spallanzani must have killed the few remaining microbes.

(k) What conclusion would you draw from Spallanzani's experiment?

The Source of Needham's Error

Because Needham only examined a few drops from the beef broth immediately after boiling, he missed the few microbes that were not killed. Imagine finding just a few cells in a 500-mL flask of beef broth.

(l) If only a few microbes remained in the flasks after Needham heated them, how could you explain that the flasks were filled with microbes a few weeks later?

Accepting a New Theory

Like many people, scientists do not accept change easily. New theories are often opposed, even if they are supported by experimental evidence. So, some scientists constructed arguments to dismiss Spallanzani's experiment. These critics suggested that sealing the flasks prevented the "active principle" in the air from reaching the broth so microorganisms could be created.

(m) What problem is created when fresh air gets into the flask?

Louis Pasteur and the End of the Spontaneous Generation Theory

The final blow that demolished the theory of spontaneous generation was delivered by the great French scientist, Louis Pasteur (1822–1895). In 1864, Pasteur had a glass-worker make special flasks called swan-necked flasks, as shown in **Figure 5**. Broth was placed in a flask and boiled to destroy the microbes. Fresh air entered the flask as the flask cooled; however, microbes were not carried into the broth from the surrounding air. The microbes were trapped in the curve of the swan-necked flask.

(n) What conclusion can be drawn from the observation that the flask appeared clear both immediately after heating and three weeks later?

(o) Is this conclusion supported by the rest of the observations of the experiment?

As a finale, Pasteur tipped the broth in one of the flasks, allowing it to run into the

Figure 5

Pasteur's experiment

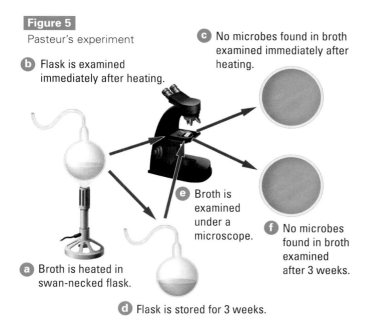

b Flask is examined immediately after heating.

c No microbes found in broth examined immediately after heating.

e Broth is examined under a microscope.

f No microbes found in broth examined after 3 weeks.

a Broth is heated in swan-necked flask.

d Flask is stored for 3 weeks.

curve of the swan-necked flask (see **Figure 6**). The broth here quickly became cloudy.

(p) Compare the colour of the two flasks, for procedure A and procedure B, in **Figure 6**.

(q) Why are microbes found in a sample of the broth from procedure B but not from procedure A in **Figure 6**?

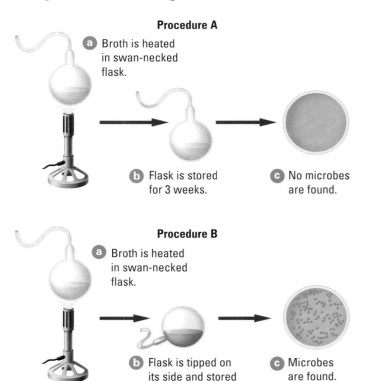

Procedure A

a Broth is heated in swan-necked flask.

b Flask is stored for 3 weeks.

c No microbes are found.

Procedure B

a Broth is heated in swan-necked flask.

b Flask is tipped on its side and stored for 3 weeks.

c Microbes are found.

Figure 6

Pasteur's second experiment

The Cell Theory

The theory of spontaneous generation was thus replaced by the third, and last, part of the modern cell theory, which states that all cells come from preexisting cells. This explains how life perpetuates itself.

The complete cell theory is
- All living things are composed of one or more cells.
- The cell is the functional unit of life.
- All cells come from preexisting cells.

Understanding Concepts

1. What is spontaneous generation?

2. What variable was Redi attempting to control in his experiment?

3. Identify one variable in Needham's ②A experiment.

4. What were two major differences between Needham's and Spallanzani's experiments?

5. Examine Needham's experiment. After two weeks of storage, why were more microbes found in the unsealed flask than in the sealed flask?

6. Use Pasteur's experiments to explain why controls are important.

7. The modern cell theory states that all cells come from preexisting cells. What evidence have you seen to support this theory?

Exploring

8. Suggest what evidence would have to be collected to prove that cells come from other cells. How would you gather that evidence?

9. Repeat Needham's experiment to determine ④B the minimum boiling time required to kill all the microbes in the beef broth. Have your teacher check your written procedure for safety before beginning.

Challenge

For centuries, people believed in spontaneous generation. People still have misconceptions about reproduction. How could you identify those misconceptions for your display?

The Importance of Cell Division

Have you ever peeled the skin from your shoulder after a sunburn? Imagine your terror if new cells had not replaced the dead skin cells you pulled off. Imagine what you would look like if every scratch or blemish on your skin remained. Cells come from preexisting cells through the process of cell division. Throughout your entire life, you will rely on cell division to replace dead or damaged cells in your body (**Figure 1**).

Functions of Cell Division

Healing and Tissue Repair

Healing and tissue repair are important functions of cell division. A related function is the replacement of dead cells (**Figure 2** and **Figure 3**). You don't go through life with all the same cells you had at birth. Every second, millions of your body cells are injured or die. If the remaining cells did not reproduce, your body would gradually shrink in size and eventually die.

Growth

A more obvious function of cell division is to increase the number of cells. As the number of cells in an organism increases, so does the size of the organism. Growth of all living organisms depends on cell division. Human growth begins with the division of a fertilized egg cell. All multicellular organisms also rely on cell division to grow.

Most cells are small and of a relatively constant size. Instead of dividing, why don't cells simply continue to increase in size? The reason is that the relationship of the surface area of the cell membrane to the volume of cytoplasm is very important. As a cell grows, the volume of cytoplasm increases faster than the surface area of the cell. All essential substances enter and exit the cell through the cell membrane. If a cell became too large, there would not be enough exchange of materials through the cell membrane to sustain it.

Also, the distance of the nucleus, which controls all of the cell's activities, from all parts of the cytoplasm must be kept small so that messages can be relayed efficiently. In short, cell division allows an organism to grow, while still maintaining a cell size that keeps the organism healthy.

Figure 1

Normal activities remove millions of skin cells that must be replaced daily. It has been estimated that approximately 50% of the dust in a furnace filter is dead human skin cells.

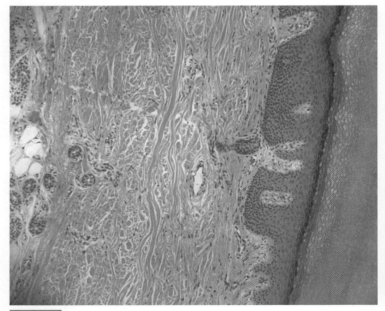

Figure 2

The outer layer of skin is made up of dead cells that become twisted and bent out of shape as they dry. Replacing these dead cells is an important function of cell division.

Reproduction of Organisms

Another very important function of cell division is that it perpetuates life. This is most obvious in the case of unicellular organisms, like bacteria. Cell division in unicellular organisms creates two new organisms. Cell division is also fundamental to reproduction of multicellular organisms.

Unanswered Questions

Cell division is one of the most studied, yet least understood areas of biology. Through painstaking hours of observation, scientists have collected a great deal of information about cell division. Yet despite all they have observed, many questions remain unanswered. How do cells know when to divide? The formation of calluses on your hands after a few days of working in the garden provides evidence that the rate of cell division can be altered. How and why is it altered? Why does a fertilized egg cell divide so rapidly after fertilization? Why do the cells that give rise to red blood cells divide at enormous rates, but brain cells rarely divide in adults? These are some of the questions that still need to be answered as the science of cell division continues to evolve.

Did You Know ❓

In the human body, red blood cells live a mere 120 days; white blood cells anywhere from 1 day to 10 years; and platelets, the cells that help blood clot, only about 6 days.

Understanding Concepts

1. Why is cell division important?

2. Provide evidence that suggests that not all cells in your body divide at the same rate.

3. Imagine two cubic cells, one with sides of 1 mm, and one with sides of 2 mm. For each cell, calculate

   (a) the total surface area

   (b) the volume

   (c) the surface area/volume ratio

   Using these calculations, explain why cells have to divide as an organism grows.

Making Connections

4. At one time, doctors transfused blood from younger individuals to the elderly. They believed that the younger blood would provide the elderly with more energy. Do older people actually have older blood? Support your answer.

5. Why might scientists want to get mature nerve cells to divide?

Exploring

6. Research reasons for the different rates of replacement of red blood (3A) cells, white blood cells, and platelets.

⬤ Challenge

In your display, should you present the questions scientists are still investigating that may affect decisions about reproductive technology?

Figure 3

Cell division is not limited to animals. The oldest living cells in a 300-year-old redwood tree are no more than 30 years old. The inner portion of a tree trunk is composed mainly of dead cells. Although dead, these cells still function to transport water up the tree.

Cell Division

All cells come from preexisting cells through cell division. Cell division, then, is how life is perpetuated (**Figure 1**). The approximately 100 trillion cells in your body began as a single, fertilized egg cell. This cell divided into two cells. Then each of these divided into two cells, and so on, until they formed the complete, functioning, multicellular organism that is you.

The Cell Cycle

Cells alternate between stages (phases) of dividing and not dividing. The sequence of events from one division to another is called the **cell cycle**, shown in **Figure 2**. For most cells, the cell division phase is a small part of this cycle. The stage between cell divisions is called interphase. During interphase, the cell takes in nutrients, such as sugars, and produces building materials, such as proteins. These materials are used by the cell for energy, growth, and repair of damaged parts. After a period of rapid growth, the cell prepares for division by duplicating its chromosomes within the nucleus. It is critical that the genetic material is duplicated before cell division. The chromosomes contain all the necessary information for all cell functions—including cell division! Each new cell will need a copy. After the genetic material is copied, there is another period of growth and preparation for cell division.

Figure 1
On a cellular level, reproduction is one cell becoming two.

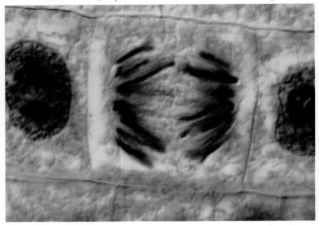

a A plant cell dividing

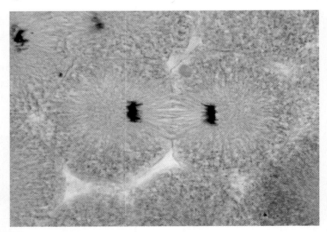

b Cell division in a whitefish embryo

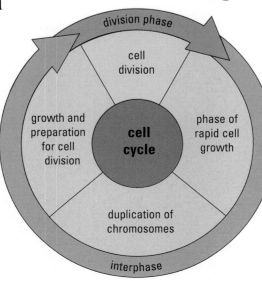

division phase

cell division

growth and preparation for cell division

cell cycle

phase of rapid cell growth

duplication of chromosomes

interphase

Figure 2
The cell cycle. The circle represents the entire life of a cell, which can be divided into two major phases: interphase and the cell division phase.

Mitosis and Cytokinesis

Despite great differences among living things, most cells show remarkable similarities in the way they divide. Cell division occurs in very simple, unicellular forms of life, such as the bacteria, as well as in complex, multicellular organisms, such as humans. In all cases, the initial mother cell divides into two identical daughter cells, as shown in **Figure 3**.

Cell division involves the division of nuclear materials and the sharing of the cytoplasm, which includes the organelles. During cell division, the duplicated chromosomes, copied during interphase, divide and move to opposite ends of the cell. This process of dividing nuclear material is called **mitosis**.

Cell division continues with the separation of the cytoplasm and its contents into equal parts. This process is called **cytokinesis**. This process begins before mitosis is complete. About half of the cytoplasm, containing about half of the organelles, goes to each daughter cell.

Cytokinesis differs in animal and plant cells. In animal cells, the cell membrane pinches together in the middle, separating the cytoplasm into equal parts and creating two new cells. In plant cells, a new cell wall forms along the middle, creating two new cells.

Together, mitosis and cytokinesis result in cell division, or the production of two new daughter cells. As the daughter cells grow during interphase, they make additional cytoplasm and organelles and eventually reach the size of the parent. The daughter cells will also use the single chromosomes to synthesize a duplicate set of genetic material.

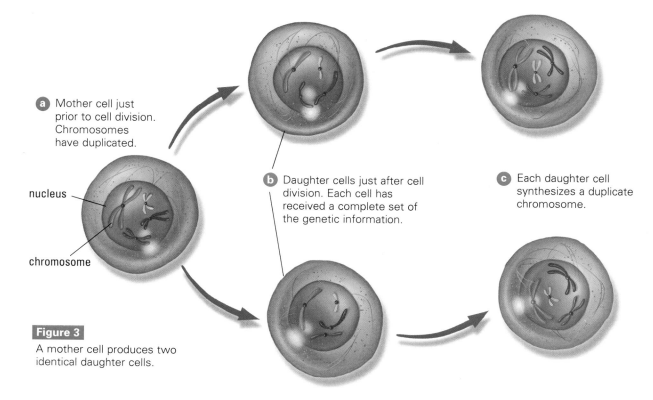

a Mother cell just prior to cell division. Chromosomes have duplicated.

nucleus

chromosome

b Daughter cells just after cell division. Each cell has received a complete set of the genetic information.

c Each daughter cell synthesizes a duplicate chromosome.

Figure 3

A mother cell produces two identical daughter cells.

The Phases of Mitosis

Figure 4 shows an animal cell dividing. To help describe the events of mitosis, scientists have divided the process into several phases. However, you must remember that the process is a continuous one. Think of each phase as a snapshot taken at a particular moment during cell division.

1. Interphase

During **interphase** the cell grows then prepares for cell division by duplicating its genetic material. Another growth phase readies the cell for division.

2. Prophase

In **prophase**, the individual chromosomes, now made up of two identical strands of genetic information, shorten and thicken. They become visible with the use of a light microscope. The nuclear membrane appears to fade when viewed under the microscope; in effect, it is dissolving.

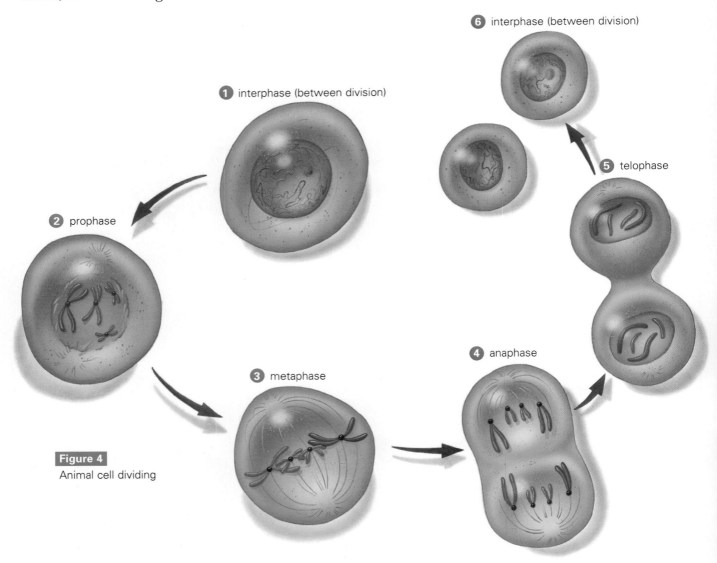

Figure 4
Animal cell dividing

6 interphase (between division)

1 interphase (between division)

2 prophase

5 telophase

3 metaphase

4 anaphase

3. Metaphase

In **metaphase**, the double-stranded chromosomes line up in the middle of the cell.

4. Anaphase

During **anaphase**, each chromosome splits. The two halves move to opposite poles of the cell. If anaphase proceeds correctly, each of the daughter cells will have a complete set of genetic information.

5. Telophase

During **telophase**, the chromosomes reach the opposite poles of the cell and a nuclear membrane begins to form around each set. Cytokinesis begins. The cytoplasm and organelles separate into roughly equal parts, and the two daughter cells are formed.

6. Interphase

The daughter cells begin growth and duplication of genetic material.

Try This — A Dynamic Model of Mitosis

Although mitosis is described in stages, the process of cell division is continuous. To help you understand this process, work with a partner to build a dynamic model in which chromosomes can be moved to show the events of cell division. To keep your model simple, use only two to four chromosomes. In your model be sure that you are able to line up the chromosomes in the centre of the cell, and that the single strands are able to move to opposite ends of the cell as they do during anaphase.

Challenge

One way to analyze and present data is to use graphs, such as a pie graph. How will you design the questions in your survey so the data are easy to analyze and present?

Understanding Concepts

1. Describe the cell cycle. What happens during interphase?

2. Why is the duplication of the nuclear material necessary during the cell cycle?

3. How do the new cells formed during cell division compare with the initial cell?

4. List and describe four phases of mitosis.

5. A normal human cell has 46 chromosomes. After the cell has undergone mitosis, how many chromosomes would you expect to find in each cell?

6. Cells alternate between phases of dividing and
(7A) not dividing. The sequence of events from one interphase to the next is called the cell cycle.

   (a) Describe the differences between the two cell cycles in **Figure 5**.

   (b) Which cell cycle represents a cell of an embryo or fetus and which a cell in an adult? Give your reasons.

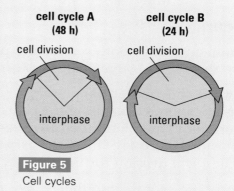

Figure 5
Cell cycles

Making Connections

7. Speculate about how modern advances in microscopes would also have advanced our understanding of cell division.

8. X rays and other forms of high-energy radiation can break chromosomes apart. Physicians and dentists ask women if they are pregnant before taking X rays. Why don't they want to X ray pregnant women?

Reflecting

9. Draw a sketch of your body. Under the sketch, list areas of the body where you think cell division is most rapid. Why do you think cells from these areas divide most rapidly? Check your answer once again at the end of the chapter.

Observing Cell Division

In the previous sections you learned why and how cells divide. In this activity you will have an opportunity to view and compare plant and animal cells during mitosis. You will examine prepared slides of the onion root tip and the whitefish embryo to identify cells that are dividing. Because prepared slides are used, these cell divisions have been "frozen in time." You will not be able to watch a single cell divide from prophase to telophase.

Materials
- microscope
- lens paper
- prepared microscope slide of an onion root tip
- prepared microscope slide of a whitefish embryo

Procedure

1 Obtain an onion root tip slide and place
5B it on the stage of your microscope.

2 View the slide under low-power magnification. Focus using the coarse-adjustment knob. Find the cells near the root cap (**Figure 1**). This is the area of greatest cell division for the root.

3 Centre the root tip and then rotate the nosepiece to the medium-power objective lens. Focus the image using the fine-adjustment knob only. Identify a few dividing cells.

✎ (a) How can you tell if the cells are dividing?

4 Rotate the nosepiece to the high-power objective lens. Use only the fine-adjustment knob to focus the image. Locate and observe cells in each phase of mitosis. Use the photographs of cells dividing, shown in **Figure 2**, to help you. Don't worry if what you see does not look exactly like the photographs.

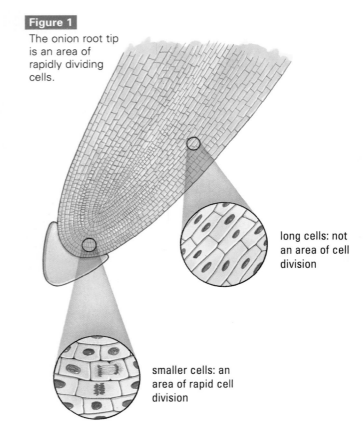

Figure 1
The onion root tip is an area of rapidly dividing cells.

long cells: not an area of cell division

smaller cells: an area of rapid cell division

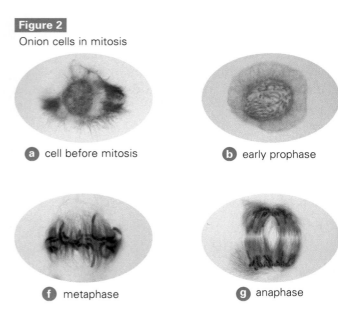

Figure 2
Onion cells in mitosis

a cell before mitosis

b early prophase

f metaphase

g anaphase

✏️ (a) Draw and title each of the phases that
6C you see. Label chromosomes if they are
visible. It is important to draw and
label only the structures that you see
under the microscope.

5 Return your microscope to the low-power
objective lens and remove the slide of
the onion.

6 Place the slide of the whitefish embryo on
the stage. (An embryo is an animal in the
very early stages of its development.)
Focus the slide using the coarse-
adjustment knob.

7 Repeat steps 3 and 4 for the whitefish cells.

✏️ (a) Draw and title each of the phases that
you see. Label the chromosomes if they
are visible.

8 Compare your diagrams with those of
other students in your class. Assist each
other in locating phases or cell structures.

9 Return your microscope to the low-power
objective and remove the slide of
the whitefish embryo. Put away your
microscope and return the slides to
your teacher.

Understanding Concepts

1. Why were plant root tip cells and animal embryo cells used for viewing cell division?

2. Explain why the cells that you viewed under the microscope do not continue to divide.

3. Compare the appearance of the dividing animal cells with that of the dividing plant cells. You may wish to use a table to list the differences and similarities.

4. If a cell has 10 chromosomes, how many chromosomes will each cell have following cell division by mitosis?

5. Predict what might happen to each daughter cell if all of the chromosomes moved to only one side of the cell during anaphase.

Exploring

6. Search the Internet for pictures of cells that
3A are dividing. What additional information about cell division can be gained by studying these pictures?

Challenge

Modern microscopes are linked to computers that can measure and count cells much faster than humans. Will this technology improve? What place could it take in your story or play?

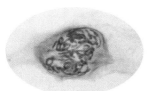

c prophase

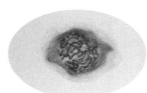

d late prophase

e transition to metaphase

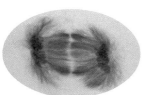

h telophase

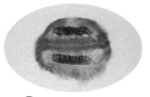

i late telophase

j cells after mitosis

Determining the Rate of Cell Division

A great deal of money is spent on agricultural chemicals used to increase plant growth. To find out if this is an efficient use of money, scientists must be able to analyze the effectiveness of these chemicals. One way to do this is by studying the rate of cell division. In this investigation, using prepared slides, you will be able to observe and determine the rate of cell division yourself.

Question

Can you tell how fast an organism is growing?

Hypothesis

1 Write a hypothesis for this experiment. **(4A)**

Materials

- microscope
- lens paper
- prepared microscope slide of an onion root tip
- prepared microscope slide of a whitefish embryo

Procedure
Part 1: Onion Root Cells

2 Obtain an onion root tip slide and place it **(5B)** on the stage of your microscope.

3 View the slide under low-power magnification. Focus using the coarse-adjustment knob. Locate the area of cell division, immediately above the root cap.

4 Centre the root tip and then rotate the nosepiece to the medium-power objective lens. Focus the image using the fine-adjustment knob.

5 Count 20 cells that are next to one another in the onion root tip. Determine which of those cells are dividing.

(a) In your notebook, record the number of cells dividing.

(b) Of the 20 cells, calculate the percentage that is dividing. For example, if you found 8 cells dividing:

Percentage of cells in prophase = $\frac{8}{20} \times 100\% = 40\%$

6 Examine two other areas of the onion root tip to determine if the division rate is the same in those areas.

(a) Construct and title a table to record the division rate of the three areas of the root tip.

(b) Draw a small diagram of the onion root **(6C)** tip and show the approximate location of each area you selected.

7 Return your microscope to the low-power objective lens and remove the slide of the onion root tip.

Part 2: Whitefish Embryo Cells

8 Place the slide of the whitefish embryo on the stage. Focus the slide using the coarse-adjustment knob.

(a) Predict whether the onion root tip or the whitefish embryo will have a greater percentage of actively dividing cells. Give reasons for your prediction.

9 Centre the whitefish embryo and then rotate the nosepiece to the medium-power objective lens. Focus the image using the fine-adjustment knob. Under medium-power magnification, repeat steps 5 and 6 for the whitefish cells.

10 Return the nosepiece to the low-power objective lens and remove the slide of the whitefish embryo.

Part 3: Making a Cell Division Clock

11 Replace the slide of the onion root tip under your microscope, and focus using the coarse-adjustment knob.

12 Under high-power magnification, locate 20 cells that are dividing. Identify the phase each cell is in. (Do not count cells in interphase.) You may have to search for enough dividing cells by moving the slide.

(a) Make a chart like the one below and enter the number of cells you found in each phase.

Phase	Number of cells	Percentage of total in phase
prophase	?	?
metaphase	?	?
anaphase	?	?
telophase	?	?

(b) Calculate the percentage of cells that are in each phase of division and include that number in your table.

13 With your data, you can now construct
(7B) a clock for cell division. It actually takes between 12 h and 16 h to complete one mitosis, but for the sake of simplicity, assume it takes 12 h.

(a) Calculate the number of hours spent in each phase by multiplying the percentage of cells in each phase by 12 h. If you found 40% of the 20 cells in prophase, for example:

Time spent in prophase = 40% × 12 h = 4.8 h

(b) Draw a clock and indicate the amount of time spent in each phase of cell division.

Analysis and Communication

14 Analyze your observations by answering the following questions:

(a) Which areas of the onion root tip have the fastest cell division rate?

(b) Were there any differences in the cell division rates of the three areas of the whitefish embryo? What do you conclude from this?

(c) Which has the greater percentage of dividing cells, the whitefish embryo or the onion root tip? What does this indicate about the cell division rates of the plant root tip and the animal embryo?

Making Connections

1. Why might someone be interested in determining the cell division rate of a plant or animal?

2. Herbicides are chemicals designed to kill weeds. Some herbicides, such as 2,4,D and 2,4,5,T, make plant cells divide faster than normal. Why would these herbicides kill weeds? (Hint: Think about what cells do when they are not dividing.)

3. What important part of the cell cycle is missing from your cell clock?

Exploring

4. Choose one of the cell division rates you
(7B) calculated in this investigation. Draw a pie graph showing the percentage of cells involved in cell division and those not involved.

Reflecting

5. Mitosis is described in phases because this makes it easier to talk about. But the cells do not stop between each of the phases. What observations have you made that suggest that mitosis is a continuing process?

Career Profile

Using Law in Science

Harriet Simand was 20 years old, a healthy student just back from a summer in Europe, when she went for a routine medical checkup. That is, routine until she found she had a rare kind of vaginal cancer known as clear-cell adenocarcinoma. The cancer was so advanced she needed an immediate hysterectomy.

Simand's cancer had been caused by a drug known as DES: diethylstilbestrol. She had never taken it herself; it had been prescribed to her mother, and to thousands of other pregnant women, to prevent miscarriage.

With little background in science, Simand soon learned a great deal about chemistry, especially about pharmaceuticals and their effect on people. She found that between 200 000 and 400 000 women took DES in Canada before the drug was banned in 1971. Side effects include breast cancer among mothers who took the drug, and premature births and fertility problems in their daughters. There may be problems with their sons as well, but this area has not been studied fully.

Simand decided to obtain a law degree so that she could work to change the laws concerning pharmaceuticals. Since getting it, she has worked to publicize the dangers of DES. The first job is to find women who were exposed, a difficult job because "drug companies have never invested any money in tracking people down. We are still finding people who didn't know they were exposed, who don't even know what DES is... [But] when the cancer is caught early it has a higher cure rate."

> "Even without a medical degree you can become an expert in your own health. You still need to inform yourself enough to know what questions to ask the doctor, because doctors can't read every single journal."

Exploring

1. In what other areas could a knowledge of law and science be used side by side?

2. There are other pharmaceuticals that have caused harm to fetuses. Choose one to research. Present your findings to your class.

Challenge

In your story or play, one of your characters will face a decision. Could the decision have some negative results? What might they be?

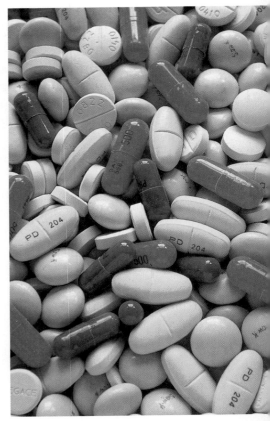

Reproduction and Cell Division

Reconsider the modern cell theory.
- All living things are made up of one or more cells.
- The cell is the functional unit of life.
- All cells come from preexisting cells.

   Cell division, the process by which cells come from preexisting cells, is the process that perpetuates life and allows species to continue. Just as cells reproduce as part of the cell cycle, living organisms reproduce as part of their life cycle.

   Organisms of all species reproduce. They may reproduce sexually or asexually. In **asexual reproduction** a single organism gives rise to offspring with identical genetic information. The cells of the human body, other than those found in female ovaries and male testes, reproduce asexually by mitosis. Most single-cell organisms, such as bacteria, and some multicellular organisms use asexual reproduction to produce offspring.

   In **sexual reproduction**, genetic information from two cells is combined to produce a new organism. Usually, sexual reproduction occurs when two specialized sex cells unite to form a fertilized egg called a **zygote**. **Figure 1** compares asexual and sexual reproduction at the cell level.

   Note that some organisms use both methods of reproduction. For example, bacteria reproduce mostly in an asexual process called binary fission, which is basically cell division as you have learned it. However, bacteria are also able to exchange genetic information in a form of sexual reproduction. Similarly, most plants reproduce sexually, in the process that results in seeds, but many also reproduce asexually in various other ways, which are shown in **Figure 2**.

Figure 1

Asexual and sexual reproduction of cells. Notice that when a cell reproduces asexually, the parent cell becomes two identical daughter cells. In sexual reproduction two specialized cells fuse.

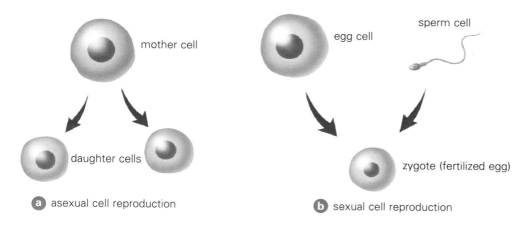

a asexual cell reproduction

b sexual cell reproduction

Figure 2
Various types of asexual reproduction

a binary fission

In **binary fission**, the organism splits directly into two equal-sized offspring, each with a copy of the parent's genetic material. Binary fission is a common type of reproduction in single-celled organisms, such as these intestinal bacteria.

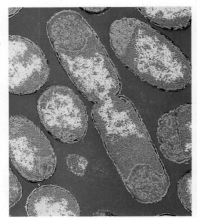

b budding

In **budding**, the offspring begins as a small outgrowth from the parent. Eventually, the bud breaks off from the parent, becoming an organism on its own. Budding occurs in some single-cell organisms, such as yeast, and in some multicellular organisms, such as the hydra shown here.

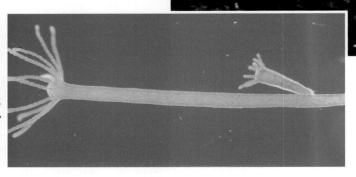

Try This How many divisions will it take?

Many single-cell organisms reproduce by binary fission. How many organisms would there be after five divisions? Copy and complete the chart to help you answer this question.

Number of divisions	Number of organisms
0	1
1	2
2	4
3	?
4	?
5	?

1. What pattern relates the number of divisions to the number of resulting organisms?

2. Use the pattern to predict how many organisms are produced after 10 divisions.

3. How many divisions are required to produce over a million organisms?

4. The single, fertilized cell from which you began divided to produce the many cells that make up your body. Estimate how many cell divisions it took.

c fragmentation

In **fragmentation**, a new organism is formed from a part that breaks off from the parent. Many types of algae and some plants and animals can reproduce this way. If a starfish is cut through its central disk, each section will develop into a new starfish that contains identical genetic information.

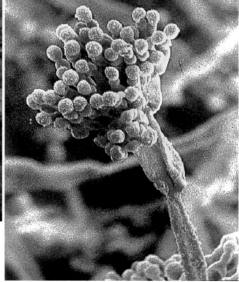

d spore formation

In **spore formation**, the organism undergoes frequent cell division to produce many smaller, identical cells called spores. The spores are usually housed within the parent cell. Many spores have a tough, resistant coating that allows them to survive after the parent cell dies. The penicillium mould shown here reproduces by forming spores. Each spore can develop into a mature organism.

e vegetative reproduction

Many plants, such as spider plants, strawberries, and the quaking aspen shown on page 139, make use of **vegetative reproduction**. They produce runners that can develop into another plant. Each runner from this spider plant can develop into another plant with identical genetic information.

Understanding Concepts

1. How is asexual reproduction different from sexual reproduction?

2. Why must the genetic material of the cell be duplicated before cell division begins?

3. How is the zygote, produced by sexual reproduction, different from daughter cells, produced by asexual reproduction?

4. **(a)** Describe briefly five types of asexual reproduction.

   (b) Choose one type of asexual reproduction. Explain how a plant nursery could make use of it.

Making Connections

5. Identify the type of asexual reproduction in each of the following situations:

   (a) A multicellular algae is struck by a wave. The algae breaks up and each new piece grows into a new organism.

   (b) A new tree begins to grow from the root of a nearby tree.

   (c) A small cell begins to grow on the outside of another cell. Eventually, it breaks away from the larger cell and continues to grow.

Reflecting

6. What advantages might an organism have that can reproduce asexually? Make a list of the advantages. Add to your list or modify it as you progress through this unit.

Challenge

Could humans attempt to reap the benefits of asexual reproduction? Could an asexual reproduction technology play a role in your play or story?

Calculating Population Growth Rates

You've created a clock to measure how long cells in an onion root tip spend in each stage of reproduction. Because those cells were dead when you observed them, you couldn't measure their rate of reproduction. Also, the cells in the onion tip reproduce asexually, but they also work together with other cells in a multicellular organism—their rate of reproduction varies according to where they are in the root tip. What about unicellular organisms? What is their rate of reproduction? What factors affect that rate?

In this investigation you will be observing paramecia, unicellular organisms that live in fresh water. Under most conditions, they reproduce asexually by binary fission. You will learn to estimate the population of paramecia in a culture using a sampling technique. In addition, you will change the environment of the organisms. After several days of observation, you will be able to calculate the rate at which their population grows.

Question

How do populations of paramecia change over time?

Hypothesis

1 Write a hypothesis to predict how a population of paramecia will change under different conditions

Materials

- apron
- light microscope
- transparent ruler (mm marks)
- paramecium culture
- microscope slide
- cover slip
- rice grains
- medicine dropper

Procedure
Part 1: Determining the Field of View

2 Measure the field of view with the low-
(5B) power objective lens in place.

✎ (a) Record the field diameter under the low-power objective lens in millimetres.

3 Measure the field of view for the medium-power objective.

✎ (a) Record the field diameter under medium-power objective lens in millimetres.

4 Determine the field of view under high-power magnification.
- Calculate the ratio of the magnification of the high-power objective lens to that of the low-power objective lens.
- Use the ratio to determine the field diameter under high-power magnification.

$$\text{Field diameter (high power)} = \frac{\text{field diameter (low power)}}{\text{ratio}}$$

✎ (a) Calculate the field diameter of the high-power lens. Show your calculations.

Part 2: Determining the Number of Paramecia

5 Using a medicine dropper, place a drop from the paramecium culture on a microscope slide and add a cover slip. Using a ruler, measure the diameter of the wet mount preparation.

✎ (a) Record the diameter. Calculate and record the area of the wet mount.

Step 5

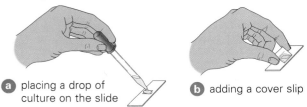

ⓐ placing a drop of culture on the slide   ⓑ adding a cover slip

6 Place the slide on the microscope stage and examine for paramecia first under low-power and then medium-power magnification.

    (a) Describe the appearance of the paramecia.

    (b) Do any of the paramecia appear to be dividing?

    (c) Draw any cells that you believe are (6C) undergoing cell division.

7 Using medium-power magnification, estimate the number of paramecia in three different fields of view.

    (a) Record the number of paramecia seen in each field.

    (b) Calculate the average number of paramecia in a field of view.

Part 3: Environmental Factors Affect Growth Rate

8 Measure 10 mL of the culture and pour it into a glass or plastic container. Add a few grains of rice to the container. The rice will serve as a food source.

Step 8

10-mL culture — rice — rice added — original culture

9 Check the paramecia cultures over the next 10 days and determine the population by the sampling technique described in Part 2. Compare the two cultures: with rice and without.

    (a) Construct a data table to show populations every day.

    (b) Record your daily observations in the data table.

Analysis and Communication

10 Analyze and summarize your results by completing the following:

    (a) Identify the control used in Part 3.

    (b) Graph the changes in the population (7B) of the paramecia in both the control and experimental groups. Plot time along the x-axis and population along the y-axis.

    (c) Does the population grow at a constant rate? Give reasons for your answer.

    (d) Extrapolate from your graph the estimated number of paramecia that would be found if the experiment were extended to 20 days.

    (e) Calculate the number of paramecia in 1 mL (20 drops from the medicine dropper = 1 mL). Use the following formula to help you:

$$\text{Number of paramecia in one drop} = \frac{\text{area of the wet mount} \times \text{no. of paramecia per field of view}}{\text{area of field of view (medium power)}}$$

    (f) Calculate the population growth by using the following formula:

$$\frac{(\text{final population/mL}) - (\text{original population/mL})}{\text{Time}}$$

Making Connections

1. Did the populations change? If so, why?

2. From your investigation, what do you think paramecia need to live and reproduce?

Exploring

3. Different groups in your class could test different numbers of rice grains. Is there an "optimum" nutrient level?

4. Design and conduct an investigation to test (2A) another environmental factor that might increase the rate of population growth.

Challenge

In both the survey and the display, you must present data. How could you use graphs to improve your presentation?

Hormones for Cell Growth and Division

What causes some plants, like the ones in **Figure 1**, to grow full and bushy while others grow tall and thin? Why do some animals grow faster and larger than others? Understanding the factors that either promote or inhibit cell growth and cell division is important for scientists who are working to increase food production.

Figure 1

Scientists don't really know why some cells divide more frequently than others. However, they do know that cells communicate with one another using chemical messengers. These messengers, called **hormones**, are produced in cells in one part of the body and can affect cells in other parts. Some hormones trigger cells to grow or divide.

Plant Growth Hormones

Plants produce a variety of growth hormones. For example, when one side of a plant is not exposed to light, hormones called auxins collect on the dark side. These hormones signal the cells to grow so they become longer. As the cells grow in length, the plant bends toward the light, as shown in **Figure 2**.

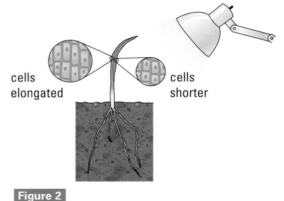

cells elongated   cells shorter

Figure 2

Grass seedlings bend toward the light.

As the name suggests, cytokinins are plant hormones that stimulate cell division. Cytokinins released from the roots promote cell growth and division in the buds on the side of a plant. This causes the plant to grow wider.

Often, cytokinins and auxins work in opposite ways. For example, auxins produced at the top of a plant inhibit the growth of the buds on the side. This causes the plant to grow up. Horticulturists have known about the interaction of auxins and cytokinins for years. By removing the top buds from a plant, auxin production is reduced, and the plant slows its upward growth and becomes bushier. Apple growers prune the tops of their trees, making them bushier. Low, bushy apple trees mean more fruit, easier picking, and less bruising of mechanically picked apples.

Animal Growth Hormones

Animals also have hormones that affect the growth and division of their cells. Growth hormone (GH) is produced in the pituitary gland and carried by the blood to all areas of the body. However, GH affects some cells, such as bone, muscle, and cartilage cells, more than others. Stimulated by GH, they divide rapidly to produce more cells, which then grow and make a bigger organism.

The effects of human growth hormone are particularly noticeable when it is produced in abnormal amounts. Low production of GH during childhood can result in dwarfism, while high secretions can result in gigantism, as shown in **Figure 3**.

Abnormally low or high secretions of growth hormone can result in dwarfism and gigantism.

Understanding Concepts

1. What are hormones and why are they important to the survival of multicellular organisms?

2. What environmental stimulus could cause the release of a hormone?

3. Identify two hormones of plants and explain what each does.

4. What is animal growth hormone? What does it do?

5. Calluses form when cells in the skin layer divide and grow rapidly to protect cells below. How does this indicate that chemical signals stimulate cell division?

Making Connections

6. What do you think might happen if you added cytokinins to the soil of your houseplants?

7. What could a plant nursery worker do to make young plants bushier?

Exploring

8. Some farmers give "growth (3A) enhancers" to some of their (3B) livestock, particularly beef cattle. Many consumers are concerned that this practice could cause health problems in humans. Research both sides of this topic. Use the Internet and talk to people in the agriculture industry to find out what growth enhancers are and how they work.

Try This — Response to Sunlight

Set up an experiment, as shown in **Figure 4**, to study a plant's response to sunlight. Predict what you think will happen to the plant as you observe it for several days. Record your observations. In which parts of the plant do the cells elongate? (2A)

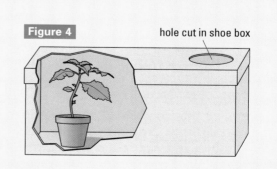
Figure 4

hole cut in shoe box

Measuring Plant Growth

Selecting the plants best suited for a particular environment, whether it's a field, an orchard, or a garden, requires an effective way of measuring growth rates. In this investigation, you will mark a growing root and then measure it to determine the area of the root where cell division took place at the most rapid rate.

Question

1 Write a question for this experiment. **4A**

Hypothesis

If a plant grows, then its cells must be undergoing cell division and cell growth.

Materials

- germinating seedlings, approximately 4 cm long
- 250-mL beaker
- elastic band
- ruler (mm)
- paper towel
- petri dish
- permanent marker
- nylon thread
- glass plate or cardboard sheet
- cross section of tree
- prepared microscope slide of a woody stem

Procedure

2 Prepare a growth chamber by lining a 250-mL beaker with damp paper towel. Add approximately 20 mL of water to the beaker. Place a petri dish cover over the beaker.

Step 2

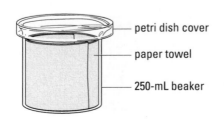
petri dish cover
paper towel
250-mL beaker

3 Select 5 seedlings that are reasonably straight. Place them on a dry paper towel to remove excess water.

Step 3

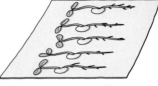
select germinating seedlings

4 Place a piece of nylon thread on a sheet of paper. Run a permanent marker pen along the thread until the thread has picked up the ink. Use the thread and a ruler to lightly place a mark every 1 mm along the root of each seedling.

Step 4

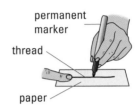
permanent marker
thread
paper

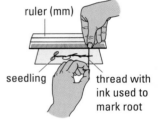
ruler (mm)
seedling
thread with ink used to mark root

(a) Why is it important to use permanent ink rather than water-soluble ink?

5 Cover a glass plate or small piece of cardboard with a paper towel. Use an elastic band to gently hold the 5 seedlings in place on the paper towel.

Step 5

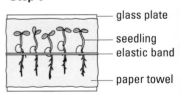
glass plate
seedling
elastic band
paper towel

6 Stand the plate with the seedlings in the growth chamber, cover the chamber, and leave it in a dark place for 48 h.

Step 6

petri dish cover
seedlings
beaker

SKILLS HANDBOOK: **4A** Asking Questions and Hypothesizing

7 Make a data table such as **Table 1** to record your observations. Section 1 is the section closest to the root tip.

Table 1	Root Growth in Seedlings				
Section number	Distance (mm) after 48 h				
	Seedling 1	Seedling 2	Seedling 3	Seedling 4	Seedling 5
1	?	?	?	?	?
2	?	?	?	?	?

8 After 48 h, gently take one of the seedlings **(6B)** and measure the distance between each pair of ink marks on the root. Remember that your original marks were 1 mm apart. Replace the seedling on the plate.

✎ (a) Record your measurements in your table.

Step 8   After 48 h

section 1
section 2
section 3
section 4

9 Repeat step 8 for the remaining seedlings. Return the seedlings to the growth chamber.

✎ (a) Record all measurements in your table.

10 Repeat steps 8 and 9 after 72 h, and again after 96 h.

✎ (a) Record all measurements in your table. Note any changes in the root sections that show the most growth.

11 In addition to increasing in length, roots and stems also grow thicker. This type of growth is referred to as secondary growth. Because little growth occurs during the winter months, the secondary growth can be identified as bands of cells, called annual rings. Examine the cross section of a tree.

✎ (a) Estimate the age of the tree in **Figure 1**.

Analysis and Communication

12 Analyze your results by completing the following:

(a) Which section of the root showed the most growth?

(b) Make a graph to show which section of **(7B)** the roots grew the most after 48 h. Plot the average length of each section on the *y*-axis and the section number along the *x*-axis.

(c) Construct a graph that shows the change in growth rate over time.

(d) Why was most of the growth located in one area?

(e) Summarize in a paragraph why you used 5 seedlings in this experiment instead of only one.

(f) What can you conclude about the relationship between organism growth, cell growth, and cell division?

(g) Write a hypothesis that explains why some annual rings are thicker than others.

Exploring

1. Repeat the experiment after removing different lengths of root tip from a number of seedlings. Compare the growth rate of these seedlings with normal seedlings.

2. Design a method for determining the growth **(4B)** rate of leaves and stems. Plant some of the remaining seedlings used in this activity in vermiculite and test your procedure.

Figure 1

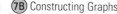

Cell Division and Patterns of Growth

Humans grow most quickly during the nine months before and the first three months after their birth. However, not all of their body parts grow at the same rate. In this case study, you will investigate some of the differences in growth rate as a human body matures.

Growth of the Body

Examine **Figure 1** showing changes in body proportions from the fetus to the adult. The individual diagrams are not drawn to the same scale. However, proportions are indicated by eight different segments, each of which represents one-eighth of the total body size.

Figure 1

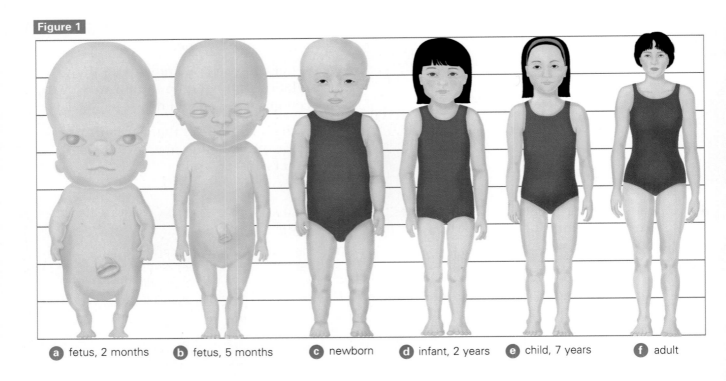

a fetus, 2 months   b fetus, 5 months   c newborn   d infant, 2 years   e child, 7 years   f adult

(a) Which parts of the body appear to grow the most between a two-month-old fetus and an infant?

(b) Which parts of the body appear to grow the most between infancy and adulthood?

(c) Which parts of the body grow the least during each time?

(d) Speculate about why an infant's head is so large in comparison to the rest of its body.

Growth of Organs

Examine the graph in **Figure 2** showing the rates of growth of the brain, heart, and body. The graph shows that at the age of two, the masses of the brain and heart have doubled, whereas the mass of the body is almost four times the mass at birth.

(e) By how many times has the body mass increased by age 19? By how many times has the heart increased in mass?

(f) At what approximate age does the brain reach its maximum mass?

(g) How does the growth of the heart compare with that of the brain?

(h) Would you expect the change of mass of the heart and body to continue at the same pace after 19 years of age? Explain your answer.

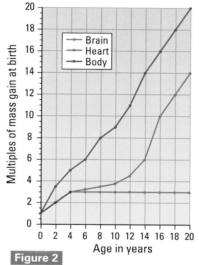

Figure 2

Growth rates of the brain, heart, and body

Growth of Bones

Figure 3 shows changes in the growth rate of the foot and shin bone (tibia).

(i) Which body part (foot or shin bone) grows faster?

(j) Plot a graph that shows the difference between the growth rate of the foot and that of the shin bone (tibia).

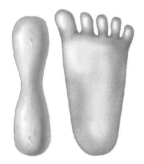 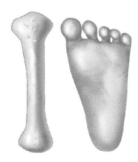 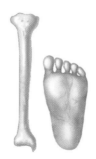

Figure 3

The bones of the foot and tibia do not grow at the same rate.

Understanding Concepts

1. Where in your body would you expect to see the highest rate of cell division? Explain your answer.

2. Based on the information in this activity, what can you conclude about the growth of your brain?

Making Connections

3. The graph of the growth rates of the brain, heart, and body is taken from data collected from a large group of people. Why do scientists compile data from many people rather than just record the information from a single individual?

Exploring

4. How large would the picture of the adult be if it were drawn to the same scale as the infant? Assume that the size of the adult's head is approximately twice that of the infant's head. The following steps may help your calculations:

   • Use a ruler to take a horizontal measurement of the infant's head and the adult's head.

   • Calculate the size of the adult's head, if drawn to same scale as the infant's head.

   • Calculate the size of the adult, if drawn to same scale as the infant.

Reflecting

5. What evidence can you draw from your own growth patterns to suggest that all parts of your body do not grow at the same rate? If possible, use photographs of yourself at different ages as evidence.

Slowing Down Aging

People are living longer. The life expectancy (the average lifespan) for Canadians born in 1991 was 75 years for men and 81 years for women. In 1951, the life expectancy was only 66 years for men and 71 years for women. According to Statistics Canada projections (**Figure 1**), the percentage of the population that is older than 65 years is going to grow every year for decades (**Figure 2**). People have always aged, but there was little that could be done about it. In the near future, that may change. The question "Why do cells age?" is central to a growing field of research.

Aging and Cell Division

The answer to the question of aging seems to have a lot to do with cell division. The oldest organisms on Earth, trees such as the giant redwood trees on the west coast of North America, started growing around the same time people figured out how to use iron to make axes—about 3000 years ago. Yet even in these incredibly ancient trees, the oldest living cells are only 30 years old. The same is true of the cells in your body—very few of the cells you were born with are still around. They have died and been replaced by their descendants. Red blood cells live only about 120 days; your skin cells are replaced by the hundreds of thousands daily. Few cells exist for the entire life of most plants and animals.

Figure 1

Population projections for the years 2001 and 2016 show that the number of older people will increase.

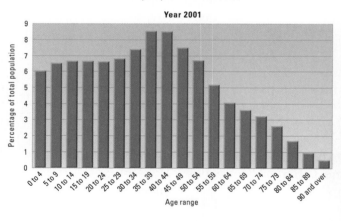

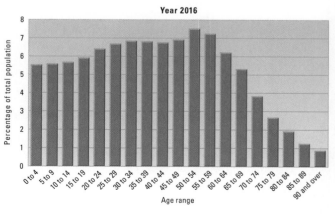

Figure 2

To stay alive, you rely on cell division. But research on chicken heart cells grown in tissue culture indicates that there's a limit to the number of times cells can divide. For those immature heart cells, the maximum was 50 times. Other types of cells have different limits, but once they have reached their maximum, they can no longer divide. If a neighbouring cell dies or is injured, they cannot reproduce to replace it. This may be at the core of aging. Cells die and are not replaced. Injuries are not repaired. Because of accumulating damage, the function of organs slows. Eventually the damage becomes so serious that an organ or an organ system can no longer function, and the person dies.

Fighting Age

Is there a way around the biological clock? Scientists are exploring this question. A 1990 study indicated that injections of growth hormone (GH) could slow aging. According to the research, the injections in older people increased muscle development and caused fat to disappear. In many ways, GH seemed to reverse decades of aging.

There may be a cost. Researchers warn that the long-term effects of GH injections have not been studied, and the hormone may not be for everyone. Scientists and non-scientists remain doubtful about the potential of GH. Since 1990, many scientists have begun research on other substances that may reverse the aging process.

 Should we be fighting nature?

Statement

Hormones or drugs should not be used to reverse or even slow the processes of aging.

Point

- The idea of reversing aging presents many difficulties. First, the cost would be immense. GH is expensive. At current prices, injections of GH for a 70-kg man would cost about $14 000 per year. Only the richest in society would be able to pay for such treatments.

Counterpoint

- Expense has no bearing on the issue. Cosmetics are a billion-dollar business. The money spent on hormone or other treatments for aging would also produce economic benefits. People would work longer and generally experience a better quality of life.

What do you think?

- In your group, discuss the statement and the point and counterpoint above. Write down additional points and counterpoints that your group considered.
- Decide whether your group agrees or disagrees with the statement.
- Search newspapers, a library periodical index, a CD-ROM directory, and, if available, the Internet for information on drugs or hormones used to slow aging.
- Prepare to defend your group's position in a class discussion.

Key Expectations

Throughout this chapter, you have had opportunities to do the following things:

- Explain the cell theory and illustrate its contributions to the concept of cell division. (5.1, 5.3, 5.5, 5.8)
- Explain mitosis and its significance. (5.4, 5.5, 5.6)
- Describe representative types of asexual reproduction. (5.8, 5.9)
- List the advantages of asexual reproduction. (5.8)
- Investigate the processes of cell division, and organize, record, analyze, and communicate results. (5.6, 5.7, 5.9, 5.11)
- Formulate and research questions related to cell division and communicate results. (5.4, 5.6, 5.8)
- Use a microscope to identify cells undergoing division. (5.6, 5.7, 5.9)
- Use a microscope to observe an organism undergoing fission, and design and conduct an investigation into cell division. (5.9)
- Predict the number of cells produced in a given amount of time. (5.8)
- Describe the historical development of reproductive biology, including the role of the microscope. (5.1, 5.3)

- Provide examples of how advances in cell biology will affect human populations. (5.10, 5.13)
- Describe Canadian contributions to research and technological developments in genetics and reproductive biology. (5.1)
- Explore careers that require an understanding of reproductive biology. (Career Profile)

KEY TERMS

anaphase	Golgi apparatus
asexual reproduction	lysosome
binary fission	metaphase
budding	mitochondrion
cell cycle	mitosis
cell membrane	nucleolus
cell wall	nucleus
centriole	organelle
chloroplast	prophase
chromosome	ribosome
cilia	sexual reproduction
cytokinesis	spore formation
cytoplasm	telophase
endoplasmic reticulum	vacuole
flagellum	vegetative
fragmentation	reproduction
gene	

Reflecting

- "All living things undergo cell division. Cell division is essential for the perpetuation of life." Reflect on this idea. How does it connect with what you've done in this chapter? (To review, check the sections indicated above.)
- Revise your answers to the questions raised in Getting Started. How has your thinking changed?
- What new questions do you have? How will you answer them?

Understanding Concepts

1. Make a concept map to summarize the material that you have studied in this chapter. Start with the words "cell division."

2. Use the diagram in **Figure 1** of plant and animal cells during cell division.
   (a) Identify each of the cells as either plant or animal cells.
   (b) Identify the phases of cell division.

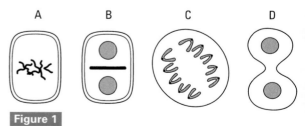

A   B   C   D

Figure 1

3. What is the cell cycle?

4. What is interphase and why is it important for the process of cell division?

5. Compare mitosis in plant and animal cells.

6. What evidence can you provide that suggests that not all cells divide at the same rate?

7. Why is the duplication of genetic material important for cell division?

8. How did the experiments performed by Spallanzani and Pasteur support the cell theory postulate that states: "All cells come from preexisting cells"?

Applying Skills

9. Three groups of seedlings were placed in growth chambers. Each growth chamber received 10 mL of a different nutrient solution. The root lengths of the seedlings were measured over a five-day period. The data in **Table 1** were obtained:

Table 1

Time (days)	Root Length (mm)		
	Solution X	Solution Y	Solution Z
0	2	2	2
1	2	4	4
2	3	6	9
3	4	10	14
4	4	12	18
5	5	15	28

(a) Graph the results obtained by plotting days on the *x*-axis and root length on the *y*-axis.

(b) Provide a conclusion from the data given.

10. The data in **Table 2** were collected from two different fields of view of hamster embryo cells. The number of cells found in each phase of cell division was recorded. It took 660 min to complete one cell cycle from the beginning of one interphase to the beginning of the next.

Table 2

Cell phase	Area 1	Area 2	Total cell count	Time for each phase
interphase	91	70	?	?
prophase	10	14	?	?
metaphase	2	1	?	?
anaphase	2	1	?	?
telophase	4	4	?	?

(a) Copy the table in your notes and complete the calculations.

(b) Using the data provided, draw a circle graph of the cell cycle.

11. An experiment measured the rate of growth of a seedling root. Lines were marked on the root 1 mm apart. After 48 h, the root appeared as shown in **Figure 2**. Examine the results and draw a conclusion.

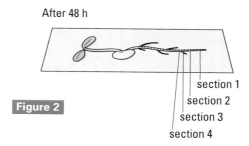

After 48 h

section 1
section 2
section 3
section 4

Figure 2

Making Connections

12. Many times science is presented as a compilation of facts; however, there are a great many things that are not known about cell division. Make a list of unanswered questions that have been introduced in the chapter. What things don't we know about cell division?

13. Irradiation can break apart chromosomes. The diagram in **Figure 3** shows the effects of irradiation on cells undergoing metaphase. Food companies sometimes irradiate fruit and vegetables to improve shelf life. How does irradiation help preserve food?

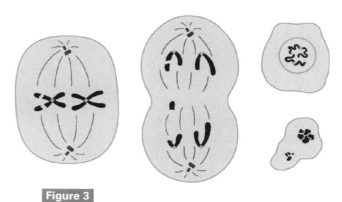

Figure 3

14. No nucleus is found in the outermost layer of skin cells that covers your body. A moisturizer claims to restore and rejuvenate these cells.

(a) Would these skin cells be capable of producing other skin cells?

(b) How would you go about testing the claim?

A Closer Look at Cell Division

1 DNA is the amazing, dynamic, genetic material found in the chromosomes of a cell. It contains all of the information that determines how cells function and respond to their environment. DNA is one of the very few molecules capable of duplicating itself. But DNA is still just a chemical. How does your hereditary material differ from that of a tree or a frog?

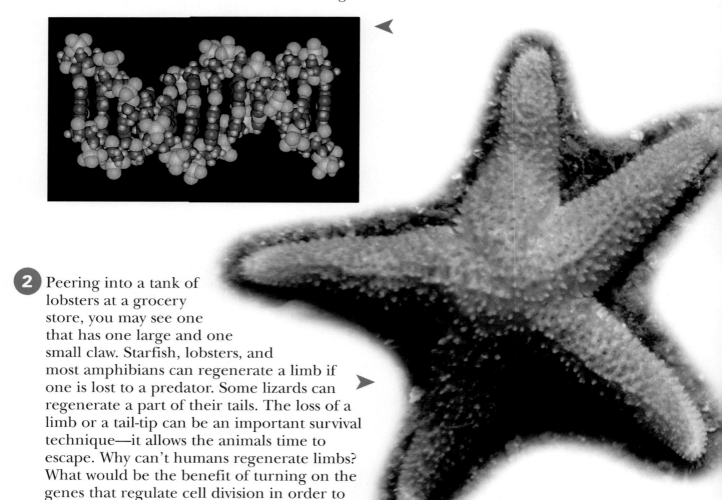

2 Peering into a tank of lobsters at a grocery store, you may see one that has one large and one small claw. Starfish, lobsters, and most amphibians can regenerate a limb if one is lost to a predator. Some lizards can regenerate a part of their tails. The loss of a limb or a tail-tip can be an important survival technique—it allows the animals time to escape. Why can't humans regenerate limbs? What would be the benefit of turning on the genes that regulate cell division in order to regenerate tissues, organs, or an entire limb?

3 The plot of the 1997 movie, *Multiplicity*, was based on cloning. The busy hero enlisted the help of scientists to make duplicates of himself so that he would be able to spend more time with friends and family. Does cloning actually work the way it was depicted in the movie? Have people ever been cloned? What ethical issues surround the cloning of humans, plants, and other animals?

Reflecting

Think about the questions in **1**, **2**, **3**. What ideas do you already have? What other questions do you have about how cells divide? Think about your answers and questions as you read the chapter.

Try This Cloning a Potato

Locate a potato that has many "eyes." Using a knife, cut the potato into several pieces. Some of the potato chunks should contain eyes, others should have none. Plant each piece in potting soil in the school laboratory. Label each piece. (You could also plant them in your garden or a school garden.)

1. Do any of the pieces of the potato grow? If so, which ones?

2. What is an "eye" of a potato tuber?

DNA: The Genetic Material

What makes you different from a mouse or a dog? Why don't you have whiskers and walk on four legs? The answer is that your body was shaped by the genetic information in each of your cells. Based on that, you might expect that your genetic information would be obviously different from that of a mouse or a dog. The packaging is certainly different. A look into one of your cells as it divided would reveal 46 chromosomes. A dog has 78 chromosomes, and a mouse has 40. However, a closer look at the chromosomes reveals something very surprising. All chromosomes, whether in mice, dogs, eagles, fish, toads, or humans, are composed of the same chemical. This genetic chemical is called **deoxyribonucleic acid** (**DNA** for short).

DNA provides the directions that guide the repair of worn cell parts and the construction of new ones. It describes how cells will respond to changes in their environment and to messages from other cells. This information is sent from DNA in the nucleus to the organelles in the cytoplasm using chemical messengers.

The Code Inside the Chromosome

The surprise that the genetic material is the same chemical in all living things is explained by a closer look at DNA. The molecule has an interesting structure. It is made up of a series of chemicals called nitrogen bases, held in a long, winding helix (**Figure 1**). These nitrogen bases are used like letters or characters in a simple code. Computers use a two-character code (1 and 0) to store information. English uses a 26-character code called the alphabet. Other languages may use many more or fewer characters.

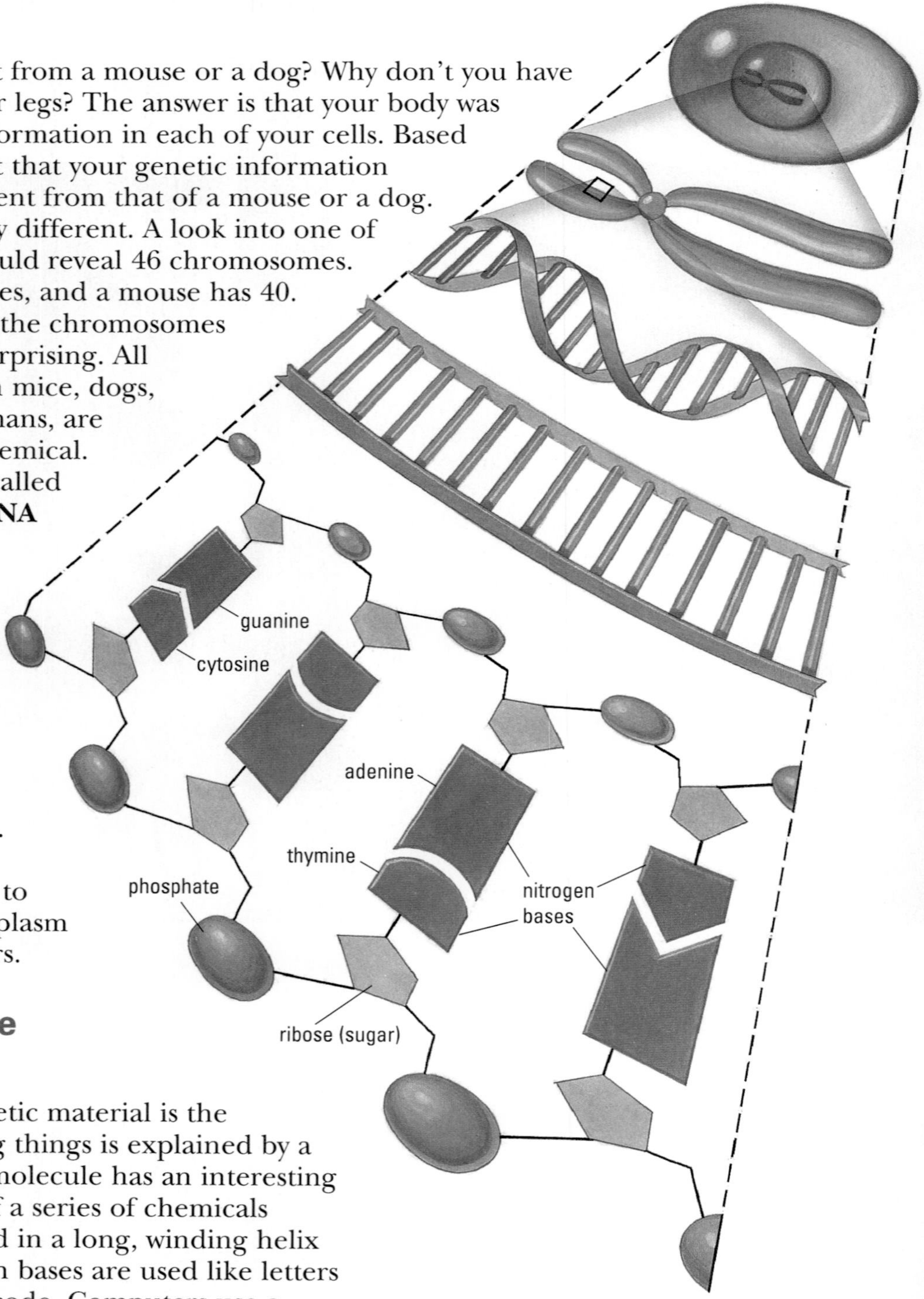

Figure 1

DNA molecules are shaped like a twisted ladder. Sugar and phosphate molecules make up the sides of the "ladder," while nitrogen bases form the "rungs." The nitrogen bases form pairs—adenine always pairs with thymine and cytosine always pairs with guanine. The sequence of the bases makes each DNA molecule different from any other DNA molecule.

Labels in figure: guanine, cytosine, adenine, thymine, phosphate, ribose (sugar), nitrogen bases

(a) In the early 1920s, shortly after World
War I, smoking became fashionable for
men. Hypothesize about why lung
cancer rates did not increase until the
1950s.

(b) Suggest a reason why no comparable
increase in lung cancer in women
occurred during the same period.
Explain.

(c) Using what you know or can find out
about the current smoking habits of
Canadians, predict trends in lung
cancer over the next 10 to 20 years.

(d) Compare the trends for males and
females between 1980 and 1995.

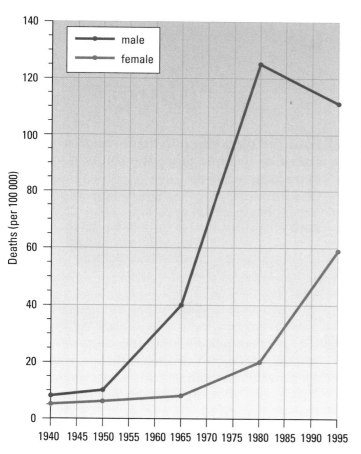

Figure 2

Lung cancer deaths in Canada

Understanding Concepts

1. What is a tumour?

2. Describe the difference between benign and
malignant tumours.

3. Calculate the amount of carcinogenic tar
absorbed by a smoker in one week. The
following data will help you with your
calculations:
- Assume that the smoker smokes
  10 cigarettes a day.
- There are 20 mg of tar in most cigarettes
  (1000 mg = 1 g).
- Approximately 25% of the tar is released in
  the form of smoke or is exhaled. The
  remaining 75% is absorbed by the smoker.

Making Connections

4. The money spent on cancer treatment
continues to grow every year. One politician
has suggested that cancers caused by
inappropriate lifestyle choices should be given
a lower priority for treatment.

(a) Identify cancers that could be reduced by
changing lifestyles.

(b) Comment on the politician's statement.
Should health care money be used first to
treat people who have not contributed to
their problem?

(c) Who would decide who gets treatment
and who doesn't?

Challenge

Statistics is a set of mathematical tools
that can be used to compare data that
have been organized into categories. How
will you use statistics in your survey report
or your display?

Inhibiting Cell Division

Have you ever noticed that very little will grow under a pine tree? Pine trees, like many species of plants, produce chemicals that inhibit the growth of plants that would otherwise compete with them for available nutrients.

Researchers looking for a cancer cure are very interested in these chemicals because they selectively destroy only the most rapidly dividing cells, and cancer cells divide continually and rapidly. Treating cancer with such chemicals is called chemotherapy.

In this investigation, plants are chopped up in a blender to obtain an extract. The extract is poured on germinating seedlings, and root growth is monitored. The root tip is the area of most rapid cell division.

Question

(4A) **1** Write a question for this investigation.

Hypothesis

If plants produce chemicals that inhibit cells that divide rapidly, then they should affect cells near the root tip.

Materials

- scissors
- pine needles (or ragweed or bitter vetch leaves)
- blender
- elastic band
- cheesecloth
- 250-mL beaker
- various germinating seeds, such as canola, tomato, carrot, lettuce, and radish
- 2 petri dishes
- wax pencil
- forceps
- 8 filter papers
- distilled water
- 10-mL graduated cylinder
- clear tape
- hand lens or dissecting microscope

✋ Caution: Scissors are sharp; handle with care. Students who are sensitive to pollen should breathe through a disposable mask.

Procedure

2 Using scissors, cut the needles from a small pine branch and place the needles in a blender. Add 50 mL of distilled water. Do not put the stem in the blender. Ragweed leaves or bitter vetch leaves may be substituted for the pine needles. Place the top on the blender and turn the blender on for approximately 2 min.

Step 2

3 Pour the blender contents through a cheesecloth filter. Store the solution, or extract, in a 250-mL beaker.

Step 3

elastic band —
cheesecloth —

4 Using a wax pencil, label the bottoms (the smaller pieces) of two petri dishes: one C for control; the other E for experimental. Label both petri dishes with the name of the seeds used by the group.

5 Using forceps, place 4 filter papers into each of the petri dishes. Add just enough distilled water to the petri dishes to moisten the filter paper. Replace the lid.

Step 5

6 Count 20 seeds and, using forceps, transfer 10 to each petri dish. Spread the seeds out in each of the petri dishes.

Step 6

7 Using a 10-mL graduated cylinder, add 4 mL of distilled water to the seeds marked C, the control. Make sure the water covers the seeds. Measure 4 mL of pine or other plant extract. Transfer it to the petri dish labelled E, experimental.

(a) Why is it necessary to add 4 mL of distilled water to the control?

(b) Why do you add the distilled water to the control first?

Step 7

8 Keeping each petri dish flat, seal the edge with clear tape and place the dishes in a dark area.

Step 8

(a) Speculate about why the petri dishes are sealed.

9 After 24 h examine the seeds with a hand lens.

(a) Select 3 or 4 seedlings from each petri dish and draw diagrams of the developing roots.

(b) Prepare a data table and record your
6D observations.

10 Repeat step 9 and question (a) after 48 h and 72 h.

11 After 72 h, select 5 seeds from each petri dish and measure the length of the shoots and roots.

(a) Record your observations in your data table.

(b) Calculate the mean length of growth for both the root and the shoot.

Analysis and Communication

12 Analyze your results by answering the following questions:

(a) Which seeds are most affected by the extract?

(b) Why were 10 seeds used for each test rather than one seed?

(c) Why was a control used for the experiment?

(d) Write a summary paragraph to explain how the pine extract could be used as a herbicide to kill weeds.

Making Connections

1. How could a growth-inhibiting substance found in the needles affect root growth?

2. What advantage might natural herbicides have over herbicides created in a laboratory?

3. Explain why scientists involved in cancer research might be interested in chemicals that affect cell division.

4. Ideally, chemotherapy should affect only cancer cells, without injuring normal cells. Why is this difficult to accomplish? (Hint: Remember that cell division is a normal process taking place in the body.)

5. Chemotherapy often causes hair loss. What reason can you suggest for this?

Exploring 4B

6. How does pine extract affect pine seedlings? Design an experiment to find the answer.

7. Design and carry out an investigation to test a variety of plants to determine which produces the best growth-inhibiting extracts.

Regeneration

On a construction site, a saw jumps while cutting a board, severing the finger of a worker. The worker, along with the finger wrapped in ice, is taken to the hospital. In the operating room, blood vessels, bones, muscles, nerves, and skin are reattached. With some luck, the finger will regain most of its original function.

Now imagine a very different scenario: the wound is cleaned and sealed, and a tiny finger begins to regrow. Months later, the tiny finger has grown to almost the same size as the other fingers on the hand. Today this scenario is science fiction. Humans are unable to regenerate fingers. **Regeneration** is the ability to regrow a tissue, an organ, or a part of the body. Tissue regeneration in humans is limited to tissues such as the blood, bone (to repair a fracture), and the outer layer of the skin (to heal a wound). Your liver and kidneys also have a limited ability for regeneration. What factors allow human skin cells, but not nerve cells, to divide and repair themselves?

Regeneration and Specialized Cells

Pass a living sponge through a cheesecloth and the cells that remain alive can join together again as a new sponge. Cut a sponge in two, and two sponges grow. Planarian flatworms can be cut in two, and each piece makes a new, identical flatworm (**Figure 1**). Through cell division, both sponges and planarians can reproduce an entire organism asexually through **fragmentation**.

Animals with little cell specialization, such as sponges and planarians, are capable of regeneration. Cells in the body of a sponge resemble each other. The planarian's cells are only slightly more specialized, and it has almost the same ability to regenerate.

As animals become more complex, greater cell specialization is needed. Cells assume specific tasks, but with greater cell differentiation comes a loss of regenerative capacity. Earthworms cannot reproduce by fragmentation and can only regenerate certain parts of the body.

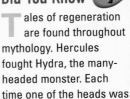

Did You Know?

Tales of regeneration are found throughout mythology. Hercules fought Hydra, the many-headed monster. Each time one of the heads was lopped off, two grew back in its place.

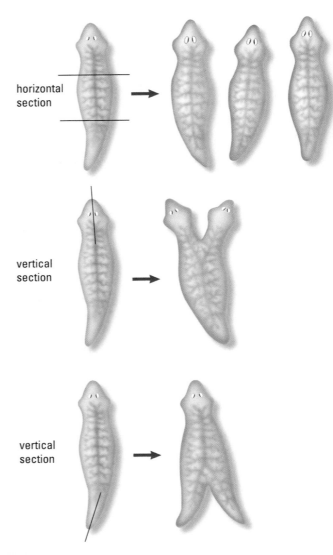

horizontal section

vertical section

vertical section

Figure 1

When cut across, a single planarian can become multiple copies by fragmentation. After an incomplete longitudinal cut through the body, the animal can regenerate an extra head or tail.

Increasingly more complex animals follow the same trend. As more cells become specialized, the ability to grow "replacement parts" is decreased. Salamanders regenerate limbs, and lizards grow severed tails, but neither can be sliced in half and have both parts survive. Why do more complex animals lose their ability to regenerate?

A clue can be found by looking at the wound of a severed limb of a starfish (**Figure 2**). Cells in the area where the limb was severed appear as an irregularly shaped ball. Microscopic examination shows little cell specialization. These cells closely resemble cells of an embryo. They are called stem cells and are able to divide many times and then specialize into cells of bone, cartilage, muscle, or nerve.

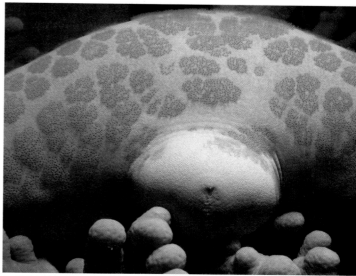

Figure 2

This starfish has lost one of its arms. The new arm will grow where the mass of unspecialized cells (white) has started to divide.

Specialization and DNA

The nucleus of every cell in your body contains a complete code for all of your characteristics and all cellular functions. In other words, every cell contains identical DNA. Why, then, are a skin cell and a nerve cell different? Each of these cells uses different parts of their DNA: they are directed by different genes. Specialized cells use only small amounts of their genetic information. For example, a nerve cell no longer needs the instructions to make muscle protein or to produce skin pigment. The genes that are not needed in a specialized cell are "turned off." Most specialized cells also turn off the genes that would allow them to reproduce.

Salamanders overcome the problem of cell specialization by "turning back the clock." Mature, specialized cells become like immature stem cells. As rapid cell division produces a new limb, the new cells specialize into cells of bone, nerve, and so on.

Stem Cells in Humans

Human skin, bone marrow, and some other tissues contain stem cells. That helps explain how we can easily grow new skin cells, but not new nerve cells. If specialized cells could "turn on" the segments of genetic information that allow them to divide, they could work like stem cells and once again begin dividing. This would mean that nerve cells could be restored following a stroke or spinal cord injury. Recent research suggests this may soon be possible.

Understanding Concepts

1. How does fragmentation differ from regeneration?

2. Why are less complex animals better able to regenerate than more complex animals?

3. What are stem cells?

4. What advantage is gained by changing specialized cells to act like stem cells?

Exploring ③Ⓐ

5. Some hospitals are giving new parents the option of saving some of the cells of their baby's umbilical cord. These are relatively unspecialized cells. Research this development and prepare a brief report on its possible benefits and drawbacks.

6. Research neurophysiologist Dr. Albert Aguayo and his work on the regenerative capacity of nerve cells.

Transplant Farms

Some people require an organ transplant—a kidney, eye cornea, liver, heart—to replace an organ that isn't functioning well. In 1996, approximately 25 000 North Americans received such a transplant. However, more than twice that number remained on waiting lists. One estimate from the United States indicates that 4000 people died that year waiting for an organ.

Organs are taken from donors after accidental death, as long as the organs are still healthy. However, the supply of organs has never come close to meeting demand.

To solve the problems created by the shortage of organ donors, scientists researched two different options. One option is to harvest genetically altered organs from farm animals such as pigs (**Figure 1**). Transplanting a normal pig kidney or liver into a human is almost impossible. The human immune system recognizes the pig cells as foreign invaders and responds by "rejecting" them. White blood cells attack the foreign tissue, and its blood supply is cut off— the organ starves and dies.

This is where the understanding of DNA is vital. Scientists believe that the rejection problem can be solved by transferring human genes into the nuclei of pig embryo cells. As the cells of the pig divide, they copy the

Figure 1

One solution to the problems created by organ shortages is to transplant organs from animals such as pigs. By inserting human genes into the cells of the pig embryo, scientists believed that the pig's organs would not be rejected by the human body.

human DNA as if it were their own. Amazingly, the human genes work in the pig's cells. The cells use the transplanted human genes to provide the information to make proteins, the structural components of cells. And the human genes cause the production of human proteins, not pig proteins, within the pig's cell. The combination (hybrid) cell, containing pig and human genes, has both pig and human cell characteristics.

Once transplanted, the human's immune system no longer identifies these pig cells as "foreign." The human proteins produced by the hybrid cell fool the human's immune system into believing that the organ or tissue is from a human and not a pig.

Issue   ## The Ethics of Transplant Farming (8B)

Statement

Pig farms should be used to grow organs and tissues for transplants.

Point

In some parts of the world, the shortage of healthy organs has created a thriving business opportunity. The illegal purchase of organs, such as kidneys, from the poor of India and South America is well documented. It has been estimated that in 1998, more than 2000 poor people in India sold one of their kidneys. In February 1998, the *New York Times* reported that the FBI had arrested two men in New York City who were selling organs taken from Chinese prisoners who had been executed. Regulated transplant farms would reduce the demand for illegal human organs.

Counterpoint

Transplanting organs from pigs sounds like a good solution, but by making pig cells "somewhat human" another problem is introduced. What happens if a virus, now found only in pigs, enters the human body because it contains hereditary material from pigs? These new hybrid cells could create a gateway through which harmful viruses would move from pigs to humans.

What do you think?

- In your group, discuss the statement, the point, and the counterpoint above. Write down additional points and counterpoints that your group considers.
- Decide whether your group agrees or disagrees with the statement.
- (3A) Search newspapers, a library periodical index, a CD-ROM encyclopedia, and, if available, the Internet for information on this issue.
- (3B) Prepare to defend your group's position in a class discussion.

Producing Plants Without Seeds

Often, a home gardener forgets to remove an onion or a potato from the garden in autumn. The following spring, a new onion or potato plant pushes up through the soil. Onions and potatoes are examples of plants that store food. As the air temperature cools and the sunlight hours decline, these plants become less active. Although the top of the plant dies in the fall, it leaves behind one or more food-containing organs. As spring approaches, the soil warms, and new shoots grow from the food-storage organs (**Figure 1**). These plants are growing without seeds through vegetative reproduction.

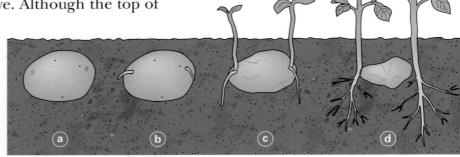

Figure 1
The potato plant dies in the fall, leaving the potato tubers (not roots) in the ground. The tubers sprout again in the spring, producing more plants. This is vegetative reproduction.

The new plant, like all products of asexual reproduction, is genetically identical to its parent. If the parent grew well in the shade or in dry conditions, so will the new plant. For this reason, people who grow plants often prefer to use asexual reproduction instead of seeds to produce new plants. They can be confident that the new cloned plants will have the same desirable characteristics as the parent.

Figure 2 shows several methods to produce plants asexually. Although the diagrams illustrate houseplants, the same methods can also be used to propagate a variety of food-producing plants such as potatoes, herbs, and fruit trees.

Did You Know ?
The word "clone" comes from the Greek word *klon*, meaning a plant cutting or twig.

Try This **Examining a Potato Tuber**

Examine a new potato and one that has been left in a dark cupboard for one month. Identify the eyes of the potato. Describe the differences between the two potatoes. Examine the old potato. Where are the new shoots getting their food?

Caution: Exercise care when using a knife.

Using a knife, cut the potato open and scrape some of the inside cells with a toothpick. Smear the collected cells across a microscope slide.

Put on gloves and safety goggles.

Caution: Iodine irritates eyes and skin and may stain.

Prepare a wet mount by adding a drop of iodine and then a cover slip. Starch turns a blue-black colour in the presence of iodine. Examine the slide under low- and medium-power of a light microscope. 5B

1. Draw diagrams of a few cells, as seen 6C under medium-power magnification.

2. Describe the appearance of the starch granules found in the cells.

SKILLS HANDBOOK: 5B Using the Microscope 6C Scientific Drawing

Figure 2
Summary of asexual reproduction techniques

a Stem cuttings
Recommended for: geraniums, coleus, begonias, dieffenbachia, impatiens, philodendrons, holly, ivy, jade, most vines.

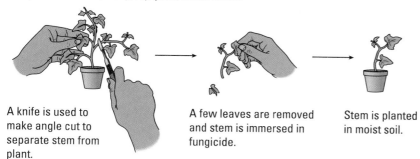

A knife is used to make angle cut to separate stem from plant.

A few leaves are removed and stem is immersed in fungicide.

Stem is planted in moist soil.

b Leaf cuttings
Recommended for: African violets, peperomia, and some begonias.

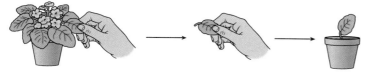

Leaf is removed and immersed in fungicide.

Leaf is placed in moist soil.

c Leaf section cuttings
Recommended for: most begonias, sansevieria.

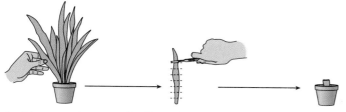

Leaf is removed.

Leaf is cut into sections and immersed in fungicide.

Section is placed in moist soil.

d Root division
Recommended for: most plants with fibrous roots, such as ferns, spider plants, peperomia.

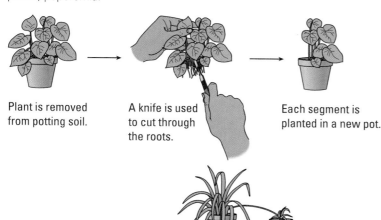

Plant is removed from potting soil.

A knife is used to cut through the roots.

Each segment is planted in a new pot.

e Runners
Recommended for: strawberries, geraniums, raspberry, ivy, spider plants, philodendron.

Runner is placed in new potting soil and kept moist.

Making Connections

1. What advantages does asexual reproduction of plants have for a farmer? Imagine that there is a dramatic climate shift during the growing season. What disadvantage would asexual reproduction have?

Exploring

2. A single apple tree can produce
3A several varieties of apples if twigs from different varieties of trees are grafted onto one root stock (**Figure 3**).

(a) What types of plants are commercially produced by asexual reproductive techniques?

(b) How long have these techniques been used?

(c) Outline some of the economic benefits associated with this technology.

Figure 3

3. Investigate the greenhouse
3A industry in Ontario. What kinds of plants are produced, and how are they propagated?

Reflecting

4. Speculate about how seedless grapes or oranges reproduce.

Challenge

Figure 2 presents information through the use of graphics. How could you use graphics in your display?

Cloning from Plant Cuttings

Some plants naturally reproduce asexually when a portion of the plant, such as a stem or leaf, breaks off and drops to the ground. Roots develop on the broken part and penetrate the soil. The broken part becomes a new plant. Willow trees spread along streams when ice storms break fragments from the original tree.

Cuttings from the seedless-grape vine have been grafted to grape vines all over the world. You wouldn't be able to eat seedless grapes if it were not for cloning. Because no seeds are produced, conventional methods of reproduction are not possible.

In this investigation, you will grow a plant from a cutting, thereby making a clone of the original plant.

Question

1 Write a question that this investigation
4A answers.

Hypothesis

If plants can be cloned, then offspring can be produced by cuttings.

Materials

- lab apron
- knife
- coleus plant (or alternative)
- small beaker or jar
- potting soil
- flower pot (or alternative)
- root stock
- appropriate twig (for grafting)
- grafting tar
- string

🖐 Caution: Exercise care when using a knife.

Procedure
Part 1: Stem Cuttings

2 Using a knife, carefully cut off the tips of three coleus stems. Include two or three leaves on each stem.

Step 2

3 Place each piece into a small beaker containing water. Water must cover as much of the stem as possible without covering any of the leaves. You may need to support the stems to keep them upright.

Step 3

4 Place the beaker in a sunny place and check daily to maintain the water level.

✏ (a) Make and record observations.

5 After the roots appear on the stem, allow an additional week's growth and then transplant the cutting into a flower pot filled with moist potting soil.

Step 5

✏ (a) Record the growth rate of the plant.

SKILLS HANDBOOK: **4A** Asking Questions and Hypothesizing

Part 2: Grafting

6 Use the illustration below as a guide for tree grafting. Grafts must be from a similar species. For example, different varieties of apple trees can be grafted together. Do not attempt to graft the branch of a pear tree with that of an apple tree.

Step 6

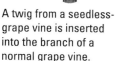

A twig from a seedless-grape vine is inserted into the branch of a normal grape vine.

The grape vine is cut back.

Area is sealed to prevent infection.

✋ Caution: Exercise care if using pruning shears.

7 Using string, tie the graft in place. Encase the graft with grafting tar to prevent the entry of bacteria or fungi.

✏️ (a) Record your observations over several weeks.

Analysis and Communication

8 Analyze your results by answering the following questions:

(a) What evidence suggests that coleus has the ability to regenerate parts of the plant lost to injury?

(b) In what ways would the new coleus resemble the parent plant?

(c) Suggest two simple ways you can prove that the roots from the cutting are growing.

(d) If the grafts did not "take," the grafted branch will die. If your graft took, when did you know the graft was successful? If it did not, suggest what you could do differently that would make success more likely.

9 Summarize your evidence to support or disprove the hypothesis.

Making Connections

1. A hailstorm can shred most plants. After a hailstorm, what advantage would a coleus plant have over plants that cannot reproduce vegetatively?

2. If a McIntosh branch is grafted onto a red delicious apple tree, would the branch continue to produce McIntosh apples?

3. Seedless grapes are genetic mutations. Unlike other grapes, they do not reproduce by seeds. Explain why grafting is used to establish new vineyards of seedless grapes.

4. Explain why someone might want to graft a number of different varieties of pears onto one tree.

Exploring

5. Grow a pineapple clone using the method shown in **Figure 1**.

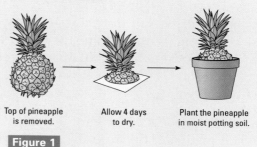

Top of pineapple is removed.

Allow 4 days to dry.

Plant the pineapple in moist potting soil.

Figure 1

6. Research the use of clones for food (3A) production. Outline the positive and negative outcomes of this practice.

🔵 Challenge

Cloning of food plants is a technology that you could present in your display. What would the issue be? Do you think most people realize that the apples, grapes, and oranges they eat come from clones? What other misconceptions might be common?

The information that is included in a question can affect the answers to that question and following questions. How will you factor this into the design of the questions in your survey?

Cloning

When people think of cloning, they often imagine a sinister scientist working in a dungeonlike laboratory, far from the peering eyes of the public. Human clones are often portrayed as robots that are exact duplicates of the original. A person aged 25 years meets his or her 25-year-old clone. Even the scars acquired during childhood are found on the clone. Such images distort science and are actually impossible. Why?

You may be surprised to know that clones have always been with us. **Cloning** is a natural process, repeated daily in nature. The vast majority of organisms on our planet produce exact duplicates of themselves by asexual reproduction, using the process of binary fission. The mother cell divides into two identical daughter cells, or clones, as shown in **Figure 1**.

The parent organism does not always split in two to produce offspring. Some organisms reproduce by budding (**Figure 2**), and others by producing runners (**Figure 3**).

Only One Parent

Cloning is the process of forming identical offspring from a single cell or tissue. Because the clone originates from a single parent, it is genetically identical to that parent. Therefore, cloning is referred to as asexual reproduction.

Figure 1
Paramecia reproduce by means of binary fission. One cell divides to form two identical cells.

Figure 2
The small outgrowth from the parent hydra's body is called a bud. Eventually the bud will break off and form a separate, but identical, organism.

Figure 3
Strawberry plants produce runners that produce new plant clones. Unlike plants that grow from seeds, these plants have the advantage of relying on the parent for nutrients. This gives the clone a head start.

Cloning a Plant from a Single Cell

Frederick Stewart caused great excitement when he revealed a plant he grew from a single carrot cell in 1958 (**Figure 4**). He isolated a group of cells from the area of rapid cell growth near the root tip and placed them in a culture medium with a plant growth hormone that promoted cell division. A mass of cells began to grow. After a few days, parts of the cell mass were transferred into another culture medium that did not contain the growth hormone. Cell division within the new culture slowed, and the cells had time to change shape and start to specialize before their next cell division. Once the cell mass began to specialize into cells that form the root, stem, and leaves, the plant was transferred into soil.

Carrots, ferns, tobacco, petunias, and lettuce respond well to cloning, but grass and legume families do not. No one really knows why. The secret seems to be hidden somewhere in the genetic makeup of the plant. Each cell contains the complete complement of DNA from the parent. Yet some cells specialize, becoming roots, stems, or leaves. Leaf cells use only certain parts of their genetic material, while root cells use other segments. Huge sections of the genetic information remain "turned off" in specialized cells. The trick in cloning plant cells appears to be in delaying specialization.

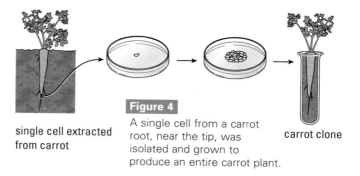

single cell extracted from carrot

Figure 4
A single cell from a carrot root, near the tip, was isolated and grown to produce an entire carrot plant.

carrot clone

Cloning Animals

Animal cloning experiments were also being conducted. Working with the common grass frog, Robert Briggs and Thomas King extracted the nucleus from an unfertilized egg cell by inserting a fine glass tube into the cytoplasm (**Figure 5**). The cell without the nucleus is called an "enucleated" cell.

Next, a nucleus from a frog embryo in the early stages of development was extracted and then inserted into the enucleated cell (**Figure 6**). The egg cell with the transplanted nucleus began to divide, much like any normal fertilized egg cell, and grew into an adult frog. Not surprisingly, the adult frog displayed the characteristics of the frog that donated the transplanted nucleus, not those of the frog that donated the unfertilized egg cell. Analysis proved that the adults were clones of the frog that donated the nucleus.

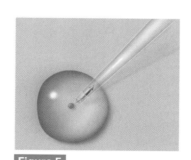

Figure 5
A small glass tube, called a micropipette, is used to remove the nucleus from a cell.

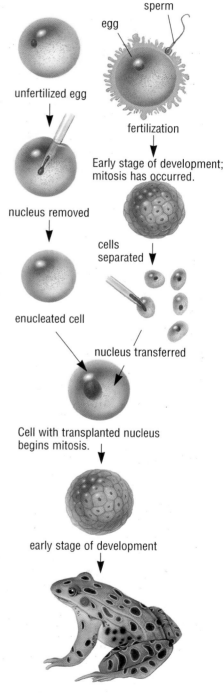

sperm

egg

unfertilized egg

fertilization

nucleus removed

Early stage of development; mitosis has occurred.

enucleated cell

cells separated

nucleus transferred

Cell with transplanted nucleus begins mitosis.

early stage of development

adult frog

Figure 6
The nucleus is transferred from one embryo cell into an egg cell.

However, different results were obtained when the researchers took a nucleus from a cell in a later stage of development. With this nucleus, the cell did not grow and divide. The scientists concluded that something was missing from the nucleus used. The nucleus of a cell in the later stages of development, unlike those of cells of an earlier stage, had begun to specialize. Some regulatory mechanism must have "turned off" some of the genes as the cell began to specialize.

Mammal Clones by Embryo Splitting

Mammalian cells have been cloned using a similar technique to that of frogs. Scientists were successful only when they took the nuclei from embryo cells up to the eight-cell stage of development (**Figure 7**). After the eight-cell stage, the cells began to specialize.

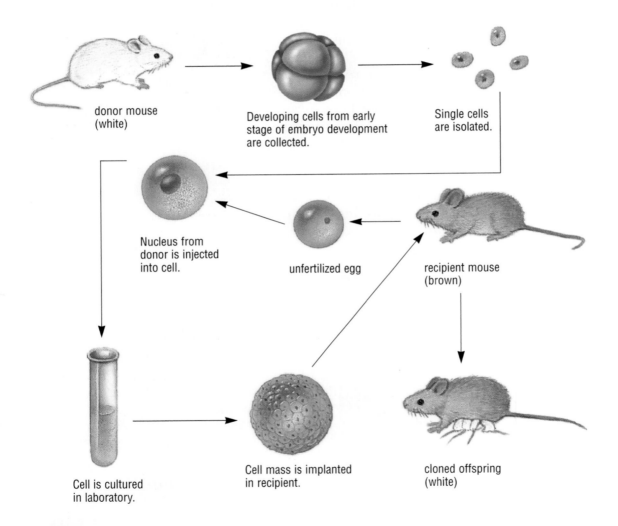

donor mouse
(white)

Developing cells from early
stage of embryo development
are collected.

Single cells
are isolated.

Nucleus from
donor is injected
into cell.

unfertilized egg

recipient mouse
(brown)

Cell is cultured
in laboratory.

Cell mass is implanted
in recipient.

cloned offspring
(white)

Figure 7
The cloned offspring appear identical to the donor parent.

The Dolly Revolution

The scientific world was rocked with the appearance of a sheep named Dolly. Prior to Dolly, scientists believed that cells that had specialized had lost their ability to give rise to new organisms. Dr. Ian Wilmut, of the Rosalind Institute in Scotland, proved the accepted rule wrong by using the genetic material from an adult cell. Some cells from the udder of an adult Finn Dorset sheep were grown in a cell culture and starved for a few days. Then a nucleus was extracted and placed into an enucleated egg cell from a Poll Dorset sheep. Finally, the dividing embryo was placed into the womb of a third sheep. The offspring, named Dolly, looked nothing like the birth mother. She carried the genetic information identical to the Finn Dorset adult—Dolly was a clone (**Figure 8**).

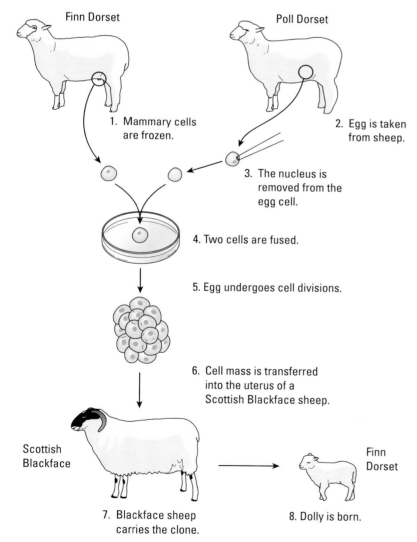

1. Mammary cells are frozen.
2. Egg is taken from sheep.
3. The nucleus is removed from the egg cell.
4. Two cells are fused.
5. Egg undergoes cell divisions.
6. Cell mass is transferred into the uterus of a Scottish Blackface sheep.
7. Blackface sheep carries the clone.
8. Dolly is born.

Finn Dorset

Poll Dorset

Scottish Blackface

Finn Dorset

Figure 8

Dolly could claim three different sheep as mothers. In fact, her genetic mother died before Dolly was born.

Understanding Concepts

1. What is binary fission?
2. How are plants cloned?
3. What is an enucleated cell?
4. Describe how nuclear transplants are used to clone animals, such as frogs.
5. If the nucleus were extracted from a human adult cell and placed into an enucleated egg, how could you distinguish the cloned individual from the original?
6. Dolly was not the first cloned animal, nor was she the first mammal clone. What made her cloning so special?

Making Connections

7. Imagine that farmers were able to easily clone any animal in their herd or flock. What might be the benefits for food production? Would there be any disadvantages? Explain.
8. What ethical issues can you list that relate to cloning? Write a paragraph sharing your own perspective on one of these issues.

Challenge

For a while, Dolly was "hot" in the media. What is hot now? Newspapers, TV news, and magazines may be good sources of ideas for your Challenge.

Movies, TV shows, comic books, and novels may also be sources of misconceptions about reproductive technologies. How do these misconceptions affect how people think about cloning? How would you address this in your display?

Chapter 6 Review

Key Expectations

Throughout this chapter, you have had opportunities to do the following things:

- Explain the importance of DNA replication for cell division and for the survival of an organism. (6.1, 6.5, 6.9)

- Describe representative types of asexual reproduction (6.5, 6.7, 6.8, 6.9)

- Explain cloning in plant and animals as examples of asexual reproduction. (6.7, 6.8, 6.9)

- Identify factors that may lead to cancer-causing mutations, and define cancer as abnormal cell division. (6.2, 6.3)

- Investigate how plants are able to reproduce without seeds, and organize, record, analyze, and communicate results. (6.8)

- Investigate questions related to the causes and treatments of cancer, and organize, record, analyze, and communicate results. (6.3, 6.4)

- Formulate and research questions related to DNA and its role in cell division, and communicate results. (6.1, 6.2, 6.5, 6.6, 6.7, 6.8)

- Provide examples of how developments in reproductive biology have affected plant and animal populations. (6.6, 6.7, 6.9)

- Describe Canadian contributions to research and technological developments in genetics and reproductive biology. (Career Profile)

KEY TERMS

benign	fragmentation
cancer	malignant
carcinogen	mutation
cloning	regeneration
deoxyribonucleic acid (DNA)	tumour

Reflecting

- "The way in which a cell functions and divides is determined by genetic information contained in its nucleus." Reflect on this idea. How does it connect with what you've done in this chapter? (To review, check the sections indicated above.)
- Revise your answers to the questions raised in Getting Started. How has your thinking changed?
- What new questions do you have? How will you answer them?

Understanding Concepts

1. Make a concept map to summarize the material you have studied in this chapter. Start with the words "deoxyribonucleic acid."

2. Why is DNA replication important for the survival of life on Earth?

3. Use **Figure 1** to answer the following questions:

    (a) Describe the process shown.

    (b) Indicate why the process is essential to all living things.

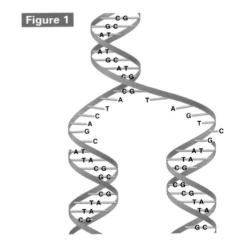

Figure 1

4. Describe a DNA molecule and identify its three chemical components.

5. How does DNA replicate?

6. Briefly describe the process of DNA fingerprinting.

7. Describe two methods that permit the cloning of plants.

8. Describe the process of nuclear transplants during animal cloning.

9. What evidence can you provide that suggests that not all cells divide at the same rate?

10. Explain why a salamander can regenerate a limb but a human can not.

11. What are stem cells?

12. In what ways does a cancerous cell differ from a normal cell?

13. List three factors that cause or contribute to the development of cancer.

14. What changes in lifestyle could reduce the incidence of cancer?

15. Why was the birth of Dolly considered to be such ground-breaking research?

16. Why might a gardener want to graft branches from different trees to a single stem?

17. Extracts from plants, such as pine needles, prevent the growth and division of cells in the root tip of a germinating seed. Explain how plant extracts serve as a model for understanding how chemotherapy drugs work to destroy cancer cells.

Applying Skills

18. The spinal cords of various mice were severed. All of the mice lost control over the lower limbs of their bodies. Researchers transplanted cells from the spinal cord of a mouse embryo into the severed spinal cord of each adult mouse. They noticed that the embryonic nerve cells began to grow in the adult mouse. After a while, some of the muscle control to the lower limbs had returned.

    (a) What was the purpose of the experiment?

    (b) How would you set up a control for the experiment?

    (c) Identify the independent and dependent variables.

    (d) Provide an explanation for the results obtained in the experiment.

    (e) Do you think the experiment was justified? What practical application might this procedure present for society?

19. A scientist treats cells that produce cartilage with a drug that reduces their activity. The cells become less specialized and begin to act like stem cells. The cartilage-producing cells begin to divide at a rapid rate.

    (a) Why would the scientist want to make mature cells behave like cells that aren't specialized?

    (b) What practical application would this experiment have?

20. A manufacturer claims that its sunblock has a rating greater than 15, and therefore provides added protection from skin cancer. How would you set up a controlled experiment to check this claim?

21. A research team studied the growth rate of a type of cancer cell in mice. Every 2 days for 60 days, they counted the number of cells in an area of 1 mm². Which of the following graphs represents their data? Explain your answer.

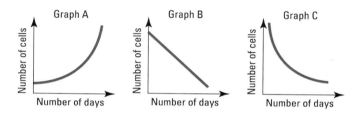

Making Connections

22. In 2105, politicians decide that all food plants will be reproduced by tissue cloning. Predict some of the potential problems.

23. In the movie *Multiplicity*, genetic engineers make many duplicate models of the movie's hero, played by Michael Keaton. A number of comedic situations are created because his wife is unable to distinguish him from the clones. Explain the scientific flaw in the plot.

24. How does society react to the research about cloning? What legal and ethical issues are raised by this research?

25. A project called the human genome project is one of the largest research initiatives ever undertaken. The object of the project is to identify the position of every human gene along each of our 46 chromosomes. Make a list of some possible benefits of this project. How might the research be used?

Sexual Reproduction and the Diversity of Life

Getting Started

1 During the summer months it is rare to find a male aphid. Female aphids reproduce asexually and give birth to female aphids. In fact, the young aphids are born already pregnant.

However, something quite incredible happens in the autumn. Some females become males, and the aphids can then reproduce sexually. What advantages do aphids gain by reproducing two different ways? Why do plants and animals reproduce in so many different ways? Are there advantages to reproducing sexually rather than asexually?

2 In many plants, sexual reproduction occurs in the flower. Unlike humans, the male and female sex cells in many plants are found in the same individual. What are sex cells? Where are the sex cells located within the flower? Where are they found in male and female humans? Do any animals contain both male and female sex cells?

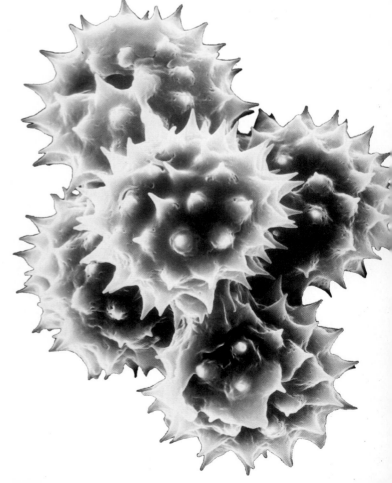

You may have heard of genetic screening. Various procedures are available to expecting parents to discover if a developing fetus has a genetic disorder, or what sex the child will be. Doctors can determine the answers to such questions simply by looking at the chromosomes of one cell. How do they do this? What are they looking for? How do chromosomes influence the development of organisms? What are the ethical implications of such tests?

Reflecting

Think about the questions in ❶, ❷, ❸. What ideas do you already have? What other questions do you have about sexual reproduction? Think about your answers and questions as you read the chapter.

 **Examining Pollen**

Pollen are the male sex cells of the flower. Like the sperm cells of animals, pollen cells fertilize egg cells. Collect pollen from a flower by removing a stamen (the pollen-producing organ). Using a paint brush, sweep some of the pollen into an open petri dish that contains a 10% sucrose solution. Check the pollen grains at the end of the class and 24 h later. You could use pollen from several different flowers, using a separate petri dish for each one.

✋ Caution: Students who are sensitive to pollen should breathe through a disposable mask.

1. Describe what happens to the pollen grains.

2. Draw a diagram of the changes (6C) that you observe.

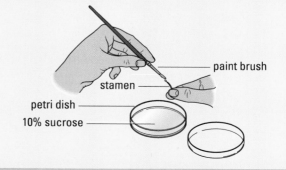

7.1 Case Study

Reproductive Strategies

In previous chapters you studied asexual reproduction. In this chapter you will study sexual reproduction. In general, less complex organisms produce offspring by asexual reproduction, while more complex organisms reproduce sexually (**Figure 1**), although there are many exceptions.

In the simplest form of asexual reproduction, a single cell, the mother cell, duplicates genetic information and becomes two daughter cells, but in all forms of asexual reproduction the offspring have the same genes as the parent. These genes will allow the offspring to meet any environmental challenges as well as the parent did, but it's not likely they would do any better.

reproduction
asexual
sexual
budding
binary fission
spore formation
fragmentation
vegetative reproduction
hermaphroditic
conjugation
separate sexes

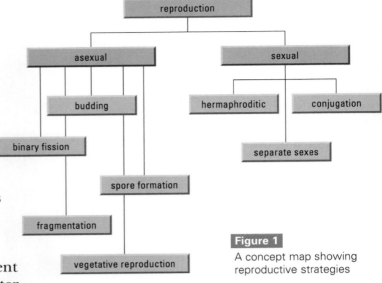

Figure 1
A concept map showing reproductive strategies

(a) In **Figure 2**, which of the cells would you label the mother and which would you label the daughter cells?

(b) In what ways would the new cells that are produced by asexual reproduction resemble the original cell?

Figure 2
Binary fission in bacteria is a form of asexual reproduction.

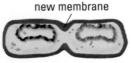

chromosome attachment site
circular chromosome
cell wall
a Bacterium
duplicate chromosome
b Bacterium replicates genetic information.
new membrane
c Circular chromosomes attach to different areas, and new membrane and wall begin to form.
d Cell membrane and cell wall cleave, and two new cells are formed.

Sexual Reproduction

Sexual reproduction is common among multicellular organisms. In sexual reproduction, genetic information from two cells is combined to form the genetic code for a new organism. In complex animals, this usually involves two specialized sex cells, a sperm and an egg, that combine to form a zygote.

Offspring are not identical to either parent, or even to each other. Sexual reproduction produces new combinations of genes that may allow organisms to adapt better to a given environment.

(c) How do the offspring produced by sexual reproduction resemble the original organisms?

(d) How is sexual reproduction different from asexual reproduction?

(e) Explain why the offspring of an organism that reproduces sexually might be able to adapt better to a new environment than offspring of an organism that reproduces asexually.

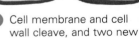

Conjugation

Most of the time bacteria reproduce asexually by binary fission (**Figure 2**), but they also reproduce sexually in a process called conjugation (**Figure 3**). In **conjugation**, two cells come in contact with each other and exchange small pieces, but rarely all, of their genetic information.

If the genetic material that is exchanged includes genes that allow the bacteria to survive in a new environment, both of the cells will now be able to survive in the environment, and so will all the descendants they create through asexual reproduction. Conjugation increases the diversity of bacteria species.

(f) Explain how conjugation provides greater diversity among bacteria.

(g) List the ways in which conjugation differs from what you know about the way humans reproduce.

Hermaphrodites

Sexual reproduction presents special problems for organisms that have restricted movement. Animals and plants that are attached to a spot, such as sponges or tomatoes, will not come in contact with another member of their species. Some burrowing animals, such as worms, may contact only a few other members of their species. If worms were either male or female, as humans are, it would restrict their chance of reproducing. One solution to these problems is hermaphroditic reproduction. An organism that creates both male and female sex cells is referred to as a **hermaphrodite**.

Individual sponges, earthworms, and tomato plants are both male and female. They contain male sex organs that produce sperm and female sex organs that produce eggs. They can reproduce with any other member of their species.

In some cases, two hermaphrodite animals can join together, and each deposits sperm into the other animal, as shown in **Figure 4**. Some animals that live in water simply release their sex cells into the water. This usually happens at a certain time of year, so all the animals of one species release their sex cells at the same time. Land plants, many of which are hermaphrodites, have a similar strategy.

(h) What advantage is gained from fertilizing the eggs of another worm, rather than the worm's own eggs?

(i) Explain how hermaphroditic reproduction helps the earthworm to reproduce.

Figure 3

Conjugation, a form of sexual reproduction, is used by bacteria to exchange genetic information. Unlike human cells, bacteria do not have nuclei. Their DNA floats in the cytoplasm. It usually consists of one large ring and several smaller ones, called plasmids.

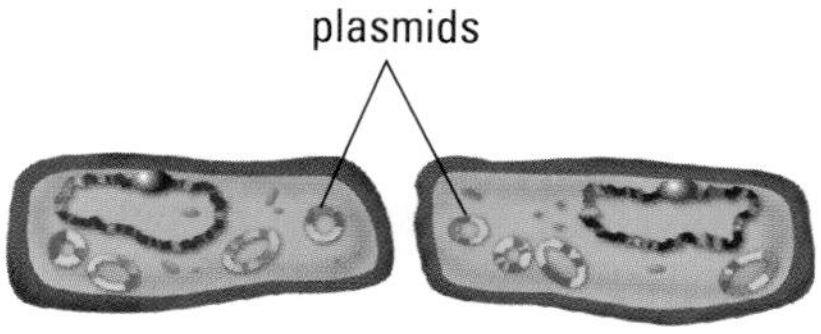

plasmids

a Plasmids are small segments of genetic information.

conjugation

b A bridge is formed between two bacterium cells, and plasmids are exchanged between the cells.

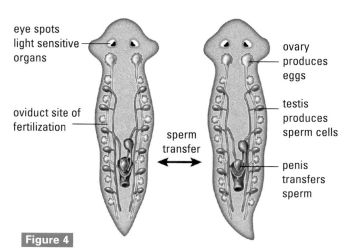

eye spots
light sensitive organs

ovary produces eggs

oviduct site of fertilization

testis produces sperm cells

sperm transfer

penis transfers sperm

Figure 4

Sperm is transferred between two flatworms. Each flatworm acts as a male by releasing sperm and as a female by receiving the sperm to fertilize egg cells. Once fertilized, the egg cells are released inside a capsule.

Separate Sexes

Most complex animals and some plants have separate sexes, males and females. Males produce sperm cells and females produce egg cells.

Of course, if an organism has separate sexes, something has to determine which individuals are males and which are females. For example, in humans this is determined by two chromosomes, called the X and Y (**Figure 5**). Females have a pair of X chromosomes in each cell (one from each parent). Males have a single X chromosome (from their mother) and a much smaller Y chromosome (from their father) in each cell.

Animals with separate sexes use one of two different methods of fertilization. For example, in fish the female releases her egg cells, and the male releases sperm. The sex cells unite outside the female's body. This process is called **external fertilization**.

In most land animals, including humans, the male deposits sperm inside the body of the female, where it fertilizes her eggs. This process is called **internal fertilization**.

(j) How do the reproductive strategies used by humans differ from those used by earthworms?

(k) How do internal and external fertilization differ?

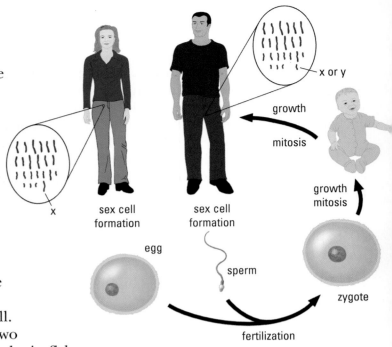

Figure 5

Humans contain organs that produce either egg cells or sperm cells.

Chromosome Number and the Formation of Sex Cells

Human cells contain 46 chromosomes. Imagine what would happen if a human cell containing 46 chromosomes fertilized another cell with 46 chromosomes. The resulting cell would contain 92 chromosomes! If cells with 92 chromosomes united, the following generation would have 184 chromosomes, and so on. For sexual reproduction to occur, there must be a way to reduce the number of chromosomes. This is why sex cells are formed. Sex cell formation differs from normal cell division.

The process that forms sex cells is called **meiosis**. During meiosis the chromosome number is reduced by half. A human cell containing 46 chromosomes undergoes meiosis to produce sex cells that have 23 chromosomes. The 46-chromosome number is referred to as the diploid chromosome number. It is written as $2n$. The 23-chromosome number is referred to as the haploid chromosome number and is given the symbol n.

The union of a haploid sperm cell and a haploid egg cell creates a diploid zygote.

(l) Which cells in **Figure 5** would contain 23 chromosomes? 46 chromosomes?

Aphids: Both Asexual and Sexual

If you look at the underside of new leaves or the buds of plants in the summer, you may observe a mass of tiny green or brown organisms. These are aphids, which survive by sucking the juices from the plant. You may notice large aphids surrounded by little aphids—this is evidence of an unusual life cycle (**Figure 6**). During the summer, female aphids reproduce asexually, giving birth to live female aphids. The newborn aphids contain unfertilized eggs that start developing into female aphids. It's as if summer aphids are born pregnant. This process allows their population to grow explosively if they find a good source of food.

(m) Explain why the offspring in the summer are clones of the mother.

(n) Would it be possible for a female aphid to produce a male aphid by asexual reproduction?

In the fall, when the days get shorter and evenings cooler, some of the female aphids become male. It's not really known how this change occurs, but the change allows the aphids to switch strategies. Now they reproduce sexually, and the females lay eggs that survive over the winter. These offspring, produced through the fertilization of sperm and egg, carry a combination of genetic information from both parents.

(o) What advantage do aphids gain from reproducing sexually?

(p) Speculate as to why aphids reproduce asexually in the summer and sexually in the fall.

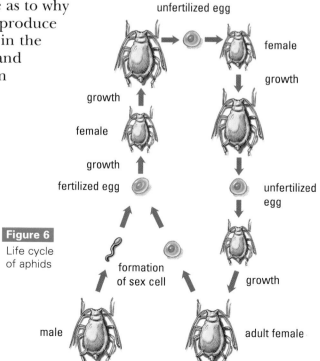

Figure 6
Life cycle of aphids

Understanding Concepts

1. What is a zygote?
2. Describe the process of conjugation.
3. What is hermaphroditic reproduction?
4. Give three examples each of animals that reproduce using
   (a) internal fertilization
   (b) external fertilization
5. In terms of chromosomes, how do female mammals differ from male mammals?
6. A muscle cell from a mouse has 22 chromosomes. How many chromosomes would you expect in
   (a) an unfertilized egg cell?
   (b) a zygote?
   (c) a brain cell?
   (d) a sperm cell?
7. What evidence do you have that suggests that people do not reproduce by conjugation?

Making Connections

8. What advantage is gained by bacteria that can reproduce both asexually and sexually?
9. What is the advantage of hermaphroditic reproduction in
   (a) plants?
   (b) animals?

Meiosis

Organisms that reproduce sexually show a greater range in their characteristics than those that reproduce asexually. Because the male and female sex cells come from different individuals in most species, sexual reproduction ensures a recombination of genes. Offspring carry genetic information from each parent. That may explain why you have thick hair like your father's, while your brother has thin hair like your mother's.

Although you may look more like one parent than the other, you receive the same amount of genetic information from each parent: a set of 23 chromosomes. Your father gives you a chromosome that contains genes that code for the thickness of your hair, but so does your mother. Each of the 23 chromosomes that you receive from your father is matched by one of 23 chromosomes from your mother. Chromosomes that are similar in shape, length, and gene arrangement are called homologous chromosomes (**Figure 1**). Your appearance is determined by the way the genes from your homologous chromosomes interact.

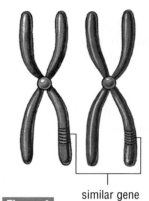

similar gene

Figure 1

Homologous chromosomes carry genes that code for the same trait, in the same position on the chromosome. One of these chromosomes came from the mother, the other from the father.

Stages of Meiosis

Organisms that reproduce sexually contain two types of cells. Cells that reproduce only by normal cell division and mitosis are called **somatic cells**. Skin cells and muscle cells are examples. When these cells divide, each of the daughter cells is identical to the mother cell and has the

Meiosis I

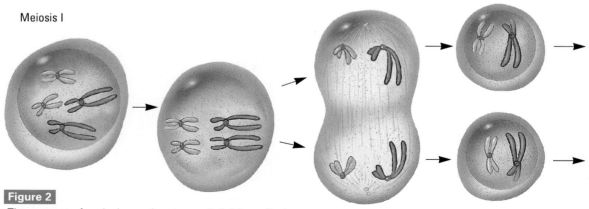

Figure 2

The process of meiosis requires two cell divisions. During the first division, the chromosome number is reduced. During the second division, the chromosomes divide and move to opposite poles.

same number of chromosomes. However, this is not true of the cells that produce sex cells. **Reproductive cells** produce sex cells that contain only half the number of chromosomes through the process of meiosis.

Meiosis involves two cell divisions that produce four haploid cells (**Figure 2**). During the first division, called meiosis I, homologous chromosomes move to opposite poles of the dividing cell. During this division a diploid cell (2n) becomes two haploid cells (n).

In the second phase, meiosis II, the chromosomes are divided.

Try This A Dynamic Model of Meiosis

In Chapter 5, you may have built a model of mitosis. To help you understand the differences between meiosis and mitosis, work with your partner to modify your model so it can show the events of sex cell formation. Be sure that you can show both meiosis I and meiosis II with your model.

Understanding Concepts

1. How do somatic cells and reproductive cells differ from each other? How are they similar?

2. What are homologous chromosomes?

3. Describe the two divisions of meiosis.

4. Use **Figure 3** to compare meiosis and mitosis.

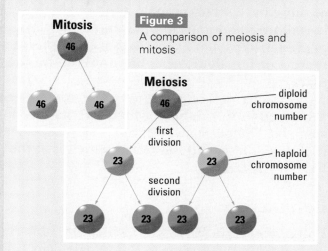

Figure 3
A comparison of meiosis and mitosis

5. Why is meiosis necessary?

6. A dog has 78 chromosomes in each somatic cell. How many chromosomes would you find in each of its sex cells?

7. Do homologous chromosomes have the same number of genes? Explain why or why not.

8. Do homologous chromosomes have identical genes? Support your answer.

Meiosis II

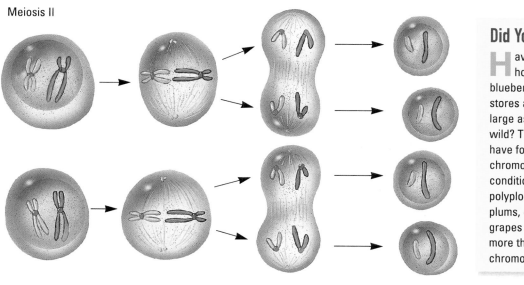

Did You Know

Have you ever noticed how some blueberries in grocery stores are almost twice as large as those found in the wild? These blueberries have four sets of chromosomes. The 4n condition is called polyploidy. Other fruits like plums, cherries, and grapes can also have more than the usual 2n chromosome number.

Reproduction in Flowering Plants

A plant's flowers contain its reproductive cells. Many plants have both male and female sex organs, and hence are hermaphrodites. Some of these have separate male and female flowers, while others have both sex organs in the same flower. A few plants, such as poplar trees, have separate sexes. Poplars have male flowers or female flowers, but not both on the same plant.

The male sex cells of the flower, the **pollen**, are produced in the **anthers**, which are the tips of the **stamens**. Female sex cells, called **eggs**, are located in a structure called the **ovary**, which is at the base of the **pistil**.

Pollination

Pollination is the process by which the pollen is moved from the anther to the female eggs cells and fertilizes those cells. In some plants, the pollen can fertilize the eggs in the same flower. However, in many plants the pollen has to travel to another flower before it can fertilize an egg cell. Wind often carries pollen. Many people are bothered by wind-carried pollen, such as from ragweed or oak. If you have hay fever, it may well be caused by pollen.

Plants that have brilliantly coloured, fragrant flowers, and a sugary nectar attract animals to pollinate them. Most pollinators are insects, but bats and hummingbirds also pollinate flowers. As the animal crawls into the flower to collect the nectar, pollen from the anther falls on its body. When it moves to the next flower in search of more nectar, some of the pollen brushes off onto the pistil. Animals may deposit pollen over a wide area.

Once a pollen grain arrives at the pistil, it may unite with and fertilize an egg.

Seed and Fruit Formation

The fertilized eggs (zygotes) of the flower become the seeds. The petals slowly shrivel and fall from the plant. In some species, the ovary surrounding the zygotes develops into the fruit. Fruits help protect and disperse the seeds.

There are many different types of fruits, from pea pods to apples. Many birds and animals eat fruits and scatter the seeds, often in their droppings. The life cycle of one plant, the tomato, is shown in **Figure 1**.

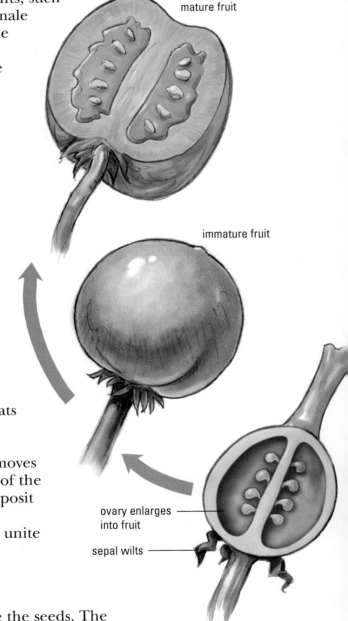

mature fruit

immature fruit

ovary enlarges into fruit

sepal wilts

Figure 1

The life cycle of the tomato plant. After the pollen has fertilized the female eggs, the ovary develops into the fruit.

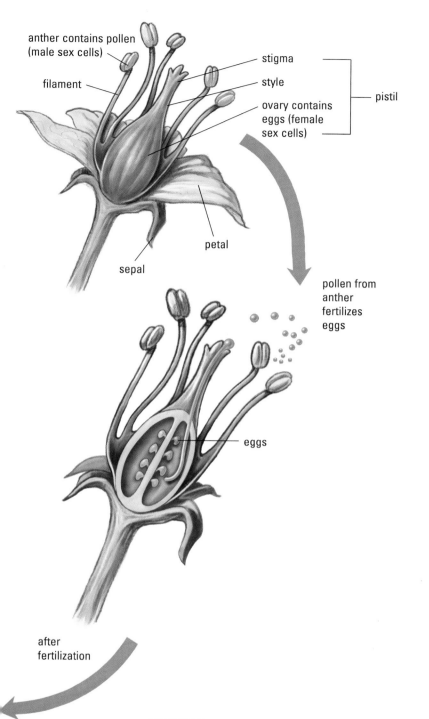

anther contains pollen
(male sex cells)

filament

stigma

style

ovary contains
eggs (female
sex cells)

pistil

petal

sepal

pollen from
anther
fertilizes
eggs

eggs

after
fertilization

Understanding Concepts

1. Many plants are hermaphrodites. Explain.

2. Name the male and female sex cells of the flower.

3. What is pollination? How does it occur?

4. Explain how fruit is formed.

5. Many seeds have specialized adaptations for dispersal. Some seeds have structures that allow them to be carried by the wind. Other seeds can cling to animal fur. Examine the seeds in **Figure 2**. Do you think they are carried by the wind or animals? What special structures do each of the seeds have that help them travel away from the parent?

6. Speculate about how fruit formation helps plants spread their seeds. Use several different fruits as examples.

Exploring

7. Have you ever noticed that hummingbirds are attracted to red trumpet-shaped flowers? The long beak and hovering flight of the hummingbird make it ideally suited for getting nectar from these flowers. In the process, they pollinate the plants. Provide other examples of how flowers and animals have developed special structures to ensure pollination.

Figure 2

How might these seeds be dispersed from the parent plant?

 **a** poppyseed capsule

 **b** burdock

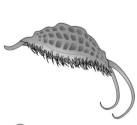 **c** unicorn capsule

 **d** maple

 **e** dandelion

SKILLS MENU
○ Questioning   ● Conducting   ● Analyzing
● Hypothesizing   ● Recording   ● Communicating
○ Planning

Flower Anatomy

In this investigation you will examine different flowers to see which structures they have in common and which are different. You will also look at the functions of the parts of a flower.

Question
How do flowers differ from each other?

Hypothesis

1 Create a hypothesis for this question.

Materials
- whole flowers (various kinds)
- coloured pencils
- small paint brush
- hand lens or dissecting microscope
- tweezers
- small knife

✋ Caution: Exercise caution when using a knife. Students who are sensitive to pollen should breathe through a disposable mask.

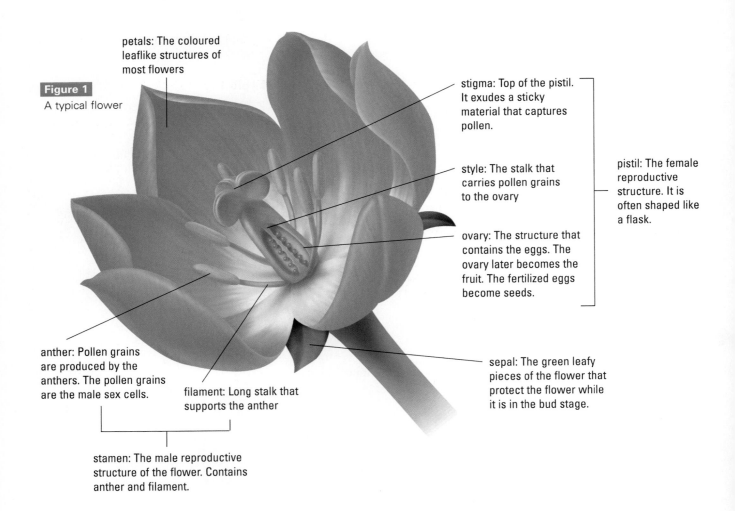

Figure 1
A typical flower

petals: The coloured leaflike structures of most flowers

stigma: Top of the pistil. It exudes a sticky material that captures pollen.

style: The stalk that carries pollen grains to the ovary

ovary: The structure that contains the eggs. The ovary later becomes the fruit. The fertilized eggs become seeds.

pistil: The female reproductive structure. It is often shaped like a flask.

anther: Pollen grains are produced by the anthers. The pollen grains are the male sex cells.

filament: Long stalk that supports the anther

stamen: The male reproductive structure of the flower. Contains anther and filament.

sepal: The green leafy pieces of the flower that protect the flower while it is in the bud stage.

SKILLS HANDBOOK: 6D Creating Data Tables

Procedure

2 Examine **Figure 1** and read the descriptions of the various parts. Not all flowers are alike. Your flowers may differ from the diagram.

3 Obtain two flowers and examine them closely with a hand lens or under a dissecting microscope.

✎ (a) Record the number of petals, sepals, ⑥ⁿ and stamens on each flower. Leave space on your data table for the data from more flowers.

Step 3

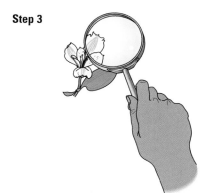

4 Compare the number of petals, sepals, and stamens on your flowers with those of similar and different flowers being examined by other groups.

(a) Do similar flowers have the same number of petals, sepals, and stamens?

(b) Do different flowers have the same number of petals, sepals, and stamens?

5 Remove a few adjoining petals and sepals. Examine the inside of the flower carefully.

(a) How many pistils do you see?

(b) Draw a diagram of your flowers and ⑥ᶜ label the parts. Identify which parts are the female reproductive system and which are the male reproductive system.

Step 5

6 Carefully cut vertically through the pistil of each flower.

(a) Draw a diagram of the inside of the pistil and label the structures.

✋ Caution: Exercise care when using a knife.

Analysis and Communication

7 Compare the flowers you examined by answering the following questions:

(a) Why do you think the flowers were ③ᴬ different? (Hint: Find out what type of flower each is, how it is pollinated, and how its seeds are dispersed.)

(b) Why do you think the similar parts of your flowers were similar?

(c) Did any flowers you examined lack one or more of the parts shown in **Figure 1**?

(d) What happens when the pollen reaches the egg?

(e) In what part of the flower do seeds form?

(f) How do insects and birds help with the process of pollination?

Exploring

1. Bees collect nectar from plants. Predict which part of the flower contains nectar. Give reasons for your prediction.

2. Explain why unripe fruit is often green and bitter, whereas ripe fruit is usually sweet and colourful.

Reproduction of Plants for Food

Imagine your school lunchroom with twice as many people but the same amount of food. The lunchroom would have two problems: the amount of space and the amount of food for each person. The problems facing world agriculture are similar.

Approximately 6 billion people inhabit 145 000 000 km² of planet Earth. Scientists estimate that the population will grow to about 10 billion within 50 years. The most dramatic increases will occur in areas already identified as crowded. Increased food production is essential to feed the extra people, but only about 11% of Earth's land is suitable for growing crops (**Figure 1**). Many experts suggest that technology will probably not be able to increase the area of agricultural land. Unfavourable climatic conditions make tundra areas and deserts inappropriate for growing food.

Figure 1
Of the world's 13.1 billion ha, only about 1.4 billion ha can be used for cropland.

Strategies for Increasing Food Production

Selective Breeding

Plants with desired characteristics are identified and crossbred with other plants that have different desirable characteristics. During **crossbreeding**, pollen is taken from one plant and used to fertilize the eggs of another. Some of the offspring will have the best characteristics of both parents, but others may have the worst characteristics. All the seeds are planted, and the plants showing the desired characteristics are bred again. After several generations of **selective breeding**, all the offspring have the desired characteristics.

Figure 2
Marquis wheat was developed by crossbreeding Red Calcutta and Red Fife wheat.

Sir Charles Saunders, a Canadian scientist, developed Marquis wheat by crossbreeding the Red Calcutta and Red Fife varieties of wheat. Red Fife was once the most common variety of wheat in Canada. Its hardy qualities were well suited for the Canadian prairie, but it matured slowly, and therefore was susceptible to frost damage. Red Calcutta wheat was less hardy and produced poor yields, but developed more quickly. Saunders crossbred the two varieties and selected the offspring that showed the characteristics he wanted. His new variety, Marquis wheat, had the good yields and hardy qualities of Red Fife, but matured approximately two weeks earlier (**Figure 2**).

Cloning

In cloning plants for food, cuttings are taken from a plant with desired characteristics (**Figure 3**). Because cloning is a type of asexual reproduction, all of the offspring are exact duplicates of the parent.

Figure 3

Grafting

Many fruit trees are grown by grafting, rather than planting from seed. Some varieties produce excellent roots but poor fruit, while other varieties have excellent fruit but their roots are susceptible to winter damage. Branches from trees with excellent fruit are grafted onto trees with excellent roots. The resulting tree produces good fruit and survives the winter well.

The Risks of Low Diversity

Historically, humans have used about 700 different species of plants. Today we rely heavily on only about 20 species. Wheat, rice, barley, corn, and sorghum are the most important crops in the world. Although these food species are reproduced sexually, they have been bred carefully so that each plant is very similar to the others, to ensure a good yield.

When large areas of agricultural land are planted with just one variety of food crop, it is called a monoculture. This has been a good strategy for farmers because the variety chosen can be perfect for the soil and the climate. However, because there is little diversity among the plants, few of the plants would be resistant to a new pest or disease, making the crop vulnerable.

What can make this situation even worse is that most land is not well suited for monocultures of cereal grains, such as wheat and barley. The soil requires added nutrients. In some areas, a few seasons after planting, the soil no longer supports the growth of crops. Have long-term needs been sacrificed for short-term gains?

Converting a natural ecosystem to an artificial ecosystem has many implications. Should we drain a marsh to plant rice? Should we irrigate a desert to plant wheat or cotton? Once it has been done, and the native plants have been destroyed, can it be undone?

Understanding Concepts

1. Why is increasing food production such a great concern?

2. How might grafting and selective breeding help increase food production?

3. How do cloning and grafting limit biodiversity?

Making Connections

4. Why might it be dangerous to reduce the number of plants used by people?

5. Many farmers grow a single crop on large areas of farmland. What are the advantages and disadvantages of this practice?

Reflection

6. Many people living in Central and South America have cut down forests to increase cropland (**Figure 4**). Explain why such a practice provides only a short-term solution to food shortages.

Figure 4

The tropical rain forests hold the greatest diversity of life on the planet. Many plants and animals living there have not been "discovered" by scientists. Unfortunately, they are being destroyed as the forests are burned to clear land for crops.

Challenge

Farmers may know more about reproductive technologies for food plants than people who live in cities. Having that knowledge could influence their opinions. Surveys may give inaccurate results if the sample of people is not broad enough. How will you decide who to interview with your survey?

Sex Cell Development in Males

In the last few sections you have been looking at the production of sex cells and the formation of the zygote in plants. Animals also have sex cells, but what do they look like, and where and how are they produced?

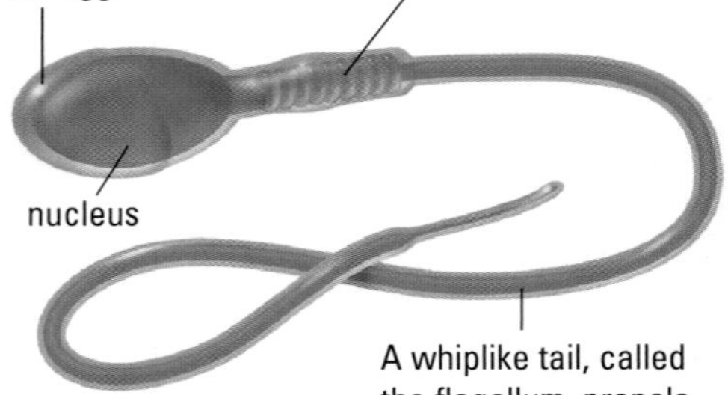

Structure of Sperm

The sperm cell, shown in **Figure 1**, is well designed for its purpose. Built for motion, the sperm cell is streamlined. The sperm cell carries little excess weight; a small amount of cytoplasm surrounds the nucleus. Although reduced cytoplasm is beneficial for a cell that must move, it also presents a problem. Limited cytoplasm means a limited energy reserve.

An entry capsule is found on the head of the sperm cell. The capsule is packed with chemicals that allow the head of the sperm cell to enter the egg.

Energy-producing organelles, the mitochondria, are located in the body, next to the flagellum.

nucleus

A whiplike tail, called the flagellum, propels the sperm cell.

Figure 1
Human sperm cell

Sperm Production and Development

The **testis** (plural testes) is the primary reproductive organ of the male mammal. It is responsible for producing and nourishing sperm as they mature. The insides of the testes are filled with tiny, twisting tubes. These **seminiferous tubules** are lined with reproductive cells that produce sex cells by meiosis (**Figure 2**). Human reproductive cells contain a full complement of 46 chromosomes. Mature human sperm cells contain 23 chromosomes.

As the sperm cells develop from the immature, shapeless form, they begin to grow a flagellum and reduce the amount of cytoplasm inside the cell. As the sperm mature, they move slowly into the epididymis, an organ that lies near the testis (**Figure 2**), where they complete their development.

Sperm are not built to last. Those not released for reproduction die within a few days and are replaced by newer cells from the testes. Scavenging white blood cells scour the epididymis, removing older sperm cells and those that have died.

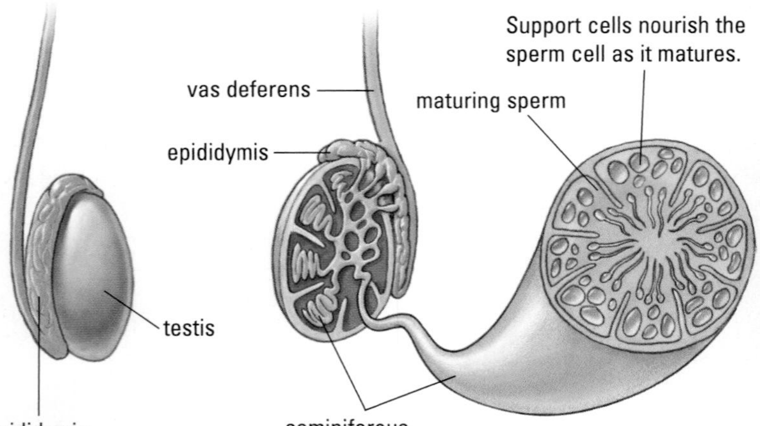

vas deferens

epididymis

Support cells nourish the sperm cell as it matures.

maturing sperm

testis

epididymis

seminiferous tubules

Figure 2
Male sex cells are produced in the testes.

Fertilization

In mammals, fertilization is internal. Each egg is fertilized by one sperm. The sperm uses its store of chemicals to enter the egg, but the egg permits only the head to enter. The body remains outside. As soon as one sperm has entered, the egg puts up a barrier that other sperm cannot break through. Once inside the egg, the sperm's nucleus merges with the nucleus of the egg. In humans, the fertilized egg has 46 chromosomes: 23 from the egg, and 23 from the sperm.

Hormones and Male Sex Cell Production

Up until week seven following fertilization, human male and female embryos are identical in appearance. Then, a chemical messenger—a hormone—is sent from the brain to stimulate the development of sexual structures.

After birth, the reproductive organs of both males and females produce low levels of sex hormones. These sex hormones continue to influence the development of male and female characteristics. However, reproductive organs are not capable of producing mature sex cells until puberty.

Puberty is a period of rapid growth and sexual maturity. It usually takes place in humans between 9 and 15 years of age. During that time, humans display a wide variety of physical changes. In the males, the amount of testosterone, a hormone secreted by the testes, increases. Testosterone stimulates sperm development.

Two other hormones are important in sperm development. Both of these hormones are released by the pituitary gland in the brain. Luteinizing hormone (LH) causes special testes cells to produce testosterone. Follicle stimulating hormone (FSH) acts directly on the reproductive cells in the testes, causing them to divide and produce sperm. FSH also causes the reproductive cells to absorb testosterone. **Figure 3** shows how these hormones interact in a feedback system.

Testosterone secretion and sperm production usually continues for the rest of the male's life. Males over 90 years old have been known to father children.

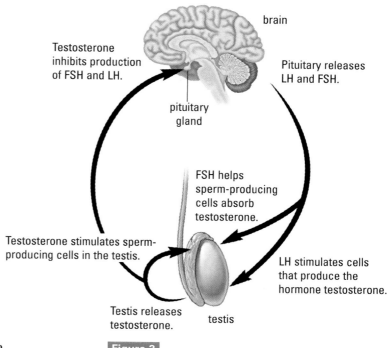

Figure 3

Hormones in the male reproductive system

(Labels in figure: brain; Testosterone inhibits production of FSH and LH.; Pituitary releases LH and FSH.; pituitary gland; FSH helps sperm-producing cells absorb testosterone.; Testosterone stimulates sperm-producing cells in the testis.; LH stimulates cells that produce the hormone testosterone.; Testis releases testosterone.; testis)

Understanding Concepts

1. What features of the sperm cell make it well suited for movement?
2. What are seminiferous tubules?
3. Describe the development of sperm.
4. What are the functions of testosterone?
5. Describe the differences between mature sperm cells and reproductive cells.
6. Why do sperm die only a few days after they are produced?

Sex Cell Development in Females

The female reproductive system in mammals (**Figure 1**) is more complicated than that of the male, because it has an extra responsibility. Not only must the female produce sex cells (eggs), but she must also nurture the embryo once an egg is fertilized.

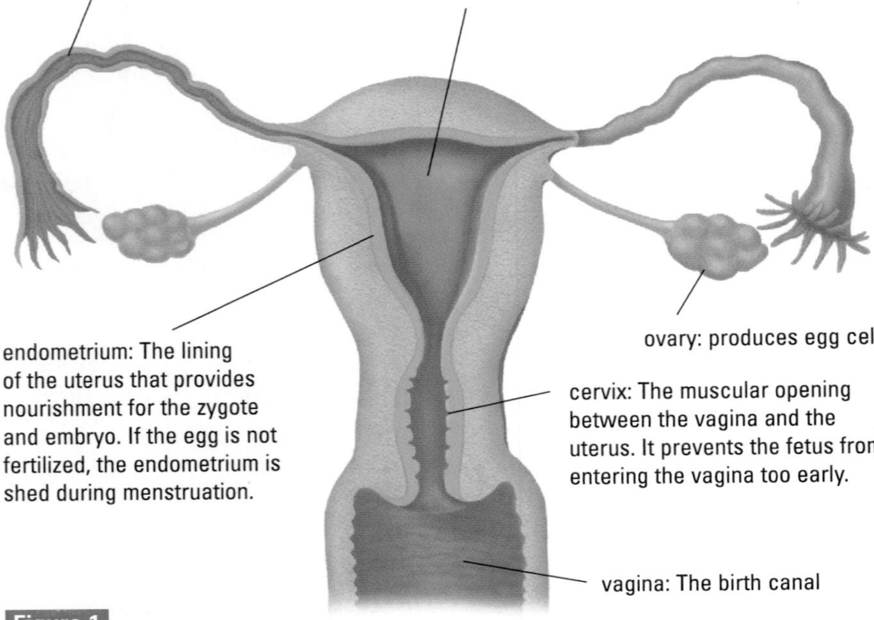

Structure of the Egg

Like the sperm, the egg is well designed for its purpose. The female sex cell is much larger than the sperm, and it is packed with nutrients, so that when it is fertilized it can divide rapidly. It is capable of building a barrier after fertilization to prevent sperm from entering. Unlike males, who manufacture millions of sperm cells every day, human females usually develop only one egg at a time.

oviduct: The site of fertilization. An egg leaves the ovary and travels through the oviduct to the uterus.

uterus (or womb): The organ in which the fertilized egg (zygote) becomes embedded. Within the uterus, the zygote develops into an embryo, which develops into a fetus.

endometrium: The lining of the uterus that provides nourishment for the zygote and embryo. If the egg is not fertilized, the endometrium is shed during menstruation.

ovary: produces egg cells

cervix: The muscular opening between the vagina and the uterus. It prevents the fetus from entering the vagina too early.

vagina: The birth canal

Figure 1

Frontal view of the female reproductive system

Production and Development of the Egg

The primary reproductive organ of the female is the **ovary**. It is responsible for producing and developing egg cells. The ovary contains small groups of cells, called **follicles**. The follicles are composed of two types of cells: a reproductive cell that produces the egg and nutrient-producing cells. The nutrient-producing cells provide energy-rich chemicals to the developing egg.

Unlike the testis, the ovary does not generate new egg cells throughout adult life. The human ovary contains about 400 000 immature follicles at puberty, and the number continually decreases. Hundreds of the follicles begin to develop during each reproductive cycle, but usually only a single follicle is allowed to reach maturity in each cycle (**Figure 2**). The other follicles that had been developing deteriorate and are absorbed into the ovary.

As the follicle develops, the reproductive cell, which contains 46 chromosomes, undergoes meiosis. It forms one large, mature egg cell with 23 chromosomes, plus three tiny cells that disintegrate. Meanwhile, nutrient cells surrounding the egg-producing cell have been dividing to nourish the future egg. The nutrient cells surrounding the egg cell develop into a fluid-filled cavity. When the egg is ready, the ovary wall bursts and the egg cell is released into the **oviduct**. The release process is called **ovulation**.

Nutrient cells that remain within the ovary are transformed into a tissue called the **corpus luteum**. The corpus luteum secretes hormones essential for pregnancy. If pregnancy does not occur, the corpus luteum degenerates after about 10 days, and the cycle that leads to ovulation begins once again.

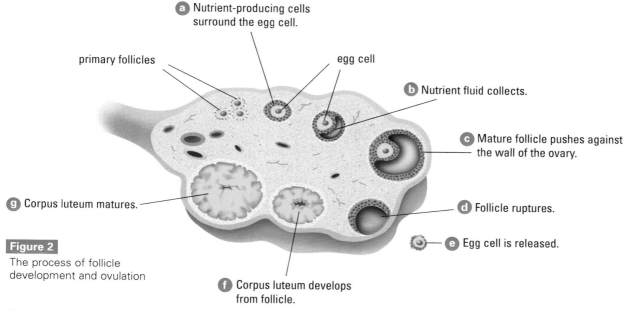

a Nutrient-producing cells surround the egg cell.

primary follicles

egg cell

b Nutrient fluid collects.

c Mature follicle pushes against the wall of the ovary.

g Corpus luteum matures.

d Follicle ruptures.

e Egg cell is released.

Figure 2
The process of follicle development and ovulation

f Corpus luteum develops from follicle.

Pregnancy

The oviduct moves the egg toward the **uterus**, the organ responsible for nourishing an embryo as it grows. If the egg has been fertilized by a sperm cell while moving through the oviduct, the zygote begins to divide by mitosis, becoming an embryo. The embryo embeds in the thick lining of the uterus, called the **endometrium**. The endometrium supports the developing embryo.

If the egg is not fertilized, the endometrium is shed in a process called **menstruation**.

The Role of Hormones

In females at puberty, the pituitary gland in the brain secretes two hormones, FSH and LH, which stimulate the changes that take place in the female's body. During puberty external and internal reproductive organs develop their mature form, and development of follicles can begin.

Estrogen

FSH also triggers the development of follicle cells in the ovary (**Figure 3**). As they develop, they begin to secrete another hormone, called estrogen. Estrogen is released into the blood and carried to the cells of the body. In the uterus, estrogen encourages the endometrium to thicken, in preparation for the zygote. As the follicles develop, estrogen concentrations in the blood increase.

Progesterone

Once the follicle cells begin to develop, LH is released into the blood. LH causes ovulation and the formation of the corpus luteum (**Figure 3**). The corpus luteum secretes both estrogen and another hormone, progesterone. Progesterone continues to stimulate the development of the endometrium and prepares the uterus for an embryo. In addition, progesterone inhibits further ovulation.

If no embryo is embedded in the endometrium, the corpus luteum stops making estrogen and progesterone. The endometrium stops developing and instead begins to break down. Menstruation starts, and a new set of follicles can develop in the beginning of the next reproductive cycle.

Understanding Concepts

1. Where are egg cells produced?
2. What is the site of fertilization?
3. In which structure does the embryo develop?
4. What are the two types of follicle cells found in the ovary? Explain the function of each of the cell types.
5. Describe ovulation.
6. How is the corpus luteum formed? Explain its function.

Making Connections

7. If the ovaries of a woman are removed, would she be able to give birth to a baby? Support your answer.
8. Ectopic pregnancies occur when the embryo becomes implanted in the oviduct rather than the uterus. Speculate about some of the difficulties that would be experienced by such a pregnancy.

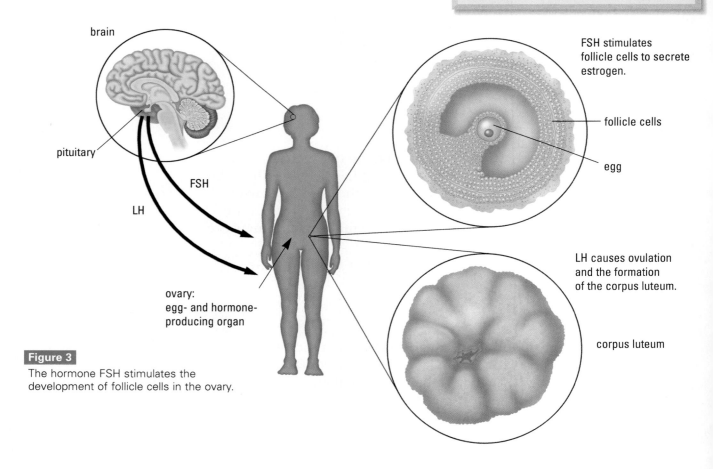

brain

pituitary

FSH

LH

ovary: egg- and hormone-producing organ

FSH stimulates follicle cells to secrete estrogen.

follicle cells

egg

LH causes ovulation and the formation of the corpus luteum.

corpus luteum

Figure 3

The hormone FSH stimulates the development of follicle cells in the ovary.

Fertility Specialist

For more than 30 years, Dr. Albert Yuzpe has been a trailblazer in the field of reproductive medicine. After obtaining an M.Sc., with a major in medical research/endocrinology, from the University of Western Ontario, Dr. Yuzpe trained as a medical doctor specializing in obstetrics and gynecology. He also became a fellow of the Medical Research Council of Canada, focusing on drugs that cause the ovaries to release eggs for fertilization. He helped develop laparoscopy—a kind of exploratory surgery that involves making a small incision in the body and inserting a tiny flexible microscope, or laparoscope. Less invasive (requiring a smaller cut) than other kinds of surgical techniques, it is used to diagnose a wide variety of ailments.

Medicine has enabled me to take my knowledge and experience abroad.

Throughout his career, Dr. Yuzpe has continued to teach and do research. He has been chief of Reproductive Medicine at University Hospital in London, Ontario, and a professor at Western's Faculty of Medicine, where he is now Professor Emeritus. He has taught courses in reproductive medicine and worked as a consultant for the World Health Organization in several countries.

Dr. Yuzpe is cofounder of the Genesis Fertility Centre, Inc., in Vancouver, where he now works with a team of specialists assisting patients with fertility problems. "The most rewarding thing is when people get pregnant," but the most difficult part of his job is to help some couples accept the fact that they simply cannot get pregnant.

Teaching has been a very important part of Dr. Yuzpe's career. His proudest moments come from seeing "the evolution of people I had a hand in training."

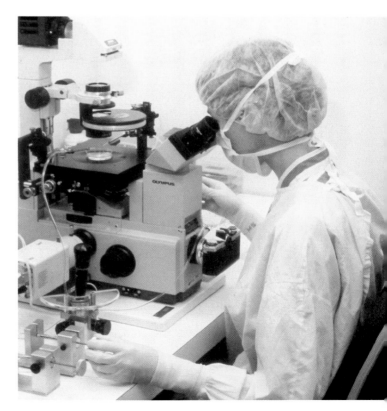

Exploring

1. The training to become a specialized medical doctor is long and demanding. Choose a specialty and research the various stages of training required.

2. Research the development of the laparoscope and its benefits.
   (3A)

3. Why is it important for university teachers to be researchers and vice versa?

Hormones and the Reproductive Cycle

The female reproductive cycle, also called the menstrual cycle, continues through a woman's life from puberty until about the age of 50 years, when no more follicles remain in her ovaries, so no more eggs can develop. This stage of life is referred to as menopause. Until menopause, however, the menstrual cycle (**Table 1** and **Figure 1**) repeats on average every 28 days, although this number varies widely in women, and may change over the life of the woman. The menstrual cycle can be divided into four distinct events: flow phase, follicular phase, ovulation, and a luteal phase.

(a) A man into his 70s or 80s can still produce sperm cells. Would a woman of the same age also be able to produce egg cells? Give your reasons.

(b) If a woman reaches puberty at age 13 and reaches menopause at 50, calculate the number of egg cells she will ovulate. (Assume one egg cell per ovulation and no pregnancies.)

(c) Why would it be impossible for a woman to give birth to that many children?

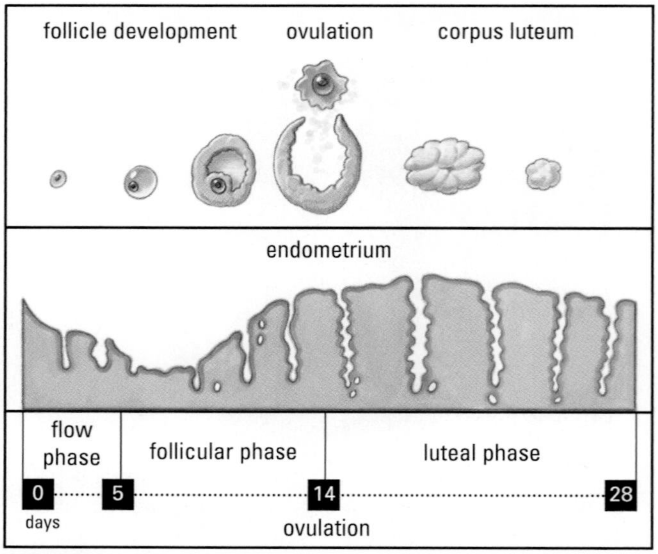

Table 1	Summary Table: Female Menstrual Cycle		
Phase	Description of events	Hormone produced	Days
flow	Menstruation		1–5
follicular	Follicles develop in ovaries. Endometrium is restored.	Estrogen produced by follicle cells	6–13
ovulation	Egg cell bursts from ovary.		14
luteal	Corpus luteum forms, endometrium thickens.	Estrogen and progesterone produced by corpus luteum	15–28

Flow Phase

The flow phase is marked by the shedding of the endometrium, or menstruation. This is the only phase of the menstrual cycle that can be determined externally. For this reason, the flow phase is used to begin the menstrual cycle. It lasts about five days.

(d) Where is the endometrium located?

(e) What is the function of the endometrium?

Follicular Phase

During the follicular phase, the follicles within the ovary develop and begin to secrete the hormone estrogen. As follicles continue to develop, estrogen levels in the blood increase. In a 28-day cycle, the follicular phase normally takes place from days 6 to 13.

(f) In which organ do the follicle cells develop?

Ovulation

Ovulation, the shortest phase of the menstrual cycle, marks the end of the follicular phase. The egg bursts from the ovary and begins to travel through the oviduct to the uterus. About four or five days after ovulation, if the egg has not been fertilized, it dies.

(g) Explain why progesterone levels only begin to rise after ovulation.

Figure 1
This graph summarizes how events in the ovary and uterus are related during the menstrual cycle.

Table 2

Event	Hormone responsible	Days of the cycle
a. follicle is stimulated	?	?
b. endometrium begins to develop	?	?
c. ovulation	?	?
d. endometrium continues to develop	?	?
e. menstruation	?	?

Table 3

Days	Temperature (°C)	
	Ovulation occurs	No ovulation
5	36.4	36.3
10	36.2	35.7
12	36.0	35.8
14	38.4	36.2
16	37.1	36.1
18	36.6	36.0
20	36.8	36.3
22	37.0	36.3
24	37.1	36.4
28	36.6	36.5

The Luteal Phase

The development of the corpus luteum in the follicle marks the beginning of the luteal phase. Estrogen levels begin to decline when the egg leaves the ovary but are restored somewhat when the corpus luteum forms. The corpus luteum secretes both estrogen and progesterone. Progesterone continues to stimulate the endometrium and prepares the uterus for an embryo. In addition, progesterone inhibits further ovulation to prevent another egg being released during pregnancy.

(h) Birth control pills contain high concentrations of progesterone. Explain how they prevent pregnancy.

(i) What might happen if an egg were released into the oviduct during a pregnancy?

Understanding Concepts

1. Copy **Table 2** in your notes and use it to summarize the events that take place in the menstrual cycle.

2. Body temperature was monitored during the menstrual cycle of two women. One ovulated and the other did not.

    (a) Graph the data provided in **Table 3**. Plot
    (7B) changes in temperature along the *y*-axis (vertical axis) and the days of the menstrual cycle along the *x*-axis (horizontal axis).

    (b) Assuming the menstrual cycle is an average 28-day cycle, label ovulation day on the graph.

    (c) Describe changes in temperature prior to and during ovulation.

    (d) Compare body temperatures with and without a functioning corpus luteum.

3. **Figure 2** shows changes in the thickness of the
   (7A) endometrium throughout the menstrual cycle.

    (a) Identify the events that occur at the times labelled X and Z.

    (b) Identify the letter-label in which follicle cells produce estrogen.

    (c) Identify the letter-label in which the corpus luteum produces estrogen and progesterone.

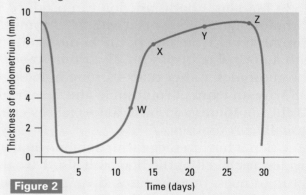

Figure 2

4. Predict the consequences if a 46-chromosome, immature cell were released from the ovary rather than a 23-chromosome egg.

Challenge

If the questions are too personal, an interview about reproductive issues may touch on some pain in a person's life. How can you craft your survey questions so they prompt answers about the issue but also protect your respondents?

Atypical Meiosis

Most processes of the body can go wrong, including cell division. If errors occur during division of a somatic cell, such as a skin cell or a liver cell, it may not harm the organism, which has many other cells. However, if something goes wrong during meiosis in a reproductive cell, the resulting embryo is in serious trouble: all of its cells will be affected.

Nondisjunction (failure to separate) is an error that occurs when two homologous chromosomes move to the same pole during meiosis (**Figure 1**). As a result, one of the daughter cells is missing a chromosome, while the other has an extra chromosome. Cells that lack genetic information or have too much information will not function properly.

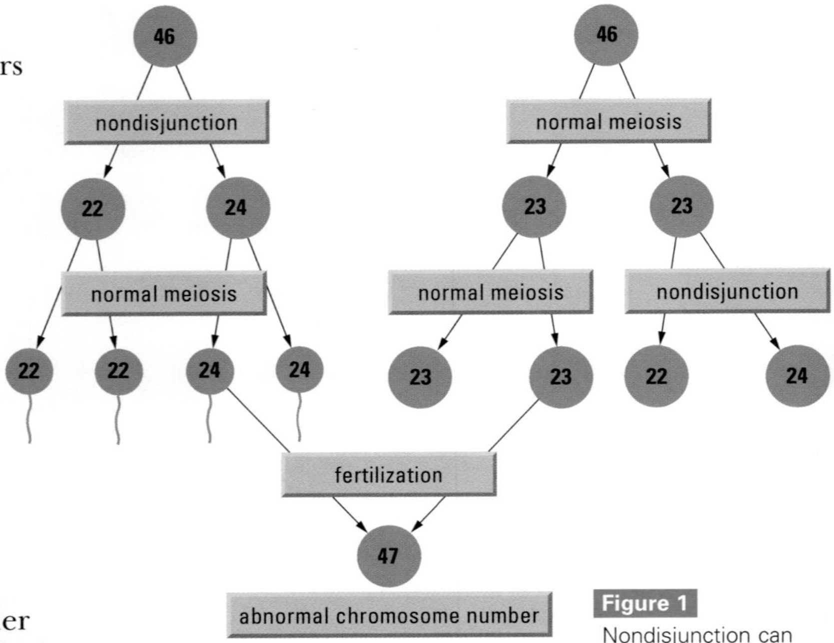

In humans, nondisjunction can produce sex cells with 22 or 24 chromosomes. The sex cell with 24 chromosomes has both chromosomes from one of the homologous pairs. If that sex cell joins with a normal sex cell with 23 chromosomes, the zygote will have 47 chromosomes, rather than 46. One homologous pair of chromosomes will have an extra chromosome and be a triplet. If a sex cell containing 22 chromosomes joins with a normal sex cell, the resulting zygote will have 45 chromosomes.

When a fertilized egg with an abnormal chromosome number divides, each cell of the body will also have the abnormal number of chromosomes. In most cases, the zygote dies, because there is either too much or not enough information in each cell. In a few cases, a person with a nondisjunction disorder is born.

Figure 1
Nondisjunction can occur at meiosis I or meiosis II stage.

Nondisjunction Disorders

Down syndrome is a genetic disorder produced by nondisjunction (**Figures 2** and **3**). Most people with Down syndrome have an extra chromosome for chromosome pair 21—too much genetic information. Children with Down syndrome may have a mental disability, although a wide range of mental abilities is possible.

Turner syndrome occurs when sex chromosomes undergo nondisjunction in either the male or the female parent, producing a female with a single X chromosome instead of two (**Figure 4**). Females

Figure 2
It has been estimated that one baby in 600 is born with Down syndrome. Researchers believe that the nondisjunction occurred when the egg was produced.

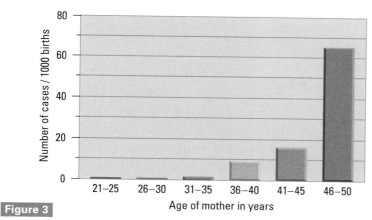

Figure 3

The incidence of birth of babies with Down syndrome increases with the age of the mother.

with Turner syndrome do not mature sexually and are shorter in height. About one in every 10 000 births is a Turner syndrome baby.

Klinefelter syndrome is also caused by nondisjunction in either the sperm or egg. The child inherits two X chromosomes plus a single Y chromosome. The child is a male, however, as he enters puberty, he begins producing high levels of female sex hormones. Males with Klinefelter syndrome cannot father children. This disorder occurs in about one in every 1000 births.

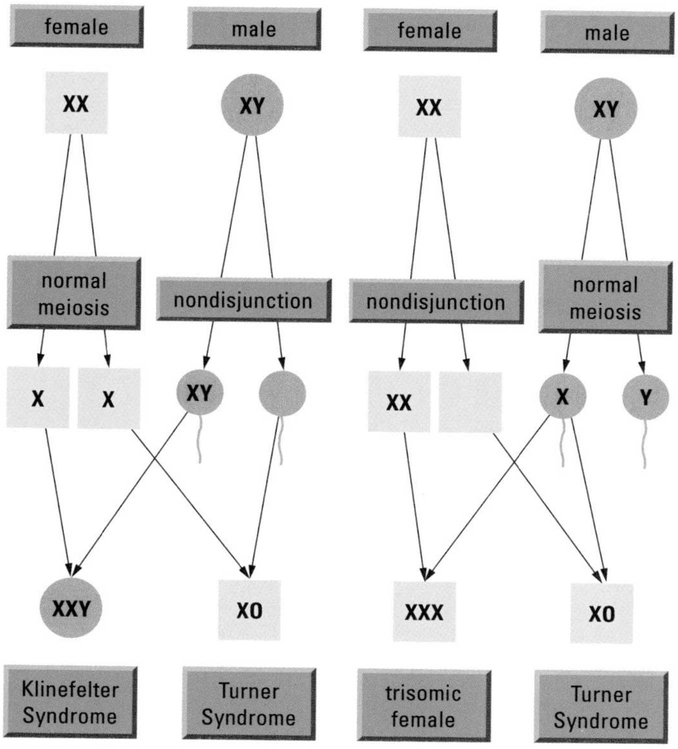

female	male	female	male
XX	XY	XX	XY
normal meiosis	nondisjunction	nondisjunction	normal meiosis
X X	XY	XX	X Y
XXY	XO	XXX	XO
Klinefelter Syndrome	Turner Syndrome	trisomic female	Turner Syndrome

Figure 4

Disjunction disorders that involve sex chromosomes

Understanding Concepts

1. What causes nondisjunction?

2. Draw diagrams to show how nondisjunction could occur during

   (a) meiosis I

   (b) meiosis II

3. Explain how a human cell could have 47 chromosomes.

4. If a zygote contains 45 chromosomes, how many chromosomes would you expect to find in nerve cells as they develop? Give reasons for your answer.

5. Use the diagram of nondisjunction in **Figure 5**.

   (a) How many chromosomes would be found in cells "A" and "B"?

   (b) Which cells in the diagram have a normal diploid chromosome number?

Making Connections

6. Provide a hypothesis that might (4A) explain why older women are much more likely to give birth to a child with Down Syndrome.

Exploring

7. Using library or Internet sources, (3A) research another nondisjunction disorder, such as Patan or Edward syndromes.

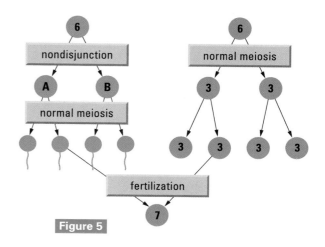

Figure 5

Genetic Screening

Many genetic disorders, such as Down syndrome, can be detected before the baby is born. By using ultrasound (**Figure 1**), physicians can locate the position of the developing fetus in the uterus of the mother. Using a second technique called amniocentesis, the doctor uses a syringe to draw fluid from a sac that surrounds the fetus. The fluid, called amniotic fluid, contains cells from the fetus. By treating the cells with special stains, the tiny chromosomes can be made visible for microscopic examination.

The physician uses a camera mounted on the microscope to take a picture of the chromosomes. The chromosomes in the digital image are identified and arranged in a karyotype chart (**Figure 2**). The karyotype compares the number, size, and shape of homologous chromosomes. A chromosome count of 47 would indicate a nondisjunction disorder. By comparing the homologous chromosomes, the specific disorder can be identified. This whole process is called genetic screening.

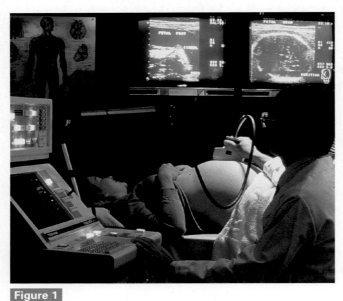

Figure 1

Sound waves are used to create an image of the fetus.

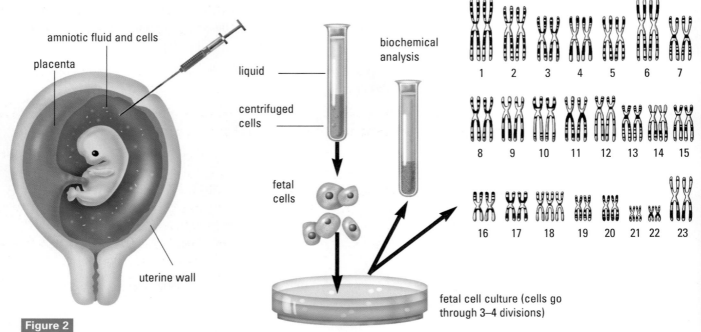

Figure 2

A karyotype is created using fetal cells drawn from the amniotic fluid.

This entire procedure of determining whether a fetus has a genetic disorder, like many other advancements in reproductive technology, raises many ethical questions. Might diagnosing genetic disorders prior to birth lead to pregnancies being terminated? Although most often the screening confirms that the baby does not carry a genetic disease, some disorders are identified. The screening also allows the physician to determine the sex of the baby. Some people fear that the technique could be used to select the sex of the offspring.

 Issue — **Should genetic screening be widely available?** 8B

Statement

Genetic screening before birth should be used on very rare occasions, if at all.

Point

- Diagnosis of genetic disorders before birth could make people consider terminating pregnancies if "imperfect" children were identified.
- In China, ultrasound scanning is used to identify male children. Because the male child has been preferred and because of state restrictions on the number of offspring, many pregnancies have been terminated because the mother carried a female child.

Counterpoint

- In most cases, genetic screening confirms that the baby is all right. This reduces worry about the pregnancy. In addition, identifying a disorder doesn't mean that the pregnancy will be terminated. It may mean that the parents are better prepared.
- Screening for gender preference has not been a problem in Canada. In Canada, there is no legal limit on the number of offspring.

What do you think?

- In your group, discuss the statement and the points and counterpoints above. Write down additional points and counterpoints that your group considered.
- Decide whether your group agrees or disagrees with the statement.
- Search newspapers, a library periodical index, a CD-ROM 3A encyclopedia, and, if available, the Internet for information on genetic screening.
- Prepare to defend your group's position in a class discussion. 3B

Human Karyotypes

Scientists are still working to discover the causes of nondisjunction disorders. Irene Uchida, a Canadian scientist then working at the Children's Hospital in Winnipeg, did research that indicates that radiation may have an effect. Her study suggested that women who were exposed to X rays were more likely to have a child with a nondisjunction disorder. This finding may explain why the incidence of nondisjunction disorders increases according to the age of the mother. The older a woman is, the more likely it is that she will have been exposed to radiation.

Dr. Uchida used karyotype charts in her work. A karyotype chart is a picture of the chromosomes from a cell, arranged in homologous pairs. The chart can be used to diagnose nondisjunction disorders.

In this activity you will analyze a karyotype chart and create your own chart. Then you will diagnose the nondisjunction disorder of the cell described in the chart.

Materials
- human karyotype plate
- scissors
- blank paper
- transparent tape

Procedure

1 Examine the karyotype chart in **Figure 1**.

(a) Which characteristics are used to arrange the chromosomes in pairs?

(b) In what ways does chromosome pair 2 differ from pair 20?

(c) What is the sex of the individual whose chromosomes are shown in the chart?

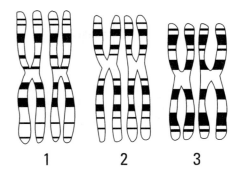

1   2   3

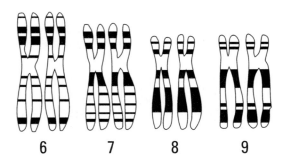

6   7   8   9

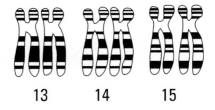

13   14   15

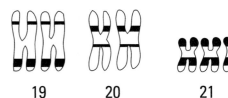

Figure 1
A karyotype chart

19   20   21

2 Obtain a karyotype chart from your teacher and carefully cut out each of the chromosomes.

3 Match the paired chromosomes and prepare a karyotype chart by taping the chromosomes to the paper. Use the karyotype chart in **Figure 1** for reference.

(a) How many chromosomes did you find?

(b) What is the sex of the individual?

4 Study **Table 1** and **Figure 1**.

(a) Using the information shown in the karyotype chart, diagnose the nondisjunction disorder.

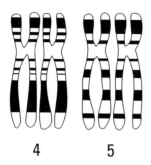

4   5

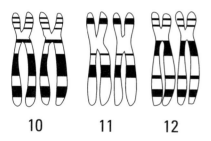

10   11   12

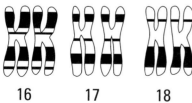

16   17   18

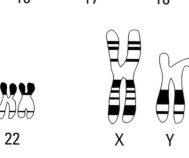

22   X   Y

Table 1		
Chromosome abnormality	**Syndrome**	**Effect**
XO	Turner	sterile female, 45 chromosomes, short stature
XXY	Klinefelter	sterile male, 47 chromosomes, often long arms and legs
three #21	Down	47 chromosomes, distinctive facial features

Understanding Concepts

1. How could a karyotype be used to identify gender?

2. Radiation can cause chromosomes to break apart. Would karyotyping be useful in identifying X-ray damage? Explain how.

3. Would the diagnosis of Turner syndrome in a single cell mean that every cell of the body contained 45 chromosomes? Explain your answer.

4. Is it possible for two people who have Down syndrome to have a child that doesn't have the syndrome? Explain your answer.

5. Would it be possible to produce a baby with 48 chromosomes? Explain your answer.

Making Connections

6. More males than females suffer from colour blindness. Speculate why females with Turner syndrome have a similar incidence of colour blindness as males do.

Chapter 7 Review

Key Expectations

Throughout this chapter, you have had opportunities to do the following things:

- Describe and give examples of types of sexual reproduction. (7.1, 7.3)

- Compare sexual and asexual reproduction. (7.1)

- Describe the production, structure, and function of sex cells. (7.3, 7.4, 7.6, 7.7)

- Describe how hormones influence development of sperm. (7.6)

- Describe how hormones influence development of the eggs and ovulation. (7.7, 7.8)

- Investigate the reproductive structures of flowering plants, and organize, record, analyze and communicate results. (7.4)

- Formulate and research questions related to reproductive technologies, and communicate results. (7.9, 7.10)

- Provide examples of how developments in reproductive biology have affected the identification of genetic disorders, and discuss the social implications of these developments. (7.9, 7.10, 7.11)

- Provide examples of how developments in reproductive biology have affected food production. (7.5)

- Describe Canadian contributions to research and technological developments in genetics and reproductive biology. (7.5, 7.11, Career Profile)

- Explore careers that require an understanding of reproductive biology. (Career Profile)

KEY TERMS

anther	oviduct
conjugation	ovulation
corpus luteum	pistil
crossbreeding	pollen
egg	pollination
endometrium	reproductive cells
external fertilization	selective breeding
follicle	seminiferous tubules
hermaphrodite	somatic cells
hormone	sperm
internal fertilization	stamen
meiosis	testis
menstruation	uterus
ovary	

Reflecting

- "Sexual reproduction creates a diversity of species." Reflect on this idea. How does it connect with what you've done in this chapter? (To review, check the sections indicated above.)
- Revise your answers to the questions raised in Getting Started. How has your thinking changed?
- What new questions do you have? How will you answer them?

Understanding Concepts

1. Make a concept map to summarize the material you have studied in this chapter. Start with the words "sexual reproduction."

2. Using the life cycle of aphids, distinguish between sexual and asexual reproduction.

3. Explain how sexual reproduction provides genetic diversity.

4. Classify the following as either sexual or asexual reproduction:
   (a) A small piece of a cactus breaks off the original plant, falls to the ground, and begins to grow.
   (b) Pollen from a male poplar tree fertilizes sex cells on a female poplar tree.
   (c) Two earthworms each produce sperm and eggs and fertilize each other. Eggs are laid.
   (d) A flatworm is cut in half and grows into two flatworms.

5. How can two brothers with the same parents have different hair colour?

6. Compare meiosis with mitosis. Copy **Table 1** in your notebook and place a check mark in the correct box.

Table 1

Description	Meiosis	Mitosis
produces four cells	?	?
daughter cells are clones of mother cell	?	?
haploid cells produced	?	?
skin cells undergo this cell division	?	?
formation of sex cells	?	?
two stages of cell division	?	?

7. Describe how pollen cells fertilize egg cells in a flower.

8. Draw the flower shown and label the parts. Indicate which are male or female parts and which are neither.

9. Draw a diagram that shows how a child could inherit three number 21 chromosomes.

Applying Skills

10. Fruit flies normally have 8 chromosomes. **Figure 1** shows the events of meiosis.
    (a) In which parent did nondisjunction take place?
    (b) How many chromosomes would be found in zygotes D, E, and F?
    (c) What is happening during process X?
    (d) Which zygote would most likely be healthy?

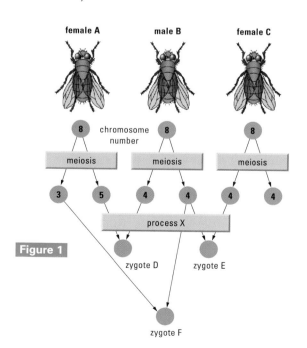

Figure 1

11. Use **Figure 2** to answer this question.
    (a) How many chromosomes were in the sperm cell?
    (b) Explain how this sperm cell could be produced from a cell that had 46 chromosomes.
    (c) How many chromosomes would be found in each cell following mitosis?

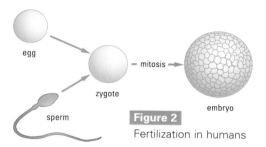

Figure 2
Fertilization in humans

Making Connections

12. Use **Figure 3** to answer the following questions:
    (a) At which stage would you find a haploid chromosome number?
    (b) What occurs at stage 3?
    (c) Would the change that occurs between stages 3 and 4 occur because of mitosis or meiosis? Explain why.
    (d) How many homologous pairs of chromosomes would be found in the baby if it was a female?

Figure 3

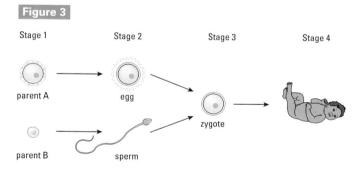

13. What advantage does cross-pollination provide flowering plants?

14. A male has cancer in one testis and has to have it removed. Speculate about how this will affect his ability to father children.

15. If a couple is planning to have a baby, on which days of a 28-day menstrual cycle is the female most fertile?

Getting Started

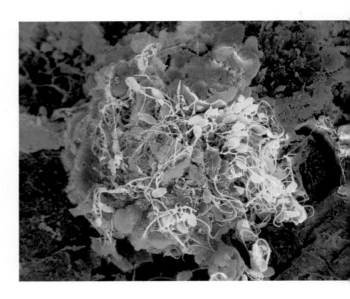

1 Only one egg cell and one sperm cell are needed to form a zygote. In fact, as soon as the egg is fertilized, it forms a barrier to keep all other sperm out. But both plant and animal males produce millions of sperm cells. Why are so many produced if only a single sperm is required?

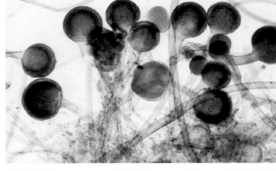

2 Organisms develop in so many different ways. Look at the photographs and list some of the strategies for development that are represented. What are the advantages of each strategy? Can you identify any possible disadvantages? What are they?

3 The female sea turtle comes ashore once every two or three years to nest and lay her eggs. Some species travel as much as 3000 km from their birthplace to their feeding grounds. Then, they return to the beach where they were born. They dig holes, lay their eggs, cover them up with sand, and return to the ocean, abandoning the eggs. Why do the turtles travel so far to reproduce? Why do they lay so many eggs? The sea turtle produces many eggs, while humans usually produce a single offspring. What are the advantages of each strategy?

Reflecting

Think about the questions in **1**, **2**, **3**. What ideas do you already have? What other questions do you have about zygotes and how they develop? Think about your answers and questions as you read the chapter.

 **Observing the Growth and Development of Brine Shrimp**

Brine shrimp, related to shrimps and lobsters, have a simple life cycle. Eggs hatch into larvae. As they develop to adults, they shed their outer skeleton approximately 12 times. With each shedding, the brine shrimp not only get larger but also add body segments. Close examination will reveal that the number of hairs on their legs increases as they age.

Place 100 mL of a 4% salt solution into a 250-mL beaker. Add approximately 0.1 g of shrimp eggs to the water. Using a hand lens, examine the water for brine shrimp eggs each day until they hatch.

1. Describe the appearance of the eggs.

2. Construct a data table. How many days did it take for the eggs to hatch?

3. Describe the appearance of the brine shrimp as they develop.

 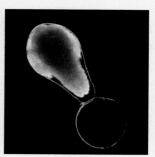

Survival and Development of Organisms

For a species to survive, its offspring must survive. Different organisms use very different strategies and combinations of strategies to ensure the survival of their offspring (**Figure 1**). Nevertheless, some general categories can be identified.

Key Strategies for the Survival of Offspring

- Some organisms produce a zygote that remains in suspended animation (i.e., does not develop) until environmental conditions become favourable. Water, food, and/or warmth can activate growth and development.

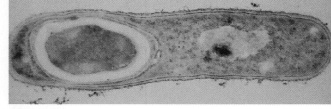

Figure 1

strategies for growth and development of offspring

spore | seed | egg | young develop in womb

marsupial mammal | placental mammal

- Some organisms wrap their developing zygote—called an **embryo**—in a food package. This food supports immediate development, making the organisms less dependent upon environmental factors.
- In some organisms, the embryos develop within the adult organism, protected from unfavourable environmental conditions.
- The adults of some organisms protect and nourish their offspring after birth.

Spores

A **spore** is a reproductive body encased within a protective shell. If environmental conditions become unfavourable for life, the organism may produce spores that enter a state of suspended animation. Once conditions become favourable again, the spores germinate, and the organisms enter a growth phase.

Many different bacteria form spores, as shown in **Figures 2** and **3**, with each cell forming a single spore. The spore contains genetic material surrounded by a tough cell wall, which is resistant to heat, drying, boiling, and radiation. When conditions improve, the wall breaks down and an active bacterium emerges.

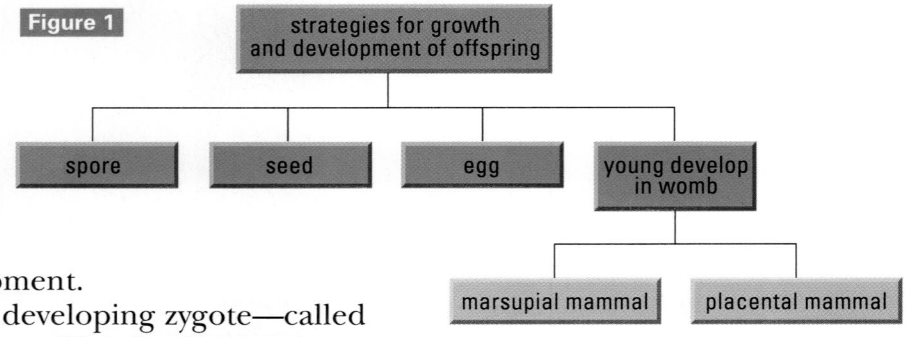

Figure 2
An electron microscope photo of a spore inside a bacterial cell

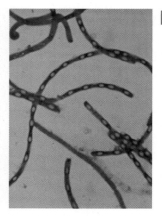

Figure 3
Anthrax is a disease caused by the spore-forming bacteria, *Bacillus anthracis*. Each of the rodlike filaments is a separate organism. The white, oval disks that appear in each of the filaments is a spore. The deadly disease affects sheep, cattle, and people. Spores can lie on the ground for many years, becoming active only when they enter an animal.

Fungi reproduce by way of tiny spores that are easily dispersed by wind. Once environmental conditions become favourable, cells in the spore begin to divide and grow.

A spore from the fungus that grows into bread mould can sit on a counter for months. If the spore finally lands on a food source, such as a sandwich, it germinates. If the sandwich sits in a locker for a few weeks, the conditions are ideal for growth. The warmth and moisture cause the fungus to grow into thousands of threadlike structures. Each thread develops a spore case, which can release thousands of spores (**Figure 4**).

Moss and fern plants also reproduce by way of spores (**Figures 5 and 6**). The spores grow into structures that produce male or female sex cells.

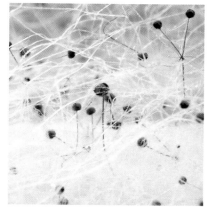

Figure 4
The black spots on the bread mould are spore cases.

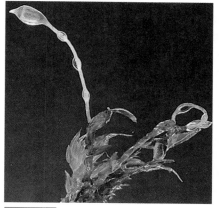

Figure 5
The orange structure is a ripe spore case of a moss plant.

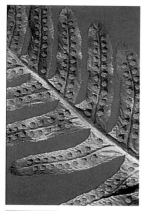

Figure 6
Spore cases are the brown structures found under the leaves (fronds) of fern plants.

Seeds

A seed contains the plant embryo wrapped in a protective package that contains food. Unlike spores, which must wait until environmental conditions become favourable, seeds bring nutrients to their environment and can get a head start on growth.

Needle-producing trees, such as pine, spruce, and fir, have exposed seeds in a conelike structure, as seen in **Figure 7**. Because the seeds are not protected, they are often described as "naked seed coats." Flowering plant seeds are enclosed and protected inside a fruit, formed by the ovary of the flower (**Figure 8**).

Figure 7
These pine seeds are not protected.

Figure 8
The apple seeds are protected within the fruit.

Eggs

The vast majority of animals lay eggs. An egg includes a zygote and some food, plus some mechanism for protection (e.g., a shell or a jellylike substance). Some organisms fertilize their eggs internally (e.g., insects, birds) before a tough shell is secreted. Others, usually aquatic animals (e.g., fish, frogs), fertilize the eggs externally.

The tapeworm, a parasite that lives in the intestines, produces thousands of eggs in each egg case (**Figure 9**). As an infected animal defecates, some of the egg cases break and eggs are released. A resistant coating prevents the eggs from drying up and sticks to blades of grass or other plant material. If another animal eats the grass, it swallows the eggs. In the intestines, they develop into adult tapeworms that can grow up to 20 m long.

For the tapeworm, the strategy of laying a large number of eggs is important, because only a few of the eggs will actually find their way into the body of another animal. Many other animals produce thousands of eggs for this same reason: few of their offspring survive to adulthood. Most of the others are eaten.

Reptile and bird eggs are especially well adapted for surviving on land. Their "amniotic" egg, shown in **Figure 10**, provides a self-contained environment for the developing embryo.

The spiny anteater and duckbill platypus are the only mammals that lay eggs (**Figure 11**). Like birds, they incubate the eggs outside the body. However, they nourish the young from mammary glands that secrete milk directly onto the fur.

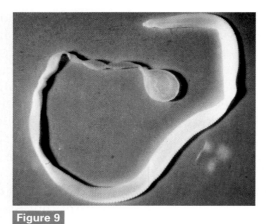

Figure 9

Each egg case produced by this tapeworm will contain thousands of eggs.

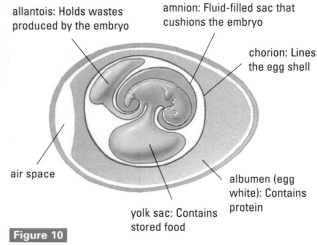

allantois: Holds wastes produced by the embryo

amnion: Fluid-filled sac that cushions the embryo

chorion: Lines the egg shell

air space

yolk sac: Contains stored food

albumen (egg white): Contains protein

Figure 10

The amniotic egg is a protective, self-contained environment.

Figure 11

The spiny anteater (left) and the duckbill platypus (right) are mammals that lay eggs.

Marsupial Mammals

Marsupials (e.g., kangaroos, koalas, and opossums) are born tiny and immature (**Figure 12**). After emerging from the uterus, they crawl into a pouch and attach to the nipple

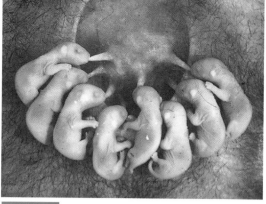

Figure 12

Opossums are marsupial mammals. Inside their mother's pouch, each of the young has its own nipple.

of a mammary gland. Once they have grown too big to be carried, the offspring leave the pouch but may return for additional milk as long as the mother allows.

Placental Mammals

Most mammals, including humans, belong to a group called placentals (**Figure 13**). Their young develop in the uterus or womb. After the embryo implants in the endometrium, blood vessels from the mother and embryo grow side-by-side and form an organ called the **placenta**. Although the two blood supplies remain separate, oxygen and nutrients diffuse from the mother's blood into the embryo's blood. Wastes from the embryo diffuse into the mother's blood and are carried away to be excreted. The umbilical cord connects the embryo and the placenta.

In general, the offspring of placental mammals develop more slowly than the young of other animals and require greater long-term care. The mother provides the young with milk and protects them from danger.

Figure 13

Wildebeest are placental mammals. A young wildebeest must begin running shortly after it is born. If it doesn't, it might fall prey to hyenas.

Understanding Concepts

1. Identify an organism that uses each of the following strategies to ensure survival of their offspring. Describe the strategy more fully and explain why it is used.

   (a) The zygote remains in suspended animation until environmental conditions become favourable.

   (b) The zygote is wrapped in a food package.

   (c) The young develop within the adult organism.

   (d) The adult protects and nourishes the offspring after birth.

2. Identify three organisms each that produce spores, seeds, and eggs.

3. Compare reproduction by spores and seeds by listing the advantages and disadvantages of each strategy.

4. Compare the development of the zygotes of a bird, a frog, and a human, by listing the advantages and disadvantages of each strategy.

Making Connections

5. In Australia, the introduction of (3A) placental mammals has often caused problems for resident marsupial mammals. Research the impact on the ecosystem and hypothesize about why marsupial mammals aren't able to compete with placental mammals that have similar needs.

Challenge

How does changing the way animals reproduce affect the other species in the ecosystem? Consider this question as you think about your Challenge.

Examining Plant Embryos

One successful strategy for reproduction is the production of seeds. A seed is actually an embryo with packaged food. The food, usually in the form of starch, provides nourishment for the early growth of the young plant until it begins to manufacture its own food. The food in seeds also feeds humans around the world. Half the world's food comes from the seeds of just three plants—wheat, rice, and corn.

In this investigation iodine is used to identify starch. Starch turns a blue-black colour when iodine is added.

Question

Do seeds carry food?

Hypothesis

1 Write a hypothesis for this experiment.

Materials

- safety goggles
- gloves
- apron
- dry bean seeds
- bean seeds soaked in water for 24 h
- dissecting needle
- hand lens or dissecting microscope
- iodine solution
- paper towels
- coloured pencils
- corn seeds soaked in water for 24 h
- single-edged razor blade

Procedure

2 Obtain a dry bean seed and one soaked in water for 24 h. Use **Figure 1** to help you identify the various parts of the seed.

(a) Compare the appearance of the soaked and dry bean seeds.

(b) Suggest a reason for at least one difference you observe.

3 Use your fingers to gently remove the seed coat from the soaked bean seed. Using a dissecting needle, pry the two sections of the seed apart. Locate the embryo and examine it with a hand lens or dissecting microscope.

(a) Draw a diagram of the embryo. Label **6C** the cotyledon, the hypocotyl, the epicotyl, and the radicle.

⛔ Caution: Exercise care when using the sharp dissecting needle.

4 Devise a test to determine which parts of **4B** the embryo contain starch. Test the embryo for the presence of starch.

(a) Describe the test you used for starch.

(b) On your diagram, use a coloured pencil to mark any areas where you obtained a positive test for starch.

(c) What function does the starch serve?

⛔ Caution: Iodine irritates eyes and skin and may stain.

5 Obtain a corn seed (kernel of corn) that has been soaked in water for 24 h. Lay the corn seed down with the embryo facing upward. Use **Figure 2** to help you identify the various parts of the seed. Carefully examine the seed.

(a) How does the corn seed differ from the bean seed?

(b) One side of the corn seed is lighter. This is the location of the embryo. Draw the embryo and label your diagram.

  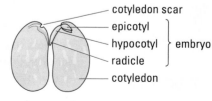

bean seed (side view)   bean seed (front view)   bean seed (split open)

cotyledon scar
epicotyl
hypocotyl } embryo
radicle
cotyledon

Figure 1
A bean seed

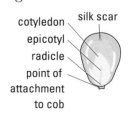

cotyledon   silk scar
epicotyl
radicle
point of attachment to cob

Figure 2
A corn seed

whole kernel (front view)   whole kernel (side view)

SKILLS HANDBOOK: **6C** Scientific Drawing   **4B** Designing a Procedure

6 Using a razor blade, cut the seed in half lengthwise to expose the interior.

Step 6

side view

🖐 Use caution when using the razor blade.

7 Test various areas of the corn seed for starch using the same technique you used in step 4.

(a) Which areas have the greatest amount of starch? Use a coloured pencil to indicate these areas on your diagram.

Analysis and Communication

8 Analyze your observations by answering the following questions:

(a) Bean seeds are classified as "dicotyledons" and the corn seeds are classified as "monocotyledons." On the basis of your observations explain the meaning of the prefixes "mono" and "di."

(b) Your teacher will supply you with a diagram of a germinating bean seed, similar to **Figure 3**. Using a red coloured pencil, colour the radicle in each of the successive pictures. Repeat the procedure by colouring the hypocotyl with a blue pencil and the epicotyl with a green pencil.

Figure 3
A germinating bean seed

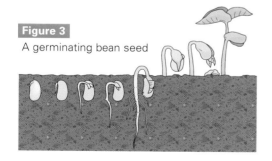

(c) Your teacher will supply you with a diagram of a germinating corn kernel, similar to **Figure 4**. Using a red coloured pencil, colour the radicle in each of the successive pictures. Repeat the procedure by colouring the hypocotyl with a blue pencil and the epicotyl with a green pencil.

Figure 4
A germinating corn seed

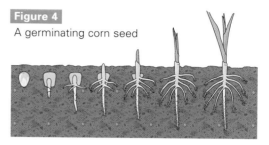

(d) Examine the diagram of the germinating bean seedling and describe what happens to the cotyledon as the seedling grows.

(e) Why does the bean plant no longer need the cotyledon?

(f) Examine the diagram of the germinating corn seedling. Explain why the endosperm gets smaller as the plant grows.

Exploring

1. Find out more about the importance of a
(2A) cotyledon by doing this experiment. In a tray of potting soil, plant three rows of bean seeds. In row 1, plant the entire seed. In row 2, remove one of the cotyledons and then plant the seeds. In row 3, remove both cotyledons and plant the embryos only. Measure and compare the growth rates of the plants.

SKILLS MENU
- Questioning
- Conducting
- Analyzing
- Hypothesizing
- Recording
- Communicating
- Planning

Germinating Seeds

Everything a new plant needs is found inside the seed: the embryo and a packaged food supply. Its protective coat resists the cold and prevents drying for many months or maybe even years. Each seed is specially adapted for specific environmental influences. Each embryo "knows" when it is time to start growing (germinating).

Most of the world's food crops are annual plants. Harvesting crops means taking the seeds and growing them anew each year (**Figure 1**). Knowledge of factors affecting seed germination is vital to food production.

In this investigation, you will design an experiment to determine how various environmental factors affect seed germination. Because not all seeds are the same, your conclusions must be restricted to the seeds you are studying.

Figure 1
Seeds, like this wheat grain, are one of Canada's most important products.

Question

1 Write an overall question for this
(2A) investigation.

Hypothesis

2 Write a hypothesis for your group's research project.

Materials

- safety goggles
- apron
- gloves
- radish and tomato seeds
- bean and lettuce seeds
- other materials as required

Experimental Design

3 You will work in research teams. Each team will work on one of the questions in **Table 1**. Pairs of teams working on the same question will use either radish and tomato seeds or bean and lettuce seeds, and then share data.

4 Work with your team to design an
(4B) experiment to investigate your problem. Work with other teams to ensure that you will be able to exchange data later. Read the following suggestions before you begin:

- Make sure that your seeds do not dry out.
- Check your seeds for mould that will slow growth and eventually kill the seeds. Remove seeds that show signs of mould. Always use clean tweezers to handle seeds, to reduce the risk of introducing mould.

Table 1	The Research Questions
A	How does temperature affect the germination and growth of seeds? Consider testing a warm environment, room temperature, and a low temperature.
B	Does light affect germination and growth of seeds? Consider using different light sources (artificial and natural) and a dark area.
C	How do plants grow in acidic conditions? Use different concentrations of acetic acid (vinegar) in water.
D	How do plants grow in basic conditions? Use different concentrations of dilute sodium hydroxide in water as the base.

  Caution: Sodium hydroxide is corrosive. Wear gloves and safety goggles when using; avoid contact with skin.

SKILLS HANDBOOK: (2A) Controlled Experiments (4B) Designing a Procedure

5 Your design must include a complete list of materials and equipment and describe any safety precautions needed.

Wear safety goggles and a lab apron for the entire procedure. Even if your procedure does not require you to work with an acid or base, other people in the class will be. The entire work area must be safe.

6 You must identify your independent and dependent variables and describe how you will control other variables.

7 Your design must include a description of how you will measure your variables. Include sample tables for recording data.

8 Present your design to your teacher for approval.

Procedure

9 Conduct your investigation.

Analysis and Communication

10 Analyze your results by answering the following questions:

(a) What conclusions can you draw from your experiment?

(b) What other experiments might you need to perform to test your conclusions?

(c) If possible, present your data in graph form.

(d) How did your results compare with those of the other groups with your research problem? with your types of seeds?

(e) Can you draw any conclusions from the class data?

11 Prepare a complete written report. **8A**

Making Connections

1. Did you notice the different lengths of the **7B** seedlings? With enough data, you can determine the range of growth for a particular plant at a particular time. Construct a graph to represent the data in **Table 2**. What conclusions can you draw from the data? Why aren't all of the seedlings the same length if the environmental conditions are the same?

Table 2

Length of shoot (mm) after day 2	Number of plants
10	1
12	3
14	6
16	14
18	22
20	15
22	7
24	3
26	1

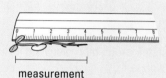

measurement

2. Why might scientists be interested in the answers to the questions your class researched?

Reflecting

3. If your team repeated the experiment, what would you do differently? Why would you change your approach?

Challenge

In this investigation you had to work in a team to design and conduct an investigation. What have you learned from this experience that might help you when you are working on your challenge?

Eggs and Embryo Development

Life begins as a single unfertilized egg which, when fertilized, becomes a zygote. As the zygote becomes an embryo, it undergoes many cell divisions by mitosis. Specific genes are turned "on" and others are turned "off" as the number of cells increases. Cells begin to differentiate into various tissues, and organs begin to develop.

Although animals vary greatly as adults, their initial stages of development are surprisingly similar. By observing the development of a frog embryo, you can begin to appreciate how a human embryo develops.

Materials

- small bowl
- unfertilized chicken egg
- pencil
- fertilized frog eggs (if available—spring only)
- microscope slide
- tweezers
- prepared slides: frog embryology set
- light microscope

Procedure

1 Fill a small bowl halfway with water. Gently crack the shell of an unfertilized chicken egg around its centre. Place the egg in the small bowl and gently break open the shell to release the contents.

Step 1

2 What you see is a single, large cell (the yolk) surrounded by the egg white. Carefully examine the egg cell. Use **Figure 1** to help you identify and locate the structures. You may have to gently turn the egg yolk to see the blastodisc. Also, examine the shell to find the egg membranes and air space.

Figure 1

The egg shell is fused to two cell membranes all around the egg. At the airspace at the blunt end, the two membranes are separated. Note the chalaza that connect the egg cell with the two ends of the egg. They probably act as shock absorbers.

albumen (egg white): Contains protein secreted by the walls of the oviduct

blastodisc: Small white spot under the cell membrane, contains the nucleus of the egg cell

shell: Hard, but permeable to atmospheric gases and various chemicals

air space

shell membranes

yolk: Contains protein and fat

chalaza: Two cordlike structures of thickened protein at either end of yolk

(a) Often, the appearance of a structure provides clues to its function. Create a summary table similar to **Table 1**. Record the appearance and predict a function for each structure.

Table 1

Structure	Appearance	Predicted function
shell	?	?
shell membranes	?	?
air space	?	?
yolk	?	?
albumen	?	?
blastodisc	?	?

3 Using the tweezers, gently touch the yolk. Note the thin membrane that surrounds it.

   (a) Gently puncture the membrane and describe what happens.

4 If living frog eggs are available (**Figure 2**), ⑤ᴮ place a single egg on a microscope slide and observe it with a microscope using low power.

   (a) Describe the appearance of the frog egg.

5 Obtain prepared slides of developing frog embryos and examine them using the low power of the microscope. Use **Figures 3**, **4**, and **5** to help you identify what you see.

   (a) Make sketches of the slides you viewed. ⑥ᶜ Make sure to label your diagrams.

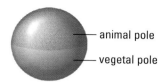

— animal pole
— vegetal pole

Understanding Concepts

1. Unlike human mothers, chickens and frogs supply no food or oxygen to developing embryos after their eggs are laid.

   (a) How do these embryos get food?

   (b) How do they get oxygen?

2. How is the chicken egg different from the frog egg?

3. Why do frogs lay so many more eggs than chickens?

Exploring

4. The chalaza may help rotate the blastodisc to the top of the yolk. What is the advantage of having the blastodisc on the top of the egg?

Figure 3

After fertilization, the zygote undergoes several cell divisions, called cleavages, during which all the cells divide at the same time. ⓐ The first cleavage divides the zygote into two cells. ⓑ The next cleavage occurs perpendicular to the first and divides the zygote into four cells. The third cleavage divides the mass into eight cells.

ⓐ first cleavage, 2.5 h

ⓑ second cleavage, 3.5 h

ⓒ eight-cell stage, 4.5 h

   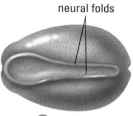

neural folds

ⓐ blastula, 12 h

ⓑ gastrula, about 20 h

ⓒ yolk plug stage, 32 h

ⓓ embryo, 48 h

Figure 4

About 8 h after fertilization, the zygote, now called an embryo, resembles a raspberry. ⓐ The 64-cell stage is a hollow ball of cells. ⓑ In the next stage, some cells migrate toward the inside of the hollow mass. ⓒ A round structure called a yolk plug can now be seen protruding from the inner layer. ⓓ After 48 h, a groove called the neural fold appears along the outer surface of the embryo. Cells in this area will develop into the spinal cord and brain.

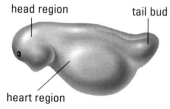 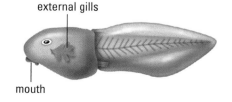

head region   tail bud

external gills

heart region

mouth

ⓐ heartbeat, 4 days

ⓑ circulation in gills, 6 days

Figure 5

By the fourth day after fertilization the embryo begins to look like a tadpole. ⓐ On day 4 beating of the heart is visible. ⓑ After 6 days, blood can be seen moving through the developing gills.

Human Conception and Pregnancy

During human intercourse, about 150 to 300 million sperm cells travel through the vagina into the uterus. However, only a few hundred actually reach the oviducts and the egg (**Figure 1**).

During fertilization (**Figure 2**), also called **conception**, the head of the sperm cell penetrates the cell membrane of the egg. Even though several sperm become attached to this membrane, only a single sperm cell penetrates the egg.

Did You Know ?

Sperm cells can survive as long as five days in the oviduct of the female. The egg cell is capable of surviving only 48 h after ovulation if it is not fertilized.

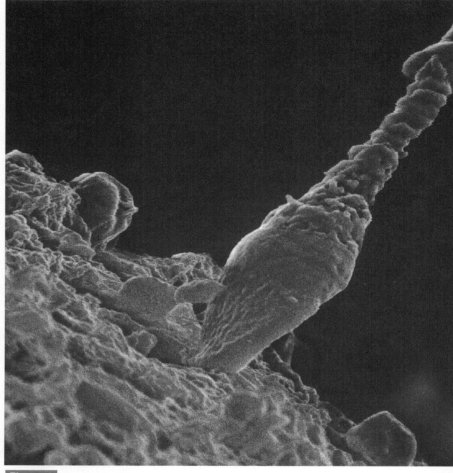

Figure 1
Human sperm cell and egg cell

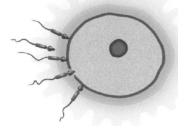

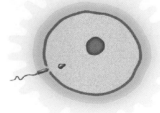

Figure 2
Fertilization

a Sperm cells attach to the egg cell. A single sperm cell penetrates the cell membrane.

b The mitochondria and flagellum of the sperm cell are pinched off by the cell membrane of the egg cell.

c The nucleus of the sperm cell finds the nucleus of the egg cell, and the 23 chromosomes from the sperm cell combine with the 23 chromosomes from the egg cell.

That sperm cell's nucleus combines with the egg's nucleus. The mitochondria and flagellum of the sperm cell are pinched off by the egg's cell membrane.

Within hours after fertilization, tiny hairlike cilia that line the oviduct move the zygote into the uterus. As it travels, it undergoes mitosis (**Figure 3**). Around the fourth day, a 16-cell mass enters the uterus where it floats freely for about two days (**Figure 4**). The dividing cell mass, which now has more than 100 cells and is a hollow sphere, implants in the wall of the endometrium. It is now called an embryo.

Embryonic blood vessels and maternal blood vessels in the endometrium combine to form the placenta. Through the placenta, nutrients and oxygen diffuse from blood vessels of the mother into blood vessels of the embryo, while wastes diffuse from the embryo's blood into the mother's. The mother disposes of the wastes.

After about three months of pregnancy, the placenta begins producing estrogen and progesterone. High levels of progesterone, first from the corpus luteum and later from the placenta, prevent further ovulation. This means that once a woman is pregnant, she cannot conceive again until after the birth.

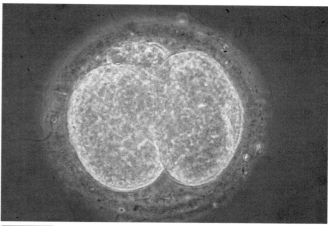

Figure 3

After conception, the zygote begins to divide rapidly. Here it has completed the first division.

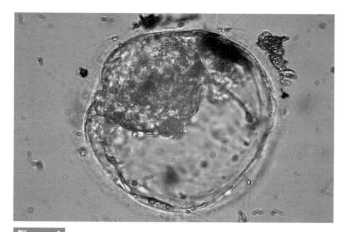

Figure 4

Cell division continues through the six days from conception to implantation.

Understanding Concepts

1. Describe the journey of the egg from ovary to uterus. How long does it take the fertilized egg to travel to the uterus?

2. How does the placenta form? What is its function?

3. The placenta begins secreting progesterone and estrogen after three months of pregnancy. Progesterone is responsible for the following functions:
   • maintains the glandular tissue of the endometrium;
   • inhibits ovulation;
   • inhibits contractions of the uterus.

   (a) Predict what would happen during pregnancy if the placenta became damaged and could not maintain progesterone levels. Provide reasons for your prediction.

   (b) Why don't pregnant women conceive again later in their pregnancy?

Human Reproductive Technology

Technology has a long history of medical inventions that have enabled people to live fuller lives. Crutches and splints were likely the first such inventions, allowing more mobility. Eyeglasses, developed in the mid-1600s, were one of the first technologies to assist a faulty organ. However, no technology has raised the interest and concern of people the way that reproductive technology has. Reproductive technology is not only changing the way that babies are born but the very laws that determine parenthood and responsibility. As research related to reproductive technologies continues to develop, controversial issues will arise. These issues will require us to debate how issues and personal beliefs relate to their use.

Many people are unable to have a baby. There are many reasons for this: a man may not produce enough sperm, a woman's oviducts may be blocked, the person's brain may not secrete enough hormones, a woman's follicles may not function properly. Some of these people go to fertility doctors in order to conceive a child.

Using Reproductive Technology

Fertility Drugs

Fertility drugs simulate the action of hormones from the pituitary (**Figure 1**). The drugs stimulate follicle development in the ovary, which makes it more probable that one or more egg cells may be released. Because fertility drugs increase the probability that multiple eggs will be released, the chance of having a multiple birth also increases.

Cytoplasmic Transfer

In cytoplasmic transfer, the cytoplasm from an egg from a younger woman is transferred into the egg cell of an older woman. It is believed that the transfer reduces the probability of genetic defects following fertilization.

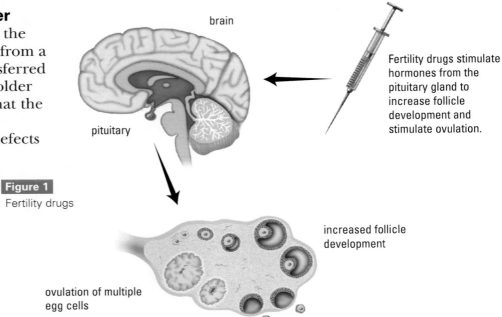

brain

Fertility drugs stimulate hormones from the pituitary gland to increase follicle development and stimulate ovulation.

pituitary

increased follicle development

ovulation of multiple egg cells

Figure 1
Fertility drugs

Intrauterine Insemination

Sperm cells are transferred directly into the oviducts of the woman following ovulation in intrauterine insemination (**Figure 2**). Normally, most sperm cells die as they travel through the uterus to the oviduct. Insemination ensures that sperm cells reach the egg in greater numbers.

Gamete Intrafallopian Transfer

Gamete intrafallopian transfer involves the sperm and egg being inserted in the oviduct (or Fallopian tube). The technique is believed to increase the chances of successful fertilization by bringing sperm and egg cells together.

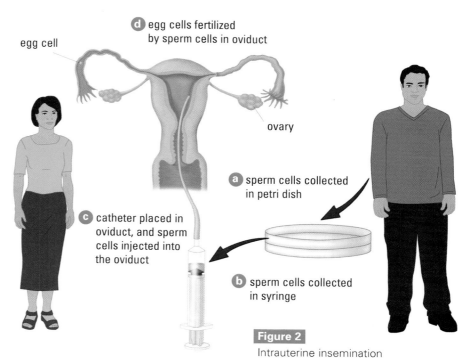

d egg cells fertilized by sperm cells in oviduct

egg cell

ovary

a sperm cells collected in petri dish

c catheter placed in oviduct, and sperm cells injected into the oviduct

b sperm cells collected in syringe

Figure 2

Intrauterine insemination

In Vitro Fertilization

The first step of in vitro fertilization (**Figure 3**) is the use of hormones to prepare the ovaries for ovulation. During ovulation, a physician inserts an instrument, called a laparoscope, into the woman's abdomen. A light in the laparoscope enables the physician to locate the ovary. A suction apparatus extracts mature eggs from the ovary. The eggs are placed in a glass petri dish and fertilized by the partner's sperm. Following a brief incubation period, one or more embryos are transferred into the uterus by a small catheter. If at least one of the embryos implants, a baby will be born nine months later.

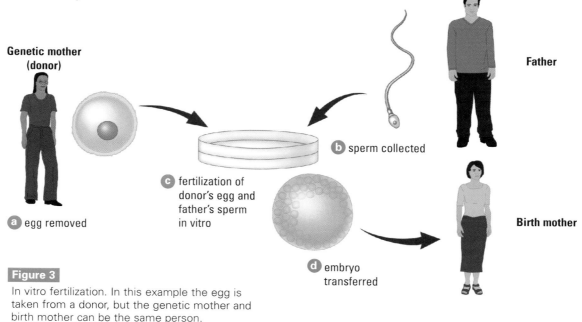

Genetic mother (donor)

Father

a egg removed

b sperm collected

c fertilization of donor's egg and father's sperm in vitro

d embryo transferred

Birth mother

Figure 3

In vitro fertilization. In this example the egg is taken from a donor, but the genetic mother and birth mother can be the same person.

Egg Freezing and Egg Donations

Fertility drugs are employed to initiate multiple ovulations. Although a single egg may be fertilized, excess eggs are frozen. At a later date these eggs can be thawed and fertilized. Sometimes zygotes are frozen after fertilization. Some of the eggs or zygotes could be implanted into the same mother at a later date, or donated to another woman who either had no eggs in her ovary or was unable to ovulate.

Embryo Transfer

A woman with a defective cervix or uterus may ask another woman to give birth to her genetic child through embryo transfer (**Figure 4**). In this case, the egg from the first woman is combined with the sperm of her partner. Fertilization occurs in vitro. The zygote is transferred to a "surrogate" mother who carries the baby to term and then returns it to the genetic parents. This technology raises legal and ethical questions.

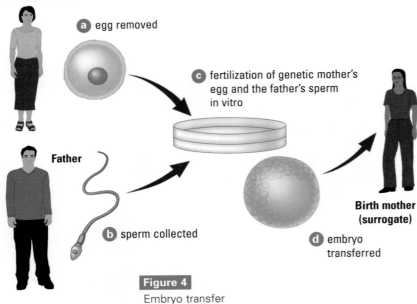

Genetic mother

a egg removed

c fertilization of genetic mother's egg and the father's sperm in vitro

Father

b sperm collected

Birth mother (surrogate)

d embryo transferred

Figure 4
Embryo transfer

Understanding Concepts

1. Explain why the name "test-tube baby" is not accurate.

2. Why might in vitro fertilization be used?

3. Identify one disadvantage of using fertility drugs.

4. Reorder the random list of statements to illustrate the correct sequence of steps performed during in vitro fertilization.
   - Eggs are placed in a petri dish.
   - Eggs are extracted from the ovary.
   - Eggs are fertilized by sperm.
   - Hormone fertility drugs are given to the woman.
   - Embryo is transferred to the uterus.
   - Embryos are incubated.

Making Connections

5. A baby could have "five different parents" if some of these reproductive technologies were used. What legal and moral issues would this raise?

6. Reproductive technologies create controversy because of the ethical questions involved. Discuss issues related to each of the reproductive technologies described in this section. Prepare a summary paragraph outlining your point of view.

Challenge

You may want to create a scenario (a short "real-world" story) for each issue in your survey. This will allow the person being interviewed to better understand the meaning of the questions. If you chose to address issues involving the technologies in this section, what scenarios could you write?

For your display, you will need a central issue. What issues can you see in the technologies presented in this section?

The future technology in your story or play may be one of the technologies in this section. How will the characters in your story or play address the issues that relate to it?

Working to Improve Women's Health

Madeline Boscoe wears many hats: she is both the executive director of the Canadian Women's Health Network and the director of the Women's Health Clinic in Winnipeg.

Boscoe started her career in community-based health education programs geared toward women, particularly around family planning and prenatal care. This experience led her to a decision to become a registered nurse, training at the Vancouver General Hospital School of Nursing. Although today she uses little of her nursing training ("I learned most of my work on the job"), she feels that some background in nursing or a related area of medicine is important for a job in the health care field.

In order to do her job, Boscoe draws on her communication skills: writing, conflict resolution, policy analysis, and strong "people skills, including networking." She advocates for better public policies in women's health care, ensuring that women have enough information to make informed choices with respect to their health. This information might include counselling or participating in experimental treatments.

Health advocacy, Boscoe believes, starts with looking at the health of the whole person, including economic health.

My work is about making a contribution to a better future.

Exploring ③A

1. Research the background and training required to become a registered nurse.

2. Research which careers require a nursing training.

3. Might poverty be the enemy of good health? Write a report describing your views.

4. Choose one type of medical treatment that might be suggested by a doctor, and imagine that it is being offered to you. What kinds of information would you need to know before you decided whether or not to give consent for the treatment? What ethical issues would you need to consider?

Comparing Embryo Development

Embryology is the study of the development of an organism before hatching or birth. During the late 1800s, scientists noted a striking similarity between the embryos of different groups of vertebrate animals. Embryologist K. E. von Baer once wrote that because he had not labelled two similar embryos when he received them, he could not identify whether they were the embryos of lizards, birds, or mammals. Later, biologists suggested that embryos are similar because all vertebrates have a common ancestor. This does not mean that birds evolved from lizards, or mammals from birds; it means that the young forms of these organisms resemble the young of related species.

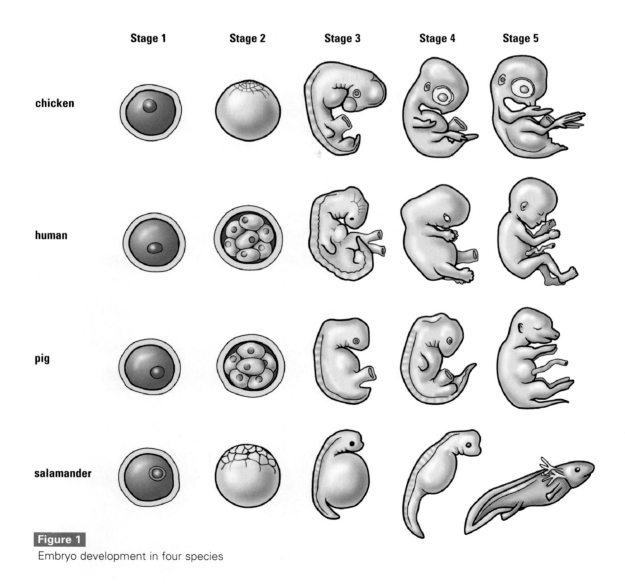

Figure 1

Embryo development in four species

Examine the embryonic development of the chicken, human, pig, and salamander shown in **Figure 1**.

The first stage of embryonic development is the fertilized egg, or zygote. The eggs of the different animals are very similar in shape and appearance. Small differences can be detected in the size of the eggs and the volume ratio of nucleus to cytoplasm. The second stage shown follows cell division of the zygote. This stage illustrates how a single cell is now a ball containing several cells.

(a) Which cell mass is most similar to that of the salamander?

(b) In what way does the cell mass of the human embryo differ from that of the chicken?

Embryos that develop in a uterus have a small yolk sac early in their development, but lose it later.

(c) Identify the yolk sacs of each embryo in stage 3 of **Figure 1**.

During stage 3, tiny gill slits can be seen in the sides of the neck. In fish, gill slits develop into gills. In mammals, they develop into parts of the ear. The backbone and tailbone can also be seen. The yolk sac in the egg-laying animals has shrunk as the animal grows. An umbilical cord is now visible in mammal embryos.

(d) Identify the gill slits, backbone, and tailbone in each embryo.

During stage 4, more differences can be identified among the organisms. Limb buds and a primitive brain and eyes can be seen.

(e) Identify the brain and eye for each embryo.

(f) Identify changes in the yolk sac of the salamander from stages 3 to 4. What has caused the change?

(g) Identify the limb buds in each embryo.

During stage 5, the embryos begin to look like the adult form. An ear can be seen in most of the embryos. Specialized structures such as the chicken's beak and wings, and the human's fingers, can now be seen. At this stage, a mammalian embryo is called a fetus.

(h) Which of the organisms spends time in a larva form?

(i) Which organism loses its tailbone?

Understanding Concepts

1. At which stage of development can you differentiate between the embryos of the chicken and the human? Explain your answer.

2. At which stage of development can you differentiate between the embryos of the pig and the human? Explain your answer.

3. Suggest an animal whose embryo would resemble a human embryo even more than the pig embryo. Provide reasons for your selection.

The Human Embryo

Your life began as a single cell about the size of the dot on an "i." In a mere 280 days (nine months), you grew into a baby about 50 cm long and with a mass of about 3 kg, composed of trillions of cells. Never again will you experience growth at such a rate, nor the number and diversity of developmental changes.

After the embryo implants in the endometrium of the uterus, it grows rapidly. By the fourth week of pregnancy the yolk sac, which has no nutrient value in humans, develops beside the embryo (**Figure 1**). A membrane, called the **amnion**, develops into a fluid-filled sac that insulates the embryo, protecting it from infection, dehydration, impact, and changes in temperature.

Blood vessels from the embryo and the mother's endometrium form the placenta. Nutrients and oxygen diffuse from the mother's blood into the blood of the embryo. Wastes diffuse in the opposite direction, moving from the embryo to the mother. The **umbilical cord** connects the embryo with the placenta.

Figure 1

Formation of the membranes that protect the embryo

First Trimester

Human pregnancy can be divided into three **trimesters**, or three-month stages. The first trimester lasts from fertilization to the end of the third month (**Figure 2**). By the end of the first month, the embryo is 7 mm long, 500 times larger than the fertilized egg. The four-chambered heart has formed, a large anterior brain is visible, and limb buds with tiny fingers and toes have developed. A tail and gill arches, characteristics of all vertebrates, can be seen.

By the end of the second month, the cartilage of the embryo's skeleton begins to be replaced by bone and the embryo is now called a fetus. Most of the body parts have formed. The facial features continue to develop as the fetus now goes into a growth phase. Arms and legs begin to move and a sucking reflex can be seen.

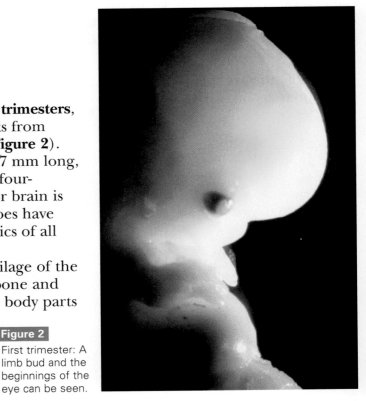

Figure 2

First trimester: A limb bud and the beginnings of the eye can be seen.

Second Trimester

By the second trimester (**Figure 3**), the 57-mm fetus moves enough to be felt by the mother. All the organs have formed and, like other mammals, soft hair covers the entire body.

Early in the fourth month the fetus begins to swallow amniotic fluid and hiccup. A month later, it may suck its thumb. By the sixth month, eyelids and eyelashes form. Most of the cartilage in the skeleton has been replaced by bone.

If the mother goes into labour at the end of the second trimester, there is a chance that the 350-mm, 680-g fetus will survive.

Did You Know ?

When a mother smokes, the embryo receives many of the harmful chemicals through the placenta. Nicotine can constrict the blood vessels to the placenta, reducing the supply of oxygen to the embryo. This may explain why mothers who smoke give birth to small babies.

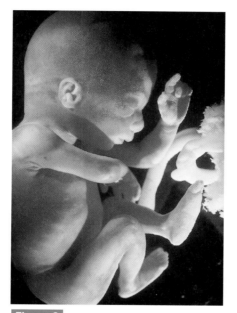

Figure 3

Second trimester: All of the organs and facial features are present, but the organs may not be fully functional.

Third Trimester

During the third trimester (**Figure 4**), from the seventh month until birth, the fetus grows rapidly. The organ systems, established during the first two trimesters, begin to function properly. All that remains is to increase body mass. The average baby is approximately 530 mm long and has a mass of 3400 g at birth.

Understanding Concepts

1. Explain why some people claim that you have to swim before you learn to walk.

2. What function does the following serve?
   (a) the placenta
   (b) the umbilical cord

3. Differentiate between an embryo and a fetus.

4. Make a chart of the changes that occur during each trimester.

5. What is the function of the amnion?

Making Connections

6. The circulatory system is the first system to function in the embryo. Why do you think this happens?

Exploring

7. What are some of the early signs
   (3A) of pregnancy? Research pregnancy tests and how they work.

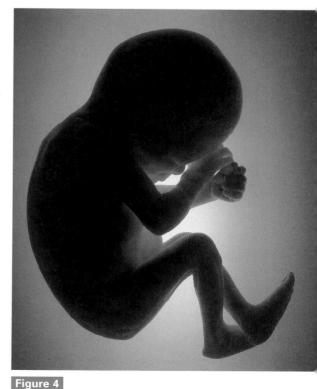

Figure 4

Third trimester: The fetus grows rapidly.

Fetal Alcohol Syndrome

The placenta acts as a barrier, preventing the mother's blood cells and large molecules from entering the circulatory system of the fetus. Smaller molecules, such as nutrients and oxygen, do move across the placenta (**Figure 1**). Unfortunately, some harmful agents, such as alcohol, are also small enough to cross the placenta. When a mother drinks alcohol, it crosses the placenta and enters the blood of the embryo.

The effects on the embryo are the same as those on the mother: alcohol depresses the functioning of the nervous system. Alcohol is also a poison. Like other poisons, it is broken down by the liver. Unfortunately, the liver of an embryo is not fully developed until the last few months of pregnancy, so alcohol cannot be broken down quickly. This means that alcohol remains harmful much longer in the embryo than it does in the mother. Not only can alcohol kill many embryonic cells, it may also change the genetic information in some cells, which can produce a mutation.

Fetal alcohol syndrome (FAS) is a host of birth defects associated with alcohol consumption by the mother. Symptoms of fetal alcohol syndrome include mental disability, abnormal facial features, central nervous system problems, behavioural difficulties, and growth deficiencies. It causes facial deformities such as a small head, thin upper lip, and small jaw bone. FAS can be mild or serious, depending on the amount of alcohol consumed by the mother.

It has been estimated that 60% to 70% of alcoholic women who become pregnant give birth to babies suffering from FAS. And evidence suggests that the problem is getting worse. A woman need not be an alcoholic to have an FAS child.

Figure 1

Alcohol and other drugs can pass from the mother's blood across the placenta into the embryo.

One study indicated that four times as many pregnant women admitted to "frequent" drinking in 1995 than in a similar 1991 study. Among 1313 pregnant women, 3.5% said they had an average of seven or more drinks a week or had consumed five or more drinks on at least one occasion in the previous month. According to the March of Dimes, alcohol is the most common cause of fetal damage in the country and the leading cause of preventable mental disability.

 Issue **FAS: A Preventable Problem?** (8B)

Statement

Pregnant women should be required to have blood tests on a regular basis to monitor drinking problems.

Point

- FAS is the third most common reason for babies being born with mental disability in Canada and the United States. (The most common reasons for mental disability are genetic defects, which are not preventable.) Heart defects and defects of the nervous system are most common in FAS children. Any measure we can take to reduce this toll is acceptable.
- Despite a growing awareness that avoiding alcohol prevents FAS, about one-fifth of pregnant women continued to drink even after they learned they were pregnant. Education is not enough.

Counterpoint

- Most birth defects occur during the first three months of pregnancy, when the organs are forming. Should women also be monitored to ensure that they have a well-balanced diet, don't gain too much weight, or don't become depressed? Scientific studies have also linked many other factors to birth problems.
- The suspension of rights for any individual is a serious matter. All people would hope that mothers would recognize their responsibility, but legislation is not the answer. Changes in attitudes are best accomplished through education.

What do you think?

- In your group, discuss the statement and the points and counterpoints above. Write down additional points and counterpoints that your group considered.
- Decide whether your group agrees or disagrees with the statement.
- (3A) Search newspapers, a library periodical index, a CD-ROM directory, and, if available, the Internet for information on FAS and other preventable birth defects.
- (3B) Prepare to defend your group's position in a class discussion.

Birth

About nine months after the embryo was implanted, contractions of the muscles of the uterus signal the beginning of **parturition**, the process of birth (**Figure 1**). The cervix begins to dilate, or open up. The membrane surrounding the baby is forced into the vagina (also called the birth canal). Usually, the amniotic membrane breaks and amniotic fluid lubricates the canal. This event is often referred to as the "breaking of the water." Once the cervix has dilated enough, uterine contractions push the head into the birth canal followed by the rest of the body. In human babies, the head and shoulders are widest. Once they are free of the birth canal, the baby slips out easily. The baby is born. A short while later, the placenta is expelled from the uterus.

Immediately after birth, the baby still remains attached to the mother by the umbilical cord. Once the blood flow through the umbilical cord becomes restricted, the baby must begin breathing on its own. The umbilical cord is cut and tied off to prevent bleeding (**Figure 2**). Within a few weeks, the dead tissue from the umbilical cord dries and falls off the baby's abdomen, leaving the navel.

Hormones play a vital role in the birthing process. Prior to labour, the hormone relaxin is produced by the placenta. Relaxin causes the ligaments in the pelvis to loosen. This provides a more flexible passageway for the baby during birth. As labour starts, the hormone oxytocin is secreted by the pituitary gland. Oxytocin causes strong uterine contractions, which push the baby into the birth canal.

After the placenta is expelled, secretion of estrogen and progesterone stops. The levels of these two hormones in the mother's blood drop. If the mother does not breast feed, her menstrual cycle begins again within a few months after birth.

Hormones and Lactation

At the beginning of puberty, estrogen stimulates breast development. During pregnancy, high levels of estrogen and progesterone prepare the breasts for milk production. Each breast contains about twenty lobes of glandular tissue, each supplied with a tiny duct that carries fluids toward the nipple (**Figure 3**).

Figure 1

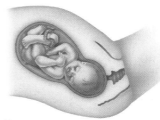

a cervical opening starts to enlarge

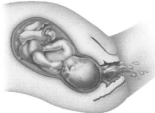

b amniotic sac breaks and fluid flows out

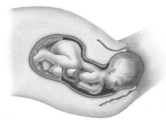

c uterine muscles contract to push the baby out

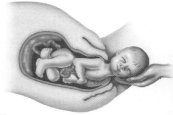

d baby emerges from the birth canal

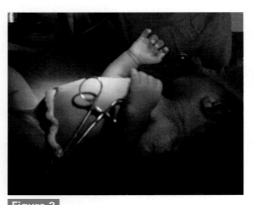

Figure 2

The umbilical cord no longer functions after birth. It dries up and falls off.

At birth, the mother's pituitary gland in the brain secretes prolactin, another hormone (**Figure 4**). Prolactin stimulates the glands in the breast to begin producing fluids. Initially, the fluid produced is called colostrum. Colostrum contains sugar and proteins but lacks the fats found in breast milk. It also helps develop the baby's immune system. A day or two after birth, the prolactin stimulates the production of milk.

Milk production is stimulated by the baby's sucking action and the removal of milk. Many North American mothers who breast-feed prefer to end breast-feeding once their youngster begins developing teeth. In some countries, especially those with few available sources of protein, breast-feeding may continue for four or even five years.

Milk and the Mother

Milk production greatly increases the energy and nutrient requirements of the mother. Human milk contains about 50% more lactose (milk sugar) than cow's milk does. At the height of lactation, a woman can produce as much as 1.5 L of milk each day. A mother producing that much milk would need about 50 g of fat and 100 g of lactose each day to replace her losses. In addition, a breast-feeding mother needs 3 g of calcium phosphate each day. In order to have enough calcium and phosphate in her milk, the woman's body can start to take those minerals from her bones. Failure to replace the needed minerals results in a progressive deterioration of the mother's skeleton and teeth.

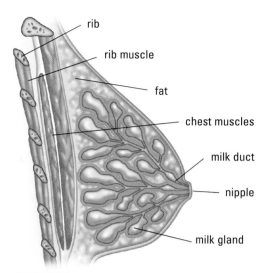

Figure 3

The mother's breast contains glands that begin producing milk shortly after the baby's birth.

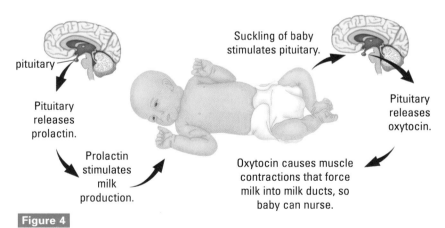

Figure 4

Milk glands, the baby, and hormones work together to ensure there is milk when the baby is hungry.

pituitary

Pituitary releases prolactin.

Prolactin stimulates milk production.

Suckling of baby stimulates pituitary.

Pituitary releases oxytocin.

Oxytocin causes muscle contractions that force milk into milk ducts, so baby can nurse.

rib
rib muscle
fat
chest muscles
milk duct
nipple
milk gland

Understanding Concepts

1. Summarize the four stages of labour shown in **Figure 1**.

2. Explain the functions of relaxin and oxytocin.

3. How does a sucking action stimulate the production of breast milk?

4. Occasionally a physician gives a pregnant woman who is past her delivery date an injection of oxytocin. What is the injection expected to do?

Making Connections

5. Why is it important for nursing mothers to eat a well-balanced diet?

Exploring

6. What is colostrum and why is it important for the newborn baby to drink it?

Chapter 8 Review

Key Expectations

Throughout this chapter, you have had opportunities to do the following things:

- Identify advantages and disadvantages of various strategies for the development of a zygote. (8.1)

- Describe the events of fertilization, the stages of embryo development, and the stages of birth. (8.4, 8.5, 8.7, 8.8, 8.10)

- Examine the structures within a seed and an egg that enable adaptation to a variety of environments, and organize, record, analyze, and communicate results. (8.2, 8.3, 8.4)

- Formulate and research questions related to reproductive technologies, and communicate results. (8.8, 8.9)

- Provide examples of technologies that help produce a baby, and discuss the ethical implications of using such technologies. (8.6)

- Explore careers that require an understanding of reproductive biology. (Career Profile)

KEY TERMS

amnion	placenta
conception	spore
embryo	trimester
parturition	umbilical cord

Reflecting

- "Different organisms use very different strategies, and combinations of strategies, to ensure the survival of their offspring." Reflect on this idea. How does it connect with what you've done in this chapter? (To review, check the sections indicated above.)

- Revise your answers to the questions raised in Getting Started. How has your thinking changed?

- What new questions do you have? How will you answer them?

Understanding Concepts

1. Make a concept map to summarize the material you have studied in this chapter. Start with the word "zygote."

2. Provide examples of organisms that reproduce by way of
   (a) spores
   (b) seeds
   (c) eggs
   (d) zygotes developing within the parent

3. In plant reproduction, what advantages do seeds have over spores?

4. Draw and label the following diagram of a seed and indicate the function of the labelled parts.

5. Where is the food found in a seed? How is that food used by the plant? by the human population?

6. In the human female reproductive system, identify the
   (a) organ that is the site of implantation of the embryo
   (b) tissue that provides nourishment for embryo
   (c) tissue that secretes estrogen after ovulation
   (d) site of egg development and estrogen secretion

7. Trace the path of an egg from the ovary through fertilization to implantation. What happens to the egg if it is not fertilized?

8. Explain why pregnancy is not possible once a woman reaches menopause.

9. Draw and label a diagram of an egg and explain the function of each of the structures.

10. What causes fetal alcohol syndrome, and what are some of its symptoms?

11. Explain the functions of the hormones oxytocin and relaxin during birth.

Applying Skills

12. A group of science students decided to test different nutrients to determine which would speed the growth of brine shrimp. The following procedure was followed.

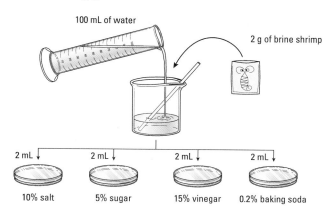

(a) Identify the independent and dependent variables.

(b) Write a hypothesis for the investigation.

(c) Identify any problems with the experimental design.

(d) How would you correct the problems you identified in (c)?

13. Some students decided to do an experiment to determine the effect of temperature on seed germination. The four groups, called A, B, C, and D, proposed different procedures, shown in **Figure 1**.

(a) Identify the question being investigated.

(b) Which procedure was the best experimental design to test the question?

(c) Identify any errors in the other procedures.

(d) Identify the independent and dependent variables in procedure C.

(e) What question is being investigated in procedure D?

(f) Identify the variables that have been controlled in procedure D.

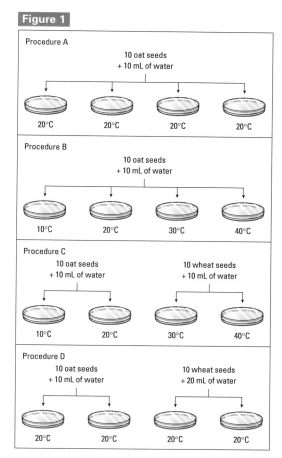

Figure 1

Making Connections

14. Compare in vitro fertilization with the normal course of events in human reproduction.

15. Indicate the advantages for each of the following reproductive strategies:

(a) A bacterium forms a spore with a resistant coating.

(b) An opossum produces eight embryos, but only six find their way into the pouch and attach to a nipple.

(c) The wildebeest gives birth to young that begin running next to the mother a few minutes after birth.

(d) The whooping crane lays two eggs. The first chick to hatch breaks the other egg. Only the first chick will survive.

(e) A parasitic worm produces hundreds of thousands of eggs. The eggs are released with the solid wastes of the host animal.

16. When the rabbit was introduced into Australia it quickly took over much of the habitat of the wombat (a marsupial). What makes the rabbit more successful than the wombat?

Challenge

Society and Reproductive Technology

When Dolly the cloned sheep was born (**Figure 1**), it set off an explosion of debate about the ethics of reproductive technology.

Scientists are making rapid advances in their understanding of reproduction, both in animals and plants. Out of those discoveries are rising new technologies that allow infertile people to have children, that allow genetic material from one organism to be transplanted into another, that allow doctors, drug designers, and seed makers to probe into reproductive cells and change them. Advances in science and technology are happening so rapidly that many people do not have access to the most recent information. Can an individual make a responsible decision about using these technologies without fully understanding their impact?

1 A Survey on Reproductive Technology

What do people believe about reproductive technology? What do they find acceptable and unacceptable? You will conduct a survey to determine public opinion on two or three issues surrounding reproductive technology, and then analyze and report your results. Are there differences in opinion based on age or gender beliefs? What other factors are involved?

Figure 1

Dolly was the first mammal to be cloned using non-reproductive cells.

Design and conduct a survey that includes:

- Brief descriptions of two or three issues around reproductive technology that you have chosen for study. The issues may be related to human reproduction or reproduction of food plants.
- A set of questions for each issue that probes the beliefs of the people you will survey.
- Completed questionnaires.
- A report that analyzes the data you have collected.

2 A Public Information Display

Before individuals can make responsible decisions about reproductive technology, they must have a clear understanding of the issues. You will prepare an information display that clearly presents an issue in reproductive technology in a way that is informative. Your display should also deal with any misconceptions that may exist in the minds of the public.

Design a display about an issue in reproductive technology in video, audio, or multimedia that includes:

- Basic information about a reproductive technology, presented in a way that is easy to understand. The display may deal with a technology used in humans or other animals or plants.
- Some of the misconceptions around the technology, and ways that help individuals deal with them.
- Different perspectives on the issue that involves the technology.
- A report explaining how you decided what information to include and why you chose to present it as you did.

3 A Futuristic Short Story or Play

The pace of development of new technology is increasing. Will the incredible rate of discovery and development continue? If it does, what kinds of reproductive choices and decisions will future generations face?

Write a short story or play related to the use of reproductive technology in the future that includes:

- A description, using scientific terms, of how the future technology works.
- Characters or situations that illustrate different perpectives on the issues related to the reproductive technology chosen.

Assessment

Your completed challenge will be assessed according to how well you:

Process

- understand the specific challenge
- develop a plan
- choose and safely use appropriate tools, equipment, and materials when necessary
- conduct the plan applying technical skills and procedures when necessary
- analyze the results

Communication

- prepare an appropriate presentation of the task
- use correct terms, symbols, and SI units
- incorporate information technology

Product

- meet established criteria
- show understanding of concepts, principles, laws, and theories
- make effective use of materials
- consider legal and ethical issues
- address the identified situation/problem

Unit 2 Review

Understanding Concepts

1. In your notebook, write the following sentences, filling in the blanks with the word or phrase that correctly completes the sentence. Use the words from this list: meiosis, mitosis, placenta, hermaphroditic, mutation, clone, interphase, metaphase, and conjugation.

   (a) This phase of mitosis, called __?__, is identified by chromosomes lining up in the middle of a cell.

   (b) The stage between divisions, called __?__, is marked by rapid growth, the duplication of genetic material, followed by another period of growth.

   (c) The division process in which sex cells are formed is referred to as __?__.

   (d) The division process in which a single cell divides into two identical daughter cells is referred to as __?__.

   (e) If the nucleus is removed from one egg cell and replaced with the nucleus from an embryo, the offspring is a(n) __?__ of the cell that donates the nucleus.

   (f) A change in a cell's genetic information is a(n) __?__.

   (g) Animals that contain both male and female sex organs are said to be __?__.

   (h) Bacteria exchange genetic information by way of plasmids in a process called __?__.

   (i) The organ responsible for nutrient exchange between a human mother and fetus is said to be the __?__.

2. Indicate whether each of the statements (a) to (j) is TRUE or FALSE. If you think the statement is FALSE, rewrite it to make it true.

   (a) The larger the organism, the larger is the size of its cells.

   (b) If a fertilized egg from a mouse has 22 chromosomes, you should expect 22 chromosomes in the muscle cell of the same mouse.

   (c) When plants such as strawberries reproduce by sending out runners, they reproduce without sex cells.

   (d) If a sheep were cloned you would expect the offspring to be the same sex as the parent or original.

   (e) All of the cells in the human body divide at the same rate.

   (f) Human sperm cells have half as many chromosomes as unfertilized egg cells.

   (g) Multiple human sperm cells fertilize a single egg cell simultaneously.

   (h) Cancer cells divide at a faster rate than normal cells.

   (i) Female aphids can give birth to male aphids through asexual reproduction.

   (j) Fertilization of the human egg takes place in the ovary and the baby develops in the oviduct.

For questions 3 to 6, write the letter of the best answer in your notebook. Write only one answer for each question.

3. The purpose of meiosis is to

   (a) produce offspring

   (b) duplicate DNA

   (c) produce daughter cells with the same chromosome number as the mother cell

   (d) produce sex cells with half the chromosome number as the mother cell

4. A human sperm cell contains

   (a) 23 chromosomes of which two are X chromosomes

   (b) 46 chromosomes of which two are Y chromosomes

   (c) 23 chromosomes of which one is X or Y

   (d) 46 chromosomes of which one is X or Y

5. Normal chimpanzee cells have 48 chromosomes. Which one of the following statements about chimpanzee cells is correct?

   (a) muscle cells have 24 chromosomes

   (b) a zygote would have 96 chromosomes

   (c) a cell following meiosis I would have 24 chromosomes

   (d) a cell following meiosis I would have 48 chromosomes

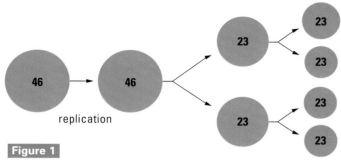

Figure 1

6. The process in **Figure 1** would occur in the
   (a) brain
   (b) heart
   (c) liver
   (d) ovary

Use the information below to answer questions 7 and 8.

A science class decides to design an experiment to test the effect of pH on seed germination.

Problem: How does pH (acidity) affect the growth of germinating seeds?

Materials:

• various pH buffer solutions

• 4 petri dishes

• 40 seeds per group

• ruler

Design: Germinating seeds were placed in petri dishes. Different pH solutions were provided. The length of the seedling was measured daily.

Three groups submitted different proposals, as shown in **Figure 2**.

7. The best proposal was provided by
   (a) group 1, because a wide range of pH was used and various amounts of solution were added.
   (b) group 2, because everything was kept constant.
   (c) group 3, because only pH was changed.
   (d) group 1 and group 3 are equally good, because a wide range of pH was used.

8. The appropriate independent and dependent variables for the problem stated above would be:
   (a) independent variable = amount of solution used,
   dependent variable = type of seeds.
   (b) independent variable = number of petri dishes used,
   dependent variable = pH solutions used.
   (c) independent variable = different pH solutions used,
   dependent variable = length of seedlings.
   (d) independent variable = same pH solutions used,
   dependent variable = length of seedlings.

9. The nutrient starch would be found, in greatest quantity, in the area of **Figure 3** labelled
   (a) 1   (d) 4
   (b) 2   (e) 5
   (c) 3

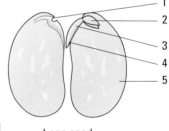

Figure 3   bean seed

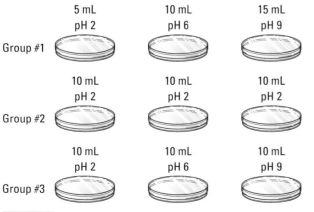

	5 mL pH 2	10 mL pH 6	15 mL pH 9	20 mL pH 12
Group #1				
	10 mL pH 2	10 mL pH 2	10 mL pH 2	10 mL pH 2
Group #2				
	10 mL pH 2	10 mL pH 6	10 mL pH 9	10 mL pH 12
Group #3				

Figure 2

Use **Figure 4** below to answer questions 10 and 11.

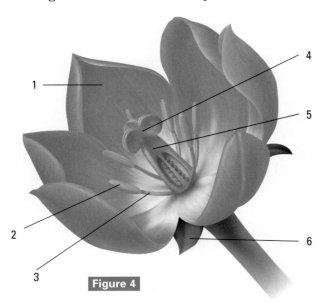

Figure 4

10. The male parts of the flower can be identified as structures

(a) 1 and 6

(b) 1 and 2

(c) 2 and 4

(d) 5 and 6

(e) 2 and 3

11. The female structure that receives pollen from an insect during cross-pollination is labelled

(a) 1

(b) 2

(c) 4

(d) 5

(e) 6

Applying Skills

12. John Needham boiled flasks containing nutrient meat broth for a few minutes in order to kill the microbes. The broth appeared clear after boiling. The flasks were then tightly sealed and left for a few days and the murky contents were examined under a microscope. The broth was teeming with microorganisms.

(a) Does this mean that the broth had spontaneously created microorganisms? Give your reasons.

(b) What changes would you make to Needham's experimental procedure before accepting his data?

13. A scientist wanted to determine if age affects body mass. The scientist hypothesized that body mass increases with age. To test this hypothesis, five people from each age group were selected at random. Their body mass was recorded. The results are provided in **Table 1** below.

Table 1

Age group	Average body mass (kg)
20 - 29	60 kg
30 - 39	65 kg
40 - 49	72 kg
50 - 59	75 kg
60 - 69	68 kg

The scientist concluded that the older a person becomes, the greater their body mass. Critique the experimental design used by the scientist. What additional information would you want to collect before accepting the conclusion?

14. Use **Figure 5** to answer the following questions.

(a) In which phase of the cell cycle does growth occur?

(b) In which phase of the cell cycle do double-stranded chromosomes become single-stranded chromosomes?

(c) In which phases of the cell cycle is the cell undergoing cell division?

(d) In which phase of the cell cycle is the cell duplicating genetic information?

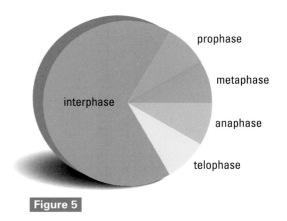

Figure 5

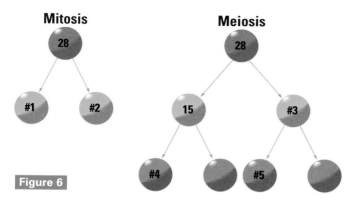

Mitosis

Meiosis

Figure 6

15. **Figure 6** compares mitosis and meiosis in a cell that contains 28 chromosomes.

   (a) How many chromosomes would be found in cell 1?

   (b) How many chromosomes would be found in cell 3?

   (c) If nondisjunction did not occur, how many chromosomes would be found in cell 3?

   (d) How many chromosomes would be found in cell 4?

   (e) Which cells shown in the diagram above would have a haploid chromosome number?

16. **Figure 7** outlines the events that take place during the female menstrual cycle.

   (a) Use **Figure 7** to describe the events of the female menstrual cycle during the flow phase, follicular phase, and luteal phase.

   (b) If a woman wanted to become pregnant, which time during the cycle would be best for fertilization?

   (c) Which hormone is produced by follicle cells?

   (d) Which hormone(s) is produced by the corpus luteum?

   (e) How would the diagram differ if a woman reached menopause?

Making Connections

17. Describe the effects of both parents using drugs and alcohol on the conception and development of their fetus.

18. Many people exposed to nuclear radiation from the Chernobyl nuclear disaster developed tumors. Some of these cancers are the result of chromosomal damage. High intensity radiation causes chromosomes to break apart, and small fragments become scattered throughout the nucleus.

   (a) Why would the fragmentation of chromosomes affect cell division?

   (b) Suggest a method that can be used to detect these changes in chromosomes.

19. In intrauterine insemination, sperm cells from the donor or partner are transferred by way of a catheter into the oviducts of the woman following ovulation.

   (a) Suggest one reason why this technique might be employed.

   (b) Provide one reason why someone might object to this technology.

20. Fertility drugs can be employed to initiate multiple ovulations, and excess eggs can be frozen. At a later date these eggs can be thawed and fertilized. Some of the eggs could be implanted back into the same mother at a later date or given to another woman who either had no eggs in the ovary or was unable to ovulate.

   (a) Suggest one reason why this technique might be employed.

   (b) Provide one reason why someone might object to this technology.

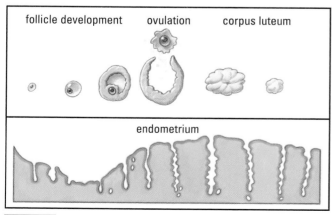

follicle development   ovulation   corpus luteum

endometrium

Figure 7

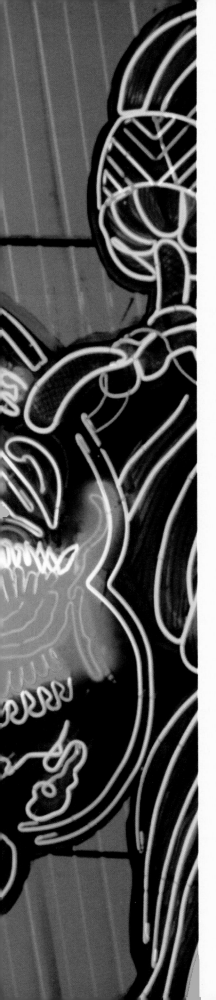

Electricity

Unit 3 Overview

Electricity is an integral part of our lives. It is fundamental to every atom and molecule, and to many of the conveniences we enjoy in our daily lives. What exactly is static electricity, and how does the operation of a photocopier depend on it? Why do we use electrical energy as the main energy source in our homes? How is electricity produced, and what new ways are being developed to produce it more efficiently? What kinds of electric circuits do we use in our homes, and why? How can we promote the safe and sustainable use of electrical energy?

9. Electrostatics

Static electricity affects the behaviour of the particles of matter, presenting both opportunities for practical use and potential hazards.

In this chapter, you will be able to:

- investigate the properties of static electric charges, and their interactions

- explain the behaviour of charged objects in terms of atomic structure

- infer, through observation, the kinds of charges transferred when a charged object contacts an uncharged object

- explain several methods of charging and discharging objects

- identify both positive and negative situations related to static electricity and suggest ways to solve problems that might arise

10. The Control of Electricity in Circuits

The electric circuits in our homes and in many electrical appliances consist of combinations of series and parallel circuits.

In this chapter, you will be able to:

- identify and explain the functions of the components of a complete electric circuit

- compare static and current electricity

- use an analogy to describe electric potential

- use Ohm's law to describe the relationship among electrical resistance, potential difference, and current

- recognize safety issues related to the use of electric circuits

- draw, construct, and analyze series and parallel circuits, and measure electric potentials and current related to the circuits

11. Harnessing Electrical Energy

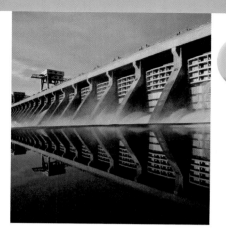

Electrical energy is often called the "in-between" form of energy because we produce it from energy sources such as fossil fuels or moving water, and then convert it into the forms of energy we need.

In this chapter, you will be able to:

- investigate the relationship among voltage drop, current, and time
- calculate the rate at which electrical energy is used
- identify renewable and nonrenewable sources of energy
- compare methods of producing electrical energy, including their advantages and disadvantages
- assess alternative sources of generating electrical energy

12. Using Electricity Safely and Efficiently

Electricity can be used thoughtfully, with due regard for safety, economy, and the environment.

In this chapter, you will be able to:

- describe and explain the distribution and operation of household electrical wiring systems
- demonstrate understanding of the importance of safety when using electricity and electric devices
- conduct a family energy audit and suggest ways to reduce energy consumption
- describe the relationship among electrical energy, power, and time interval, and calculate electrical energy using these physical properties
- calculate the percent efficiency of electrical transformations
- create a plan to use renewable energy resources to meet the needs of a dwelling, farm, or community in Ontario
- assess factors and make decisions on how to conserve energy and use it efficiently

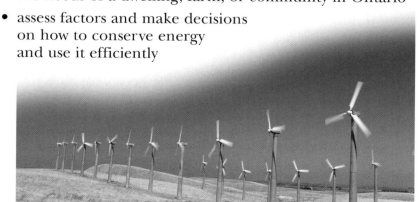

Challenge

In this unit, you will be able to... demonstrate your learning by completing a Challenge.

Using Electric Circuit Boards to Promote Energy Conservation

As you learn about electricity and continuing attempts to produce it efficiently and sustainably, think about how you would accomplish these challenges.

1 An Electric Quiz Board

Design and make an electric quiz board that asks questions about the importance of energy conservation.

2 A Display of Renewable and Nonrenewable Resources

Design a display using an electric circuit board to compare the use of resources to meet our needs for electrical energy.

3 A Consumer Awareness Display

Design a display using an electric circuit board to inform the public about how much energy is consumed by household electrical devices.

To start your Challenge, see page 388.

Record your ideas for the Challenge when you see

Challenge

Getting Started

1 Your hair stands on end after you've pulled on a sweater. You get a shock when you touch the doorknob. Your clothes stick together when you take them out of the dryer. In each case, you are experiencing the effects of static electricity. Think about times you have experienced such effects. Write down what happened, what you were doing, where it occurred— even what you were wearing. If you were walking across a room, what type of floor covering were you walking on? Did you notice the effects more often at certain times of the year?

Based on your previous knowledge of static electricity, answer the following questions:
- What do I already know?
- What do I think I know (but am not sure)?
- What would I like to know?
- Why do I need to know more?

2 The manufacturers of products that reduce the effects of static electricity (or take advantage of it) often get requests from consumers to explain how their product works. Some even print a 1-800 number or a web site on the product package so you can find out more about their products. Do you know how fabric softener sheets work? why plastic wrap clings? how photocopies are made? Make a list of some of your ideas.

3 One of the safest places to be in a lightning storm is inside a car (with the windows up, of course!). Why is this so? Brainstorm explanations of what you think provides you with protection. ➤

Reflecting

Think about the questions in **1**, **2**, **3**. What ideas do you already have? What other questions do you have about static electricity? Think about your answers and questions as you read the chapter.

Try This — Attracting Dust

With your fingernail scrape some small particles from a stick of chalk onto the edge of your hand below your little finger (or tear some paper into pieces as tiny as possible). Use some glass cleaner and a cloth rag to thoroughly polish a glass plate or one of the window panes in the room. Slowly approach the polished glass with the part of your hand covered with chalk dust. Observe what happens to the chalk dust. Now approach an unpolished piece of glass with chalk dust on the edge of your hand. Observe what happens this time. With the knowledge that you currently have about static electricity, explain what you have observed.

Investigating Electric Charges

One of the most common effects of static electricity occurs when you comb your hair on a day when the air is very dry. Your hair stands on end, and there may be a crackling noise. Static electricity causes clothes to cling together in the dryer. If, on a dry winter day, you take off an acrylic sweater that has been worn over a woollen shirt, you can hear, and sometimes even see, the discharges of built-up static electricity. You can also sense and hear its effects if you move your hand lightly over the surface of a television or computer screen after it has been on for a while.

You have considered some everyday situations in which you have experienced static electricity. Is there some way we can predict what will happen when static charges are produced? In this investigation, you will electrically charge a variety of substances, identify some properties of electric charges, and demonstrate the law of electric charges.

Question

How are uncharged and charged substances affected when they are near one another?

Hypothesis

By charging different substances and bringing them near uncharged and charged substances, it is possible to identify some of the properties of electric charges and to determine interactions that may occur between them.

Materials

- inflated balloon (with the letters A, B, and C equally spaced around the middle)
- comb
- water taps
- fur
- paper
- silk
- polyethylene strips (ebonite rods)
- acetate strips (glass rods)
- evaporating dish (and modelling clay) or support stand and stirrup

Procedure

1 Inflate the balloon. Place the part of the balloon labelled A against your hair and rub vigorously. Quickly place the A part of the balloon, and then the unrubbed B and C parts of the balloon, against several vertical surfaces. Repeat, if necessary.

✎ (a) Record your observations.

2 Rub a plastic comb on a piece of fur. First bring the comb, then the fur, near some very small pieces of paper.

✎ (a) Record your observations.

3 Rub a polyethylene strip with fur, then hold the strip near a fine stream of tap water.

✎ (a) Record your observations.

4 Repeat step 3, using the acetate strip and the silk.

✎ (a) Record your observations.

5 Rub one end of the polyethylene strip with fur. Mount the strip on an evaporating dish, or in a stirrup, as shown.

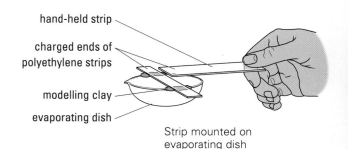

hand-held strip
charged ends of polyethylene strips
modelling clay
evaporating dish

Strip mounted on evaporating dish

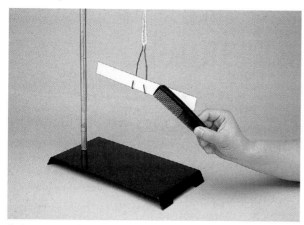

Strip mounted in stirrup

6 Rub one end of another polyethylene strip with fur. Bring this charged end of the strip close to, but not touching, the charged end of the strip on the dish.

✎ (a) Record your observations.

7 Bring the uncharged end of the polyethylene strip close to the charged end of the polyethylene strip on the dish.

✎ (a) Record your observations.

8 Repeat steps 5, 6, and 7, but substitute acetate strips that have been rubbed with silk.

✎ (a) Record your observations.

9 Charge a polyethylene strip with fur and mount it on the dish. Charge one end of an acetate strip with silk, and bring it near the charged end of the polyethylene strip on the dish.

✎ (a) Record your observations.

Analysis and Communication

10 Analyze your observations by answering the following questions:

(a) Compare the behaviour of the balloon when parts A, B, and C were placed against the wall. What does this imply about the movement of the electric charges on the surface of the balloon?

(b) What happens to uncharged objects when placed near charged objects?

(c) What happens to the force exerted between charged and uncharged objects as the distance between them decreases? Support your answer with evidence from this investigation.

(d) What evidence suggests that there are two kinds of electric charge?

(e) From your observations, what can you infer about the behaviour of objects brought close to one another if they have like charges, and if they have unlike charges?

(f) Summarize in a paragraph a test that you could perform to determine
   (i) whether an object is charged or uncharged, and
   (ii) the kind of charge present on a charged object.

(g) List the properties of electric charges you have identified in this investigation.

Making Connections

1. The hairs from a cat, dog, or rabbit stick readily to upholstered furniture, draperies, and clothing. Why does this happen? Brainstorm ways that you or others have tried to minimize the effects of "static cling" with pet hair. Invent a device that could be used to remove pet hairs from fabrics.

Exploring

2. Try combing your hair with combs made of different materials. Record what happens and develop an explanation for your observations.

Reflecting

3. Make a chart listing different situations where you have experienced the effects of static electricity. Beside each, indicate what pairs of materials you think might be responsible for producing the static electricity.

The Electrical Nature of Matter

Rubbing a balloon against your hair doesn't *create* **electric charges**. They were already there. In fact, all the atoms in matter always contain electric charges. But you aren't aware of the charges in the balloon or your hair until you make them move from their normal positions. When they are forced to move, as they are when objects of different materials are rubbed together, we say that the materials have become "charged" with "electricity." Some objects remain charged very briefly, while others, such as satellites in space when they become charged by being hit with cosmic dust particles, can remain charged for months or even longer.

As you know from your own experience, clothing, such as nylon shirts and woollen sweaters, often becomes charged with electricity when the articles rub against one another in a dryer. On many common substances the electric charge remains "static": in other words, the charge stays where the rubbing action occurred on each of the charged objects. Consequently, the phenomenon has come to be known as **static electricity**. The study of static electric charge is called **electrostatics**.

Did You Know?

Experiments by the French chemist Charles DuFay (1698–1739) showed that there were two kinds of electric charges. In 1747, Benjamin Franklin named the kind of charge we find on a polyethylene strip the "negative" charge, and that found on wool or fur the "positive" charge. Although Franklin used different materials, the charges were the same kind, and the names have remained the same ever since.

Electric Charge

A plastic comb and a woollen sweater are both electrically uncharged (neutral). Each substance has an equal number of positive and negative charges. Neither object would attract small, uncharged pieces of paper on a desk. When the comb is rubbed with the woollen sweater, however, both the comb and the sweater become electrically charged. As **Figure 1** shows, they both now attract the small uncharged pieces of paper, even though they did not touch the paper.

There are two kinds of electric charge: negative and positive. When two different neutral substances are rubbed together, one substance always becomes negatively charged while the other one becomes positively charged. For example, when the comb is rubbed with wool, the comb acquires a **negative charge**, while the wool acquires a **positive charge**. Both positively charged and negatively charged objects attract most neutral objects, including liquids and gases.

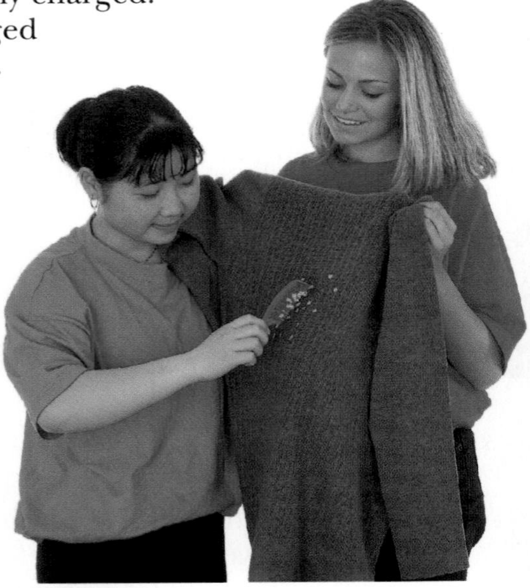

Figure 1
The pieces of paper stick only to the part of the comb that was rubbed with wool. The unrubbed part of the comb is still uncharged. Because the electrically charged particles do not move from their original positions on the comb, the electric charge is said to be static (stationary).

The Law of Electric Charges

Two objects with like charges, whether positive or negative, always repel one another. However, when a positively charged object is brought near a negatively charged object, they attract one another. This constancy of behaviour is known as the **law of electric charges**, which states: *like charges repel one another, and unlike charges attract one another*. To determine whether an object is charged and, if it is charged, whether the charge is positive or

Figure 2

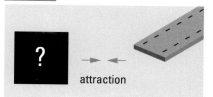

attraction

a The object being tested could be neutral or positive.

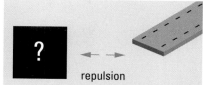

repulsion

b The object being tested must be negative.

negative, you must observe the object being repelled by an object with a known charge. Charged objects will attract both neutral objects and objects with unlike charges. On the other hand, charged objects will only repel objects with like charges (**Figure 2**).

A Model for the Electrical Nature of Matter

Scientists believe that all matter is made up of atoms containing particles that possess electric charges. Because these atoms are too small to be seen even with the aid of a microscope, a theoretical model, first proposed by Rutherford and later refined by Bohr, was developed that allows us to visualize the particles present in an atom and their relationships to one another (**Figure 3**).

This model provides a basis to explain how matter is structured and how it behaves. It helps explain the effects of electric charge and allows us to predict how charged objects will behave. The main ideas of this model, as they relate to the electrical nature of matter, are summarized in **Table 1**.

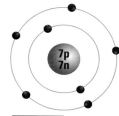

Figure 3
The Bohr-Rutherford model of a nitrogen atom

Understanding Concepts

1. **(a)** Why is the term "static" electricity used?

   (b) Describe a situation involving static electricity to explain your answer.

2. **(a)** State the law of electric charges.

   (b) Explain in detail how you could demonstrate this law.

   (c) Why does the attraction test not prove that two objects have opposite charges?

3. List the properties of electric charges identified so far.

4. Draw a labelled diagram showing (6C) the structure of the atom. Indicate the kind of charge on each of the particles that make up the atom.

Making Connections

5. A photocopier or computer printer sometimes takes in several sheets of paper at once.

   (a) Why does this happen?

   (b) What can be done to correct this problem?

Reflecting

6. **(a)** Describe what happens when charged objects are brought near uncharged objects.

   (b) Illustrate your answer with a real-life example from your list in ❶ and ❷ on page 269.

Table 1	A Model for the Electrical Nature of Matter

1. All matter is made up of submicroscopic particles called atoms.

2. At the centre of each atom is a **nucleus**, with two kinds of particles: the positively charged **proton** and the uncharged **neutron**. Protons do not move from the nucleus when an atom becomes charged (**Figure 3**).

3. A cloud of negatively charged particles called **electrons** surrounds the nucleus.
   An electron has the same amount of charge as a proton, but the kind of charge is opposite. When atoms become charged, only the electrons move from atom to atom.

4. Like charges repel each other; unlike charges attract each other.

5. In some elements, such as copper, the nucleus has a weaker attraction for its electrons than in others, and electrons are able to move freely from atom to atom.
   In other elements, such as sulfur, the electrons are strongly bound to each atom.

6. In each atom, the number of electrons surrounding the nucleus equals the number of protons in the nucleus.
   A single atom is always electrically neutral.

7. If an atom gains an extra electron, the net charge on the atom is negative, and it is called a **negative ion**.
   If an atom loses an electron, the net charge on the atom is positive, and it is called a **positive ion**.

Charging by Friction

Which rooms at school or in your home produce the worst shocks from static electricity in winter? How can you minimize or eliminate the problem? The effects of static electricity occur only when electric charges shift from their normal position on a neutral object or are transferred from one object to another. Remember, when an atom (or molecule) is electrically neutral, the positive and negative charges are equal in number and are positioned to make the atom appear uncharged. There are three ways in which objects become electrically charged: by friction, by contact, and by induction.

Charging by Friction

When plastic food wrap is smoothed and shaped to the sides of a bowl, static charges are produced that cause it to stick to the bowl. This is because electric charges can be transferred by a rubbing action or friction. **Charging by friction** causes many of the effects produced by static electricity. Large amounts of electric charge build up on clothes in a dryer because the tumbling motion is a kind of rubbing action. When someone walks across a carpet, the friction between the carpet and the person's shoes produces a charge on both the person and the carpet.

Sometimes substances rub against one another in a less obvious way. For example, when gasoline rushes out of a hose at a gas station, or when dry air rushes over the surface of a car or an airplane travelling at high speeds, large amounts of electric charge can be transferred. You can receive an electric shock when you touch the surface of a car charged this way, especially in winter. Even wearing clothes made of different materials can cause a buildup of electric charge. As the different materials rub together, each piece of clothing develops its own electric charge.

The model for the electrical nature of matter can be used to explain charging by friction (**Figure 1**). When a comb is rubbed against your hair, the comb becomes negatively charged, and the hair becomes positively

Did You Know ?

A Greek philosopher named Thales is reputed to have studied static electricity around 600 B.C. He found that when amber was rubbed with fur it attracted a small piece of cloth or leaves. The Greeks called this attraction the "amber effect."

Around A.D. 1600, William Gilbert (1540–1603), personal physician to Queen Elizabeth I, found that many other materials, including glass, also showed the amber effect. Because the Greek word for amber is "elektron," Gilbert called this effect "electric."

Figure 1

When a comb is rubbed through hair, the comb becomes negatively charged, and the hair becomes positively charged. Only electrons move during the transfer of electric charge on an atom. The protons remain in their original locations, at the centre of the atom.

a Before being rubbed together

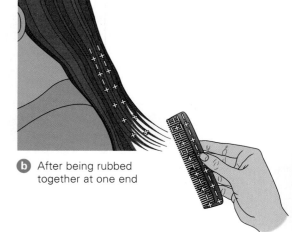

b After being rubbed together at one end

charged. In terms of the model, this occurs because the positively charged nuclei of the molecules in the comb attract electrons, including the electrons on the hair molecules, much more strongly than the nuclei of the hair molecules.

Just touching the hair with the comb allows only small numbers of both kinds of molecules to come close enough to transfer electrons from the hair to the comb. The rubbing brings many more molecules of hair into contact with molecules of the comb, allowing the transfer of significantly more electrons. The large number of electrons transferred causes an excess negative charge to build up on the comb. Because the hair loses some electrons, it becomes positively charged.

The Electrostatic Series

You can use a list called the **electrostatic series** (**Table 1**) to determine the kind of electric charge produced on each substance when any two substances on the list are rubbed together. When charging by friction occurs, the substance higher in the list (for example, acetate) always loses electrons and becomes positively charged, while the substance lower in the list (for example, silk) gains those same electrons and becomes negatively charged.

Table 1	The Electrostatic Series
acetate	Weak hold on electrons
glass	
wool	
cat's fur, human hair	
calcium, magnesium, lead	
silk	
aluminum, zinc	
cotton	Increasing tendency to gain electrons
paraffin wax	
ebonite	
polyethylene (plastic)	
carbon, copper, nickel	
rubber	
sulfur	
platinum, gold	Strong hold on electrons

Understanding Concepts

1. **(a)** Draw a series of diagrams and explain, in
   (6C) terms of the electrical model for matter, how objects become charged by friction.

   (b) What two factors affect the amount of static charge produced when you rub two different substances together? Explain why.

2. How can you use the electrostatic series to determine how two different substances will become charged when rubbed against one another?

3. **(a)** Predict what will happen when (i) acetate is rubbed with fur, and (ii) rubber is rubbed with cotton.

   (b) Explain your answer in terms of the electrical model for matter.

Making Connections

4. A silk blouse and a pair of wool socks are put into a clothes dryer. What charge will appear on the blouse when it rubs against the socks? Explain why. How would an antistatic product help?

5. Which kinds of combs are best to use in winter, plastic combs or combs made of aluminum? Explain why (in terms of the electrostatic series).

Reflecting

6. Sometimes at night if you move your feet around quickly against the sheets you can observe an interesting effect. Explain what is happening.

Challenge

When deciding where to place (set up) your electric circuit boards, what should you do to minimize the effect of static electricity on the users?

Charging Different Substances by Contact

Sometimes just shaking hands with a friend or touching your pet can have an unexpected effect: an electric shock. These little irritations are caused when electric charge is transferred from one substance to another because the substances touch each other. This is called **charging by contact**. However, charging by contact does have its uses. Part of the photocopying process is dependent on the transfer of electric charges from one substance to another.

By investigating what happens, we can develop ways to minimize or control the effects caused by charging by contact. In this investigation you will electrically charge objects by contact and determine the kind of charge transferred. A pith-ball **electroscope** (**Figure 1**) will be used to help detect and identify the kind of charge being transferred. It consists of a small, light ball suspended by a thin cotton thread. The ball moves in response to the electric forces on charged objects held near it. (Note: If a charged object touches the pith ball, it may become charged. To remove the charge, simply hold the pith ball gently in your fingers—do not squeeze.)

Figure 1

Question

How can we determine what kind of charge is transferred when a charged object contacts an uncharged one?

Hypothesis

If we know that an object possesses either a positive or negative charge, then the charge transferred by contact to an uncharged object can be determined using the law of electric charges.

Materials

- pith-ball electroscope
- polyethylene strip
- fur
- acetate strip
- silk
- evaporating dish (and modelling clay) or support stand and stirrup
- iron rod
- glass rod

Procedure

1 Without allowing the pith ball to touch the strip, bring a negatively charged polyethylene strip close to the uncharged pith ball, and then remove it.

✎ (a) Record your observations.

2 Repeat step 1 using a positively charged acetate strip.

✎ (a) Record your observations.

3 Touch the uncharged pith ball with a negatively charged polyethylene strip, remove the strip, and then approach the pith ball with the charged strip again. Then approach, but do not touch, the pith ball with a positively charged acetate strip. Remove the charge from the pith ball by touching it with your hand.

✎ (a) Record your observations.

4 Touch the uncharged pith ball with a positively charged acetate strip, remove the strip, and then approach the pith ball with the charged strip again. Then approach, but do not touch, the pith ball with a negatively charged polyethylene strip.

✎ (a) Record your observations.

5 Set up the equipment as shown below.

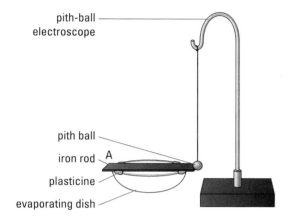

pith-ball electroscope

pith ball
iron rod A
plasticine
evaporating dish

6 Touch a negatively charged polyethylene strip against end A of the iron rod, then remove it.

✎ (a) Record your observations.

7 Recharge the polyethylene strip and bring it close to the pith ball. Discharge the pith ball by touching it with your hand.

✎ (a) Record your observations.

8 Repeat steps 5 to 7, replacing the iron rod with a glass rod.

Analysis and Communication

9 Analyze your observations by answering the following questions:

(a) What can you infer about the transfer of electric charge when a charged object is placed near, but does not touch, the pith-ball electroscope? Explain your answer.

(b) What can you infer about the transfer of electric charge when the pith ball is touched by
(i) a negatively charged object

(ii) a positively charged object
Explain your answer.

(c) Write summary statements that explain your observations in steps 3 and 4. Use diagrams to illustrate the transfer of charges.

(d) What kind of charge is on the pith ball in step 6?

(e) Explain what happened when the polyethylene strip touched the end of the iron rod in step 6?

(f) What kind of charge is on the pith ball in step 8?

(g) Explain what happened when the polyethylene strip touched the end of the glass rod in step 8.

(h) Summarize your observations in a statement by comparing the movement of electric charge in the iron rod with that in the glass rod.

Understanding Concepts

1. If you walked over a wool carpet in cotton socks, in what direction would the negative charges move when you touched a neutral doorknob? Explain your answer.

Exploring

2. Use two pith-ball electroscopes and predict, observe, and explain what happens when they are brought close together if

(a) they are charged alike;

(b) they are charged oppositely;

(c) one is charged and the other is not.

Use simple diagrams to analyze your observations.

Transferring Charge by Contact

A single spark produced by a charge transferred by contact can cause dangerous fires and explosions. Safety precautions, such as wearing boots and shoes that do not produce sparks and the use of special clothing, are required in grain elevators, flour mills, coal mines, hospital operating theatres (**Figure 1**), and some parts of oil refineries. Planes and vehicles transporting flammable materials need to have special equipment installed to prevent or control sparks produced by static electricity.

Transferring a charge by friction is difficult to avoid. Even if you are initially uncharged and walk carefully over a carpet, many electric charges are gradually transferred by the rubbing action of your shoes on the carpet fibres. But when charging by contact occurs, one object is already electrically charged. The other object may or may not be charged as well. The important factor is that there must be a difference in the amount of charge already on the two objects.

Before you touch the doorknob, you may be charged negatively, due to friction with the carpet. The doorknob is usually uncharged. When you touch it, some of the extra electrons on your body transfer

Did You Know ?

The rubbing produced by sliding across a car seat may charge you to an electric potential of 15 000 V. When you step out onto the ground, the charge is transferred by contact, but the high voltage produces only a small electric shock because the amount of electric charge is too small to be dangerous.

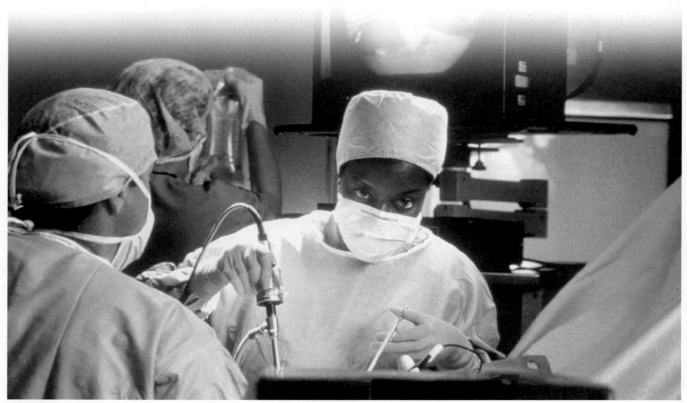

Figure 1
The clothing worn by doctors and nurses in operating theatres is woven with special fibres designed to eliminate sparks caused by static electricity.

to it. Thus, as shown in **Figure 2**, the total electric charge on your body is shared between you and the doorknob, and the charge transferred to the doorknob is also negative.

The shock produced by this kind of charge transfer can be surprising and even painful. This is because the electric charges on your body are shared with the doorknob very rapidly. In fact, usually your hand doesn't even touch the doorknob before the charges begin to transfer in the form of a spark. The electrons actually jump across the air gap between your hand and the knob just before you touch it, like a miniature lightning stroke. What could you do to prevent yourself from getting a shock? If you can't think of an answer, try again after reading more about electrostatics.

Figure 2

a Hand with negative charge approaching doorknob with no charge

b Hand touching doorknob, now both charged negatively

Understanding Concepts

1. What happens when a negatively charged object touches an uncharged pith ball on an electroscope? Use a diagram to explain your answer.

2. When an object is charged by contact, what kind of charge does the object have compared with that on the object giving the charge? Explain in terms of the model for the electrical nature of matter.

3. Why does a spark occur when a person who is charged touches an uncharged object? Would moving the hand to the doorknob very fast prevent the spark from occurring?

4. If a cat was combed with an ebonite comb, and someone else touched the cat, what charge would that person receive from the cat's fur? (See **Table 1** on page 275.) Explain your answer.

Making Connections

5. List three situations in which charging by contact can be dangerous. Explain why in each case and suggest safety precautions.

6. What makes grain elevators and flour mills among the most dangerous places to work? What precautions are taken?

Exploring

7. Talk to your friends and relatives to see if they have found ways to reduce or eliminate shocks produced by charging by contact. List some solutions.

8. What special precautions must astronauts take to guard against static electric shocks

    (a) while inside the spacecraft?

    (b) on returning to the spacecraft after a walk in space?

9. Carry out some research in your school or local ③A library, or the Internet, or talk to people you know to identify some actual cases where explosions or fires have been caused by static electricity. Write a report or present what you find to your class.

Insulators and Conductors

Insulators

We often wear several different substances, such as wool and nylon, at the same time. The fabrics rub against each other and continually become charged with static electricity. The result can be very irritating, especially during winter, because the different pieces of clothing tend to stick to each other. The static charge remains in the places where the wool and nylon rub together because they are **electrical insulators**.

An electrical insulator is a substance in which electrons cannot move freely from atom to atom. If some atoms of an insulator become negatively charged with extra electrons, these electrons remain on the same atoms until removed by a substance that exerts a stronger force on the electrons. An insulator that has positively charged atoms on its surface behaves in a similar manner. This explains the continuous buildup of static charge on furniture and glass during cleaning. Wooden furniture and glass are both electrical insulators. When you polish furniture, the electric charges remain on the surfaces and attract uncharged dust particles.

Very large amounts of charge can still build up on the surface of an insulator. Review the electrostatic series on page 275. The amount of charge that builds up depends on the relative ability of the two substances to hold on to their electrons, and how much rubbing action occurs. Paint and wax are both insulators. The surface of a car can often build up very large amounts of charge due to the air rushing over it. Most people have experienced a static shock from a car when stepping out of it after a journey.

However, even though some insulators do cause static electricity problems, they can also be very useful (**Figure 1**). Because electrons cannot be conducted *through* electrical insulators, these materials can protect us from electric shocks. The two wires carrying the electric

Figure 1

The long ceramic insulators isolate the high voltage transmission line from the metal support towers.

Table 1	Common Conductors and Insulators	
Good Conductors	**Fair Conductors**	**Good Insulators (Nonconductors)**
silver	carbon	oil
copper	nichrome	fur
gold	human body	silk
aluminum	moist human skin	wool
magnesium	acid solutions	rubber
tungsten	salt water	porcelain, glass
nickel	Earth	plastic
mercury	water vapour (in air)	wood
platinum		paper
iron		wax
selenium (in the light)		ebonite
		selenium (in the dark)

current to an electric kettle would be very dangerous if they were not covered with a plastic or rubber insulating substance. Insulators cover many household tools and appliances. Electrical cords, plugs, wall sockets, and switches are actually metal conductors covered by an insulating substance.

Conductors

It doesn't matter how hard you polish a metal tap in the kitchen or bathroom, it never builds up a static charge because metals are **electrical conductors**. A conductor is a substance in which electrons can move freely from one atom to another. If a conductor becomes negatively charged with extra electrons, they move freely (are conducted) along the conductor. When taps are charged negatively by friction, the extra electrons repel one another and are conducted away from the taps along metal water pipes to the main water supply pipe, where they transfer into the ground. Because the electric charge is conducted away as soon as it is produced, the taps remain uncharged.

Table 1 lists common conductors and insulators.

Static Electricity and Winter

The reason that problems with static electricity are much worse in winter than during other times of the year is that the cold air is so much drier and contains fewer water molecules than it does in other seasons. Dry air is an insulator and does not easily pick up charges from our body as the air molecules constantly collide against us. So, in the winter, any static charge that builds up on our clothes or on painted or polished surfaces tends to stay there. At other times of the year, the air is warmer and contains huge numbers of water molecules. Water molecules tend to pick up and transfer electric charges easily. When the air is moist, the molecules of water vapour in the air are constantly striking us all over our bodies, and these water molecules redistribute the static charges produced by friction as soon as they occur.

Challenge

What materials would you consider using when building the case for your electric circuit board to minimize the effect of static electricity for the user?

Understanding Concepts

1. Explain the difference between a conductor and an insulator, in terms of the transfer of electrons. Use diagrams to illustrate your explanation.

2. **(a)** Why does the amount of static charge continue to increase on a glass surface as you rub it?

   (b) What would eventually happen if you continued rubbing it?

3. If you charged the end of a plastic comb and then put the same kind of charge on one place on the surface of a round metal ball on an insulating stand, what would happen to the charge in each case? Explain why with the help of a diagram.

4. Why are problems with static electricity more common in winter than at other times of the year? How could any of these problems be reduced?

Making Connections

5. Look around your home, or in the family car, and identify examples where insulators and conductors are used on electrical equipment. Give reasons for their use.

6. List at least two reasons why you think plastic materials are used to cover the copper wires in electrical equipment.

Exploring

7. When people began using electricity in homes, copper wire was used. Then in the 1970s aluminum wire largely replaced copper. After a few years, aluminum was replaced by copper again. Why did each change occur? List the advantages and disadvantages of using each metal.

Discharging Electrically Charged Objects

Every time you pump gasoline into the gas tank of a car, the flow of gas through the nozzle generates large amounts of static electricity. Think what would happen if a static spark jumped from the nozzle to the car. As aircraft travel through the air, they continuously build up huge amounts of static charge on the outside surface of the plane. In both cases, if there were not a way of continuously removing the charge as soon as it was produced, there could be serious consequences. The plane's communications systems would not operate properly, and the gasoline would ignite.

If a charged object has all the excess electric charges removed, it is said to be **discharged** or **neutralized**. Several methods are used to discharge charged objects.

Grounding

The simplest way to discharge an object is to connect it to Earth itself by means of a conductor, such as a wire connected to a metal rod buried in the ground. When a charged object is connected, or **grounded**, to Earth, it shares its charge with the entire Earth. The damp soil is a fairly good conductor, and Earth is so large that it effectively removes all the excess charge from the object. All the parts of the gas pump, and everything attached to it, are very carefully grounded. As soon as the charge is generated at the nozzle, it is immediately conducted safely to ground. People who assemble sensitive electronic equipment, such as microchips to computer circuit boards, usually wear metal straps on one of their wrists (**Figure 1**). The strap is attached by a wire to a grounding system on all the benches, and they in turn are grounded to Earth. Astronauts sometimes wear similar straps, using their spacecraft as the ground.

Figure 1

Discharge at a Point

Clearly, the grounding wire is not a very practical idea for aircraft or cars. Another method, which is based on the way electric charges behave on the surface of conductors, is used to discharge airplanes. The surface shape of a charged conductor affects the rate at which it becomes discharged. Smooth, spherical shapes retain charges indefinitely, because the charges spread themselves evenly over the surface. However, conductors pointed at the ends lose charges rapidly

Did You Know

The discharge of charged particles from a point is sometimes called corona discharge. Corona discharge also occurs from the wires on high-voltage transmission lines.

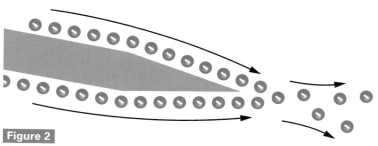

Figure 2

The electrons at the tip of the pointed rod are repelled into the air by the electrons just behind the point.

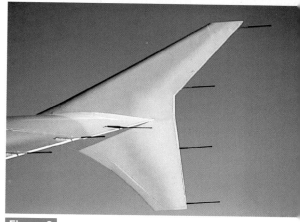

Figure 3

Static wicks on airplane wings and tail

(**Figure 2**). At the sharply curved point of a negatively charged rod, electrons repel one another so strongly that those right at the tip are actually pushed off the point in a continuous stream. This method of discharging charged objects is called **discharge at a point**.

Several different point discharge methods are used to continuously discharge airplanes. Planes, ranging in size from small executive jets to Boeing 747s, use "static wicks" (**Figure 3**), which are pointed metal rods that stick out from the movable control surfaces in the wings and tail. They allow a continuous discharge of static electricity. To prevent the discharge from travelling through the hinges on the control surfaces, flexible wires connect the movable and fixed parts of the aircraft. Aircraft maintenance workers must monitor the condition of the static wicks and test them on a regular basis. Medium-sized aircraft also use an inflatable rubber covering on the leading edge of the wings. The rubber surface is covered with a conductive cement that overlaps the metal parts and allows the electricity to leak away. These protective measures are also useful when airplanes are flying in stormy weather. The smallest aircraft lack static electric reduction devices and, therefore, are prohibited from flying when storms are in the area.

Other Ways to Discharge Objects

Over a period of time, charged objects can be discharged by simple exposure to the air. On a humid summer day, because of the number of water molecules in the air, the charge leaks away so rapidly that many of the problems caused by static electricity are not noticeable. However, on a cold, dry winter day, when the humidity is low, the charge leaks away so slowly that just combing your hair can be difficult. Other ways to discharge an object are to shine a light on it or to expose it to radioactivity.

Understanding Concepts

1. What is the meaning of the term "discharge"?

2. Why does Earth not become charged when so many electrons are constantly flowing into it as various devices are "grounded"?

3. (a) Why does the flow of gasoline through the hose and nozzle of a gas pump produce a static charge?

   (b) Why is there no static discharge at a gas station when the nozzle of the pump is brought up to the car's fuel tank opening?

Making Connections

4. Why do people working on electronic equipment have metal wires clipped to their bodies as they work?

5. Why would it not be acceptable to allow the discharge from a wing flap of an airplane to travel through the hinges attaching it to the wing?

6. On a cold, dry winter day, how would a woollen toque or the hood of a parka affect your hair? Explain.

7. How is a gasoline tank truck protected from static discharge?

8. Would an airplane made of wood be subject to a buildup of static electricity? Explain.

9. How can you avoid getting a shock from static electricity when you touch a doorknob?

Aircraft Mechanic

Gary Masse is an aircraft mechanic, pilot, and owner of an aircraft maintenance company called WCS Aviation. His interest in airplanes dates back to his early teens when he became a member of the local Air Cadet Squadron. His desire to work with planes was inspired when he talked with a mechanic for the Snowbirds.

Masse took machine shop and welding in addition to mathematics, science, and the other required areas of study in high school. He emphasizes the importance of English since aircraft mechanics must communicate accurately with others and must write clear records and reports.

After graduation, he enrolled in a two-year Aircraft Maintenance program at Canadore College. The areas of study included airframes, avionics (communications systems), power plants, federal regulations, mathematics, science, and English. Approximately 20% of the time at school was spent studying electricity.

Aircraft must be serviced frequently and thoroughly. Safety is everything. The mechanics must pay particular attention to the static electricity discharge equipment because communications systems and instruments will not work properly when static charges build up. Masse enjoys the precision and accuracy of these procedures. "Everything must be done according to the maintenance manuals." Records must be kept of all parts that are used so that every one can be traced back to its original manufacturer. In addition, all parts must be destroyed when their rated life span has passed.

Masse recommends that high school students who think that they might be interested in a career in aircraft mechanics should try to obtain summer jobs working around aircraft. And ask lots of questions.

> It is clean work with no cutting of corners and with a great deal of attention to detail.

Exploring

1. Research courses in aircraft mechanics. Where are they available, how long do they take, and what entrance requirements are there?

2. Why is electricity such a major part of an aircraft mechanic's training?

3. Servicing aircraft takes great precision and attention to detail. List 10 other careers in which these characteristics are valuable.

Induction

What do lint and dust sticking to your clothes and the operation of a pager have in common? They are both examples of the third way of charging an object, **charging by induction**. In physics, the term "induction" suggests that something is made to happen without direct contact. With the other two methods of transferring charge, charging by friction and charging by contact, the buildup or transfer of charge occurs only when the objects rub against or touch one another.

Induced Charge Separation

When an uncharged object, such as a dust particle, is charged by induction, the nearby charged object doesn't actually touch the dust particle at all. As you know, the surface of a television screen or computer monitor becomes charged after it has been operating for a while. When a dust particle is near a charged television screen, the charges on the screen cause, or induce, the electrons on the dust particle to change position slightly. This slight shift of the electrons has the effect of making the side of the dust particle facing the screen have the opposite charge to that on the screen, and the dust is attracted toward the charged television screen. This charging effect is known as **induced charge separation**. **Figure 1** shows what happens to a dust particle near a negatively charged object. Whether an object is charged positively or negatively, the dust particle is still attracted to it. A neutral object always has an opposite charge induced on the surface closest to the charged object. This is why there is an attractive force between charged objects and neutral objects.

Charging Conductors by Induction

What would your life be like if you were no longer able to watch television, listen to the radio, or talk on a cell phone? All these modern conveniences use charging by induction to operate. Previously, you learned about the induced charge separation that occurred on uncharged insulating materials, such as dust. When induced charge separation occurs in conductors, it is possible to make the electrons move from atom to atom along the uncharged conductor, rather than just shift

Figure 1

Induced Charge Separation

dust particle

a Dust particles floating in the air are usually made of substances that are insulators. They are normally neutral with both kinds of charges distributed evenly over the surface.

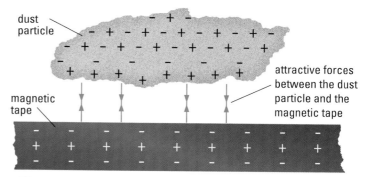

dust particle

attractive forces between the dust particle and the magnetic tape

magnetic tape

b As the uncharged dust particle comes close to the negatively charged object, the negative charges on the dust particle closest to the object are repelled and shift position slightly. The side of the dust particle closest to the object now has a positive charge induced on its surface, and it is attracted to the negatively charged object. However, it is important to realize that the dust particle is actually still neutral.

position slightly. In fact, it is possible to induce a permanent charge on a conductor.

If a negatively charged strip is brought near one end of an uncharged metal rod, the electrons at that end of the metal rod are repelled toward the other end. This movement of electrons induces a negative charge on the other end of the metal rod (**Figure 2a**). When the negatively charged strip is removed, the electrons redistribute themselves evenly along the rod again (**Figure 2b**).

However, if a conducting wire is connected from the metal rod to a "ground" such as a water tap, the rod will lose electrons, as explained in **Figure 3**. The induced charge is always opposite to that of the charged object producing the charge. For example, a negative charge can be induced on a metal rod by placing a positively charged strip near the metal rod when the conducting wire is connected to it.

In this explanation of charging by induction, the conducting wire was connected and disconnected to show that a charge can be induced on the conductor. However, in practical applications of charging by induction, the wire is usually connected permanently. The induced charge can then move onto and off the object freely whenever electric forces act on the object. Devices such as electrostatic microphones, television and car radio aerials, cell phones, and lightning rods all work on the principle of induced charges.

Using Static Electricity to Advantage

Although static electricity can be annoying and even dangerous, it also has many practical uses. Charging by induction is the principle used in the design of many devices that remove pollutants and dust from the air. It is also used as a way of coating surfaces with a variety of coverings. As you will learn (in section 9.11), the photocopier uses two of the three ways of transferring charge as it makes copies.

Pollution and Dust Control

The electrostatic air cleaners installed in homes and hospitals use static electricity. Dust and other particles are removed from the air by using the attraction between unlike charges. The dirty air is usually sprayed with positively charged ions as it passes into the air cleaner. The positive ions are attracted to the dust particles by induction. This produces positively charged dust particles that are then forced between negatively charged flat plates. These plates attract the oppositely charged dust particles, and the cleaned air passes on through. A similar method removes many kinds of smoke particles and harmful pollutants as they travel up industrial chimney stacks (**Figure 4**).

Figure 2

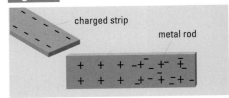

a Electrons in the metal rod are repelled by the negatively charged strip.

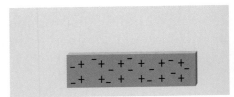

b When the negatively charged strip is removed, the electrons redistribute themselves evenly throughout the metal rod.

Figure 3

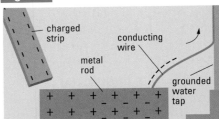

a When the negatively charged strip is brought close to the metal rod, some of the electrons repelled along the metal rod are conducted into the wire.

b If the wire is now removed from the metal rod, the metal rod will have lost some electrons and will be positively charged.

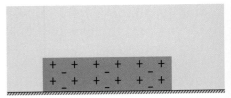

c When the negatively charged strip is removed, the remaining electrons spread evenly along the metal rod again, but the rod has a permanent positive charge.

Figure 4

Many smokestacks are equipped with electrostatic air cleaners, which remove air pollutants before they enter the atmosphere.

Coating Surfaces with Particles

An electrostatic process is also used in machines that paint objects (**Figure 5**). The tiny paint particles from the spray gun are electrically charged as they pass an electrode attached to the gun. The object to be painted is given a charge opposite to that on the paint. If the object is made of an insulating material, it is first coated or dipped in a conducting substance. The charged paint particles are attracted toward the oppositely charged object. Paint droplets that would normally have missed the object are pulled toward it by the attraction of opposite charges.

Did You Know

A completely emission-free automobile spray-painting system is now being tested. It uses dry-powder paint instead of water-based paint. Even the small amount of electrically charged paint powder that misses its target can be collected and reused, leaving no waste whatsoever.

Understanding Concepts

1. Why is a neutral dust particle attracted to a charged object?

2. Explain the difference between induced charge separation and charging by induction.

3. When an object is charged by induction, what kind of charge does the object have compared with that on the object inducing the charge? Explain.

4. It is possible to spray the back of an object, even though the spray gun is pointed at the sides. Explain why.

Making Connections

5. **(a)** The same ducts are used to distribute hot air from the furnace in winter and cold air from central air conditioning in the summer. Would dust build up on the return air flow grates more quickly in winter or summer? Explain your answer.

   (b) Would you recommend the air ducts in your house be cleaned in summer or winter?

6. Why is charging by induction important in helping to protect our environment? Describe some of the ways it is used to do so.

Exploring (3A)

7. How does an electronic air cleaner in an automobile work?

8. When was the "electrostatic precipitator" first developed? Who did it? Why is it so important to our everyday lives?

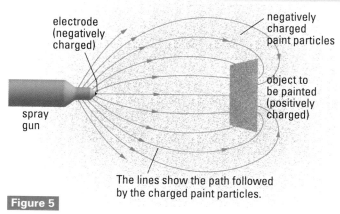

electrode (negatively charged)

negatively charged paint particles

object to be painted (positively charged)

spray gun

The lines show the path followed by the charged paint particles.

Figure 5

An electrostatic spray-painting machine

Charging by Induction

The next time you are eating nuts or cooking with different kinds of seeds, think about how they are cleaned and processed. The sorting of many kinds of seeds and the separation of nuts from their shells is done using charging by induction. Over 30 different kinds of minerals can be separated by electrostatic processes. In this investigation, you will study the two different methods of charging objects by induction. You will also develop your understanding by making predictions and testing them.

A metal-leaf electroscope consists of a small metal ball or plate connected to a metal rod (**Figure 1**). Hanging from the rod are two thin metal strips, or leaves, which are protected from air currents by being enclosed inside an insulated glass container. Because the parts of the electroscope inside the container are made of a conducting metal, electrons are able to move freely. When charged objects are brought near the ball, electrons move onto or out of the metal leaves through the rod. The resulting charge on the leaves causes them to repel each other.

Question

How can we determine the kind of charge induced on a neutral object when it is approached by a charged object (step 5)?

Hypothesis

1 Read the procedure and write a hypothesis
④Ⓐ that will answer the question.

Materials

- metal-leaf electroscope
- polyethylene strip
- fur
- acetate strip
- silk
- water taps
- insulated wire conductor
  (with alligator clips)

Figure 1

Procedure

2 Without touching the uncharged metal-leaf electroscope, approach it with a negatively charged polyethylene strip. Move the strip toward and away from the ball several times.

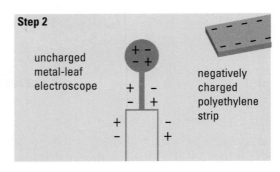

Step 2

uncharged metal-leaf electroscope

negatively charged polyethylene strip

(a) Draw the above diagram. Draw a
⑥Ⓒ second diagram to record what happens when the strip is very close to the ball.

3 Repeat step 2, using a positively charged acetate strip.

(a) Record your observations using diagrams.

4 Attach an insulated wire conductor from the rod of the metal-leaf electroscope to the water tap, as shown below.

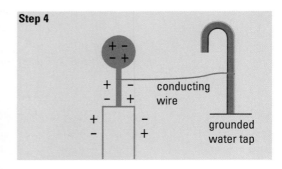

Step 4

conducting wire

grounded water tap

5 Bring a negatively charged polyethylene strip near (but don't touch) the ball on the metal-leaf electroscope. While the charged strip is near the metal ball, remove the wire conductor from the rod. Remove the charged strip.

(a) Draw the diagram in step 4, then draw diagrams to show what happens to the metal leaves (i) when the strip is brought near the ball before the conductor is removed, and (ii) after the conductor is removed and the charged strip is moved away from the ball.

6 Predict the kind of charge that has been induced on the electroscope, and state how you will test your prediction. Test your prediction, using the appropriate charged strip. Discharge the electroscope by touching the ball.

✎ (a) Record your observations using labelled diagrams.

7 Predict what will happen if steps 4 and 5 are repeated using a positively charged acetate strip. Then test your prediction.

✎ (a) Record your observations using labelled diagrams.

Analysis and Communication

8 Analyze your observations by answering the following questions and making the necessary explanations:

(a) In terms of the law of electric charges, what can you infer about the kind of charge that appears to be induced on the leaves of electroscope in step 2, and in step 3? Explain your answers.

(b) In terms of the movement of electrons, explain what happened to the electroscope in step 2, and in step 3.

(c) Why is it possible for an uncharged object to appear charged if no charge has been transferred to it?

(d) What kind of charge was induced on the metal-leaf electroscope in step 5? Comment on the validity of your predictions and testing procedure in step 6.

(e) Explain what happened in step 5, in terms of the movement of electrons.

(f) What kind of charge was induced on the metal-leaf electroscope in step 7? Comment on the validity of your predictions and testing procedure.

(g) Explain what happened in step 7, in terms of the movement of electrons.

9 Write a summary of your observations based on the following:

(a) Identify the kind of charge that is temporarily induced on the side of a neutral object closest to the charged object and explain this charge, using supporting diagrams.

(b) Identify the kind of charge on an uncharged object after it has had a charge induced on it and explain this charge, using supporting diagrams.

Understanding Concepts

1. Why are dust particles attracted to a newly polished car?

2. Explain why lint and hairs from pets stick to your clothes more readily in winter than in other seasons of the year, in terms of induced charge separation. Would you alter the device you invented in question 1 in section 9.1, based on new knowledge you have gained?

3. (a) If a dust particle in the air floated near some free electrons that had been released by an electrostatic device, what would happen? Explain why.

(b) If a charged dust particle floated near a wall or a piece of furniture, what would happen? Explain why.

4. How is charging by induction involved in the operation of cell phones, radios, and televisions? In this investigation, the wire was disconnected to show charging by induction. What is the difference with the devices mentioned above?

Exploring

5. Look back at Table 1 on page 275. In a group, brainstorm eight common materials not included in the electrostatic series list in the table. Make your own simple electroscope and, by using only three materials selected from the electrostatic series list, produce a revised electrostatic series that includes the eight items you thought of.

Lightning

Just about everyone has at least one vivid memory of a particularly violent lightning storm. Although lightning and thunderstorms seem to occur infrequently, about 2000 thunderstorms are occurring throughout the world at any given time, generating about 100 lightning strikes every second, or about 8 million strikes daily.

Actually, lightning is part of a natural process of exchanging electric charges between the atmosphere and Earth itself. Electric charges, mostly negatively charged electrons, are continuously being removed from Earth's surface by a variety of processes. Some are natural processes, such as the evaporation of water molecules, and others are related to the production of exhaust gases by vehicles and industrial activities. When thunderclouds form, huge numbers of these negative charges tend to concentrate near the bottom of the cloud. When the negative charge at the base of the cloud moves over tall objects, such as the buildings in **Figure 1**, it is sometimes close enough to return to the ground in a huge spark we call lightning.

Lightning appears to be a jagged path of white light moving toward the ground. The jagged path is caused by the electric charges moving along the path of least resistance in the air. This path is sometimes created by traces of moisture in the air, or by a concentration of positive ions. The electric charge flows in a series of steps as it finds the easiest path to the ground. The charges are more likely to move toward the tallest objects, because it shortens the path to the ground, especially if they are made of metal conductors.

Figure 1

Try This Lightning and Safety

Think about the last major thunderstorm you experienced.

1. During the thunderstorm, what activities did you continue to engage in that you now realize were placing you in danger?

2. When is the appropriate time to stop playing golf, baseball, soccer, or tennis, or to get out of a pool or lake when a thunderstorm is approaching?

3. What safety precautions should you take when indoors during a lightning storm?

4. Make a list of the actions that you will take in the future if you are engaged in an activity or are in a location that could place you in danger during a thunderstorm.

Research and review the Canada Safety Council recommendations on actions to take if you are caught in a thunderstorm. (3A)

Lightning Rods

The diagram in **Figure 2** shows how a lightning rod can protect a house from a lightning strike. A pointed metal rod is attached to the highest part of the building. A thick conductor, usually copper, is connected from the pointed rod to a metal plate buried in the ground. The plate is used to conduct the electric charges between the rod to the ground.

Lightning rods provide two kinds of protection: they help prevent lightning from striking and, if lightning does strike, they direct the charge through the conductor to the ground. To understand the first case, look at the charges indicated in **Figure 2**. The negative charge at the base of the thundercloud induces a positive charge on the things below it, including the building, the lightning rod, and the ground. The lower atmosphere always contains 1000 to 2000 positively and negatively charged ions in every cubic centimetre of "normal" air. The positive ions are repelled by the highly charged (positive) lightning rod toward the thundercloud, thereby neutralizing some of the negative charge. This can prevent a lightning strike. If, however, a lightning strike does occur, the copper conductor carries the negative charges safely to the plate in the ground.

A car is a safe place to be in a lightning storm because most of the car body is made of a metal conductor. Also, because it is usually raining, the outside of the car is wet. When lightning strikes the car, the electric charge travels over the entire body of the car and then easily crosses the short distance from the base of the car to the ground.

Did You Know?

On May 6, 1937, the huge 240-t airship *Hindenburg* burned as it came in to the landing area during a storm at Lakehurst, New Jersey. Research has revealed that the fire was caused by an electrical discharge that ignited the highly flammable coating painted on the cotton-fibre skin.

Figure 2

The protection of a building by a lightning rod

Understanding Concepts

1. Why does lightning occur?

2. Why does lightning seem to strike the tallest objects?

3. Draw and label the typical path of a lightning strike. Explain why it looks as it does. **(6C)**

4. **(a)** What are the two main ways a lightning rod protects a building?

   (b) Explain how each protection method works.

Making Connections

5. What kinds of buildings are most at risk from lightning strikes? What buildings are at the least risk? Explain why.

6. Why are golf courses, parks, and open boats particularly dangerous places to be during a thunderstorm? How would you minimize the safety risks in these public places?

Exploring

7. Use the Internet or a library to investigate the effects of lightning on the human body. Choose two of the following questions and prepare an artistic or electronic visual presentation for the class: **(3A)**
   - What are the physical signs you might experience to warn that lightning might be about to strike?
   - What parts of the body are most vulnerable to lightning damage?
   - How does lightning cause death?
   - How many people are injured or killed annually by lightning in Canada?

8. What positive results occur because of lightning strikes? Investigate any chemical reactions that might occur as the result of a lightning strike.

Interesting Insulators and Conductors

Fabric Softener Sheets

A clothes dryer is a perfect static electricity generator. When clothes made of different materials rub against one another as they tumble in the dryer, some of the molecules making up the materials lose electrons and some gain electrons. As the clothes become drier, the humidity in the dryer is lowered and the fabrics act as insulators preventing the charges from easily returning to their original locations. This favours the additional buildup of charge, and the clothes snap and crackle with static charge when we take them out of the dryer.

As sheets of fabric softener tumble around with the clothes, they act as conductors allowing the electrons to migrate among the clothes easily, lessening the buildup of charges. The molecules of the softener transfer to the fabrics and make the different materials appear more like one another. This allows the electrons to distribute themselves more evenly throughout the entire load in the dryer, and clothes can be separated without clinging to each other.

(a) If you take the clothes out of the dryer when they are only partially dry, is there any static cling? Explain your answer.

(b) Why do the clothes become charged? How could you predict which clothes would be charged (i) positively, and (ii) negatively?

(c) What two things does the fabric softener do to reduce static cling?

The Photocopier

The photocopying machine provides an interesting example of how one substance, selenium, has the unusual property of being able to function both as an insulator and a conductor, depending on how much light is shining on it. Selenium is **photoconductive**—it acts as an insulator in darkness and as a conductor when light shines on it.

A copying machine uses the scientific principle that electric charges can be removed from a surface by light (**Figure 1**). First, a special selenium-coated, flat metal plate is positively charged (step 1). Light shone through a lens then projects an image of the

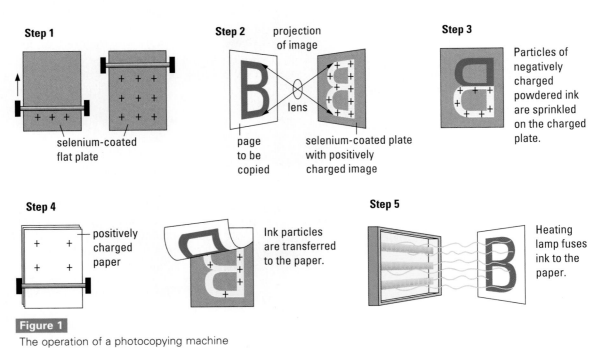

Step 1

selenium-coated flat plate

Step 2 projection of image

B

lens

page to be copied

selenium-coated plate with positively charged image

Step 3

Particles of negatively charged powdered ink are sprinkled on the charged plate.

Step 4

positively charged paper

Ink particles are transferred to the paper.

Step 5

Heating lamp fuses ink to the paper.

Figure 1
The operation of a photocopying machine

page to be copied onto the charged plate. Wherever the light falls on the selenium plate, the plate becomes a conductor and the positive charge is conducted away from the selenium by the metal plate (step 2). Only the areas coloured yellow, representing the printed material on the page, remain charged (the selenium remains an insulator). Particles of negatively charged powdered ink are then sprinkled over the plate, and these particles are attracted to the positively charged areas (step 3). Next, a positively charged piece of paper is brought in contact with the plate, and the negatively charged ink particles transfer to the paper (step 4). A lamp rapidly heats the paper, causing the ink to fuse or melt onto it, making a permanent copy of the original page (step 5).

(d) What kinds of charge transfer occur during the photocopying process? Describe each one and explain why it is used.

(e) Why is the metal plate coated with selenium?

(f) Use the Internet or a library to research (3A) who invented the photocopy process. Outline the difficulties that were faced in marketing the idea to a manufacturer.

Cell Phones and Car Radios

Every time you get a signal on a pager, talk to someone on a cell phone, or listen to the radio in the car, charging by induction is occurring on some rather unusually shaped conductors. Attached to each of these devices is a conductor called an aerial. Sometimes the aerial is so small it can be placed inside the device itself. In many cars it is attached to the back window.

Whenever a cell phone receives a signal, there is no actual wire connecting it to the source of the signal (**Figure 2**). The signals are waves of electromagnetic energy that travel from the person calling you through the air to the aerial on the cell phone. When the electromagnetic wave interacts with the conducting aerial in the cell phone, the electric charges in the conductor move by

induction in exactly the same patterns as the electromagnetic wave. This movement of electric charges is passed on to the circuits in the phone, and you can hear the voice of the person calling.

(g) Sometimes you cannot receive the signals very well. Think back to your own experiences and list the places where this problem has occurred. Suggest reasons why this happened in each case.

(h) Why are portable telephones called "cellular" phones? How do they work?

Figure 2

Chapter 9 Review

Key Expectations

Throughout the chapter, you have had opportunities to do the following things:

- Describe the properties of static electric charges, and explain electrostatic attraction and repulsion. (9.1, 9.2, 9.3)

- Describe and explain several methods of charging and discharging objects. (9.3, 9.4, 9.5, 9.6, 9.7, 9.8, 9.9, 9.10)

- Use safety procedures when conducting investigations. (9.1, 9.4, 9.9)

- Investigate static electricity, and organize, record, analyze, and communicate results. (9.1, 9.4, 9.9)

- Formulate and research questions related to electrostatics and communicate results. (9.5, 9.6, 9.8, 9.9, 9.10, 9.11)

- Charge objects using several methods. (9.1, 9.4, 9.9)

- Explain practical applications of electrostatics. (9.3, 9.8, 9.11)

- Describe common electrostatic phenomena, and suggest and assess solutions to problems related to static electricity. (9.3, 9.5, 9.6, 9.7, 9.10)

- Explore careers requiring an understanding of electricity. (Career Profile)

KEY TERMS

charging by contact	induced charge
charging by friction	separation
charging by induction	law of electric
discharge	charges
discharge at a point	negative charge
electric charge	negative ion
electrical conductor	neutralize
electrical insulator	neutron
electron	nucleus
electroscope	photoconductive
electrostatic series	positive charge
electrostatics	positive ion
ground	proton
	static electricity

Reflecting

- "Studying static electricity helps us to understand the behaviour of particles of matter, to explain common effects produced by static charges, and to become aware of both its potential hazards and its many practical uses." Reflect on this idea. How does it connect with what you've done in this chapter? (To review, check the sections indicated above.)

- Revise your answers to the questions raised in Getting Started. How has your thinking changed?

- What new questions do you have? How will you answer them?

Understanding Concepts

1. Make a concept map to summarize the material you have studied in this chapter. Start with the word "electrostatics."

2. What is
   (a) a negatively charged ion?
   (b) a positively charged ion?

3. List six materials that are electrical conductors and six that are electrical insulators.

4. Describe three methods of discharging a charged object.

5. (a) List three methods of charging a neutral object.
   (b) List two examples of situations where neutral objects are charged by each of the methods in part (a).

6. Which particles in the atom move when electric charge is transferred from one atom to another? Explain why.

7. Explain the purpose of the electrostatic series. Describe a practical example to illustrate your answer.

8. (a) What is the function of an electroscope?

(b) Name the kind of electroscope that can be recognized as being charged, just by observation. Explain why.

(c) Explain how to identify the kind of charge present on (i) a charged pith-ball electroscope and (ii) a charged metal-leaf electroscope.

9. Describe and explain, with the aid of diagrams, what happens when (a) a negatively charged and (b) a positively charged object is brought up to an uncharged pith-ball electroscope.

10. Explain how to identify the unknown charge on an object, using a pith-ball electroscope.

11. If you rub a comb in your hair and bring it close to some small pieces of paper, some of the pieces jump toward the comb, and then quickly jump off it again. Explain why.

12. Explain why an electric charge quickly builds up on the surface of furniture when it is being polished, but not on water taps.

13. What does it mean to "ground" an object?

14. When static electricity is discharged rapidly, what forms of energy can be produced? List examples to support your answer.

15. If you were given only a negatively charged strip, how could you charge a metal-leaf electroscope (a) positively and (b) negatively?

Applying Skills

16. Explain how you could determine the charge on your comb after you comb your hair. Design a test and carry it out. What is the kind of charge on the comb?

17. Consider the following interactions between various combinations of four charged pith balls, A, B, C, and D. B repels A. D attracts C and A. If D is attracted to an acetate strip that has been rubbed with silk, what are the charges on A, B, and C?

18. (a) Describe what happens when a negatively charged object is touched to a metal-leaf electroscope, and then removed.

(b) Draw a series of diagrams to show what happens to the movement of electric charges on the electroscope in part (a).

19. (a) Suppose you have a positively charged acetate strip, and two uncharged metal spheres mounted on insulating stands. Describe how to use the charged strip to charge one sphere positively and one sphere negatively by electrostatic induction.

(b) Explain what happened to the charges on the two spheres. Draw a series of diagrams to illustrate your answer.

(c) Describe how you can check the kind of charge on each sphere.

20. The Bohr-Rutherford model for the structure of the atom was developed after much careful experimentation and thought by many brilliant scientists. An interesting problem arises if you think about the structure of the nucleus itself. Based on the Law of Electric Charges, what should happen to the protons in the nucleus? What does happen? Try to develop a theory to explain why the protons behave as they do.

Making Connections

21. Why do some cars and trucks have wires and chains hanging underneath them?

22. (a) How does a lightning rod protect a house during a thunderstorm?

(b) List three safety precautions that you should take inside a house during a lightning storm.

(c) List three safety precautions to follow if you are outside during a lightning storm.

23. You want to buy a new comb to reduce the problem of flyaway hair on dry winter days. What should the comb be made of? Explain why. (Hint: Review Table 1 in 9.3 and Table 1 in 9.6.)

24. When you polish a metal ornament in the kitchen, dust particles are attracted to it. However, when you polish a nearby metal water tap, it does not attract dust particles. Explain these two situations.

25. Are all makes of cars equally safe during a lightning storm? Explain your answer.

26. Which has more need of lightning protection, a wooden barn or a steel skyscraper? Explain.

27. (a) List two major areas of application in which electrostatic devices are used.

(b) For each area, state the properties of electric charges that are used to advantage.

10 The Control of Electricity in Circuits

Getting Started

1 We use lights for so many different purposes that we tend to take them for granted. Walk through each room in the house and count the number of light sources in each room. Don't forget to include lights that might not actually be glowing or flashing at the time you look at them. There will be a number of them in the basement, garage, outside and inside the car, and outside the house.

How many different kinds of light sources that are operated with electrical energy are there in your home and in the car? To summarize your results, draw up a table that lists the type of light source (bulb), the number of each type, the operating voltage, the size, and the colour.

What proportion of the total electrical energy used in your home do you think is used up by operating light bulbs?

2 Has this ever happened to you? The big day arrives, and you finally get a gift you have been hoping for. You eagerly tear open the package, take out your gift, turn it on and—nothing happens! You look at the box and there is the answer. It reads "Batteries not included. Requires 2 AA alkaline cells."

Every day we use electrical and electronic devices that we can carry around with us. They require a portable source of electricity that comes in the form of a battery or dry cell. Look in and around your home and find devices that require batteries. Make a table in which you record the following: Device, Type of Battery, Number Required, Voltage, Rechargeable or Disposable.

Why are there so many different kinds and sizes of cells and batteries? Although we often use the terms "cells" and "batteries" interchangeably, they are not the same thing. What is the difference between them?

3 It is not by accident that electrical energy is the energy of choice for modern society. It has many advantages over the other forms of energy we could use to operate all the different appliances found in our homes. Just imagine what it would be like to have appliances in the kitchen and laundry room operated by little gasoline engines, similar to the ones used on lawnmowers.

However, from our very earliest days when we are first able to understand our parents' warnings, we are constantly made aware of just how dangerous electricity can be. There are many situations in our daily lives where there is the potential for electricity to harm us. Electrical outlets have the potential to be dangerous, depending on the room they are in. Extension cords, which can be dangerous, are continually being used to operate appliances and devices, both inside and outside the house or apartment.

What are some of the ways in which a careless person could receive an electric shock? Look around your home and list some of the safety features that have been designed into the electrical equipment you use.

Reflecting

Think about the questions in **1**, **2**, **3**. What ideas do you already have? What other questions do you have about electric circuits? Think about your answers and questions as you read this chapter.

Try This   Make a Cell

In earlier grades you may have made a cell using pieces of metal and an orange or lemon. These kinds of cells can be made with common materials available in your home. You can use such things as aluminum foil, the copper tubing used for plumbing, and the zinc on galvanized nails for the metal plates of a wet cell. You will need some wire to connect the pieces of metal together.

Try making a 3-V flashlight bulb glow with just one cell. What happens? Look at the number of dry cells in one of the larger flashlights. Try to do something similar with one or more oranges, potatoes, or some other fruit or vegetable. What arrangement did you need to light the bulb? Explain why.

The Electric Circuit

Use a flashlight, plug in an electric fan, or turn on the defroster on the back window of the car. In each case you are changing electrical energy into another form of energy. In this activity you will construct a simple electric circuit that operates safely and can be controlled. You will also determine the function of each of its parts.

Question

What are the components of a complete electric circuit?

Hypothesis

A source of electrical energy, connecting wires, a switch, and a device operated by electricity can be connected to create an electric circuit.

Materials

- dry cell or battery (with holder)
- switch
- light bulb (with holder)
- connecting wires (conductors)
- small electric motor

Procedure

1 Study the electric circuit shown in **Figure 1**.

(a) Draw the diagram. Do not draw the 5D wires connecting the parts of the circuit together.

⚡ Ensure that the switch is "open," as shown in **Figure 1**, before connecting the wires in the electric circuit. When the arm on the switch is up, it is "open," and when it is pushed down, the switch is said to be "closed."

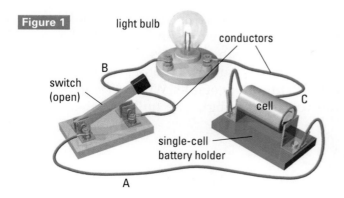

Figure 1
light bulb
conductors
switch (open)
B
cell
C
single-cell battery holder
A

2 Place the dry cell, the (open) switch, and the light bulb on your desk, in the positions shown in **Figure 1**.

3 Identify the negative terminal of the dry cell and connect a wire from it to one side of the switch.

(a) Draw a line on the circuit diagram to show which wire has been connected.

4 Connect a wire from the other side of the switch to the bulb.

(a) Draw the appropriate line on your circuit diagram.

5 Connect a wire from the other side of the bulb to the positive terminal of the cell.

(a) Draw the appropriate line on your circuit diagram.

⚡ Do not operate the circuit until it has been checked by the teacher.

6 Close and open the switch several times. Touch the light bulb.

✎ (a) Record your observations.

7 Close the switch. Disconnect and then reconnect each end of all three wires in turn. Open the switch.

✎ (a) Record your observations.

8 Close the switch. Remove the bulb from its socket, and then replace it again. Open the switch.

✎ (a) Record your observations.

9 Disconnect the wires attached to the light bulb holder, and reconnect them on the opposite sides of the holder. Close and open the switch.

✎ (a) Record your observations.

10 Disconnect the light bulb holder from the circuit and connect the electric motor in its place. Repeat steps 6 to 9 with the motor instead of the light bulb.

✎ (a) Record your observations.

Analysis and Communication

11 Analyze your observations by answering the following questions:

(a) What happens to the stored chemical energy in the dry cell when the switch is closed?

(b) What energy changes occur (i) in the light bulb and (ii) in the motor?

(c) What is the function of (i) the dry cell, (ii) the switch, (iii) the light bulb and the motor, and (iv) the wires?

(d) Which one of the four parts of the circuit can be omitted while allowing the circuit to continue to function? Why is it usually included in a circuit?

(e) List three different ways of turning the light bulb on and off.

(f) List the ways you can start and stop the motor.

(g) Would the circuit operate differently if (i) the connections on the switch were reversed and (ii) the switch were connected on the other side of the light bulb? If you aren't sure, try making the changes. Explain your answers.

(h) What happened when wire C was disconnected? Why does wire C have to be there?

(i) What effect did reversing the connecting leads have on (i) the light bulb and (ii) the motor? Explain your answer.

12 Write a summary paragraph in response to the question posed for this investigation.

Making Connections

1. Identify and describe three kinds of switches (a) in your home, (b) on electrical devices you use every day, and (c) in a car. Suggest reasons why different switches are used in different situations.

2. Identify three different kinds of electrical connecting wires used on electrical devices and appliances in your home and a car. Suggest reasons why different wires are used in different situations.

3. Think about toys with electric circuits in them that you used when you were younger or that children use now.

(a) What problems did you have with the electrical parts of the toys?

(b) How did you know how to replace the cells or batteries in the correct position or order? What did the toy manufacturer do to try to make sure you didn't put the cells in incorrectly?

(c) How has the electrical operation of toys improved in the past few years?

Exploring

4. (a) Predict what would happen if you connected the light bulb directly to the motor in the same circuit at the same time and closed the switch. Try it and comment on your prediction.

(b) How many different ways can you connect the light bulb and the motor directly to each other? What happens in each case? Explain why.

● Challenge

What components will you need for your electric circuit board?

Electricity and Electric Circuits

Static electric charge may build up to the point that it causes a discharge in the form of lightning strikes or a spark jumping from your hand to a doorknob. Whatever way it happens, electrical energy is transferred by the movement of electric charge. The movement, or flow, of electric charges from one place to another is called an **electric current**. A more detailed discussion of electric current, how it is measured, and the units it is measured in occurs on page 314.

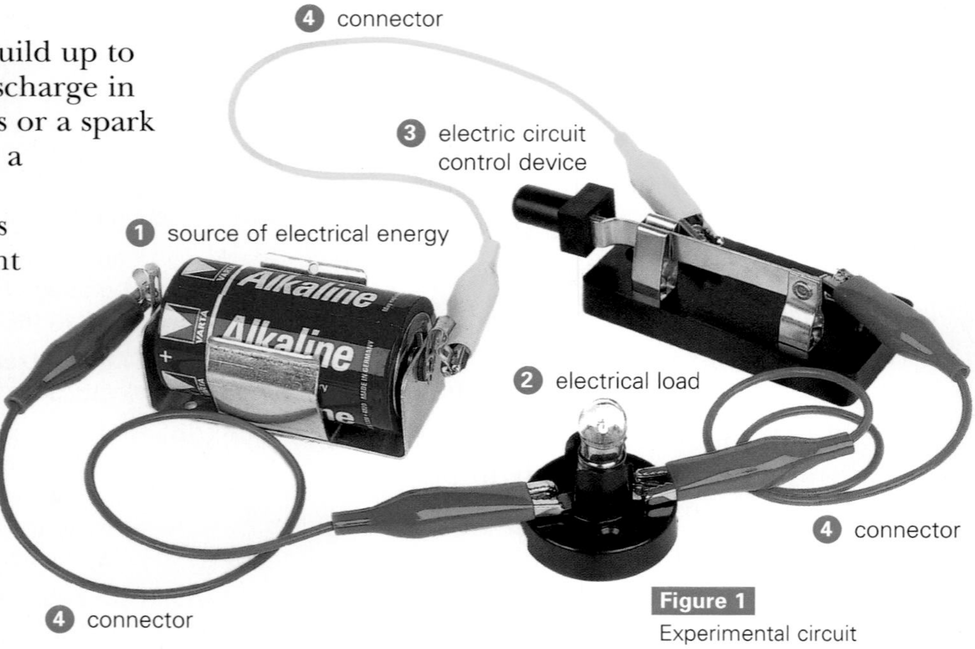

4 connector

3 electric circuit control device

1 source of electrical energy

2 electrical load

4 connector

4 connector

Figure 1
Experimental circuit

There is one very important difference between the electric current flowing during a lightning strike and that flowing through a light bulb in a flashlight. The current passing through the bulb is flowing in a controlled path called an **electric circuit**. Electric circuits are used to convert electrical energy into the other forms of energy we need.

The Parts of an Electric Circuit

A study lamp, a flashlight, and the experimental circuit shown in **Figure 1** look quite different. However, the electric circuits that operate all three of them are essentially the same. They all have the same four basic parts found in the simple electric circuit shown in **Figure 1**. These four parts are:

1. Source of Electrical Energy

Almost daily scientific and technological developments provide new ways of producing electrical energy. They range from the minute amounts of electrical energy generated for obtaining information from a computer hard drive to the large amounts produced at nuclear power stations. In between these two extremes are such sources as the photoelectric cells used in calculators, cells and batteries, portable generators, and of course wall outlets. Electrical energy is discussed in more detail in Chapter 11.

2. Electrical Load

Although the word "load" normally tends to imply something heavy, an **electrical load** is simply the name given to anything that converts

Did You Know

We often use the term "battery" instead of "cell." A battery is actually a combination of two or more cells.

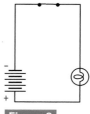

Figure 2
Schematic diagram of a closed circuit shown in Figure 1

electrical energy into whatever form of energy we need. The electrical load is actually the reason the electric circuit exists in the first place. The toaster you use at breakfast is an electrical load. What electrical loads have you used so far today?

3. Electric Circuit Control Device

The most obvious devices for controlling electric circuits are the simple switches we use in our homes, cars, and many kinds of portable electronic equipment. However, there are many more that we never see. They are often hidden inside the appliances, like the clock timer that controls the microwave oven. Many operate automatically, like the thermostat that controls the temperature of the house.

4. Connectors

One of the most amazing electric circuits is the microchip. The conducting wires, or **connectors**, used in these circuits are now so small that they are sometimes only a few atoms wide. However, whether they are the size of the wires on transmission lines or microscopic strands of wire 10 000 times thinner than a human hair, they all have the same purpose: to provide a controlled path for electric current to flow to each part of the circuit.

The words used to describe whether an electric circuit is operating often cause confusion. When a circuit is operating, and current is flowing, there is said to be a **closed circuit**. In the closed circuit shown in the photograph (**Figure 1**), the arm on the switch is connected to the other part of the switch, and the switch is said to be "on." When the arm of the switch is not connected to the other part of the switch, the switch is said to be "open" or "off" and there is said to be an **open circuit**. The electric current flows in a continuous loop from the negative terminal of the cell, through the wires, the switch, and the light bulb, and returns to the cell's positive terminal.

Electric Circuit Diagrams and Symbols

To simplify the drawing of electric circuits, a special set of symbols is used. This is much more convenient because we need to draw only one symbol for a switch, instead of different symbols for each kind of switch that exists or will be invented. Drawings of circuits using these symbols are called **schematic circuit diagrams** (**Figure 2**). (5D)

In these diagrams, the connecting wires are usually drawn as straight lines, with right-angled corners. This makes it easier to understand complicated circuits.

Understanding Concepts

1. Describe the difference between static electricity and current electricity.

2. Make a chart listing the parts of an electric circuit. State a function for each part and provide two examples.

3. In which direction does the electric charge flow around the circuit in **Figure 1**? What causes it to happen?

Making Connections

4. List four examples of electrical loads in the kitchen that convert electrical energy to (a) light energy and (b) mechanical energy. Predict which load uses the most, and the least, amount of energy.

5. List four different examples of electric control devices

   (a) in the kitchen;

   (b) the basement or laundry room;

   (c) in a car.

   Choose two devices from your list and suggest reasons for their design.

Exploring

6. What process is used to create the complex, multilayer circuit diagrams that make up the microchips used in computers? Visit the Internet sites of some of the major computer microchip manufacturers and find out how this is done. Create a flow chart describing the process.

Reflecting

7. Why are schematic circuit diagrams used rather than pictorial circuit diagrams?

Challenge

What should be included in the schematic circuit diagram for your challenge?

Electric Potential (Voltage)

Why is it safe to touch some sources of electrical energy and very unsafe to touch others? Most people know that if you touch both ends of a 1.5-V dry cell, whether it is a small AAA cell or a large D cell, you will not get an electric shock. However, everyone knows how dangerous it is to touch the terminals of a wall outlet.

The difference between the dry cell and the wall outlet is in the amount of energy that each electron receives from the energy source before moving into the electric circuit. The energy given to each electron leaving the terminal of a 120-V wall outlet is about 80 times greater than the energy given to each electron leaving the terminal of a 1.5-V dry cell. In fact, the energy of the electrons leaving the terminals of a wall outlet is great enough to cause a dangerous amount of electric current to flow through your body, giving you a severe electric shock.

Figure 1

A dry cell connected to a motor

A Model of Electric Potential

When you connect a simple electric circuit using a dry cell and an electric motor, the dry cell supplies electrical energy to the motor and causes the motor shaft to turn (**Figure 1**). However, you cannot actually see what is happening inside the circuit. To understand what is happening in a circuit, we can use an analogy to help us visualize it.

The energy in falling water has been used for thousands of years to turn water wheels. The way falling water gives up its gravitational potential energy to turn a water wheel can be compared with the way the electric charges released from a dry cell give up their energy to turn the electric motor.

In **Figure 2a**, a pump has lifted a large amount of water to the top of the building. Once the pump has lifted the water to the top, we say that the water has some stored or potential energy. If the water is now allowed to pour out on to a water wheel, the energy of the falling water can be used to turn the wheel. As long as the pump inside the building gives the water at the bottom potential energy by pumping it back up to the top, it will again be available to fall from the top of the building to keep the wheel turning.

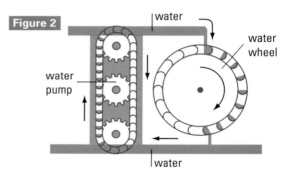

Figure 2

a The gravitational potential energy given to the water by the pump is released to turn the water wheel.

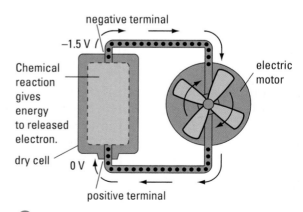

b The electric potential energy released by the chemical reaction in the cell is used to turn the electric motor.

In the same way that many molecules of water gain energy by being lifted up to the top of the building, a dry cell gives a huge number of electric charges (electrons) a certain amount of electric potential energy (**Figure 2b**). The energy each electron gains is released to it by the chemical reactions that occur. Just as the water molecules have gravitational potential energy at the top of the building, the electrons have electrical potential energy at the negative terminal of the dry cell.

Electricity from Chemical Reactions

Even if the dry cell is not connected to a circuit, the stationary electrons at the negative terminal have electric potential energy. At the instant the dry cell is made, a chemical reaction occurs, and the stored energy released by the reaction exerts a force on the electric charges and pushes a certain number of them on to the terminals of the dry cell. An excess of electrons accumulates on the negative terminal, thus giving it a negative charge, and a matching number of positive charges remain on the positive terminal.

Electrons can be released with their electric potential energy in a circuit only when the switch is closed and electrons can flow completely around the circuit. Any further chemical action that occurs in the dry cell will happen only if, for every electron that leaves the negative terminal to move into the circuit, another electron at the other end of the circuit enters the positive terminal to replace the one that left. These electrons are required to complete the chemical reaction. Whenever a current flows in an electric circuit, there is a continuous, unbroken chain of moving electrons in the circuit. As the electrons move through the circuit, they release the energy to the electrical load in the circuit.

In **Figure 2b**, the electrical energy is carried through the circuit by the electrons and is used to turn the electric motor. The energy each electron has is called the **electric potential** of the electron. Electric potential is commonly referred to as **voltage**. The SI unit used to measure electric potential is the **volt**, and the symbol for this unit is V. **Table 1** lists some sources of electric potential, with typical voltage values.

Table 1

Source of Electric Potential	Voltage (volts)
tape playback head	0.015
human cell	0.08
microphone	0.1
photocell	0.8
electrochemical cell	1.1 to 2.9
electric eel	650
portable generators	24, 120, 240
wall outlets in house	120, 240
generators in power stations	550

Understanding Concepts

1. (a) Why is it necessary for the electrons to move continuously around the circuit?

   (b) From which terminal do the electric charges flow into the circuit? Explain why.

2. (a) Define the term "electric potential."

   (b) State the SI unit and name the symbol used for electric potential.

3. Why is it possible to measure an electric potential across the terminals of a dry cell, even if electrons are not flowing into the circuit?

4. Explain, in terms of the energy of the electrons, why someone would receive a severe electric shock from a 120-V source, yet hardly notice the electric shock from a 6-V battery.

Exploring

5. Research the typical voltages (3A) generated by computer hard drives, VCRs, and tape cassette players. Record your findings as an information fact sheet for electrical equipment.

Reflecting

6. Make your own analogy using marbles to help you visualize how electrons can move continuously through a circuit.

Master Electrician

When Shelley Harding-Smith was a little girl, she was fascinated by the debris that came home from her father's various jobs as a master electrician. When she was ten, she went to work with her father. Although she didn't see any women on the job, the work appealed to her.

After a "typical high school program of language, social sciences, mathematics, and science," Harding-Smith enrolled in a three-year electrical apprenticeship program at St. Clair College, involving on-the-job training as an indentured apprentice and three months a year of classes on the college campus. She was the only woman in a class of 34 men.

After graduation, Harding-Smith first worked with a small firm doing residential, commercial, and industrial work. Attracted by the challenge of the new field of robotics, however, she returned to St. Clair to complete two years of training as a robotic electronics technician. She subsequently worked at a firm that used robots to weld automobile bumpers, then as a maintenance electrician at an amusement park, meanwhile obtaining her master electrician's certificate.

Harding-Smith started her own electrical contracting business, hired her son as an apprentice, and proudly watched him become a third-generation licenced electrician.

But the robotics field still intrigued Harding-Smith. She went to work at Chrysler Canada, at first with only two welding robots; now servicing and trouble-shooting the many robots occupies approximately 75 electricians per shift. Harding-Smith calls her work "challenging, rapid-paced, and exciting," and she enjoys the interaction with the other workers.

> Recognize the great opportunities for a career where you constantly progress, constantly grow, and are well rewarded.

Exploring

1. Check community colleges and technical schools in your area for courses in trades related to electricity. What high school qualifications are needed for admission into these courses?

2. Research and compare the advantages of working for a large company and being self-employed.

Building a Simple Wet Cell

When you began this chapter you made a table describing the many types of cells or batteries that you found around your home. No matter how many different kinds of dry cells exist, they all operate using the same principle—a chemical reaction that releases electric potential energy to the electrons. In this activity you will make your own wet cell.

Materials

- zinc plate
- copper plate
- steel wool
- 150-mL beaker
- light bulb
- light bulb holder

- voltmeter
- connecting wires
- dilute sulfuric acid
- small piece of insulating material
- small brush
- safety goggles

Sulfuric acid is corrosive. If it touches your skin, wash with cold water and inform your teacher.

Procedure

1 Put your safety goggles on. Polish the zinc and copper plates using steel wool.

2 Place the zinc and copper plates in the beaker. Place a small piece of insulating material between the two metal plates.

3 Connect the light bulb and the voltmeter to the two metal plates as shown in the schematic circuit diagram in **Figure 1**.

✎ (a) Record your observations. Record the
5E voltmeter reading.

4 Pour about 50 mL of dilute sulfuric acid into the beaker.

✎ (b) Record your observations
    (i) immediately and (ii) after 4 min.
    Record the voltmeter readings.

5 Using the brush, sweep the bubbles off the plates.

✎ (c) Record your observations. Record the voltmeter reading.

6 If plates of different metals are available, try different combinations of two plates.

✎ (d) Record the voltmeter readings. Design
6D a suitable table to display the data.

7 Return used solutions to containers designated by your teacher. Rinse, clean, and dry all equipment. Wash your hands.

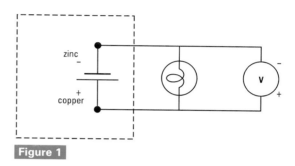

zinc
−
+
copper
V
−
+

Figure 1

Understanding Concepts

1. List the energy changes that take place (a) in the wet cell and (b) in the bulb.

2. Draw the schematic circuit diagram shown in **Figure 1**; draw an arrow on your circuit diagram to indicate the direction of current flow in the circuit.

3. (a) Why does the brightness of the light from the bulb change in step 4?

(b) How can this be overcome?

4. Explain the changes in voltage readings observed in steps 3 to 5.

5. (a) Why is it necessary to consider the positive and negative terminals of the wet cell when you connect the leads from the voltmeter to the circuit?

(b) What would happen if the voltmeter were connected incorrectly?

Reflecting

6. Compare the similarities and differences of the physical process of charging different materials by rubbing them together and the chemical process of producing electricity, using different metals in wet cells.

Electrochemical Cells

Primary Cells

There are two basic types of **primary cells**: the primary wet cell (voltaic cell) and the primary dry cell. In a primary cell, chemical reactions use up some of the materials in the cell as electrons flow from it. When these materials have been used up, the cell is said to be discharged and cannot be recharged.

The Primary Wet Cell

The primary wet cell, or **voltaic cell**, was developed in 1800 by an Italian scientist, Alessandro Volta. It is called a **wet cell** because it is made of two pieces of metal that are placed in a liquid. The metal plates, usually zinc and copper, are called **electrodes**. The liquid in the cell is called the **electrolyte**. An electrolyte is any liquid that conducts an electric current.

The zinc electrode reacts chemically with the sulfuric acid. The chemical energy released separates electrons from the zinc atoms. These electrons collect on the zinc plate, which is called the **negative terminal** of the cell. At the same time, positive charges collect on the copper plate, which is called the **positive terminal** of the cell. These electric charges remain static on each electrode. Cells discharge only when connected to a closed electric circuit.

Two major disadvantages of the wet cell are the danger of spilling the electrolyte, and the continual need to replace the zinc plate and the electrolyte.

The Primary Dry Cell

The familiar primary **dry cell** functions in the same way as a primary wet cell, but the electrolyte is a moist paste rather than a liquid (**Figure 1**). When most of the negative electrode has been used up by the chemical reaction, the electrons stop flowing, and the cell is discharged. Two common types of dry cells are shown in **Figure 2**.

The printed warnings on battery blister packs (**Figure 3**) are there for your safety. It is unsafe to recharge disposable (primary) cells and batteries. If you tried to recharge the cell, the energy supplied by the charging device would cause the cell to heat up, and the cell casing could break open, or the cell could even

Figure 1

Primary dry cells

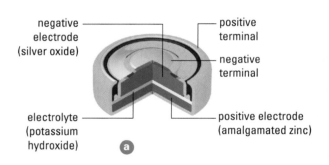

Figure 2

A cross section of two kinds of primary cells:
a a silver oxide cell and **b** an alkaline cell.

negative electrode (silver oxide) — positive terminal

negative terminal

electrolyte (potassium hydroxide) — positive electrode (amalgamated zinc)

a

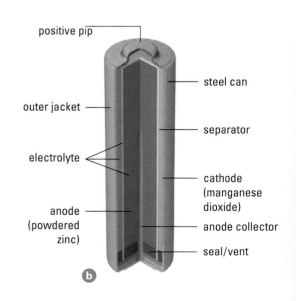

positive pip

steel can

outer jacket

separator

electrolyte

cathode (manganese dioxide)

anode (powdered zinc)

anode collector

seal/vent

b

explode. Also, because the outer casing of the cell is very tightly sealed, it is dangerous to throw a discharged cell into a fire. The gases formed inside may cause an explosion. If the cell's outer casing is punctured, corrosive liquids can leak out.

Over long periods of time, even new dry cells or batteries will gradually discharge. An expiry date is printed directly onto many dry cells. Dry cells and batteries in flashlights and lanterns, especially those used for emergencies, should be checked regularly to ensure they will operate when needed.

Secondary Cells

Unlike the single-use, disposable primary cell, a **secondary cell** can be discharged and recharged many hundreds of times. It is called a secondary cell because there are two chemical processes involved, one to discharge the cell, and another to recharge it to its original state. The secondary cell was initially developed to provide larger amounts of electrical energy economically, especially for cars, but now many different kinds of rechargeable cells are available. A car battery consists of a group of connected secondary cells.

Try This: Batteries on the Internet (3A)

Visit the web sites of some well-known battery manufacturers. Explore each of the sites to investigate different kinds of cells.

1. How many different types of primary and secondary cells can you find that are available commercially?
2. List the advantages and disadvantages of at least four types of secondary cells.

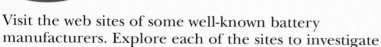

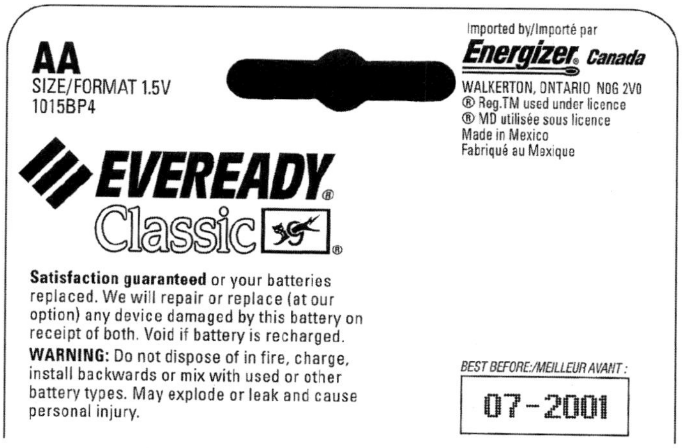

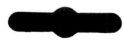

AA
SIZE/FORMAT 1.5V
1015BP4
EVEREADY®
Classic®
Satisfaction guaranteed or your batteries replaced. We will repair or replace (at our option) any device damaged by this battery on receipt of both. Void if battery is recharged.
WARNING: Do not dispose of in fire, charge, install backwards or mix with used or other battery types. May explode or leak and cause personal injury.
Imported by/Importé par
Energizer Canada
WALKERTON, ONTARIO N0G 2V0
® Reg.TM used under licence
® MD utilisée sous licence
Made in Mexico
Fabriqué au Mexique
BEST BEFORE:/MEILLEUR AVANT:
07-2001

Figure 3
This standard warning found on battery blister packs reminds us that a battery is an electrochemical device that needs to be used and disposed of with care.

Understanding Concepts

1. **(a)** What energy changes occur in an electrochemical cell when electric current flows from it?
   (b) Describe the conditions necessary in a voltaic cell to produce a steady supply of electrons.
2. **(a)** Explain why primary dry cells were developed.
   (b) Explain the difference between a primary cell and a secondary cell.
3. Why is it necessary for one electron to leave the circuit at the positive terminal of the cell every time an electron enters the circuit from the negative terminal?

Making Connections

4. How do manufacturers of devices that use batteries make sure you do not install the batteries incorrectly? Check some devices and describe the different solutions.
5. (1A) Make a safety chart on using batteries. List four safety precautions to be observed when using dry cells or batteries. Indicate what could result if each safety precaution is not observed. Use WHMIS symbols where appropriate.

Exploring

6. Which would be best for making a simple voltaic cell—a lemon or a potato? Why? Is size important?
7. Volta and Galvani were two scientists who had very different ideas about the effects caused by electricity. Carry out some research, and describe what they disagreed about, who was eventually proved to be correct, and why.

SKILLS HANDBOOK: (3A) Research Skills (1A) Safety Conventions and Symbols

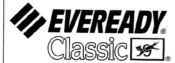

10.6 Investigation

SKILLS MENU
- Questioning
- Hypothesizing
- Planning
- Conducting
- Recording
- Analyzing
- Communicating

Batteries—Combinations of Cells

When you are on a camping trip and have only a small, single-cell flashlight to read with, how long will it last? Compare this with the amount of time for which a large lantern battery can produce light. The large battery can store much more energy than the small cell.

Dry cells can be connected together to increase the amount of energy available to operate the electrical load in the circuit. In this investigation, you will study the characteristics of each of the two basic kinds of electric circuits used to form batteries.

Question

1 Formulate a question about connecting cells in series and parallel and identifying (4A) the relationships among the variables.

Hypothesis

2 Restate the question in a testable form.

Materials

- 4 dry cells
- 4 dry-cell holders
- voltmeter
- connecting wires
- switch
- light bulb
- light-bulb holder

⚡ Be careful not to connect a wire from the positive terminal to the negative terminal on the same cell. When there is no light bulb to act as an electrical load, a connecting wire provides a "short" circuit for the electric current. When a short circuit occurs, very large currents flow from the cell that will cause it to heat up, and it may explode.

Procedure
Part 1: Connecting Cells in Series (5D)

3 Connect the positive terminal of a dry cell to the positive terminal of the voltmeter.

4 Connect the negative terminal of the cell (5E) to the negative terminal of the voltmeter.

(a) Draw the schematic diagram and record the voltmeter reading.

5 Disconnect the wire attached to the negative terminal of the cell. Connect the negative terminal of the first cell to the positive terminal of a second cell. Cells connected in this way are said to be connected in "series" (**Figure 1**).

6 Predict the voltage that will be produced by the two cells when connected in series. Reconnect the wire from the negative terminal of the voltmeter to the negative terminal of the second cell.

(a) Draw the schematic diagram, record your prediction, and then connect the voltmeter and record the reading.

7 Repeat steps 5 and 6, this time connecting a third cell to the second one.

(a) Draw the schematic diagram, record your prediction, and then connect the voltmeter and record the reading.

8 Repeat steps 5 and 6, this time connecting a fourth cell to the third one.

(a) Draw the schematic diagram, record your prediction, and then connect the voltmeter and record the reading.

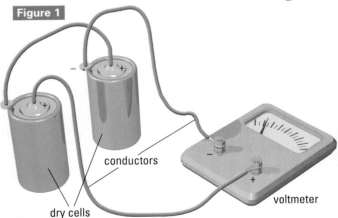

Figure 1

conductors

dry cells

voltmeter

a pictorial circuit diagram

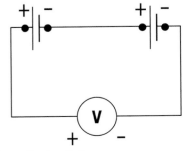

b schematic circuit diagram

SKILLS HANDBOOK: (4A) Asking Questions and Hypothesizing (5D) Drawing and Constructing Circuits

Part 2: Connecting Cells in Parallel

9 Connect the circuit as shown in **Figure 2**.

10 Close the switch, measure the voltage across the cell, and note the brightness of the bulb. Then open the switch.

✎ (a) Draw the schematic diagram and record the voltmeter reading and your observations.

11 Connect the negative terminal of the first cell to the negative terminal of the second cell, and the positive terminal of the first cell to the positive terminal of the second cell. The second cell is now connected in "parallel" with the first cell.

(a) Draw the schematic diagram.

12 Predict the voltage that will be produced by the two cells when connected in parallel, and what will happen to the brightness of the light from the bulb.

✎ (a) Record your prediction, then record the voltmeter reading and your observations.

13 Repeat steps 11 and 12, connecting a third cell in parallel with the first two cells.

✎ (a) Draw the schematic diagram, record your prediction, then record the voltmeter reading and your observations.

14 Review the observations you have made in both parts of the investigation, and organize and display them in a suitable format.

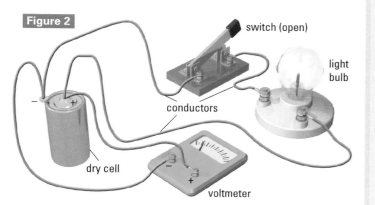

Figure 2

switch (open)

light bulb

conductors

dry cell

voltmeter

Analysis and Communication

15 Analyze your observations by answering the following questions:

(a) Comment on the validity of the predictions you made throughout the investigation.

(b) (i) What happens to the total voltage of a battery when its cells are connected in series?

(ii) What can be inferred about the electric potential of the electrons leaving the negative terminal of the battery as each new cell is added?

(c) (i) What happens to the total voltage of a battery when its cells are connected in parallel with one another?

(ii) What can be inferred about the electric potential of the electrons leaving the negative terminal of the battery as each new cell is added?

(d) What happens to the brightness of the bulb as more cells are added in parallel? Explain why.

Making Connections:

1. What changes have occurred and what features have been added in the way cells and batteries have been manufactured over the past few years? Which changes have been most useful to you? Why?

Exploring

2. How are cells connected together in car batteries? How many cells are there in most modern car batteries?

3. On a single piece of graph paper, plot a graph
(7B) of Total Voltage versus Number of Cells for the combinations of cells used in the investigation

(a) for cells connected in series;

(b) for cells connected in parallel.

From the graph determine how many cells would need to be connected in series to produce 15 V. How did you obtain your answer? Why was it possible to use this method?

Cells in Series and Parallel

Look at the collection of batteries in **Figure 1**. The little 9-V battery has six miniature dry cells in it. The big 6-V lantern battery has eight much larger dry cells. The flashlight requires three D size dry cells. All three combinations of dry cells produce different voltages. Dry cells can be connected together as a battery using two basic kinds of circuits: **series circuits** and **parallel circuits**. Which battery in the photo is connected using both kinds of circuits?

Figure 1
A 9-V battery, a 6-V battery, and three 1.5-V cells

Cells in Series

The electric potential given to a single electron by a dry cell has an absolute maximum value of slightly under 2 V. The value depends on the two materials used for the two electrodes of the cell. However, by connecting cells together in a series circuit, it is possible to obtain much higher voltages.

We can use the water analogy again. If we wanted to get twice as much energy from the same amount of water, we could simply lift the water twice as high as we did before (**Figure 2a**). In that way, it would gain twice the amount of gravitational potential energy, and then give up that energy turning the water wheel, as it fell to the bottom again.

We can achieve the same result with two dry cells by connecting them as shown in **Figure 2b**. When the switch is closed, and electrons flow around the complete circuit loop, the electrons get two boosts of energy instead of just one. Each time an electron leaves the negative terminal of cell 2 at D, another electron enters the positive terminal of cell 1 at A. When this occurs, two chemical reactions occur, one in each cell. The first reaction releases an electron at B with 1.5 V of electric potential. This means that all the electrons that go into cell 2 at C have the 1.5-V boost from cell 1. When each electron enters cell 2, a second chemical reaction takes place and a second boost occurs. The net result is that each electron leaving cell 2 to go into the circuit has a gain in electric

Figure 2

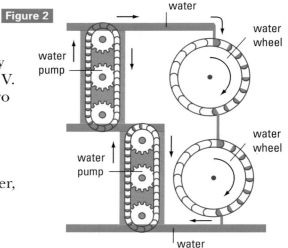

a Using two pumps connected in series the water gains twice the amount of gravitational energy and can turn two water wheels as it falls.

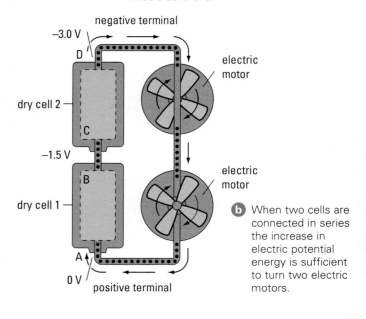

b When two cells are connected in series the increase in electric potential energy is sufficient to turn two electric motors.

potential of 3.0 V. Every time we connect a cell such that the negative terminal of one cell is connected to the positive terminal of the next one, the voltage of the combination of cells will increase by 1.5 V. Cells can be added in series indefinitely to increase the voltage of the battery.

Cells in Parallel

The cells used in many watches and calculators all have the same voltage value—1.5 V. These cells come in several different shapes and sizes. In each case, the size of the cell is based on the amount of electrical energy the device needs to operate for a reasonable amount of time (from the consumers' point of view).

There is, however, a practical limit for the size of a cell. How do you obtain more energy than can be given by just one cell? You connect the cells in parallel. Once again we can use the water analogy to explain what is happening to the electric charge from the cells (**Figure 3a**). If we have two pumping devices, each with its own supply of water, placed side by side instead of on top of one another, we can lift twice the amount of water to the top of the building and make it available to operate the water wheel. The water wheel would be able to operate for twice as long as it could with just one pump lifting the water. However, all the water would only have the gravitational potential energy from being lifted just one level.

Similarly, if we connect two cells side by side, or in parallel, with the positive terminals connected together and the two negative terminals connected together, there will be twice as many electrons available as there would be with only one cell (**Figure 3b**). Notice, however, that each electron released from the negative terminal will still only have the electrical potential energy gained from just one cell.

Understanding Concepts

1. Explain the difference between a cell and a battery.

2. **(a)** Why are cells connected in series?

   (b) Draw a schematic circuit diagram, showing five cells connected to produce the highest electric potential.

3. **(a)** Why are cells connected in parallel?

   (b) Draw a schematic circuit diagram showing four cells connected in parallel.

Exploring

4. Human and other animal cells (3A) typically produce a voltage of about 0.08 V. What is different about the way cells in humans are connected together compared with the cells in an electric eel? What are the maximum voltages and currents that can be produced by electric eels? Research these two questions and present your findings.

Challenge

Which combination of cells (if any) is appropriate for your electric circuit board?

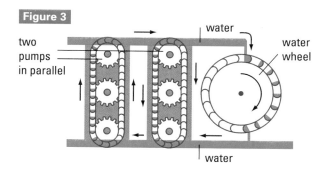

a When two pumps are connected in parallel, the water only gains the same amount of gravitational potential energy as with a single pump, but twice as much water is lifted. The pump can only turn one water wheel, but for twice as long.

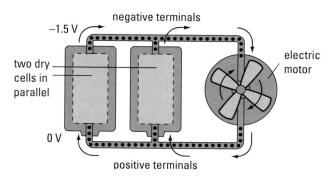

b When two cells are connected in parallel, the electric potential energy remains the same as that of a single cell, but with twice as much electric charge, the electric motor will operate for twice as long.

Figure 3

The Control of Electricity in Circuits **311**

Cells and Batteries: Costs and Benefits

Cells are a convenient, portable way of supplying electricity. They convert chemical energy directly into electrical energy. In voltaic cells the electrodes themselves are involved in the chemical reactions, so they can only supply a certain amount of energy before they are used up.

As scientists and engineers continue to design and develop new kinds of cells and batteries, they are using a variety of different substances for the electrodes and the electrolytes. Some of the substances being used in both types of commercially available electrochemical cells and batteries are listed in **Table 1**. Many other different combinations of substances are currently being investigated in research programs around the world.

A quick glance at the list of substances used in the different kinds of cells reveals that there are several areas that we should be concerned about if we wish to produce increasing numbers of cells and batteries in the future. These concerns include availability, cost, toxicity, and disposal.

Availability and Cost of Resources

Many of the substances used are quite rare and are found in specific deposits on the planet. The extraction of the elements from their minerals is an expensive process. In fact, some of the substances are so rare that it may not be practical to continue to use them, except in exceptional circumstances.

The issue of availability would be less of a problem if each cell or battery had a longer lifetime: if it were rechargeable. Rechargeable cells are now both available and suitable for most uses. The small rechargeable cells we recharge in our homes usually range from AAA to D size cells. Larger rechargeable batteries are used in cars and trucks, where they are automatically recharged by the vehicle itself. It is much more economical to use rechargeable cells than the single-use cells. However, rechargeable cells and batteries are much more expensive to manufacture and to purchase and may not be cost effective for some low-use devices such as smoke detectors.

Toxicity

Many of the substances used in cells and batteries are poisonous: lead and mercury are heavy metals, which can cause long-term health effects in a wide range of organisms; chlorine is a poisonous gas; lithium and sodium are highly reactive elements that require very careful handling. The more cells we use, the more of these substances we have around us.

Disposal

Both single-use and rechargeable cells eventually have to be replaced. Most of these cells and batteries are routinely tossed into the garbage container and dumped in the local landfill site. It has been estimated that about 50% of the mercury in our landfill sites comes from discarded cells and batteries. The uncontrolled disposal of millions of these cells, even though they are quite small, may lead to significant increases in pollution in areas around the landfill sites.

Recycling facilities are available (**Figure 1**), where used cells and batteries can be safely dismantled and their valuable components processed into a form that can be reused. However, this requires some effort on the part of the general public. Because of this, only a fairly small percentage of used cells are actually recycled. To make recycling a little more convenient, we could set up recycling boxes in stores that sell batteries.

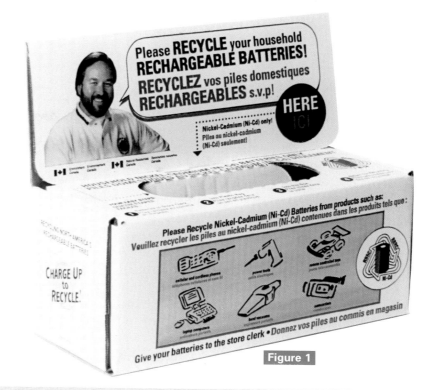

Figure 1

Table 1

Electrochemical Cells

Single Use	Rechargeable
alkaline	nickel/metal hydride
lithium/ion	nickel/cadmium
zinc/manganese	lead/acid
zinc/mercury	zinc/air

Issue Should battery recycling be mandated by law? 8B

Point

Opinion of an environmental scientist
The continued uncontrolled dumping of these toxic substances into the landfill sites over many years may lead to severe and long-term damage to the environment. If we provide a safe and convenient way to collect, process, and reuse these discarded cells and batteries we will be able to avoid significant pollution problems. We know the problem exists—we should do something about it now.

Opinion of a research scientist
If we know we will be able to recycle some of the rarer or more toxic substances, we may be encouraged to develop cells that do not use up these materials in the chemical reactions while the cell is being used.

Counterpoint

Opinion of a citizen
It will be inconvenient to have to take all the cells to the recycling centres. Besides, not all of the substances are harmful to the environment. I would have to take the cells out of some of the things I throw away, and I might not be able to take them apart.

Opinion of a store owner
The handling and storage of these discarded cells sounds dangerous, and I would have to train my staff to do this. In addition, I may have to provide a special storage area to keep them until they are collected. How am I going to be paid for providing these extra services?

What Do You Think?

Decide how you feel about this issue, and write your thoughts and reasons in the form of a position statement. Present your opinions in a letter to your local member of parliament or prepare a short speech to present at a local council meeting.

Electric Current

We have all experienced irritating electric shocks from static electricity, especially in dry, cold weather. But shocks from electric circuits are quite another matter. Every year people are injured and sometimes die from electrocution. Surprisingly small amounts of electric current can be lethal.

As you learned earlier, an electric current is made up of moving electric charges. In solids it is only the negative electric charges on electrons that move through the circuit. The positive charges on protons remain in their fixed positions in the atoms. Electric current is a measure of the rate at which electric charges move past a given point in a circuit. The metric SI unit used to measure electric current is the **ampere**. The symbol for the ampere is A. Slightly less than one ampere (1 A) of current flows through a 100-W light bulb in a lamp connected to a 120-V circuit. (**Table 1** lists the electric current required to operate some electrical loads.) Current is measured using an ammeter connected to the circuit in series.

Comparing Static and Moving Electric Charges

Static electricity is electric charge that remains in a fixed position on an insulator and distributes itself over the entire surface of a conductor. Static electric charges can be transferred by friction, contact, and induction.

Current electricity is electric charge that moves from a source of electrical energy in a controlled path through an electric circuit. The electrical energy of the moving electric charge can be converted into other desired forms of energy using a wide variety of electrical loads.

Human Response to Electric Shock

One of the most common misconceptions about electric shock concerns how much current is required to kill a person. A surprisingly small amount of current is lethal. That is one reason why it is important to read the safety precautions in the operating manuals of any kind of electrical device or equipment.

The electric potentials that cause muscle movement in the human body are produced by nerve cells and are typically about 0.08 V. When muscles are stimulated by electrochemical

Table 1

The Electric Current Ratings of Some Common Electrical Loads

Electrical Device	Electric Current (amperes)
electronic wrist watch	0.000 13
electronic calculator	0.002
electric clock	0.16
light bulb (100 W)	0.833
television (colour)	4.1
electric drill	4.5
vacuum cleaner	6.5
stove element	6.8
oven element	11.4
toaster	13.6
water heater element	27.3
car starter motor (V-8)	500.0

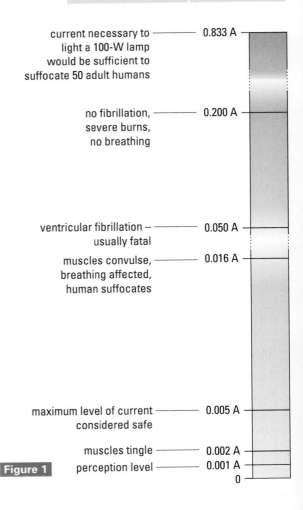

current necessary to light a 100-W lamp would be sufficient to suffocate 50 adult humans — 0.833 A

no fibrillation, severe burns, no breathing — 0.200 A

ventricular fibrillation – usually fatal — 0.050 A

muscles convulse, breathing affected, human suffocates — 0.016 A

maximum level of current considered safe — 0.005 A

muscles tingle — 0.002 A

perception level — 0.001 A

0

Figure 1

impulses from the nerve cells, the fibres in the muscle cells contract. The larger the electric current, the more strongly the muscles contract.

When a part of someone's body touches a source of electricity, and there is a complete circuit, an electric current flows through the body. If the current is large enough, the muscles in the part of the body in which the electric current is flowing automatically contract and remain contracted until the electric current ceases. The chart in **Figure 1** shows the effects produced by varying amounts of electric current. The amounts of electric current listed are average values.

Most people do not feel anything if the current is below 0.001 A, but there is a tingling sensation if the current is about 0.002 A. When the electric current is about 0.016 A, the muscles contract or convulse. This level of electric current is sometimes referred to as the "let-go threshold," because if the current is above that value, the person cannot let go of the object giving the electric shock. If the electric current is flowing from one hand to the other through the chest, the breathing muscles may become paralyzed, and the victim will suffocate unless the current is stopped.

If a current of 0.050 A or more passes through the chest, the heart muscles stop their regular pumping action and merely flutter. This fluttering of the heart muscles is called ventricular fibrillation. After a few seconds, the victim will become unconscious. The only way to stop ventricular fibrillation is to restart the pumping of the heart muscles by means of another controlled electric shock, such as you have probably seen on television or in the movies when a doctor uses defibrillator paddles on someone whose heart has stopped. This treatment must be administered as soon as possible. Electric currents above 0.200 A usually cause severe burns.

Before helping a victim of electric shock, ensure that you cannot receive a shock yourself. The electric current must be turned off, or the victim must be pulled from the danger zone with a nonconducting object, such as a piece of wood.

Understanding Concepts

1. **(a)** Define the term "electric current."
   (b) State the SI unit and name the symbol used for electric current.

2. **(a)** What kind of electric charges move through solids to form an electric current?
   (b) Why is it that only this kind of charge moves in solids?

3. **(a)** Why is it necessary to consider the positive
   (5E) and negative terminals of the ammeter when you connect leads from it to the circuit?
   (b) What happens if the ammeter is connected incorrectly?

4. Compare static and current electricity, and describe how you would use a measuring device to demonstrate at least one characteristic of each.

Making Connections

5. Why is it dangerous to try to help someone who is experiencing an electric shock? Explain what you should do if you wish to help the person.

6. **(a)** Why is it necessary to help a person who is suffering from ventricular fibrillation as soon as possible?
   (b) What treatment is necessary, and why do the medical personnel use the equipment for only a short period of time on the patient?

Exploring

7. A number of organisms stun or kill prey using
   (3A) electric shocks. Others sense danger by responding to electric fields given off by other organisms. Research these topics, using the library or the Internet, and prepare an oral report for the class.

Reflecting

8. Design a poster to inform others about why and how working with or using electrical devices can be dangerous to the human body.

Electrical Resistance and Ohm's Law

Why do we use electric circuits at all? Think about all the ways you use electricity in a typical day. Every time you use electricity, electrical energy is changed into heat, sound, light, or mechanical energy by many different kinds of electrical loads. Each electrical load actually performs a useful task for us.

There are thousands of different kinds of loads, and each has been designed to operate with a specific source of electrical energy (**Figure 1**). A digital watch or portable CD player is a load that uses a particular size and type of dry cell. An electric kettle has been designed so that the heating element (coil) inside the kettle is the correct size to heat water quickly and safely when plugged into a 120-V outlet.

Electrons are able to move easily through the atoms of a conductor. In a good electrical conductor, such as copper, the electrons lose very little energy when they collide with the copper atoms as they move through. In other materials, such as the tungsten filament in a light bulb, the electrons lose much more of their energy. As a result of the collisions with the tungsten atoms, the electric potential energy of the electrons is converted to thermal energy, and the filament becomes so hot that it glows brightly.

Figure 1
The photographs show how electricity can be transformed into heat, light, motion, or sound.

Electrical Resistance

The molecules of all types of conductors impede, or resist, the flow of electrons to some extent. This ability to impede the flow of electrons in conductors is called electrical **resistance**. Some kinds of electrical devices used in circuits are designed for this purpose and are called **resistors**. The symbol for electrical resistance is R, and the SI unit is the ohm (Ω). The resistance of a 100-W light bulb is about 144 Ω.

When electrons flow through a conductor, the electrical resistance causes a loss of electric potential (voltage). There is a "difference" in the amount of electric potential after the electrons have flowed through the conductor. Physicists refer to this loss as electric potential difference, or more simply, **potential difference**.

In 1827, the German scientist Georg Ohm (1789–1854) discovered a special relationship, now called Ohm's law, that exists between the potential difference across a conductor, such as copper wire, and the electric current that flows through it. Ohm's law states that *the potential difference between two points on a conductor is proportional (directly related) to the electric current flowing through the conductor.* The factor that relates the potential difference to the current is the resistance of the conductor or load. This very simple law is used to calculate the resistance of the load when designing many different electrical devices. Although potential difference is the correct term, the term **voltage drop** is commonly used instead. Voltage is lost or is "dropped" across the conductor.

Potential Difference = Electric Current × Electrical Resistance
  (Voltage Drop)

$$V = I \times R$$

Where potential difference (voltage drop) (V) is measured in volts (V), electric current (I) is measured in amperes (A), and resistance (R) is measured in **ohms** (Ω).

Table 1 lists the resistance of some electrical loads and the electric currents and voltages required to operate them. Ohm's law only applies to types of electrical loads called **ohmic resistors** that do not change electrical resistance with temperature.

Table 1	**Resistance of Some Electrical Loads**		
	Voltage Drop = V =	**Current** x I x	**Resistance** R
Electrical Load	**Voltage Drop (V)** (volts)	**Current (I)** (amperes)	**Resistance (R)** (ohms)
flashlight bulb	6.0	0.25	24
light bulb (60 W)	120	0.50	240
coffee grinder	120	1.20	100
food dehydrator	120	4.60	26
toaster oven	120	14.0	8.6
water heater	240	18.75	12.8

A Short Circuit

One of the warnings printed on battery blister packs states "Do not carry batteries loose in your pockets or purse." This warning is to prevent a dangerous condition known as a "short circuit." Let's suppose some dry cells get mixed up with a set of keys in a tote bag. Sometimes the positive and negative terminals of a dry cell become accidentally connected by the metal keys. There is now a complete, but very "short" circuit, with no electrical load to use up the energy from the dry cell as the current flows through the keys. In the confined space, both the keys and the dry cell may actually become hot enough to start a fire.

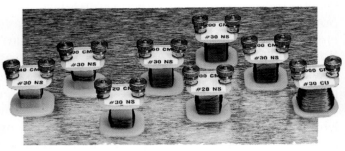

Figure 2

Many household appliances use resistors, as **Figure 2** shows. The equation for Ohm's law can be used to design electric circuits so that the resistance of the electrical load is properly matched to its energy source. For example, the tiny coil of tungsten wire (the glowing filament) inside a 100-W light bulb is just the right length so that it will glow very brightly as it uses up the electrical energy supplied by the wall outlet. Let's use Ohm's law to calculate the voltage drop across the light bulb.

Solving Science Problems Involving Formulas

The procedure below provides a standard method for solving problems.

Sample problem 1: What is the voltage drop across the tungsten filament in a 100-W light bulb? The resistance of the filament is 144 Ω and a current of 0.833 A is flowing through it.

1. **Data:** Read the problem carefully and record all given quantities, using correct symbols and units. Also, record symbols and units for the unknown quantities. (Note: Most difficulties can be traced to omissions or errors made in recording given and unknown quantities.)

   I = 0.833 A
   R = 144 Ω
   V = ? V

2. **Equation:** Write the formula(s) related to the problem. Compare the data with the formula(s). Determine how the unknown quantities can be found using the formula(s).

   $V = I \times R$

3. **Substitute:** Ensure that units for given quantities are the same as those needed for the formula. Substitute given quantities into the formula.

   $V = (0.833 \text{ A})(144 \, \Omega)$

4. **Compute:** Compute the numerical answer. Record it with the correct units.

   V = 120 V

5. **Answer:** Write an answer statement in sentence form.

   The voltage drop across a 100-W light bulb is 120 V.

Sample problem 2: An electric toaster is connected to a 120-V outlet in the kitchen. If the heating element in the toaster has a resistance of 14 Ω, calculate the current flowing through it.

Data:
V = 120 V
I = ? A
R = 14 Ω

Formula:
V = I × R

Substitute:
120 V = I × 14 Ω

Compute:
$$\frac{120 \text{ V}}{14 \text{ Ω}} = I$$

I = 8.6 A

Answer: The current flowing through the toaster is 8.6 A.

Sample problem 3: The current required to operate an electric can opener is 1.5 A. What is its resistance if the supply voltage is 120 V?

Data:
V = 120 V
I = 1.5 A
R = ? Ω

Formula:
V = I × R

Substitute:
120 V = 1.5 A × R

Compute:
$$\frac{120 \text{ V}}{1.5 \text{ A}} = R$$

R = 80 Ω

Answer: The resistance of the can opener is 80 Ω.

 **Are You Resistant?**

Most multirange meters can measure electrical resistance as well as voltage and current. Set the multirange meter to its resistance scale, and hold one of the tips of the meter leads in each hand. First, record the resistance of your body with your hands dry, and then repeat the measurement after wetting your hands. Comment on the safety aspects identified by this activity.

Understanding Concepts

1. (a) Define the term "electrical resistance."

   (b) State the SI unit and name the symbol used for resistance.

2. (a) State Ohm's law.

   (b) What does the term "potential difference" mean when applied to an electric circuit?

   (c) Explain why it is reasonable to consider the terms "voltage drop" and "potential difference" to be equivalent to one another.

3. For a given voltage drop, what would happen to the electric current through the resistance if the value of the resistance was (a) doubled, (b) halved, and (c) five times as large?

4. What is a "short circuit" and why is it considered to be a safety hazard?

5. Calculate the voltage drop across the following electrical loads:

   (a) a resistance of 500 Ω that has a current of 1.4 A flowing through it;

   (b) a resistance of 39 Ω that has a current of 0.58 A flowing through it;

   (c) a resistance of 15 000 Ω that has a current of 0.08 A flowing through it.

6. Does the wire in the electrical cord of an electric kettle have a higher or lower resistance than the heating element inside the kettle? Explain your answer.

7. A 3-V battery sends a current of 0.10 A through a light bulb. What is the resistance of the filament of the bulb?

Making Connections

8. Compared with copper, is tungsten wire a high-resistance or low-resistance metal? How does this account for how these metals are used?

Challenge

Eliminating short circuits is an important consideration when designing electric circuits. How will you avoid short circuits in your design?

Ohm's Law

Ohm's law is one of the most useful relationships in the study of electric circuits. It is used in the design of circuits that range in complexity from toasters to those used in advanced computer systems.

In Part 1 of this investigation, you will study the relationship between voltage drop and electric current for special electrical loads known as "ohmic" resistors. In Part 2 you will study the relationship between voltage drop and electric current when an incandescent light bulb is used as an electrical load.

Question

How can the relationship between the voltage drop (potential difference) across an ohmic resistance and the electric current flowing through it be determined?

Hypothesis

If we measure the changes in voltage drop across an ohmic resistor for a series of values of electric current, we can plot the experimental data on a graph and determine the relationship between the two quantities.

Materials

- low-voltage power supply (variable from 0 V–6 V)
OR
- 4 D dry cells
- holder for 4 D dry cells
- 2 different resistors (39 Ω–100 Ω) – 1-W rating minimum
- ammeter
- voltmeter
- 6 connecting wires
- switch
- bulb (6-V rating minimum)
- bulb holder

Procedure

1 Before you begin, read the procedure and analysis.

(a) Design and draw a table to record your observations in Parts 1 and 2.

Part 1: Ohmic Resistors as Electrical Loads

2 Refer to the diagram below. (Note that the connecting lead from the switch is not attached and will be moved along the set of dry cells in sequence to obtain the required voltages.)

(a) Draw the schematic diagram for the circuit.

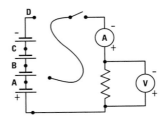

⚡ Ensure terminals are connected correctly to avoid short-circuiting the dry cells.

3 Construct the circuit you have drawn using the larger of the two resistor values (resistor 1). Do not close the switch. Ask your teacher to inspect your circuit before continuing.

4 Attach the connecting lead from the switch to the negative terminal of the battery holder at A. Close the switch and record the ammeter and voltmeter readings in the table. Open the switch.

✎ (a) Record your observations in the table.

5 Repeat step 4 three more times by connecting the lead from the switch:
(i) first to point B on the battery case;
(ii) next to point C on the battery case;
(iii) and finally to point D on the battery case.

✎ (a) Record your observations in the table for each step.

6 With the switch open, remove resistor 1 from the circuit and replace it with resistor 2. Do not close the switch. Ask your teacher to inspect your circuit before continuing.

7 Attach the connecting lead from the switch to the negative terminal of the battery holder at A. Close the switch and record the ammeter and voltmeter readings in the table. Open the switch.

✎ (a) Record your observations in the table.

8 Analyze the observations in the table for both resistors, then predict the values of voltage drop and current you expect to observe when you repeat step 5 for resistor 2.

✎ (a) Record your predicted values for voltage drop and current in the table.

9 Repeat step 5 for resistor 2.

✎ (a) Record your observations in the table in each step.

Part 2: An Incandescent Light Bulb as an Electrical Load

10 With the switch open remove resistor 2 from the circuit and replace it with the light bulb. Do not close the switch. Ask your teacher to inspect your circuit before continuing.

11 Repeat steps 4 and 5 for the light bulb.

✎ (a) Record your observations in the table in each step.

Analysis and Communication

12 Analyze your observations by answering the following questions:

(a) For each pair of values of V and I in the table, calculate the ratio of V/I and record it in the table.

(b) What quantity is represented by the V/I ratio for each resistor?

(c) In what way are the V/I values for the incandescent light bulb different from those of the resistors? Suggest reasons for these differences.

(d) On a single sheet of graph paper, plot
7B three separate graphs of V versus I for the two resistors and the light bulb. Plot I on the horizontal axis and label each of the three graph lines.

(i) From the graph line of resistor 1, record in the table the values of current for voltages of 1.0 V, 2.0 V, and 4.0 V. Then calculate the V/I ratio for these three values. What did you notice about the three V/I ratios? This ratio is known as the "slope" of the graph line.

(ii) Repeat step (i) for resistor 2. What was different about the "slope" of the graph line of resistor 2 compared with that of resistor 1?

(iii) Review your answers for (i) and (ii). Write a statement that links the concepts of slope for the V/I graph and the value of the resistors used in the investigation.

(iv) Compare the graph lines for the two resistors with that of the light bulb. What can you infer about the resistance of the light bulb as the current through it increased?

Making Connections

1. (a) What is the voltage rating of the bulb you used in this investigation?

(b) Why was it this value?

(c) What would have happened in this investigation if the voltage rating was i) 3.0 V, ii) 12 V? Explain your answers.

2. Suppose you had to replace a burned-out light bulb in a flashlight. Describe at least two ways that you could determine the correct voltage rating for the bulb.

Reflecting

3. Explain why an ammeter should be connected in series with a load, while a voltmeter should be connected in parallel with a battery or a load.

10.12 Investigation

SKILLS MENU
○ Questioning   ● Conducting   ● Analyzing
○ Hypothesizing   ● Recording   ● Communicating
○ Planning

Parallel and Series Circuits

There are many more examples of electrical loads connected in parallel than connected in series. Why is this so? When control devices such as switches are included as part of the circuit, together with the loads, you will find that most of the circuits we use are a combination of both series and parallel circuits.

Question

How can the characteristics of parallel and series circuits, including the relationships between voltage drop and current for each kind of circuit, be determined?

Hypothesis

If we measure the current flowing in and the voltage drops across the electrical loads in each kind of circuit, we can determine the relationships between current and voltage drop, and describe the characteristics of parallel and series circuits.

Materials

- 3 bulbs
- 3 bulb holders
- 4 D dry cells
- holder for 4 D dry cells

OR

- 1 6-V lantern battery
- voltmeter
- ammeter
- switch
- 12 connecting wires

⚡ Ensure terminals are connected correctly to avoid short-circuiting the dry cells.

Procedure
Part 1: Electrical Loads in a Parallel Circuit

1 Construct the circuit shown in **Figure 1a**. The ammeter will remain in the same position for the complete investigation. Ask your teacher to inspect your circuit before continuing.

(a) Draw the schematic circuit diagrams. Draw a table for your observations in Part 1 (**Table 1**).

Table 1

Supply Voltage V_{supply}	Voltage Drop across bulb V_{bulb}		Current I_{supply}		V_{supply}/I_{supply} ratio
	Predicted	Actual	Predicted	Actual	
?	?	?	?	?	?
?	?	?	?	?	?

2 Close the switch and note the brightness of the bulb. Connect the voltmeter first across the bulb and then across the battery. Open the switch.

✎ (a) Record the voltmeter readings in each case. Record the ammeter reading.

3 Connect the second bulb to the circuit, as shown in **Figure 1b**. Predict the voltmeter and ammeter readings. Repeat step 2.

✎ (a) Record your predictions and observations in the table.

4 Connect the third bulb to the circuit, as shown in **Figure 1c**. Predict the voltmeter and ammeter readings. Repeat step 2.

✎ (a) Record your predictions and observations in the table.

5 Remove one bulb from its socket, then close the switch.

✎ (a) Record your observations.

6 Open the switch, and replace the bulb in the socket.

7 Repeat steps 5 and 6 for each of the other two bulbs in turn.

Figure 1

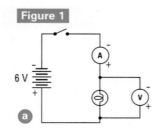
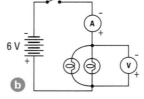
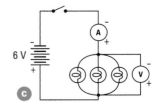

Part 2: Electrical Loads in a Series Circuit

8 Construct the circuit shown in **Figure 2a**. Ask your teacher to inspect your circuit before continuing.

(a) Draw the schematic circuit diagrams shown in **Figure 2**. Draw a table for your observations in Part 2.

9 Close the switch, and note the brightness of the bulb. Connect the voltmeter first across the battery and then across the bulb. Open the switch.

✎ (a) Record the voltmeter readings in each case. Record the ammeter reading.

10 Leaving the voltmeter connected across the first bulb, connect a second bulb in series with the first bulb, as shown in **Figure 2b**. Predict the voltmeter and ammeter readings. Close the switch.

✎ (a) Record your predictions and meter readings in the table.

11 Open the switch and connect a third light bulb in series with the other two bulbs, as shown in **Figure 2c**. Predict the voltmeter and ammeter readings. Close the switch.

✎ (a) Record your predictions and meter readings in the table.

12 With the switch closed, remove the first light bulb from its socket, then replace it in the socket. Open the switch.

✎ (a) Record your observations.

13 Close the switch and repeat step 12 for each of the other two bulbs. Open the switch.

Analysis and Communication

14 Analyze your observations by answering the following questions for both Part 1 and Part 2:

(a) How does the voltage measured across the dry cell compare with the voltage drop measured across each of the three bulbs? Explain your answers.

(b) What happens to the brightness of the light from each bulb as each new bulb is added?

(c) What can you infer about the amount of electric current flowing through each bulb as each bulb is added? Explain your answers.

(d) What can you infer about the total current flowing from the cell each time compared with the current flowing through each of the bulbs?

(e) Explain what happens when one of the bulbs is unscrewed.

(f) How many paths for current flow are there in each circuit?

(g) Calculate the V_{supply}/I_{supply} ratio.

(h) Explain why the V_{supply}/I_{supply} ratio changes as it does.

Figure 2

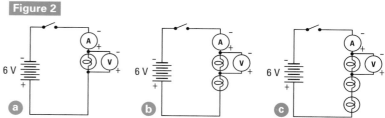

Making Connections

1. (a) Suppose 15 light bulbs were connected in series, and one bulb burned out. How could you find the defective bulb?

(b) How could you identify one defective bulb if the 15 bulbs are connected in parallel? Explain.

2. Are the electrical wall outlets in your home wired in series or in parallel? Explain.

3. (a) Why are power bars used so commonly for connecting computer systems?

(b) Are power bars examples of series or parallel circuit connections? Explain.

Electric Circuits with Multiple Loads

The cells and batteries in some electrical devices, such as calculators, simple cameras, and flashlights, operate only one electrical load at a time. However, when strings of decorative lights or the lights on a car are turned on, several electrical loads operate simultaneously (**Figure 1**). Two basic kinds of electric circuits are used to connect these loads: the series circuit and the parallel circuit.

The Series Circuit

The term "series" applies to any electric circuit in which the parts of the circuit are wired to one another in a single path. Have you ever had a set of minibulb lights that wouldn't light up when you plugged it in? These lights are connected in series and contain many bulbs in each circuit. If one bulb burns out, all the bulbs have to be checked to find the burned-out bulb. **Figure 2c** shows how the wires are connected to three bulbs in a series circuit.

We will use Ohm's law to analyze what is happening in the three-bulb series circuit. The circuit consists of a 9-V battery and light bulbs that each have a resistance of 9 Ω. In the simple circuit shown in **Figure 2a**, only one bulb is connected to a switch and a 9-V battery. When the switch is closed, the bulb lights up. A current of 1 A flows through the bulb and it glows brightly.

Figure 1

Which kind of circuit (series or parallel) would be most suitable to connect all the light bulbs on each of these trees?

Series Circuit with Multiple Loads

In **Figure 2b** a second bulb is connected directly after, or in series with, the first bulb. Notice that there is still only one single path for the current to flow through in the circuit. The current flowing through the circuit is now only 0.5 A, exactly half the current flowing with just one bulb in the circuit. The voltage drop across the two-bulb circuit is still the same 9 V, but the amount of resistance in the circuit is now doubled. This is because the two 9-Ω bulbs are in line with one another, and the resistances add together,

Figure 2

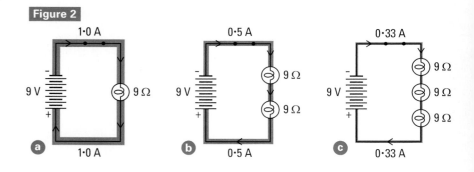

making the total resistance 18 Ω. Using Ohm's law, if the resistance in the circuit is doubled, then, for the same voltage drop across the circuit, the current will be halved.

Predict what will happen when we connect three bulbs in series, as shown in **Figure 2c**. There is still only one path through which the current can flow. The total resistance of the three-bulb circuit will now be tripled to 27 Ω. Because there is still the same 9-V battery connected to the three-bulb circuit, the current will now be only one-third of its original value—one-third of an ampere (0.33 A). Predict what the current would be if four of these bulbs were connected in series.

Characteristics of a Series Circuit

Look at the diagrams of the two-bulb and three-bulb circuits. For the two-bulb circuit, the effect on the battery is the same as if the two bulbs were replaced by just one load (resistor) with twice the resistance (18 Ω). For the three-bulb circuit, it is as though the bulbs were replaced by a single 27-Ω resistor load. As each load is added to a series circuit, the total resistance of the circuit increases as well. If the voltage remains the same, and the resistance increases, the current flowing in the circuit decreases.

The three light bulbs in **Figure 2c** are connected in a series circuit. When the switch is closed, all three bulbs produce light. When the switch is open, the three bulbs stop producing light. Notice that the three bulbs and the switch are all wired together, one after the other, to provide a single path for the electric current. The same electric current flows through the cell, the bulbs, and the switch. Another characteristic of a series circuit is that the electric current is the same in all parts of the circuit.

A burned-out bulb prevents current from flowing, just as if a switch were open. Removing a bulb from its socket has the same effect. Because there is only one path for the current to flow through, the other two bulbs will also stop glowing. If the path of the current in a series circuit is broken at any point, the current stops flowing. This is another characteristic of a series circuit.

Figure 3

The Parallel Circuit

Figure 3 shows three lights in a track-light fixture. Two of the bulbs are glowing, even though the other one remains unlit. The bulbs are connected in a parallel circuit, in which the current passes through a separate circuit to each bulb. Each separate circuit is called a **branch circuit**. Because each bulb is connected to its own branch circuit, it does not affect the other bulbs. If any one of the bulbs is removed from its socket, or the filament in the bulb breaks, all the other bulbs remain lit. **Figure 4** shows how the wires are connected to three bulbs in a parallel circuit.

Figure 4

Again, we will use Ohm's law to analyze what is happening in the three-bulb parallel circuit. We will use a 9-V battery and 9-Ω light bulbs. In the first circuit shown in **Figure 5a**, only one bulb is connected to a switch and a 9-V battery. When the switch is closed, the bulb lights up. A current of 1 A flows through the bulb and it glows brightly.

Parallel Circuit with Multiple Loads

In **Figure 5b**, a second bulb is connected beside, or in parallel with, the first bulb. Notice that the current can now flow through two separate paths or branch circuits. The two bulbs are identical, and because the voltage drop is the same in each case, the

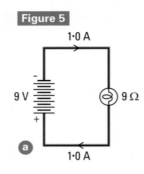 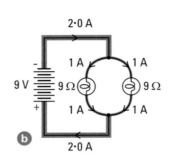 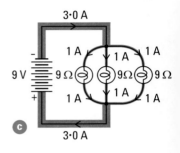

Figure 5

same current of 1 A will flow through each bulb. The current flowing from the negative terminal of the battery will now be 2 A, twice the amount it was with only one bulb. After passing through the bulbs, the two currents combine again and return to the battery. If either bulb is removed from its socket, the other one stays lit because each bulb has a separate branch circuit connected to the battery.

In **Figure 5c**, a third bulb is connected in parallel. The current can now flow in three separate paths around the circuit, so the current flowing from the battery will be 3 A, three times what it was when only one bulb was connected. One-third of the total current flowing from the battery passes through each bulb. In this circuit as well, each bulb can be switched on and off or removed from its socket without affecting the other bulbs.

Characteristics of a Parallel Circuit

Note that each time another bulb is added in parallel to the circuit, the current from the battery increases. In any parallel circuit, the total current flowing from the source of electrical energy equals the sum of all the separate branch currents in the circuit. The voltage drop across each branch circuit is 9 V, the same as that produced by the source of electrical energy.

An interesting effect of connecting loads in parallel is that the total current always increases when you add another load. Look at the diagrams for the two-bulb and three-bulb circuits. For the two-bulb circuit, the effect on the battery is the same as if the two bulbs were replaced by just one load (resistor) with half the resistance (4.5 Ω). In the case of the three-bulb circuit, the effect is the same as if the three bulbs were replaced by one load with one-third of the resistance (3 Ω). In each case the current will increase, because adding another bulb in parallel has the same effect as decreasing the effective (total) resistance

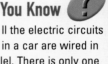

Did You Know

All the electric circuits in a car are wired in parallel. There is only one 12-V battery, but you can operate any electrical device in the car without affecting any of the others.

connected to the battery. If the voltage drop remains the same, and the resistance decreases, the current will always increase.

Large appliances in your home such as the clothes dryer, stove, and electric water heater are each connected to a separate branch circuit. On other branch circuits, as many as ten different lighting fixtures and wall outlets may be connected together in parallel. Because all the circuits are wired in parallel, any appliance, light, or wall outlet can be used without affecting the others. Almost all the electric circuits in your home are connected in parallel.

Combinations of Series and Parallel Circuits

Electric circuits often contain both series and parallel circuits that are combined together. Each parallel circuit in your home is connected to the main control panel and is controlled by a safety switch connected in series with the circuit. Every time you turn on the set of three bulbs in the bathroom fixture, you are actually using two switches in series with the three light bulbs that are connected in parallel. The first switch is the safety switch in the main control panel, and the second is the switch on the wall.

Try This Make a Circuit (5D)

Design a parallel electric circuit such that one bulb is controlled by a switch while two other bulbs are not and glow continuously. Draw the schematic circuit diagram. Construct the circuit to test your circuit design. Explain the operation of the circuit.

Understanding Concepts

1. (a) State two characteristics of (i) a series circuit and (ii) a parallel circuit.

   (b) What is meant by the term "branch" circuit?

2. What happens to the total current that flows in a

   (a) series circuit if another load is connected in series with the existing loads?

   (b) parallel circuit if another load is connected in parallel with the existing loads?

3. What effect does the change in current have on the effective resistance of the total circuit in 2 (a) and (b)?

4. Draw a schematic circuit diagram for each of the (5D) following:

   (a) Three cells are connected in series, which in turn are connected to two light bulbs, a motor, and a switch, also connected in series. A voltmeter is connected to the battery to measure its voltage.

   (b) Two cells are connected in parallel, which in turn are connected to three light bulbs connected in parallel. A switch is connected in series with just one of the light bulbs.

5. Design an electric circuit so that an electric motor is controlled by a switch. In addition, one bulb is to remain lit all the time, and another is to be lit only when the motor is operating. Draw the schematic circuit diagram. Construct the circuit to test your circuit design.

6. Try to design a light bulb that could be used in a series circuit so that, if the filament burned out, all the other lights would continue to glow.

Making Connections

7. (a) Why are the electric circuits in a house wired in parallel with one another?

   (b) The more expensive strings of coloured lights are connected in parallel. Explain why. Why are they more expensive?

Challenge

Consider which type of circuit you will need to build your electric circuit board.

Chapter 10 Review

Key Expectations

Throughout the chapter, you have had opportunities to do the following things:

- Compare static and current electricity. (10.3, 10.9)

- Explain the function of each part of an electric circuit. (10.1, 10.2, 10.4, 10.5)

- Describe the concepts of electric current, potential difference, and resistance, and their relationship to one another. (10.3, 10.9, 10.10)

- Determine resistance using Ohm's Law. (10.10, 10.11)

- Describe and compare the characteristics of series and parallel circuits. (10.7, 10.12, 10.13)

- Design, draw, and construct series and parallel circuits, and measure electric potentials and current related to the circuits. (10.4, 10.6, 10.12)

- Use safety procedures when conducting investigations. (10.1, 10.4, 10.11, 10.12)

- Investigate circuits, and organize, record, analyze, and communicate results. (10.1, 10.4, 10.11, 10.12)

- Formulate and research questions related to electric circuits and communicate results. (10.2, 10.5, 10.7, 10.8, 10.9)

- Describe practical applications of current electricity. (10.10, 10.13)

- Explore careers requiring an understanding of electricity. (Career Profile)

KEY TERMS

ampere	parallel circuit
branch circuit	positive terminal
closed circuit	potential difference
connector	primary cell
dry cell	resistance
electric circuit	resistor
electric current	schematic circuit
electric potential	diagram
electrical load	secondary cell
electrodes	series circuit
electrolyte	volt
negative terminal	voltage
ohm	voltage drop
ohmic resistor	voltaic cell
open circuit	wet cell

Reflecting

- "The electric circuits in our homes and in many electrical appliances consist of combinations of two simple electric circuits: the series circuit and the parallel circuit." Reflect on this idea. How does it connect with what you've done in this chapter? (To review, check the sections indicated above.)

- Revise your answers to the questions raised in Getting Started. How has your thinking changed?

- What new questions do you have? How will you answer them?

Understanding Concepts

1. Make a concept map to summarize the material that you studied in this chapter. Start with the words "electric circuit."

2. (a) What is a voltaic cell?
   (b) What is an electrolyte? Explain its purpose.

3. What is the major advantage of a secondary cell compared with a primary cell?

4. (a) What is an "ohmic" resistor?
   (b) Why are light bulbs not considered to be ohmic resistors?

5. State the characteristics of (a) a series circuit, and (b) a parallel circuit in terms of the electric current and voltage drops across electrical loads in each circuit.

6. (a) In what part of the circuit do the electric charges release most of their energy?
   (b) In what other parts of the circuit are very small amounts of energy released? Explain why.

7. (a) What happens when a short circuit occurs in an electric circuit?

(b) Why can a short circuit be dangerous?

8. Explain what happens when one bulb burns out in a circuit made up of six bulbs connected in series with one another?

9. Why are series circuits with more than three or four loads not very common?

10. Describe and explain what would happen in the circuit diagram shown in **Figure 1** below if
    (a) the switch is closed;
    (b) the switch is closed and light bulb 1 is unscrewed;
    (c) the switch is closed and light bulb 3 is unscrewed;
    (d) the switch is closed, and light bulb 6 is removed and replaced by a copper wire.

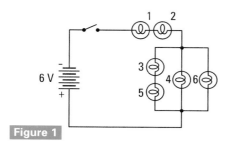

Figure 1

Applying Skills

11. Calculate the voltage drop across the following electrical loads:
    (a) A bulb that has 2.4 A flowing through it. The resistance of the bulb is 16 Ω.
    (b) A coffee grinder that has a resistance of 85.0 Ω and a current of 1.41 A flowing through it.
    (c) A current of 0.024 A flowing through a resistance of 750 Ω.

12. Determine the value of voltage indicated by the meter needle position in **Figure 2**.

Figure 2

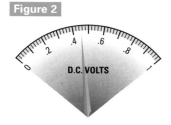

13. (a) What voltage is produced when five 1.8-V cells are connected in series? Draw a circuit diagram of this battery.

(b) What is the lowest voltage that can be obtained by connecting the five cells together? Draw a circuit diagram to show how to obtain this value.

14. What voltage is produced when three 1.3-V cells are connected in parallel? Explain why. Draw a schematic diagram.

15. (a) To measure the electric current flowing through a bulb, should an ammeter be connected in series or in parallel with the bulb? Explain your answer.

(b) To measure the voltage of a battery, how should a voltmeter be connected with the battery? Explain your answer.

16. Draw a schematic circuit diagram showing a 120-V source of electrical energy and two light bulbs connected in series, which in turn are connected to two light bulbs and a motor, all connected in parallel with each other. The complete circuit is controlled by a switch and protected by a fuse. An ammeter is connected to measure the current through the motor and a voltmeter measures the 120-V supply.

17. Design three different circuits that are combinations of series and parallel circuits, using light bulbs and switches to demonstrate your understanding of the characteristics of each kind of circuit. With your teacher's permission, construct them, have them checked, and test their operation.

Making Connections

18. List four household electrical loads and state the forms of energy each load produces.

19. List three electrical loads that use batteries with cells connected in series.

20. Make a list of devices that use batteries in your home, and identify how many cells are used in each case. Indicate whether the cells are connected in series or parallel. Display the information in a table.

21. Explain why electrical insulators are used to cover the conducting wires in electrical cords attached to appliances.

22. (a) What effect does an electric current have on human muscles?

(b) Explain what is meant by the term "let-go threshold"?

23. Are series or parallel circuits used to provide electrical energy to wall outlets? Explain why.

11 Harnessing Electrical Energy

Getting Started

1 If you use a 100-W bulb for 10 hours, you will use up 1 kilowatt hour (kW·h) of electrical energy. In 1999, in Ontario, that would cost you about $0.08. Look at the cost per kW·h for the single-use (disposable) cells in **Table 1** below.

How many times more expensive is the energy per kW·h for an AA disposable cell than the electrical energy we use in our homes? Find similar information about rechargeable cells on the Internet, and compare the costs per kW·h for rechargeable versus disposable cells.

Suggest reasons for the differences in cost per kW·h for the four sizes of disposable dry cells.

Table 1

Single Use	AAA Cell	AA Cell	C Cell	D Cell
cost per cell	$2.30	$1.89	$2.58	$2.58
cost per kW·h	$1300	$490	$260	$132

2 Sources of electrical energy range from the minute amounts produced by the magnetic tape in a portable tape player to that produced by nuclear generating stations. There are also some more unusual sources of electrical energy. This radio is powered by a large clockwork spring. You wind up the spring, and as the spring unwinds it produces sufficient energy to operate the radio for a short time. The electric bicycle is another example. The small, specially shaped batteries on these bicycles can be recharged each night.

Where or in what circumstances would these kinds of devices be most useful? List some of their advantages and disadvantages.

3 It's easy to list ways that automobiles improve our lives. Unfortunately, there are two major problems with the sources of energy used by automobiles: the fuels are nonrenewable, and their combustion causes air pollution.

Canadian engineers and scientists have developed a practical solution that can power vehicles with electricity generated from renewable fuel sources that give off only water as a byproduct. These remarkable devices are called fuel cells.

Two of the fuels that can power a fuel cell are ethanol and methanol. Why are these fuels classified as renewable sources of energy?

What other uses could an electricity-generating fuel cell system, small enough to fit into a subcompact car, have in our society? ▼

SKILLS MENU
○ Questioning   ● Conducting   ● Analyzing
○ Hypothesizing   ● Recording   ● Communicating
○ Planning

Energy Stored in Batteries

Have you ever accidentally left a flashlight switched on, or has your portable tape or CD player ever stopped because the battery is discharged? If you have ever experienced this annoying phenomenon, then you know that dry cells can only release electrical energy for a limited amount of time.

In this investigation you will study the factors related to the amount of energy released from a dry cell as it is discharged. You will need to use two circuits. Circuit 1 consists of a AAA rechargeable 1.5-V dry cell energy source that is connected to a single electrical load—a light bulb. Circuit 2 consists of an electrical energy source identical to that in Circuit 1, connected to two electrical loads identical to that used in Circuit 1. The electrical loads in Circuit 2 are connected in parallel.

Question

What factors affect the amount of energy released from a dry cell?

Hypothesis

If we observe the light bulb, measure the voltage drop across an electrical load, and the time taken to discharge the dry cell supplying the electrical energy, then we will be able to identify the factors affecting the amount of electrical energy the dry cell releases into the circuit.

Materials

- 2 fully charged rechargeable 1.5-V dry cells (AAA)
- 2 AAA dry cell holders
- 3 light bulbs (1.5-V)
- 3 light bulb holders
- 2 switches
- connecting leads
- voltmeter(s)
- stopwatch (timer)

Procedure

1 Draw the schematic circuit diagrams **(5D)** shown in **Figure 1**, and copy **Table 1** shown below to record your data.

2 Construct the circuits shown in **Figure 1**. Ask your teacher to inspect your circuits before continuing.

3 Close the switch in each circuit, and at the same time start the timer. Record the actual time when you closed the switch, and the readings on the voltmeters in each circuit. (Your teacher will instruct you on procedures to use if the discharge time of the dry cells is longer than one class.)

✎ (a) Record your initial observations, check the meter readings at regular intervals, and write notes, below the table, on any other changes that may occur.

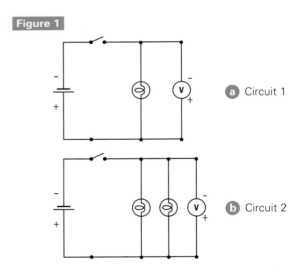

Figure 1

ⓐ Circuit 1

ⓑ Circuit 2

Table 1

	Voltage Drop (volts)	Ratio of Current in Circuit #2 compared to Circuit #1	Total Time Taken (hours)
Circuit 1	?	?	?
Circuit 2	?	?	?

Energy When You Need It Most

In some cases when a cell or battery eventually becomes discharged, it may cause a serious safety hazard. Think about the emergency lights at the critical places in the corridors of your own school or in large stores. Find out how they work. In this activity, you will identify battery-operated devices that can cause significant inconvenience or safety hazards when the battery becomes discharged. You will also identify the types of cells and batteries used in these devices, and determine the amount of energy stored in them.

Materials
- tables of data on the performance characteristics of batteries obtained from Internet web sites
- information from user manuals related to the operation, performance, and safety considerations of the devices being analyzed

Procedure

1 In a group make a list of all the commonly used electrical devices that are battery operated such as the clock shown in **Figure 1**. They can be in the home or other kinds of buildings, in vehicles, or carried with you or attached to you. From the list, identify those devices that can cause inconvenience or pose a safety hazard if the batteries become discharged at an inopportune time.

2 Carry out research on the Internet and in **(3A)** your library to obtain information about the types of cells and batteries in the devices identified in step 1, and the amount of electrical energy stored in them.

✎ (a) Design and choose a method to summarize and present the important data you have found, and the results of the calculations you will be completing.

3 Calculate the amount of electrical energy in each cell or battery using the energy equation and the information you have obtained.

✎ (a) Enter any important information and the results of your calculations in the table.

4 Consider the way that you presented your findings.

(a) Were there any limitations in using the format you chose to present all the information you found?

5 Summarize your findings.

(a) What safety recommendations will you include?

Figure 1

Making Connections
1. List four devices in which the unanticipated discharge of the battery (cell) could cause a safety hazard. Brainstorm possible solutions to overcome this hazard.

Exploring
2. List the web sites you visited. Rate their usefulness and justify your evaluation.

The Rate at Which Energy Is Used

Although you may not have previously understood what electrical power was, you have been using the units of electrical power in your everyday speech for most of your life. You probably use a 40-W light bulb in your bedside lamp. CD players usually operate on less than 1 W. Electric heaters often have much larger power ratings, which are more conveniently measured in **kilowatts** (1 kW = 1000 W), rather than watts. A typical electric space heater may have a power rating of 1500 W, or 1.5 kW. **Table 1** shows the power rating of some electrical appliances.

When we think about electrical power, we are thinking about how rapidly an appliance is using up electrical energy. A 300-W bulb uses up three times as much electrical energy as a 100-W bulb in the same amount of time. When you are in a car travelling at a speed of 100 km/h, you are travelling over the road at a rate of 100 km in one hour. Speed is a measure of the rate at which you are covering the

Table 1	Power Ratings of Common Appliances		
Appliances	**Average Power (W)**	**Monthly Energy Use (kW·h)**	**Approximate costs ($)**
air conditioner (room)	750	90–540	7.20–43.30
clothes dryer	5000	50–150	4.01–12.03
coffee maker	900	4–27	0.32–2.17
computer (monitor & printer)	600	5–36	0.40–2.89
electric kettle	1500	3–15	0.24–1.20
lighting – 60-W incandescent lamp	60	5–30	0.40–2.41
microwave oven	1000	5–20	0.40–1.60
television – colour	80	5–15	0.41–1.20
toaster	1000	1–5	0.08–0.41

Costs are based on $0.082 per kW·h.

Try This   Checking Registration Plates

If you look at the registration plate on any electrical device (**Figure 1**), you will always see at least two of the three quantities, voltage (V), current (I), and power (P) listed. This is a legal requirement of the Canadian Standards Association (CSA). It is an important part of the electrical safety regulations set up by the Canadian government to protect consumers.

Why are only two of the three quantities required to be listed? Which one is most commonly omitted? Why?

Draw the table shown below and list at least 15 electrical devices. Then locate the registration plate on a sample of each of the devices in your list, and complete the table by doing the appropriate calculation.

⚠ Only look at the registration plate if the device is disconnected to avoid danger of electric shock.

Figure 1

$$P = V \times I$$

Appliance	Electrical Power (P) (W)	Voltage Drop (V) (V)	Electric Current (I) (A)
?	?	?	?
?	?	?	?

distance. The word "rate" often involves how much a quantity is changing in a given time interval. **Electrical power** is a measure of the rate at which electrical energy is being used. The symbol for electrical power is P. The metric SI unit for electrical power is the **watt**, and the symbol is W.

Calculating Electrical Power

Electrical power can be calculated using the simple formula shown below. Just as speed is calculated by dividing distance by the time interval, electrical power is calculated by dividing electrical energy by the time interval.

$$\text{Electrical Power} = \frac{\text{Electrical Energy}}{\text{Time Interval}}$$

Symbolically,

$$P = \frac{E}{\Delta t}$$

Where E = electrical energy measured in joules (J)
P = electrical power measured in watts (W)
Δt = time interval measured in seconds (s)

Sample Problem: Calculate the power of a toaster that uses 72 000 J of energy in 50 s.

Data:
Electrical energy = 72 000 J
Time interval = 50 s
Power = ? W

Equation:
$$P = \frac{E}{\Delta t}$$
$$P = \frac{72\,000 \text{ J}}{50 \text{ s}} = 1\,440 \text{ W} = 1.440 \text{ kW}$$
The electrical power of the toaster is 1440 W (1.440 kW).

The formula for electrical power shown above is rarely used in practice because you have to measure both the energy and the time interval to be able to solve the formula. The formula normally used to calculate the power of an electrical device consists of two quantities that you have already measured many times in this unit—voltage drop and electric current:

Electrical Power = Voltage Drop × Electric Current

$P = V \times I$

This formula can be derived from what you already know:

$$P = \frac{E}{\Delta t} \quad \text{and} \quad E = V \times I \times \Delta t \quad \text{so} \quad P = \frac{V \times I \times \Delta t}{\Delta t}$$

The time interval (Δt) divides out, leaving
$P = V \times I$

Sample Problem: Calculate the power of a vacuum cleaner if the operating voltage is 120 V, and the current flowing through it when it is used is 7.90 A.

Data:
Power = ? W
Voltage = 120 V
Current = 7.90 A

Equation:
$P = V \times I$
$P = 120 \text{ V} \times 7.90 \text{ A}$
$P = 948 \text{ W} = 0.948 \text{ kW}$
The power rating of the vacuum cleaner is 948 W, or 0.948 kW.

Understanding Concepts

1. **(a)** Define the term "electrical power."

   (b) State the SI unit and symbol for electrical power.

2. What electrical quantities related to power are listed on the information plates of electrical appliances? Why aren't the energy and time listed there instead?

3. **(a)** Calculate the power rating of an electric toaster that uses 210 000 J while toasting bread for 140 s.

   (b) Calculate the power rating of a coffee grinder that operates on a voltage of 120 V. A current of 1.7 A flows in the motor.

Making Connections

4. If you wanted to plug two or more appliances into a wall outlet, how could you use the data on the appliance information plates to find out how much total current would be drawn from the outlet when they were all plugged in?

5. Locate all the electrical appliances and devices in the kitchen and calculate the total power rating. How many wall outlets would you need to plug them all in and not overload any of the 15-A circuits?

Reflecting

6. If electrical energy did not exist, what would we have had to use in its place in our homes?

Challenge

How can you make others aware of the importance of power ratings and how they relate to energy conservation?

Electrical Energy

How would your lifestyle change without electrical energy? Electricity is clean, relatively easy to transmit long distances, and can be converted into the other forms of energy we need in our everyday lives.

Because electrical energy is essential, scientists and engineers are constantly trying to develop new technologies to produce even more of it. The whole world's use of electrical energy is always increasing. In our investigations so far, we have been using only energy sources that produce relatively small amounts of electrical energy—dry cells and batteries. Now we are going to study ways to produce the huge amounts of electrical energy required to supply the constantly growing needs of our modern society.

Figure 1

Sources of Electrical Energy

The only large source of electrical energy that is produced naturally is lightning (**Figure 1**). Unfortunately, no one has yet found a way to control lightning or a way to store the huge amounts of energy that are released in a few fractions of a second. An additional problem is that it occurs quite randomly, and in some areas of the world lightning hardly occurs at all. Clearly, other energy sources must be used to produce the constant, reliable supply of electrical energy we need.

It would be preferable to use **renewable energy resources** to generate electrical energy because such resources constantly replenish themselves. Solar energy, energy from the wind, biomass, and—as long as there are no droughts—the gravitational potential energy of water stored behind a dam or from a waterfall are examples of renewable energy resources. Unfortunately, at the present time it is not possible to satisfy all the energy needs of our society using renewable resources. Instead, we use large amounts of **nonrenewable energy resources**, such as fossil fuels and nuclear fuels. One of the major disadvantages of using nonrenewable resources is that they cannot be replaced in a reasonable amount of time, such as a human lifetime.

Did You Know

After including the energy losses that occur when converting fossil fuels into electrical energy at the generating station, the ordinary incandescent light bulb converts only about 1% of the original fossil fuel energy into visible light.

Try This **Sources of Electricity**

Prepare a table with the following headings:

Source of Electricity	Renewable or nonrenewable	Advantages	Disadvantages	Ways to Improve

As we study the various ways of generating electricity, add information to this table so you will have a summary that will allow you to make comparisons and suggestions about the various ways of generating electricity.

Environmental Concerns and Sustainability Issues

A growing concern is the need to produce electrical energy without harming the environment. In addition, the economic and societal risks and benefits related to each kind of energy resource should be considered. This process applies just as much to the generation of small amounts of electrical energy in such energy sources as batteries as it does to the large-scale production in generating stations. To properly evaluate those risks and benefits, we need to consider all the major aspects of the production and use of electrical energy: where and how it is generated; where and how it is transmitted from the generating station to the place where it is used; and, finally, how it is converted into other forms of energy for different consumers.

The needs of society have to be balanced with the negative impacts of pollution produced during the generation and use of electrical energy. A particular concern is the limitation and reduction of carbon dioxide (CO_2) emissions from fossil fuels as a means of reducing global warming.

The term **sustainability**, when we are discussing electrical energy, refers to a consideration of the social, economic, and environmental aspects of its production and use, now and into the future. The production of electrical energy, now and in the future, should not place an increased burden on our environment or require the unsustainable use of nonrenewable energy resources.

In the following sections, you will learn about various aspects of the production, transmission, and conversion of electrical energy. The amounts of electrical energy will range from the extremely small, such as the amount generated when you push your debit card into a banking machine (**Figure 2**), to the huge amounts required to operate all the street lights in a big city. Think about the issues mentioned above as you develop a deeper understanding of the generation, transmission, and use of electrical energy.

Figure 2

Understanding Concepts

1. Describe some of the advantages of using electrical energy.

2. Why don't we try to capture the huge amounts of electrical energy available in lightning?

3. Compare the use of renewable and nonrenewable energy resources for producing electrical energy. List at least two examples of each kind of energy resource.

4. Define the term "sustainable" and explain its importance in terms of the production of electrical energy.

Making Connections

5. Why is the use of electrical energy increasing?

6. What negative impacts could occur as a result of increased production of electrical energy throughout the world?

Exploring

7. Use the Internet, CD-ROMs, or ③A the library to determine when and where the first mass electrification system was installed in North America. Who did it? Was this system the one adopted by all the other electrical distribution companies? Describe the system's first 20 years.

Reflecting

8. What measures has your family used to try to conserve the use of electrical energy? Begin to think about practical ways that you might be able to reduce your use of electricity. Record your suggestions so that you can share them later.

Challenge

What question would you ask to assess someone's understanding of renewable and nonrenewable energy resources and their significance in relation to sustainability?

Automobiles and the Fuel Cell

In North America, names like Ford, Stanley Steamer, and Detroit Electric were important to the early development of the automobile. Henry Ford used the internal combustion engine to drive the first automobiles his company produced, and it is still used in cars today. However, some of the early cars used steam power, like those produced by Stanley. Other manufacturers, such as Detroit Electric, used electricity. In the early 1900s, first the steam-driven automobile and then the electric automobile far exceeded the gasoline-powered car in popularity. In fact, in 1912 there were 30 000 electric cars on the road in the United States with 6000 new ones produced each year.

(a) Why did steam-driven and electric automobiles disappear from car dealerships over the years?

(b) On which continent did the internal combustion engine increase in popularity most rapidly? Explain why.

The Development of Modern Electric Cars

The size and mass of the batteries needed to allow an electric car to travel a reasonable distance before the batteries need to be recharged are problems that have been difficult to overcome. Electrically powered delivery vans, for instance, can be run within a city, but they cannot take longer trips to other urban centres. Many manufacturers are now trying to overcome this problem because the internal combustion engine produces environmental pollution and uses nonrenewable resources.

General Motors is testing a "hybrid" diesel-electric bus for city streets. The bus uses 40% less fuel and produces less pollution than diesel- or gasoline-powered buses. The wheels of the bus are driven by electric motors operated by batteries. A large set of batteries is charged by a diesel-powered engine that is about half the size of the engine normally used. The engine, which runs at a more economical rate, keeps the batteries fully charged. A city bus normally gives off the most pollution when it accelerates after its many stops. The engine in the hybrid bus runs at a constant rate, avoiding this problem.

Many manufacturers, including Daimler-Chrysler, Ford, General Motors, Honda, Nissan, and Toyota, began introducing these vehicles in the late 1990s. But, because they still use fossil fuels and produce some pollution, they are not ideal replacements for the cars we are familiar with today.

(c) Summarize the problems that must be overcome to produce a practical electrically powered automobile.

(d) What are the advantages of "hybrid" vehicles compared with those using internal combustion engines?

(e) Why are hybrid vehicles not permanent solutions to the replacement of the internal combustion engine?

Fuel Cells: A Practical Solution

Governments in Canada and the United States have passed laws that require automobile manufacturers to drastically increase the efficiency of their vehicles, while at the same time, reduce pollution.

Figure 1
A city bus powered by a fuel cell

The future for electric vehicles appears to depend on **fuel cell** technology (**Figure 1**). Most advances in this field are being made in Canada. For example, Ballard Power Systems is working with automakers to produce an energy system known as the proton exchange membrane fuel cell (**Figure 2**). The fuel cell, which operates silently, requires a continuous supply of hydrogen and oxygen to produce electricity electrochemically without combustion. In the fuel cell, the hydrogen, obtained from natural gas or methanol, is combined with oxygen from the atmosphere to produce the electricity. Water vapour and heat are the only byproducts of the process. A single fuel cell consists of a positive plate called the **anode** and a negative plate called the **cathode** which is coated with a thin layer of platinum. Electrons are produced at the anode of the fuel cell and travel as a normal electric current through the electrical load in the circuit. The protons from the hydrogen gas migrate through a proton exchange membrane inside the fuel cell that separates the two electrodes. When the protons reach the cathode, they combine with oxygen from the air and with the electrons returning from the external circuit to form water and heat. Individual fuel cells are placed together in the number needed to form a fuel cell battery.

(f) What have governments done to encourage manufacturers to produce electric automobiles?

(g) What technological breakthroughs have occurred to make it possible for a car to run solely on electricity?

(h) Why is the fuel cell preferred over the existing alternatives for driving vehicles?

A Versatile Energy Source

There are other potential uses of the fuel cell. They include electrical generators the size of refrigerators that run on natural gas and could provide electrical energy for isolated buildings. This would be especially valuable in remote communities. In addition, small portable units are being developed to operate such devices as hedge clippers and portable telephones.

Fuel cells can even be operated by the methane produced in sewage treatment plants and landfill sites around the world.

Making Connections

1. Brainstorm a list of other practical applications for fuel cell technology.

2. What other raw materials could be used to provide the hydrogen required to operate the fuel cell? Suggest how this could be useful on farms and in isolated areas.

Exploring

3. Research on the Internet and in the library to
(3A) find out more about fuel cells and other
(3B) proposed electrochemical sources of electrical energy. Is the fuel cell the best alternative for all applications? What are its disadvantages? Write a brief position paper supporting your viewpoint.

Figure 2

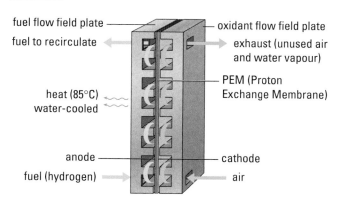

fuel flow field plate — oxidant flow field plate
fuel to recirculate — exhaust (unused air and water vapour)
heat (85°C) water-cooled — PEM (Proton Exchange Membrane)
anode — cathode
fuel (hydrogen) — air

membrane electrode assembly
flow field plate

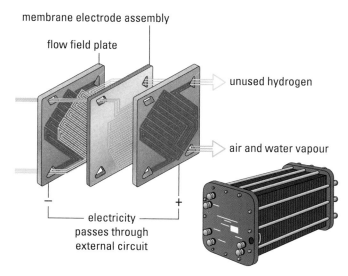

unused hydrogen
air and water vapour
electricity passes through external circuit

Energy Transformations

All forms of energy can be converted into electrical energy. Several energy transformations are needed when using fossil fuels or nuclear energy to produce large amounts of electricity. However, we have developed a number of devices that produce small amounts of electrical energy by both direct and indirect conversion. Your studies with dry cells have shown how chemical energy can be directly converted into electrical energy. Now let's investigate some energy conversions, using forms of energy other than chemical energy.

In this investigation, you will study examples of the conversion of several different forms of energy into electrical energy.

Question

How can it be demonstrated that an energy conversion from a given form of energy into electrical energy has occurred?

Hypothesis

If we can measure or identify the flow of electric charges when using only one form of energy in conjunction with an energy converter, or transducer, it can be inferred that the original form of energy has been converted into electrical energy.

Materials

- switch
- connecting wires
- photoelectric cells (photocells or solar cells)
- light bulb and holder
- light source (bright)
- barbecue lighter (equipment containing piezoelectric crystals)
- thermocouples
- voltmeter
- Bunsen burner
- microphone
- oscilloscope
- length of wire (2 m), plus cardboard tube
- ammeter
- permanent bar magnet

Your teacher may provide other materials and equipment:
- hand-operated generator or bicycle generator
- computer, with various types of sensors

Procedure

Complete each part of the investigation as instructed by your teacher.

⚡ Always connect and disconnect electrical equipment by the plug to avoid danger of electric shock.

Part 1

1 Design a circuit to measure the current flowing from the photoelectric cells through the bulb, and also measure the voltage drop across the bulb. Construct the circuit and insert the light bulb into the holder. Ask your teacher to inspect your circuit before continuing. Close the switch and shine a bright light onto the photoelectric cells. Draw a schematic diagram of the circuit.

✎ (a) Record your observations.

2 Move the light source you are using to within a few centimetres of the surface of the photoelectric cells, then slowly move it away to a distance of approximately 1 m.

✎ (a) Record your observations.

Disconnect the circuit completely before moving to the next activity.

✋ Light source may be hot. Handle with caution.

Part 2

3 Squeeze the barbecue lighter several times, using a different force each time.

✎ (a) Record your observations.

Part 3

4 Connect the wires from the thermocouple to the voltmeter.

5 Light the Bunsen burner.

✋ Tie back long hair and remove loose jewellery.

6 Hold the joined end of the thermocouple in the flame and observe the voltmeter. Move the joined end in and out of the flame.

✎ (a) Record your observations.

Part 4

7 Connect the leads from the microphone to the oscilloscope (as instructed by your teacher).

8 Speak into the microphone and observe the oscilloscope display screen. Vary the loudness of your voice and the distance of your mouth from the microphone.

✎ (a) Record your observations.

Part 5

9 Wrap the wire in a coil around the length of the cardboard tube, leaving about 15 cm at each end unwrapped to attach to the ammeter. (Ensure that if the wire is varnished, the ends of the wire have the varnish removed using sandpaper.)

10 Attach the ends of the coil of wire to the ammeter. (Your teacher will tell you which current scale to use.)

11 Move the north pole of the permanent magnet in and out of the coil of wire at a steady rate, and observe the ammeter.

✎ (a) Record your observations.

12 Repeat step 11, but move the magnet back and forth at different speeds.

✎ (a) Record your observations.

13 Predict what will happen if you repeat steps 11 and 12 using the magnet's south pole.

✎ (a) Record your predictions.

14 Repeat steps 11 and 12 using the magnet's south pole.

✎ (a) Record your observations.

Analysis and Communication

15 Analyze your observations by answering the following questions in a table:

(a) Identify the form of energy that was converted into electrical energy in each part of the investigation, and also the device used to do the energy conversion.

(b) What effect was produced when the bright light source was moved close to and away from the photocells? Explain what caused this effect.

(c) Calculate the maximum power produced by the photoelectric cells.

(d) Explain the variations in the amount of electrical energy produced when using the barbecue lighter.

(e) What temperature range is a thermocouple most suitable for measuring? Explain why.

(f) Explain the variations in electrical energy produced by the microphone. Comment on any differences between your voice patterns and that of other class members.

(g) Describe the various ways you can change the amount of electrical energy produced by moving the magnet in and out of the coil. What happens if you move it along the outside the coil?

(h) Predict what would happen if you repeated steps 11, 12, and 14 at the other end of the coil. Try it and comment on your predictions.

Making Connections

1. List two examples of the practical use of each kind of energy converter that was used in the investigation.

2. What factors might limit the amount of electrical energy produced by photoelectric cells if they were used to produce electricity for a house?

Exploring

3. Identify as many examples as possible in our everyday lives where magnetic materials or strips produce electrical energy. What must be happening for the magnetic materials to produce the energy?

Small-Scale Generation Methods

Electrical energy could almost be called the "in-between" form of energy. First, other forms of energy are converted into electrical energy in electrical generating stations. Then it is converted to other forms of energy for a variety of uses. For instance, an electric kettle heats water, a television produces both light and sound energy, and an electric fan produces mechanical energy that is used to turn a fan blade.

Electrical energy has many positive features: it is clean; it can easily be transmitted large distances; it can travel through conductors around corners, underground, and even underwater; and it is safe to use if the proper precautions are taken. However, perhaps its major advantage is that it can be produced from all other forms of energy and can then be converted back into all of them, except nuclear energy and the fossil fuel form of chemical energy.

Producing Electrical Energy Directly from Other Forms of Energy

There are five basic forms of energy that can be used to produce electrical energy directly from the source of energy. They are mechanical, thermal, sound, radiant (light, radio, microwave, solar), and chemical energy.

Mechanical Energy

Mechanical energy can be directly converted into electrical energy. When certain types of quartz crystals are subjected to stress by being squeezed or stretched, small amounts of electrical energy are produced. This effect is called the piezoelectric effect. The quartz crystals used in watches produce a very small electrical signal that controls the operation of the watch. Piezoelectric crystals are also used in some types of microphones and for producing the spark from a barbecue lighter.

Thermal Energy

Thermal energy can be converted into electrical energy using a device called a thermocouple. The thermoelectric effect occurs when two different metals are joined together at the ends, as shown in **Figure 1**, and the junction of the two metals is heated. Thermocouples are usually used to measure higher than normal temperatures.

> ### Did You Know ?
> Earthquakes in Japan have produced lightning in areas where the ground contains large amounts of quartz. Lightning bolts several metres long have been observed shooting out of the ground as the quartz crystals have been subjected to huge changes in pressure during the earthquake.

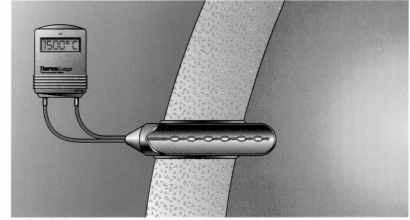

Figure 1

Sound Energy

Every time you use the telephone, you are producing electrical energy as you speak into the mouthpiece. Microphones of one kind or another are so much a part of our lives that we just take them for granted. The music we listen to and the movies and television shows we watch all required the conversion of sound energy to electrical energy when they were being recorded. Special microphones called geophones are used by geologists to investigate the transmission of vibrations through the ground when they are prospecting for minerals.

Radiant Energy

Whether you call them photovoltaic cells, photoelectric cells, solar cells or just photocells, they are becoming increasingly important as a versatile source of electrical energy. **Photoelectric cells**, like the ones in **Figure 2**, convert light energy directly into electrical energy. When light strikes the surface of certain materials such as silicon, electrons are released and produce an electric current. In effect, the photoelectric cell is a light-energized cell. However, instead of using chemical energy to provide the electric charges with energy, as a cell normally does, the photoelectric cell uses light energy.

Large flat surfaces covered with photoelectric cells are used to provide the electrical energy required by spacecraft and satellite equipment. Photoelectric cells are used as barcode readers at supermarket checkout counters, for operating automatic doors, and for automatically switching on streetlights at dusk. Your CD player uses a photoelectric cell to detect the variations in the intensity of laser light reflected from the pitted surface of the disk. Currently, photoelectric cells cost too much for use as a large-scale generator of electrical energy. However, costs are decreasing, and their performance is constantly improving. As renewable energy resources become increasingly important, photoelectric cells will become one of the major sources of electrical energy.

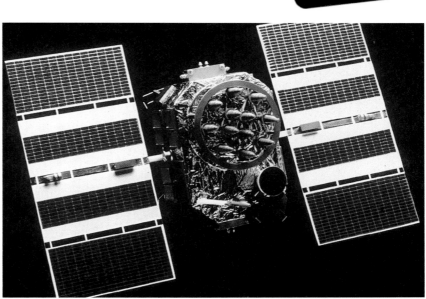

Figure 2
Photoelectric cells

Chemical Energy

You have already become familiar with two devices that convert chemical energy into electrical energy—the voltaic cell and the fuel cell. The two major types of voltaic cells, primary cells and secondary cells, were discussed in Chapter 10. More information about some of these very versatile energy sources is outlined on the next page.

The Primary Cell: A Disposable Energy Source

Many wristwatches today are digital and use small, specially designed dry cells. There are actually dozens of different-sized, special purpose dry cells available. The devices in **Figure 3** use dry cells made of special materials to suit each special requirement. Many electrically operated devices, such as wristwatches, calculators, and cameras, contain computer chips or special circuits that will operate correctly only if the voltage of the battery stays at a certain value. When the voltage drops below the critical value, the circuit will not operate. Special batteries have been designed so that the voltage remains constant until it is almost completely discharged, then within a few minutes, the voltage drops so low that the device just stops working. The original miniature batteries for these devices contained mercury, but their disposal is hazardous to the environment. Newer batteries are made with zinc and silver oxide.

Figure 3

The Secondary Cell: A Rechargeable Energy Source

The secondary cell was developed to provide larger amounts of electrical energy more economically than the primary cell. A secondary cell can be discharged and recharged hundreds of times. Although new types of secondary cells have been developed in recent years, the **lead-acid cell** is still a useful source of portable electrical energy. It has been used for many years in car batteries. This cell was developed by Gaston Plante in 1859. To recharge a secondary cell, the chemical change is reversed by connecting the cell to a source of electrical energy, such as a battery charger (or the alternator in a car). As the cell is recharged, the electrical energy from the charger changes the electrodes back to their original state. The electrical energy is converted to chemical energy. Most modern lead-acid storage batteries are sealed to reduce loss of electrolyte by evaporation. The electrolyte is usually in the form of a gel.

Lead-acid batteries are not able to store sufficient energy to operate a practical electric car. Instead, some companies are working with nickel-metal hydride (NiMH) batteries that produce about twice as much energy as lead-acid batteries, are much lighter, and have a life cycle that would allow them to last as long as the car does. In addition, they have no hazardous materials to be disposed of and have superior cold-weather performance. However, they do not operate as well as the lead-acid battery at high temperatures, do not provide as much acceleration performance, and are very expensive.

Smaller Rechargeable Batteries

One of the most reliable portable sources of electrical energy, the nickel-cadmium cell (NiCd), may be charged and discharged hundreds of times while still providing a constant voltage. In cost per hour of use, it is perhaps the most economical secondary cell. Nickel-cadmium cells are used in portable devices such as pocket calculators, electric shavers, electronic flash units for cameras, personal tape players, portable TVs, and many kinds of power tools. Nickel-cadmium cells are less powerful than alkaline cells, do not store well, and will lose about 1% of their charge every day they are not used. They also rely on the world's limited supplies of cadmium and require special disposal because the materials are toxic. Other more powerful (and more expensive) rechargeable cells include those made of nickel-metal hydride, lithium ion (Li-ion), and rechargeable alkaline cells.

Producing Electrical Energy Indirectly from Other Sources of Energy

Another group of small-scale electrical generators requires the use of moving magnets or electromagnets to produce electricity. The movement of magnetic particles when a clerk swipes a credit card, or when signals are produced in a computer from a spinning magnetic hard drive or floppy disk produces tiny amounts of electrical energy. As the magnetic tape moves past the recording heads in tape cassette players and VCRs, varying amounts of electrical energy are also generated.

Fossil fuel-powered electrical generators are available in a wide range of power ratings. These devices require several different energy conversions. The chemical energy in the fuel is converted to heat energy which makes the internal combustion engine turn. Then the spinning shaft of the engine moves magnets or coils of wire in the electrical generator to produce the electrical energy. Building contractors routinely use small, gasoline-powered generators to begin new buildings if the normal electricity supply is not available. When the severe ice storm occurred in Quebec in the winter of 1998, a portable gas generator was often the only reliable source of electrical energy for several weeks for many households. Larger diesel-powered generators are used to produce energy to keep large buildings such as hospitals operating on an emergency basis and to operate diesel-electric locomotives (**Figure 4**).

Figure 4

Harnessing Electrical Energy

Understanding Concepts

1. Why can electrical energy be called the "in-between" form of energy?

2. List some of the advantages of electrical energy.

3. Identify the five forms of energy that can be used to produce small amounts of electrical energy, and identify a device that converts each form of energy into electricity.

4. **(a)** Identify the major advantages and disadvantages of producing electricity by photoelectric cells and by using batteries.

   (b) Which method of producing electrical energy has the least impact on the environment?

Making Connections

5. **(a)** Describe four practical ways photoelectric cells are used.

   (b) Why are they not used to provide the electrical energy needed for our homes?

6. Why are portable fossil fuel-powered generators so useful? What recent developments may be used to replace the smaller versions of these generators?

Exploring

7. Read reference material related to (7B) the discharge characteristics of cells and batteries. On a graph of voltage versus time, draw a line to show how the voltage of an ordinary dry cell changes as the cell discharges. On the same graph, draw a line to show how the kind of dry cell that is used to operate the computer in a camera or a wristwatch would need to discharge.

Large-Scale Sources of Electrical Energy

About 89% of the energy resources used in Canada are nonrenewable. The remaining 11% is the renewable energy available from falling or running water. Over 80% of the nonrenewable energy resources are in the form of fossil fuels. In order of abundance, oil, natural gas, and coal are the three major fossil fuel resources in Canada. When the chemical energy in fossil fuels is converted into electrical energy, several complete energy conversions are required. The sequences of energy conversions required to produce electrical energy using both nonrenewable and renewable resources are shown in **Table 1**.

Table 1

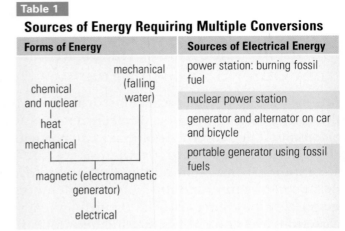

Sources of Energy Requiring Multiple Conversions

Forms of Energy	Sources of Electrical Energy
chemical and nuclear → heat → mechanical → mechanical (falling water) → magnetic (electromagnetic generator) → electrical	power station: burning fossil fuel
	nuclear power station
	generator and alternator on car and bicycle
	portable generator using fossil fuels

Nonrenewable Resources

Fossil-Fuel Generating Stations

In a generating station using any of the three forms of fossil fuel as the energy source, the fuel is burned, and the chemical energy released is used to heat water and produce steam (**Figure 1**). In turn, the high-pressure steam is used to turn a set of fanlike wheels called turbines. As the turbine wheels spin, they turn an electromagnetic generator that finally produces electrical energy. In 1997, Ontario obtained about 17% of its electricity from generating stations using fossil fuels.

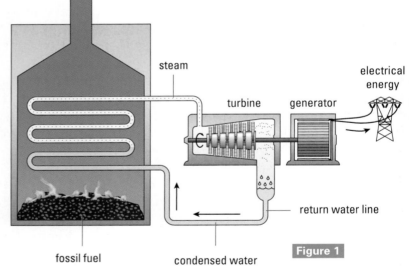

steam • turbine • generator • electrical energy • return water line • fossil fuel • condensed water

Figure 1

   Although transmission lines carry electrical energy across the country, there are communities for which it is not economical to provide electricity in this way. Such communities, often use large diesel generators, like the ones mentioned in section 11.8.

Nuclear Generating Stations

Nuclear energy is also used to generate electrical energy in several parts of Canada. The process is essentially the same as that used with fossil fuels. The basic difference is that, instead of using the chemical energy in the fuel to heat the water, the energy used is that released from nuclear reactions. In 1997, about 48% of Ontario's electrical energy production was supplied by nuclear generating stations.

Did You Know ?

In 1890, the first windmill to produce electricity began operating in Denmark.

Efficiency of Energy Conversions

When any energy conversion occurs, only part of the original energy is converted into the new form of energy. The waste energy usually ends up as thermal energy. The comparison between the amount of useful energy produced and the original amount of energy used is called **efficiency**. We will discuss efficiency in much more detail in Chapter 12. To conserve energy, it is important to try to make each stage of the production of electrical energy as efficient as possible. **Table 2** lists the efficiency of energy conversion for the different methods of producing electrical energy. A considerable amount of the original energy in the resource is lost in the processes required to produce the electricity at the generating station. Depending on the type of fuel used, only 20% to 26% of the energy contained in the original fuel is converted into electrical energy at the generating station.

Table 2

Electrical Conversions from Nonrenewable Resources

Fuel Used for Generation	Percentage of Energy from Fuel Available as Electricity
nuclear	20.0
petroleum	20.2
natural gas	20.7
coal	22.2
advanced natural gas turbine	26.4

Renewable Resources

Table 3 shows how much of our electrical energy comes from each of the major energy resources. Only 25% is provided by a renewable resource. It is important that we develop large-scale alternative ways of generating electricity from sources that will not run out.

Table 3	Electricity Sources in Ontario	
Source of Electricity	Number of Stations	Percentage of Contribution to Total Supply
hydroelectric	69	25
fossil-fuelled	6	17
nuclear	5	48
independent producers or purchased from other utilities	—	10

Hydroelectric Generating Stations

Canada is fortunate to have many locations where it is possible to produce electrical energy by using the gravitational potential energy of falling water. In 1997, about 25% of Ontario's electricity was produced by hydroelectric generating stations. A number of different designs are in operation. Many generating stations store the water behind a dam, which releases the water through special tubes called penstocks, to turn the turbine wheels below (**Figure 2**). Some, such as the one at Niagara Falls in Ontario, do not store the water at all. Rather, some of the water falling over the falls is diverted through special tubes to turn the water turbine wheels at the bottom and produce the electrical energy. A third method involves channelling fast-flowing water in a river through a set of turbine wheels to produce the electricity. The generators in the third method are usually used for producing relatively small amounts of electrical energy. The development of small, independent, hydroelectric generating stations has been increasing in recent years.

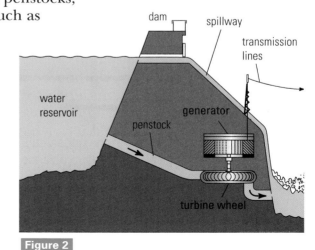

Figure 2

Cross section of a dam

Other Renewable Resources

Figure 3

As the need for an increasing supply of electrical energy becomes more urgent, scientists and engineers are creatively developing new and better sources. Perhaps one of the most promising for widespread use is the photoelectric cell. Continuing research is rapidly improving the efficiency of these devices, as well as lowering the cost of making panels large enough to mount on the roof of a house. The amount of electrical energy used in a home varies from about 20 kW·h to over 80 kW·h per day. Some of the photoelectric panels have a power rating of about 0.25 kW. So if one panel receives four hours of direct sunlight per day, it can produce about 1 kW·h of electricity. One of the major problems with the use of photoelectric cells is the wide variety of weather conditions that exist across Canada, especially in winter.

In fact, most alternative energy sources are somewhat limited in their use because they depend on the conditions at specific locations. Look at the different ways of generating electricity shown in **Figure 3**. Most of these sources are located in places with somewhat unique geographic conditions. However, usually at least one of the alternatives can be applied in almost any location, and by combining the alternatives with the more conventional sources of electrical energy, it is possible to provide a reliable supply of electrical energy.

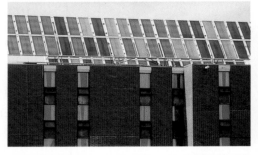

Constant Production of Electrical Energy

One of the challenges of electrical energy is that it is difficult to actually store it once it has been produced. When you recharge a battery, you are converting electrical energy from an external source back into chemical energy in the battery. Electrical energy itself is not stored in the battery. An electrical device called a capacitor can store small amounts of electrical energy. However, there is no practical way to store the huge amount of electrical energy produced at a generating station.

Because electrical energy produced at a generating station cannot be stored, it must be used up at the same rate as it is generated. As you learned earlier, the rate at which electrical energy is used is called electrical power. Thus, generating stations are often called "power" stations. The output of power stations is usually measured in megawatts (MW). The technical staff at generating stations are constantly adjusting the production of electrical energy to match the amount of energy being used by everyone all across Canada.

One of the most important aspects of any energy conversion used to produce electrical energy is the

efficiency of the process, which may be so low that it is not economically worthwhile. **Table 4** provides information on the efficiencies of various alternative energy sources.

Transmission of Electrical Energy

From generating stations, electrical energy is transmitted along the wires you see suspended from the tall transmission towers that form a huge network across Canada. Often the wires connected between the generating station and the home where it is eventually used are tens, if not hundreds, of kilometres long. Even if the wires are thick and are good conductors of electricity, that length of wire will still have some electrical resistance. When the current flows through the wire, some of the electrical energy will be converted to thermal energy due to the resistance.

The electrical energy losses due to resistance depend on the amount of current flowing in the wires. To reduce the losses, the electrical energy is transmitted using as small a current as possible. By using a special device called a transformer, the voltage of 20 000 V produced at the generating station is increased to over 500 000 V. The increase in the voltage causes a corresponding decrease in the current flowing in the wires and still allows the same amount of energy to be transmitted. However, even when the electricity is transmitted at these high voltages, there are still losses due to the resistance of the wires. Depending on the length of the transmission wires, the total losses due to the transmission of the electricity can amount to about 9% of the energy that left the generating station.

At the other end of the transmission line, the voltage is decreased to 240 V and 120 V for use in your home. So, even before the electrical energy gets into your home, some of the original energy from the energy source was lost producing the electricity, then more of it was lost as it was transmitted to your home. When the electrical energy is used in your home, even more losses occur due to the efficiency of the appliance using the energy.

Table 4

Electrical Conversions from Renewable Resources

Renewable Source of Electricity	Percentage of Energy Converted to Electricity
solar cells with concentrator lenses	30
single crystal solar cells	23
crystalline silicon solar cells	17
amorphous silicon solar cells	7
electrochemical cells	65
wind generators	30
hydroelectric generators	20

Understanding Concepts

1. **(a)** List the energy conversions that need to occur when electrical energy is produced from fossil fuels.

   (b) What are the advantages and disadvantages of using fossil fuels?

2. Why is the location of the place where the electricity is generated so important for most kinds of alternative renewable energy resources?

3. Why is the efficiency of the energy conversion from the different energy resources important?

4. Explain why electrical energy needs to be generated constantly.

Making Connections

5. Suggest reasons why Ontario produces more electricity with nuclear energy than with fossil fuels.

6. Why is it important to continue developing smaller, more efficient ways of generating electricity?

Exploring

7. Carry out research using the
   (3A) Internet and other sources of
   (3B) information to assess the different methods of producing electricity by nonrenewable energy resources. Which methods seem to be favoured? Has there been a change in perception about which method is most effective in the long term?

Reflecting

8. How might the electrical energy needs of your area be met 50 years from now?

Challenge

Identify alternative sources of energy and the advantages and disadvantages of each.

Using Renewable Resources

As we have seen, electricity can be generated in a variety of ways from renewable resources. In this activity, you will explore some of the alternative ways that electricity can be generated and compare the relative efficiencies of these methods. Remember that the concept of sustainability must be considered in any plan that is developed.

Materials
- Internet access
- CD-ROM encyclopedias and other resources
- media articles and programs
- library resources

Procedure

1 In groups, use a search engine or the
3A electronic catalogue at a library to research one alternative method of generating electricity.

2 With one group acting as a panel, evaluate the alternative methods of generating electrical energy presented by the various groups.

3 Include in the research:
- a description of the method of generating electricity
- a short history
- most practical sites for installation
- advantages of the method
- disadvantages of the method
- efficiency of the method

4 The group should make a presentation,
8A including data, to show why their method of generating electricity is the one that should be chosen instead of other methods.

5 Graphs and charts should be used to show
7B the efficiency of the method chosen.
6D

Figure 1

Exploring

1. What is the primary source of electricity in your community at present?

2. Which method of generation of electricity from renewable resources is most practical in your community? Explain your answer.

3. Where could such a facility be located in your community? How would the facility affect the environment?

4. What impact would the other methods of alternative generation of electricity have on the environment in your community?

5. List some locations in Canada where the other methods of electricity production would be practical.

Challenge

Record the alternative methods of generating electricity presented in order of efficiency of use.

Electrical Engineer

Like many young people in southern Ontario, Karen Cheung discovered science at the Ontario Science Centre. It was an exhibit about the electric light bulb that particularly fascinated her, as she realized just what a huge impact electricity has made on our lives.

Science came easily to Cheung at school, where she particularly enjoyed the technical subjects. Knowing that employment prospects were good in the high-technology careers, she decided to major in Applied Science when she obtained a Bachelor's degree in Electrical Engineering at university. She followed this up with a master's degree in Applied Science, majoring in Communications.

Cheung then joined the legions of highly-qualified young graduates looking for work in the well-paid communications industry. She didn't have long to wait. She was soon hired by Nortel Networks as a software engineer, designing and developing software for use in communication applications such as phone switches and voice mail. Cheung finds her job challenging but rewarding. She especially enjoys digital signal processing (processing of digital images and signals) because it ties in well with her university training. She also gets satisfaction from writing code that works well, and that meets the needs of customers.

What are her plans for the future? She hopes to stay with Nortel Networks for some time, diversifying her skills by working on a variety of products, and assuming more responsibility.

Following a career you love is a gift to yourself.

Exploring

1. List at least 20 different occupations that, like Karen Cheung's, involve computer skills.

2. Not all software engineers have exactly the same academic background as Cheung. Research what other university degrees are recommended for similar careers.

3. Canada is a world leader in the telecommunications industry. Do an Internet search to find out what high-tech developments Canadian companies have made recently.

Bridging the Energy Gap

One fact is certain in Ontario. We are going to need to generate increasing amounts of electrical energy far into the foreseeable future. Our population and our economy are growing, which means that we will need more electricity. In this chapter, you have learned about energy resources currently being used to generate electrical energy. As our need for electrical energy grows, we are faced with a difficult choice. What energy resources should we use to provide the extra electrical energy? There are two basic options: we can build new generating stations to produce electricity using the three major energy resources we presently use, or we can try to use alternative energy sources.

Today's Energy Sources

Most of our electrical energy is presently produced from the two kinds of nonrenewable resources—fossil fuels and nuclear fuel—and one renewable resource—water operating hydroelectric generating stations. It is not likely that we can provide sufficient additional energy by using new hydroelectric resources because most of the hydroelectric sites that can produce large amounts of electricity are already being used. So, which of the two nonrenewable energy resources would be preferable for new generating capacity in the near future?

Both options have their problems. Fossil fuels are a finite resource and cause environmental problems both during mining and burning. Nuclear fuels are also a finite resource, and their use requires continuous careful maintenance. Furthermore, we have not yet found a satisfactory way of disposing of the spent fuel rods from the nuclear reactors and the waste materials at the mines where the uranium is produced.

Renewable Energy Sources

Several renewable energy sources—thermal, photoelectric, wind, biomass, and fuel cells—could perhaps be used to produce reasonably large amounts of electrical energy in Ontario. Geothermal and solar thermal energy sources are not practical alternatives at present. Photoelectric, wind, and biomass energy sources could be used to generate significant amounts of electrical energy, but the relative cost of producing electricity with each of these alternatives is currently too high for them to be economically viable (**Table 1**). For example, producing electricity with photoelectric cells is 22 times more expensive than hydroelectricity.

None of the renewable energy sources can currently be manufactured and operated cheaply enough to make them a viable alternative in the foreseeable future.

The logical conclusion, then, is that we will have to build new generating stations that use the same nonrenewable energy resources as our existing generating stations.

Table 1	Relative Cost of Energy Sources	
Energy Source	Relative Cost (compared with hydroelectric)
hydroelectric	1
fossil fuels	2 – 5
nuclear fuel	2 – 5
geothermal energy	4 – 5
wind	4 – 5
biomass	5 – 7
photoelectric cells	22

Issue Fossil Fuels or Nuclear Power? (8B)

If we have to use nonrenewable energy resources, which is the best choice?

1. (3A) In small groups, research the issue using the Internet, your library, and any other sources available.

2. (3B) Try to classify the information you obtain into scientific, technological, and societal categories wherever possible.

3. Identify all the factors (short term and long term) related to the issue, and determine the position of your group. Consider the viewpoints of other people who would be affected by the consequences of your group's conclusion if it were carried out.

4. Write a one-page position paper, clearly stating your group's opinion and the reasons for it. Your group should also state any major problems related to the group's position.

Challenge

How will you incorporate your group's opinion and reasons into your information on energy consumption?

Chapter 11 Review

Key Expectations

Throughout the chapter, you have had opportunities to do the following things:

- Describe the relationship among electrical energy, electrical power, voltage drop, current, and time interval, and determine energy and power using the formula E = PΔt. (11.1, 11.2, 11.3, 11.4)

- Compare methods of producing electrical energy, including their advantages and disadvantages. (11.5, 11.6, 11.7, 11.8, 11.9, 11.10, 11.11)

- Describe and analyze electric circuits, and measure voltage drop and current values related to circuits. (11.1, 11.3, 11.7)

- Use safety procedures when conducting investigations. (11.1, 11.7)

- Investigate circuits, and organize, record, analyze, and communicate results. (11.1, 11.3, 11.7)

- Formulate and research questions related to electrical energy (and power) and communicate results. (11.3, 11.5, 11.6, 11.8, 11.9, 11.10, 11.11)

- Describe practical applications of current electricity. (11.6, 11.8)

- Explore careers requiring an understanding of electricity. (Career Profile)

KEY TERMS

anode	lead-acid cell
cathode	nonrenewable energy
efficiency	resources
electrical energy	photoelectric cell
electrical power	rated capacity
energy	renewable energy
fuel cell	resources
joules	substainability
kilowatt	watt
kilowatt hour	watt hour

Reflecting

- "Electrical energy is often called the 'in-between' form of energy because we produce it from energy sources such as fossil fuels or moving water, and then convert it into the forms of energy we need." Reflect on this idea. How does it connect with what you've done in this chapter? (To review, check the sections indicated.)

- Revise your answers to the questions raised in Getting Started. How has your thinking changed?

- What new questions do you have? How will you answer them?

Understanding Concepts

1. Make a concept map to summarize the material that you have studied in this chapter. Start with the word "energy."

2. Identify two devices that produce electrical energy from

   (a) sound energy

   (b) chemical energy

   (c) radiant energy

3. Identify three devices that use magnetic strips or magnetic particles to produce electrical energy.

4. Why are rechargeable cells called "secondary" cells?

5. Why are batteries made of nickel hydride preferred to lead-acid batteries?

6. List some of the energy sources that could be used to provide a source of hydrogen to operate a fuel cell. What is the other substance required to operate a fuel cell, and how is it obtained?

7. (a) Describe how a fuel cell operates.

   (b) What substances are produced as byproducts as a result of the operation of the fuel cell?

   (c) What impact do these byproducts have on the environment?

8. Why is such a high proportion of the electrical energy used in Ontario produced from nonrenewable energy resources?

9. What alternatives for producing electrical energy are currently available for small isolated communities?

10. Suggest reasons why most of the original energy in the fuel is wasted in the production of electrical energy.

11. What two factors contribute to energy losses in the transmission of electrical energy?

12. What is the factor in the equation for electrical energy that usually determines whether it is more practical to use joules or watt hours as the unit of energy for a particular calculation? Why?

13. A 6-V battery is being used to provide energy to two light bulbs connected in parallel. How will removing one of the light bulbs from its socket affect the time it will take for the battery to completely discharge? Explain why.

Applying Skills

14. (a) Calculate the energy released (in joules) from a 9-V battery that operated an alarm bell for 5 min. A current of 0.15 A flowed through the bell.

    (b) Calculate the energy released (in joules) from a 12-V car battery as it operated a starter motor. The current flowing through the starter motor was 350 A and the time the motor operated was 7.5 s.

15. Calculate the amount of energy used in each part of question 14 in units of watt hours instead of in joules.

16. (a) Calculate the power rating of a coffee maker that operates on a voltage of 120 V. A current of 5.7 A flows through the heater in the coffee maker.

    (b) Calculate the power rating of an electric kettle. A current of 12.5 A flows through the heating element and the operating voltage is 120 V.

17. Calculate the amount of energy used in each part of question 16 in kilowatts instead of in watts.

18. Design and draw a circuit using four dry cells and between one and four light bulbs. The cells and the bulbs can be connected in any desired circuit combination. Then design and draw two other circuits. Circuit 1 should discharge the four cells in one quarter of the discharge time of the original circuit. Circuit 2 should discharge the four cells in two times the discharge time of the original circuit. In each case explain why your circuits will discharge the cells in the required time. Think about the entire question before designing the first circuit.

19. Based on the information on solar panel power output listed on page 352, how many panels would be required to produce 11 kW·h of electrical energy per day if there were only three hours of direct sunlight available?

Making Connections

20. What form of energy is commonly used to alert you to the fact that the battery in a device is almost discharged? Why is this form of energy used?

21. (a) List three practical uses for photoelectric cells.

    (b) If photoelectric cells could be mass-produced at a low price, in what practical ways could they be used?

22. Which types of dry cells contain toxic materials? What would be the most responsible way to dispose of these cells? Why?

23. How can you determine the power rating of any electrical device used in Canada?

24. (a) What environmental problems can be caused by each of the various methods used to produce electrical energy?

    (b) How can these problems be eliminated or minimized?

25. (a) Predict which of the three ways of providing energy to drive cars will eventually become the most used. Explain why.

    (b) Why is the fuel cell considered to be preferable to the internal combustion engine for operating cars?

26. Suppose you had to choose whether to use a calculator operated by a battery or a set of photoelectric cells. What are the advantages and disadvantages of each type of calculator?

27. Compare using nonrenewable with renewable energy sources for producing electricity. What are the advantages and disadvantages of each?

12 Using Electrical Energy Safely and Efficiently

Getting Started

1 Every year tragic accidents take place because people become careless near exposed electrical wiring. People have been electrocuted while moving ladders around their house when cleaning windows or painting the walls. Another hazardous place is a marina or harbour where sailboats and dinghies are wheeled around on small trailers. The danger occurs when they are moved past overhead power lines on the way through the boat yard and down to the water.

Why do these kinds of accidents happen? Make a list of tasks involving ladders that could be hazardous with regard to overhead electrical wires.

What other areas, both in the home and outside, present the danger of severe electric shock?

2 The lights and the motor in this ceiling fan are controlled by the switch on the wall. The lights are controlled by a pull switch. Another pull switch controls the motor that turns the fan at three different speeds, and it can be turned on and off independently of the lights. A third pull switch reverses the direction of the motor turning the fan.

How are the lights, the switches (including the wall switch), and the motor connected in order to operate as described?

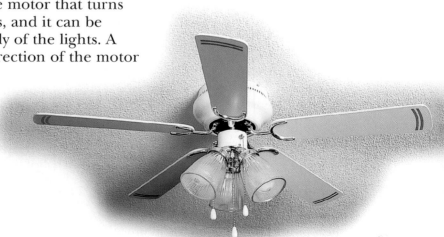

3 In 1945, engineer John Eckert and physicist John Mauchly at the University of Pennsylvania designed and supervised the building of the first large-scale programmable computer, ENIAC (Electronic Numerical Integrator and Computer). It weighed 27 t, contained 19 000 vacuum tubes, occupied the floor space of 8 average suburban houses, and required 175 kW of electrical power. By contrast, a typical desktop computer today requires only 230 W, while a laptop can operate on as little as 22 W.

ENIAC could perform 5000 calculations per second. Today's personal computers are 10 000 times faster, and the processing rate doubles approximately every 18 months. Just think of the difference in cost for a single calculation!

How many times more power was required by ENIAC compared with a typical laptop computer?

In what ways have the electrical appliances and devices you use daily changed to use less energy? What can each of us do to help conserve electrical energy?

Try This   The Fan Circuit

Design a circuit for the ceiling fan described in **2**. Select the appropriate components from the equipment you have used in the investigations so far, and construct and test your circuit. Why do you need a special kind of switch to reverse the direction of the motor?

Reflecting

Think about the questions in **1**, **2**, **3**. What ideas do you already have? What other questions do you have about using electricity? Think about your answers and questions as you read the chapter.

Safety in the Home

Although electricity is a clean, convenient energy source, we need to treat it with respect. As noted in section 10.9, electricity can cause severe injuries, and even death, if used carelessly. All equipment and appliances we use are designed to strict standards of safety.

Think of all the different kinds of electrical equipment you use in your home each day without giving your personal safety a thought. Electrical equipment is designed so that if some part breaks, or a wire becomes frayed or damaged, the user has as much protection as possible. In this investigation, you will study some common electrical devices to identify and explain their operation and structure with regard to the safe use of electrical energy.

Question

How can we become aware of the safety features of the devices we use to control electrical energy in the home?

Hypothesis

By observing the structure and function of electrical devices used, we can determine the safety features needed to be built into their design.

Materials

- several appliances (toaster, kettle, coffee maker, hairdryer, shaver, blender, can opener, coffee grinder, etc.)
- power bar (with a surge protector)
- selection of circuit breakers
- portable ground fault circuit interrupter (GFCI)
- selection of fuses
- bimetallic strip
- Bunsen burner
- main breaker switch and fuses
- several different switches
- electrically operated safety devices (Optional in Part 3: smoke alarms, CO alarms, motion detectors)
- thermostat
- timers
- extension cord (10 m)
- wire gauze

Procedure
Part 1: Operating Circuit Breakers

1 Design a table to record the values for power (P), operating voltage (V), and current (I) for each of the appliances used in the investigation. **(6D)**

(a) Draw the table and list the appliances in your table.

2 From each appliance's nameplate, identify the listed values of power, operating voltage, and current, if available.

✎ (a) Record the values in the table.

3 Identify the maximum current rating of the circuit breaker inside the power bar.

✎ (a) Record the maximum current rating.

4 Predict the combinations of the available appliances that can be safely plugged into the power bar without operating the circuit breaker. Predict one combination of appliances that will operate the circuit breaker.

✎ (a) Record your predictions.

5 Plug a portable ground fault circuit interrupter (GFCI) device into the wall or bench outlet socket. Then plug the power bar into the GFCI outlet socket. The power bar must be switched off while you insert the appliance plugs into the power bar sockets. Test your predictions by plugging into the power bar the combinations of appliances that you chose for your predictions.

✎ (a) Record your observations.

Part 2: The Structure of Fuses and Circuit Breakers

6 Observe the different-sized fuses. Describe any relationships between the properties of the material in the fuse that melts (fuses) and the fuse rating itself.

✎ (a) Record your observations.

7 Heat the bimetallic strip with the Bunsen burner flame.

✎ (a) Record your observations.

✋ Exercise proper care when using the Bunsen burner. Allow bimetallic strip to cool on wire gauze before putting it away.

8 Observe the dismantled circuit breaker and the main breaker switch. Identify, name, and state the function of each of the labelled parts.

✎ (a) Record what you have identified.

9 Construct the circuit shown below. Have the circuit checked by your teacher. Close the switch and observe what happens.

Step 9

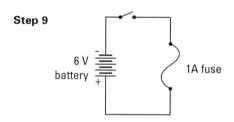

6 V battery

1A fuse

✎ (a) Record your observations.

Part 3: Control Switches and Safety Alarm Systems

10 Observe and, where possible, operate the control switches and safety alarm devices, then describe the operation of each device and identify its purpose.

✎ (a) Record your observations and comments.

Analysis and Communication

11 Analyze your observations by answering the following questions:

(a) For each appliance and device entered in the table in Part 1, calculate the missing values for power (P), operating voltage (V), or current (I). List the combination of appliances that caused the circuit breaker to open the circuit. Explain why the circuit breaker was activated.

(b) Why are circuit breakers more commonly used than fuses? How does the bimetallic strip allow the circuit breaker to operate?

(c) Explain how insulators and conductors are used in control devices and switches to provide for safe operation.

(d) Explain how the ground fault circuit interrupter operates.

12 Prepare a safety awareness poster informing others of the safety features when using electrical devices and why they are there.

Making Connections

1. Why does every appliance have a plug ending, rather than a socket ending connected to its electrical connecting cord?

2. (a) Why do people often leave some of their lights connected to special timers when they leave their homes for a period of time?

(b) How do the timers work?

(c) What problems can occur if these devices are used?

(d) How could you overcome such problems?

3. Some power bars or multiple outlets have built-in surge protectors. What is a surge protector and how does it work? Make a list of electrical devices that ideally should have surge protectors.

Distributing Electrical Energy Safely

When electrical energy is brought into your home, the voltage is reduced to 240 V. Usually three conductors, two covered with insulation and the other a bare wire, are connected from the main electrical supply wires on the poles in your area to your home (**Figure 1**). These two insulated wires are often referred to as the **live wires**. The third wire is called the **neutral wire**. When it is connected inside the building, it is literally connected to the ground itself by a special wire that is attached to the plumbing system or to a long metal stake driven into the ground. There is no voltage on the neutral wire.

Figure 1

The Energy Meter and Distribution Panel

From the place where the three wires are attached to the building, they then usually enter into a protective tube that travels down to the **electric meter**, which measures the total amount of electrical energy used in the building. The three wires then continue through the wall into the **main breaker switch** inside, where the two insulated (live) wires are each connected to a separate, special kind of switch called a **circuit breaker**.

A circuit breaker is a safety switch. If too much current flows in the circuit, the switch opens automatically. Sometimes the main breaker switch consists of a separate switch and fuse for each conductor. The circuit breaker ratings range from 60 A for a small home or cottage to over 200 A for a larger building. To completely shut off electricity from all the circuits in your home, the main breaker switch must be in the "OFF" position.

Figure 2

From the main breaker switch, the three wires pass into a metal box called the **distribution panel** (**Figure 2**). The main distribution panel is the place where all circuit breakers (or fuses in older homes) are connected to each of the separate circuits in the home.

What is the purpose of the neutral wire if it is just connected to ground? A voltage of 240 V is very dangerous. By using the neutral wire it is possible to reduce the 240 V across the two live wires down to 120 V for the circuits supplying electricity to the normal wall outlets and lights in the house. (Even 120 V is a very dangerous voltage, and causes a number of deaths by electrocution every year.) When a voltmeter is connected between the neutral wire and either of the energy-carrying (live) wires, the reading is 120 V. There are now three possibilities for connecting the circuits in the house to these three wires. The voltmeter reading between the neutral wire and one live wire is 120 V, between the neutral wire and the other live wire is also 120 V, and between the two live wires is 240 V.

If you look at **Figure 3**, you can see that there are two columns of circuit breakers in the main distribution panel. The circuit breakers in one column control circuits connected between one live wire and the neutral wire, and those in the other column control circuits connected between the other live wire and the neutral wire. All these 120 V circuits are connected in parallel throughout the home to provide electrical energy to the wall outlets and lights. The double circuit breakers control circuits connected directly to the two live wires. They are not connected to the neutral conductor at all.

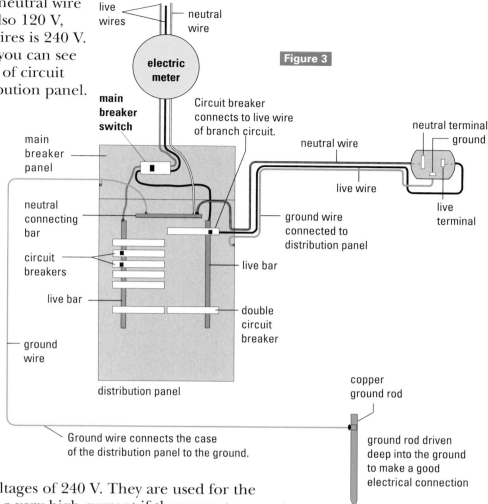

These circuits operate at voltages of 240 V. They are used for the appliances that would need a very high current if they were to operate at 120 V. By operating appliances such as electric water heaters, dryers, stoves, and ovens at 240 V, only half the current is required to supply the same amount of energy at 120 V. Why does this matter? The copper conductors used to carry the current would have to be much thicker, and more expensive, to carry the extra current needed if 120 V were used.

The Safe Use of Electricity

You may not realize it, but your personal safety is being considered every time you use an electrical device or appliance. However, most electrically operated devices are so well designed that you don't even think about safety issues when you use them.

Circuit Breakers

Safety features are built into even the simplest household circuit. Each 120 V circuit is a parallel circuit that can have only a limited number of outlets connected in parallel so as to reduce the chance of overloading the circuit too often. A circuit breaker (or fuse in some older homes) is always connected in series with the 120-V wires. This is done so that if there is a problem with any part of the circuit itself, or any device connected to it, the 120 V is immediately disconnected from the rest of the circuit. If the circuit breaker was connected to the neutral side of the circuit and a problem occurred, the 120 V would still be connected to the circuit and could still electrocute someone using a faulty device or appliance. The circuit breaker is actually in the circuit to protect your home against fire. The copper electrical cables that travel through the walls of a house to the wall outlets will overheat if the current is too high and may begin a fire inside the walls.

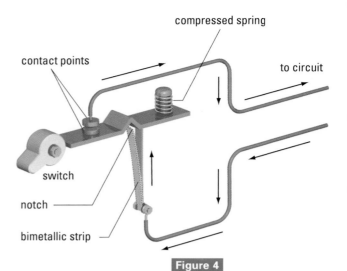

Figure 4
Circuit breaker

The operation of the circuit breaker depends on the bending action that occurs when a bimetallic strip is heated. As long as the current in the circuit is less than the maximum current allowed by the circuit breaker, the strip will not bend enough to open ("break") the circuit and stop the current flow. If a faulty appliance, or a combination of too many appliances, causes a current greater than the 15 A allowed by the circuit breaker, the bimetallic strip will bend so much that it will slip into the notch in the contact strip in **Figure 4**. A spring will push the contact strip away from the other contact that completes the circuit, and the circuit is opened. On a 240-V circuit, two circuit breakers are used, one for each of the two "live" conductors. When a circuit breaker opens the circuit, the bimetallic strip has to cool down before the switch on the circuit breaker can be reset.

Fuses

Fuses are used in cars, some electric stoves, and the distribution panels of some older homes (**Figure 5**). A **fuse** is simply a piece of material that will melt or "fuse" when it is heated to a high temperature by the current flowing through it. If the current flowing through a 15-A fuse exceeds that amount, the fuse melts, and the circuit is disconnected. Fuses are less convenient than circuit breakers because they must be replaced every time the fuse melts due to an overload. Many serious fires have occurred because homeowners have replaced burnt-out fuses with fuses of a higher rating, aluminum foil, or even pennies.

Figure 5
Fuses were popular before circuit breakers became commonly used.

Wall Outlets, Polarized Plugs, and Grounding Pins

The humble outlet found in every room has three very important safety features. First, the outlet is always a socket, not a plug. What can be done to make wall outlet sockets even safer for young children?

In more recently installed electrical wiring, the outlet socket has two different-sized slots and is said to be polarized. The narrower of the two slot-shaped terminals in the outlet is always connected to the 120-V conductor that is connected to the circuit breaker at the distribution panel. The wider slot-shaped terminal is always connected to the neutral wire in the distribution panel. Lamps and other electrical appliances have a matching set of prongs called **polarized plugs** that fit into the outlet socket. These polarized plugs are attached to lamps so that the terminal in the centre at the bottom of the lamp socket is

Figure 6
If the circuit breaker in the GFCI outlet opens the circuit, it can be reset by pressing the red button.

always attached to the 120-V conductor. Why is that a good safety feature? Also, the lamp switch is always connected to the 120-V conductor, so that when the lamp is switched off, the 120-V conductor is not connected to the lamp socket at all. In older homes where the sockets on wall outlets and plugs on appliances may not be polarized, it is possible to reverse the plug terminals when you plug in a lamp. It is a good safety precaution to unplug a lamp before adding or changing a light bulb.

In newer homes and buildings, every outlet socket has a third round hole. This is the **ground terminal** of the socket. The ground terminals on all the sockets in each circuit are connected together by a bare copper wire (called the "ground" wire) in the electrical cable that connects all the sockets to the ground terminal inside the distribution panel.

In every appliance that has an electrical cord with a three-pin plug, the ground wire in the plug is attached to the metal case of the appliance, thereby grounding it. When you are standing on the ground, your body is electrically "grounded." Sometimes when the appliance or the electrical cord becomes damaged, the live 120-V wire may contact its metal case. If there were no grounding wire, the person touching the case would get an electric shock. However, because the grounding wire has a much lower resistance than the person, a very large current flows through the grounding wire to "ground," and opens ("trips") the circuit breaker. Never remove or bend the **grounding pin** so that you can connect a three-pin plug to an extension cord or an outlet with only two socket terminals because to do so removes a very important safety feature.

Ground Fault Circuit Interrupter (GFCI)

It is possible for people to be in a situation where they are receiving an electric shock large enough to kill them, but the current passing through them is not large enough for the circuit breaker to open the circuit. Many fatal accidents have occurred in this manner, especially near or in water. A **ground fault circuit interrupter** (GFCI) outlet is a special kind of combination outlet socket and circuit breaker that prevents this from happening because it responds to very small currents. The GFCI outlet can be fitted into the space of a normal outlet anywhere in the building (**Figure 6**).

The GFCI outlet is specially designed to detect very small differences in the electric current flowing in the 120-V wire compared with the neutral wire. If the circuit is operating normally, the current in both wires should be identical. If, however, the current in the neutral wire is slightly less than the current in the 120-V wire, some of the current must be flowing elsewhere. The GFCI outlet immediately detects this difference, and its built-in circuit breaker opens the circuit within a few thousandths of a second.

Understanding Concepts

1. If the neutral wire is connected to the metal pipes in the building plumbing system, how can it be properly grounded?

2. Explain how the use of the neutral wire provides three different sources of electrical energy.

3. **(a)** Compare the advantages and disadvantages of circuit breakers and fuses.

   (b) Why would having too many outlets on one circuit cause problems?

   (c) What might happen if circuit breakers or fuses were not installed in a circuit in a house?

4. List three safety features related to wall outlets, plugs, and grounding pins, and explain how they provide protection.

Making Connections

5. What appliances are connected to double circuit breakers used in the distribution panel? Explain why.

6. Where would it be advisable to install GFCIs in and around a house? Explain why.

Exploring

7. Make a chart and list all the electrically operated devices that have been specifically designed for your personal safety in a house. Are they self-contained or are they plugged into wall outlets? How do they function? What warning system alerts you if they stop working?

Challenge

What safety features have designers included in the products that you are including in your challenge?

Practical Circuits

On a sheet of paper, make a list of all the electrical devices you or members of your family used between the time you got up and the time you left home this morning. Put a check mark beside those devices, such as a toaster, that have controls that are used directly each time. Put an X beside those, such as the refrigerator, that have electrical controls that are set once and then often continue to operate automatically for years. Each electrical device has control features for its use and safety features to protect us. All these devices consist of combinations of series and parallel circuits. Some are operated by batteries, and others by plugging into the wall outlets.

In this activity you will observe and analyze the operation of several common electrical appliances and devices. For each appliance or device, you will draw a schematic circuit diagram and describe the function of each part on the circuit diagram.
Note: Some more complex parts of a particular appliance or device may be represented by a labelled rectangle on the circuit diagram. Your teacher will indicate which parts of each appliance or device can be indicated in this manner.

Materials
- electrical appliances, devices, and operating manuals (where appropriate) supplied by your teacher

Figure 1

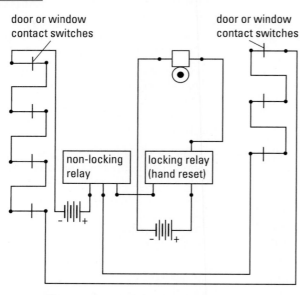

a Closed circuit alarm system

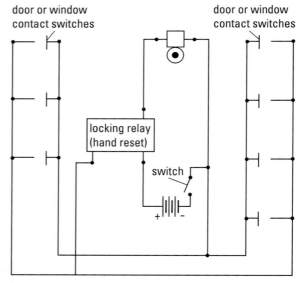

b Open circuit alarm system

⚡ Follow the instructions given by the teacher for each part of this activity. Under no circumstances should you disassemble any part of the appliance, or try to operate it in an unusual way. All safety features and procedures must be observed. If you need to operate the appliance to check how it works; you must operate it in the manner specified in the operating manual. All parts of the electric circuit diagram can be determined by simply observing the appliance, and operating (if necessary) the controls on it in the normal way.

Procedure

You will be required to follow the procedure described below for each of the electrical appliances or devices provided and specified by your teacher.

1 Observe the appliance or device. Identify all the parts that have an electrical function. Note that some of the parts will not be visible, but their presence can be inferred by the controls you operate or by the indicator lights used to show that certain functions are occurring or have finished.

(a) Write the name of the appliance, and under the heading "Electrical Components," list those parts.

2 Identify how each of the components identified in step 1 is connected to function properly. Your teacher will explain some of the more complex parts of the appliance that may be represented by a labelled rectangle.

(a) Draw the schematic diagram (by hand, 5D or by using a CAD program on a computer). Use the circuit symbols on page 545 of the Skills Handbook.

3 Describe the operation of the appliance, including any special control functions.

(a) Under the heading "Description of Operation," write the required description.

4 Identify any safety features built into the appliance, with regard to controls and also with regard to its construction.

(a) Under the heading "Safety Features," rite the required list of the safety features.

5 Repeat steps 1 to 4 for each appliance or device identified by your teacher.

Understanding Concepts

1. For each appliance or device, draw a labelled flow chart to show all the energy conversions that occur during its operation. Which forms of energy was the appliance or device designed to produce? Why were the others produced?

2. Apart from the switch, were components more commonly connected in series or in parallel in the appliances or devices? Suggest reasons for this.

3. Did the connecting cord consist of two wires or three wires? Suggest reasons why.

4. Why should you always hold on to the plug rather than the connecting cord when removing a plug from its socket?

Making Connections

5. What materials were used for the exterior case of each appliance? Why?

6. Does the instruction manual provide sufficient information on the safe operation of the appliance or device? What features would you recommend be in the instruction manual to ensure consumers read about the safe operation?

7. List any aspects of the design or operation of the appliances or devices that you think might cause safety problems.

Exploring

8. Examine the owner's manual for a car or a truck.

   (a) Make a list of the electrical circuits and the fuses that protect them.

   (b) In which circuit is the greatest number of devices protected by the same fuse?

   (c) (i) How many light bulbs are in one tail-light assembly?

       (ii) What are the functions of each bulb?

       (iii) Which bulbs are included in the operation of more than one circuit? Explain why.

⬤ Challenge

Use this list of devices in your consumer report.

Making the Most of Energy Resources

Earlier you learned that when the chemical energy in fossil fuels was converted into electrical energy, most of the original chemical energy was lost in the process. In fact, there are losses every time an energy conversion takes place. Understanding how these losses occur helps us to decide which way of producing electricity is most economical.

To provide a scientific basis for determining which way of producing electricity is most effective, scientists measure the amount of energy used to make the electricity, then measure the amount of electrical energy produced and compare the two quantities. Let's use fossil fuels as an example. The chemical energy in the fossil fuels used to make electricity is called the **input energy**. The electrical energy actually produced by the conversion is called the **output energy**. By comparing the output energy to the input energy for each way of producing electricity, we can decide which process is the most effective. The scientific term for this comparison between the output energy and input energy is efficiency. The efficiency of any energy conversion, whether it be converting nuclear energy into electrical energy in a nuclear generating plant, or electrical energy into light energy in a light bulb, can be calculated using the following formula:

$$\text{Efficiency} = \frac{\text{Useful energy output}}{\text{Energy input}}$$

Efficiency is often expressed as a percentage. The percent efficiency is simply the efficiency multiplied by 100%.

$$\text{Percent efficiency} = \frac{\text{Useful energy output}}{\text{Energy input}} \times 100\%$$

The Efficiency of Producing Electrical Energy

If you look at the percent efficiency of energy conversions for the energy sources in **Table 2** on page 351 and **Table 4** on page 353, you can see that some of the differences are significant. However, other factors need to be considered, not just the efficiency of the actual energy conversion. What would happen in dry weather or in the winter? Are there always waterfalls and rivers where we need electricity? Sometimes we are forced to use other methods for producing the electrical energy we need, even though they are less efficient.

The Efficiency of Transmitting Electrical Energy

The efficiency of the transmission lines also has to be considered. Once the electrical energy has been produced, it must be transmitted to our homes to provide us with the energy needed to cook food or wash our clothes. As mentioned earlier, as the current flows along the transmission lines, about 9% of the electrical energy produced at the generating station is lost as thermal energy (heat) to the air around the wires.

The Efficiency of Using Electrical Energy

The third area where energy losses and efficiency have to be considered is in its final use in our homes, schools, factories, and communities. Some devices are much more efficient than others. For example, an electric motor typically has a percent efficiency of over 80%, whereas an incandescent light bulb has a percent efficiency for producing light energy of only about 5%. Fully 95% of the input energy to the light bulb is released in the form of thermal energy. In some cases, the energy not converted into light is used to heat substances, such as the lamps used to heat food in restaurants. In this situation, the overall efficiency of the light would be greater than for a normal incandescent bulb. The wasted energy from the motor is also mostly released as thermal energy. What other forms of waste energy might a motor produce? The efficiency of a fluorescent light source is much higher than an incandescent bulb, as can be seen in the calculation below.

Sample Problem: Determine the percent efficiency of a 60-W fluorescent light bulb that uses 2000 J of electrical energy to produce 400 J of light energy.

Data:
Input energy = 2000 J
Output energy = 400 J
Percent efficiency = ?

Equation:

$$\text{Percent efficiency} = \frac{\text{Useful output energy}}{\text{Input energy}} \times 100\%$$

$$\text{Percent efficiency} = \frac{400 \text{ J}}{2000 \text{ J}} \times 100\%$$

Percent efficiency = 20%
The percent efficiency of a 60-W fluorescent light bulb is 20%.

If you combine the total energy losses caused by producing the energy, transmitting it to its place of use, and finally converting the electrical energy to the form we need, the overall percent efficiency for some electrical loads is disturbingly low. For example, only about 1% of the original energy in the fossil fuel is converted into light energy by an incandescent bulb.

Improving Efficiency

In some countries, such as Japan, the cost of electrical energy is much higher than in Canada. Engineers and scientists in these countries pioneered the design of electrical appliances that were much more efficient. In Canada it is now required by federal law that special "Energuide" information be attached to each appliance (**Figure 1**). This label states the amount of electrical energy per year the appliance uses. Consumers can compare the Energuide number for all the refrigerators of the same capacity and make an informed decision about which one is the most efficient model.

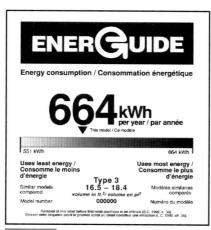

Figure 1

Understanding Concepts

1. Explain the relationship of the terms "input energy," "useful output energy," and "efficiency."

2. Suggest why it is less efficient to produce electrical energy from nuclear energy than from falling water.

3. **(a)** Calculate the percent efficiency of an electric motor that uses 15 000 J of energy to produce 11 500 J of useful energy.

   (b) Calculate the percent efficiency of an incandescent light bulb that produces 2500 J of light energy from 50 000 J of electrical energy.

Making Connections

4. How many more times efficient is a fluorescent light compared with an incandescent bulb? What would be the environmental and economic implications of using fluorescent lights exclusively?

Reflecting

5. How do you use electricity efficiently in your own home? What do you think is your family's least efficient use of electricity?

Challenge

What factors do you need to consider when deciding to choose alternative sources of energy?

How would you communicate the concept of efficiency so that its importance is understood?

Electrical Energy Use in the Home

The daily amounts of energy used in your home provide an interesting profile of your entire family's activities. Some days the family will have used much more energy than others. What factors might play a role? There are some appliances in your home that use significant amounts of electrical energy every day, regardless of what the family does. As you have already learned, the electric meter measures the amount of electrical energy used. Although the joule (J) is the normal unit for energy, the large amounts of electrical energy used in a home would require using numbers in the tens or, in some cases, hundreds of millions of joules every month. In the electric meter, the electrical energy is measured in kilowatt hours (kW·h). There are 3 600 000 J in 1 kW·h. As you can see, the kilowatt hour is a much more practical unit to use.

$1 \text{ kW·h} = 1000 \text{ W·h} = 1000 \text{ W} \times 3600 \text{ s} = 3\,600\,000 \text{ W·s} = 3\,600\,000 \text{ J}$

To understand what energy use means in terms of kilowatt hours, think about the energy used by a 100-W incandescent light bulb. One kilowatt hour (1 kW·h) of electrical energy represents 100 W of electrical power being used to operate the light bulb for 10 h.

The data in **Table 1** show the average power rating of common appliances, the estimated number of kilowatt hours (kW·h) they use per month, and the approximate monthly cost of operating them.

If possible, look at a recent electricity bill. It will be similar to the sample shown in **Figure 1**. You will see that it states the number of kilowatt hours of electrical energy used and the total cost of the energy. The electric meter records only the total amount of electrical energy used in your home. It cannot distinguish whether the energy was used for the washing machine or the toaster. However, if you want to try to conserve on your use of electrical energy, you need to know how to calculate how much energy is used by each appliance. The formulas for electrical energy and power you learned earlier can be used to complete a table like that shown in **Table 1**. The monthly amount of energy used by a clothes dryer is calculated on the next page. Instead of calculating power as we did in Chapter 11, the equation is rearranged to the form shown there.

Did You Know?

Reading the electric meter has become a little easier. Now the person who wishes to read the electric meter stops on the street near the house and activates an electronic device from within the vehicle that sends a signal to the electric meter on the wall of the house. The meter responds to the signal and automatically transmits the latest reading on the meter.

ELECTRICITY USAGE

METER NUMBER	BILLING PERIOD FROM	BILLING PERIOD TO	ELAPSED TIME	CURRENT METER READING		PREVIOUS METER READING		METER MULTIPLIER		KILOWATT HOURS USED
829046	Dec. 13 1998	Feb.11 1999	2.0 months	(27322 Actual)	–	(26370 Actual)	x	1	=	952

CURRENT CHARGES AND CREDITS

ENERGY		98.87
SUBTOTAL	98.87	
G.S.T. #85891 2572 RT		6.92
AMOUNT DUE ▶		**$105.79**
If Paid After Due Date		$110.73
Due Date		**March 5. 1999**

Energy Management Information

Reading Date	Elapsed Time (Days)	kW·h Used	Average Daily kWh	kW·h Cost ($)
Feb 11, 1999	60	952	16.1	98.87
Dec 13, 1998	57	897	15.7	94.37
Oct 17, 1998	62	1165	18.8	116.30
Aug 16, 1998	61	997	16.1	102.55
June 15, 1998	65	1307	20.7	127.91
April 11, 1998	55	1271	21.9	124.97
Feb 15, 1998	67	1311	19.6	128.24
Dec 9, 1997	56	858	15.3	91.18

Messages:
PLEASE PROVIDE YOUR DAY-TIME PHONE NUMBER SO THAT WE CAN CONTACT YOU IF NECESSARY.

Figure 1

Electrical energy = Electrical power × Time Interval

$E = P \times \Delta t$

Where E = electrical energy measured in kilowatt hours (kW·h)
P = electrical power measured in kilowatts (kW)
Δt = time interval measured in hours (h)

Sample Problem: How many kilowatt hours of electrical energy are used in one month by a clothes dryer that has a power rating of 5 kW (5000 W) and is operated for 4.5 h each week?

Data:
Electrical energy used = ? kW·h
Time interval = 4.5 h × 4 = 18 h
Power = 5 kW

Equation:
$E = P \times \Delta t$
$E = 5 \text{ kW} \times 18 \text{ h}$
$E = 90 \text{ kW·h}$
The electrical energy used by the clothes dryer in one month is 90 kW·h.

Calculating the cost of the electrical energy used is perhaps the easiest task of all. Just as you pay so much per kilogram when you buy meat or rice, you pay so many cents per kilowatt hour when you buy electrical energy. What is the current cost of one kilowatt hour of electricity where you live? It varies slightly across the country. The formula for calculating the cost of electricity is shown below. Let's calculate how much it costs to run a refrigerator for a month.

Cost = Electrical energy (kW·h) × Rate (cost per kW·h)
Cost = E × Rate

Sample Problem: Calculate the cost of the electricity needed to operate a refrigerator/freezer (500 W) for one month if it uses 75 kW·h of energy. The rate charged for electricity is $0.08/kW·h.

Data:
Energy used = 75 kW·h
Rate = $0.08/kW·h
Cost = ?

Equation:
Cost = E × Rate
Cost = 75 kW·h × $0.08/kW·h
Cost = $6.00
The cost to operate the refrigerator/ freezer for one month is $6.00.

Table 1

Appliances	Average Power W	Monthly Energy Use kW·h	Approximate costs $
air conditioner (room)	750	90–540	7.20–43.30
clothes dryer	5000	50–150	4.01–12.03
coffee maker	900	4–27	0.32–2.17
computer (monitor & printer)	600	5–36	0.40–2.89
electric kettle	1500	3–15	0.24–1.20
lighting – 60-W incandescent lamp	60	5–30	0.40–2.41
microwave oven	1000	5–20	0.40–1.60
television – colour	80	5–15	0.41–1.20
toaster	1000	1–5	0.08–0.41

Costs are based on $0.082 per kW·h.

Understanding Concepts

1. Why is the kilowatt hour used instead of the joule for measuring electrical energy used in the home?

2. (a) Calculate the energy used by a coffee maker that has a power rating of 0.900 kW and operates for 17 h.

   (b) How much energy is used by a 1.5 kW electric kettle that operates for 0.80 h per day for two weeks?

   (c) Calculate how much energy a colour television that has a power rating of 0.120 kW uses in one week if it operates 9 h per day every day.

3. Calculate the cost of operating each of the devices in question 2, if electrical energy costs $0.08 per kW·h.

Making Connections

4. List the appliances in **Table 1** that account for the greatest amount of energy use. What alternatives could you suggest to reduce the energy consumption of the appliance?

5. Is the cost for lighting in **Table 1** realistic? Explain your answer.

6. Think of several short-term and long-term ways to reduce the use of electricity in your home. Which would save the greatest amount of money? Why?

Reflecting

7. If your family was going to buy some new appliances in the near future, how important would the Energuide labels be in your decision making? What other factors would influence your purchasing decision?

The Family Energy Audit

The readings from electric meters can be used to analyze how families use electrical energy over a period of time. You now have most of the information and all the background knowledge required to develop a profile of your use of electrical energy. A quick glance at **Table 1** on page 373 indicates that about six or seven appliances account for most of the monthly cost in a typical electrical energy bill. (Remember, however, that information in the table is only for a single appliance. You may have several light bulbs, for example, operating at the same time.) There are a number of ways you can reduce energy costs if you take the time to analyze energy use in detail. Although each attempt at energy saving may not seem very significant, when all of these small savings are added up, and the savings repeated month after month, it may become possible to save hundreds of dollars a year. The savings are not only measured in terms of money. Every time you reduce the amount of energy you use, there are environmental, societal, and economic benefits.

In this activity you will collect and analyze information about the use of electrical energy in your home (or the home of a friend or relative), calculate how much it costs, and develop a practical plan to maximize the efficient use of this energy in the future.

Materials
- a set of records of the energy meter readings for your home
- data about patterns of energy use by members of the family in your home
- power ratings of a specified list of appliances in your home

Procedure

1 Your teacher will show you how to read an electric meter.

(a) Record the readings of the meters shown in **Figure 1**.

2 Predict which appliances you think your family will use most, and which will use the most energy during the monitoring period.

(a) Record your predictions.

3 To calculate the family's electricity **(5D)** consumption, check the meter reading at the same time each day for the number of days specified by your teacher.

(a) Create a data table and record your readings daily.

4 During the monitoring period, notice the family's activities, the weather, and other events that might affect the amount of electricity used.

(a) Record these activities and events.

5 List the various appliances in the home that you have been monitoring, including

Today's reading   test dial

kilowatt hours

Previous day's reading   test dial

kilowatt hours

Figure 1

when the appliances were used, and for how long.

✎ (a) Record the list of appliances and their usage.

(b) At the end of the monitoring period, summarize your usage data in a table similar to **Table 1**.

Table 1

Electrical Appliance	Power Rating (P)		Time Used (Δt)		Energy Use (E)	Cost
	(W)	(kW)	Per Day (h)	Per Month (h)	Per Month (kW·h)	Per Month ($)
?	?	?	?	?	?	?
?	?	?	?	?	?	?

6 Find out the power rating for each of the appliances.

(a) Complete the "Power Rating" column in the table, in both watts and kilowatts.

7 Calculate which appliances were used the most during the monitoring period.

✎ (a) Record your answers.

8 Share your assessments with a partner. Discuss each other's ideas and make any necessary revisions.

✎ (a) Record any changes in your assessments.

9 Calculate the amount of energy used by each appliance per month, using the formula

Energy Use (kW·h) = Power Rating (kW) × Time Used/Month (h)

(a) Complete the "Energy Use per Month" column in the table.

10 Using the current cost of electricity per kilowatt hour, calculate the cost of the energy used by each appliance per month.

✎ (a) Record the costs in the table.

(b) Write a list of the appliances, starting with the one that costs the most to operate during the monitoring period, followed by those that cost less.

11 Share your data with a partner and compare the order in which the devices appear in your lists. Note any differences that you find in your tables and discuss the relative amounts of time that the devices are used in each home.

✎ (a) Record any comments and any differences between your list and your partner's, giving reasons where possible.

(b) How did your initial predictions compare with your actual observations? Comment on any unexpected differences.

(c) Review the data in the table and suggest reasons for any unusual changes in the daily amounts of electrical energy used by your family.

12 Try to identify ways to conserve energy.

(a) Write a brief proposal to the family, making suggestions about how to reduce their electricity bill and calculating the monthly savings.

Making Connections

1. How would you convince your family members that your plan was reasonable and worth following?

2. What other resources, in addition to electricity, would be conserved if your plan were implemented?

Exploring

3. Visit some environmental web sites on the
(3A) Internet for additional suggestions about conserving electricity.

Challenge

Classify the appliances into high-, medium-, and low-energy consumption.

Efficiency, Conservation, and Convenience

Energy-Efficient Light Bulbs

Although the cost of the energy used by a common household incandescent light was one of the smallest amounts in **Table 1** on page 373, it is important to remember that the cost was for only one 60-W bulb. When you add up all the costs for every light bulb in the house, lighting becomes one of the major expenses for electrical energy. It is easy to forget this as we go from room to room and leave the lights on.

What kinds of lights are used in stores, school, factories, and offices? Fluorescent lights are the standard light source almost everywhere (**Figure 1**). Compact fluorescent bulbs are four times as efficient at producing light as incandescent light bulbs. For example, a 15-W fluorescent bulb produces the same amount of light as a 60-W incandescent bulb.

Energy-saving compact fluorescent bulbs were introduced in the late 1980s, but only 9% of households use them, while 87% of the 52.3 million bulbs used annually are still incandescent. The first energy-saving bulbs had some shortcomings, and compact fluorescent bulbs are still not suitable for all uses. When they operate, they may interfere with remote controls for TVs, VCRs, and stereos as well as with cordless phones. Although most of the problems have been largely resolved, another major obstacle to widespread use remains—their initial cost. A compact 25-W fluorescent bulb, costing about $23, replaces a 100-W incandescent bulb costing about $0.50.

(a) A 25-W compact fluorescent bulb lasts 10 000 h. If a 100-W incandescent bulb lasts 1000 h, how many are needed to last as long as one compact fluorescent bulb? What is the cost of purchasing this many incandescent bulbs?

It is immediately obvious that the compact energy-saving fluorescent bulb costs much more to purchase than the number of incandescent bulbs needed to last 10 000 h. However, an accurate comparison also needs to include what it costs to operate each type of bulb.

(b) If electricity costs 8 cents per kW·h, what is the cost of using one 25-W compact fluorescent bulb for 10 000 h?

(c) If each 100-W incandescent bulb lasts 1000 h, how much does it cost to operate it for that time at 8 cents per kW·h?

(d) How much would it cost to operate the number of 100-W incandescent bulbs required for 10 000 h of light?

Now you see the 25-W compact fluorescent bulb costs much less to operate than the 100-W bulbs over the same time period. But we still need to include the enormous difference in the bulbs' prices.

(e) What does it cost to buy and use one 25-W compact fluorescent bulb compared with the cost to buy and use the required number of incandescent bulbs for 10 000 h of light? Which type provides the better value?

Because the cost of electricity varies with location, it takes from two to four years to recoup the cost of compact fluorescent bulbs in reduced energy bills. However, if everyone in Canada used the bulbs, 3.17 billion kW·h of electricity would be saved annually.

How Much Could Be Saved If...?

Determine the number and the power of each of the light bulbs in use in your bedroom, living room, kitchen, bathroom, and hallways.

Use **Table 1** to decide which compact fluorescent bulb could be used to replace each bulb with a different power rating, determine the cost of each size of bulb, and calculate the total cost per 10 000 h for all the fluorescent bulbs required in the five areas above. Visit a hardware, home centre, discount, or department store to determine the cost and expected life of the incandescent bulbs. Remember to include the cost for each of the incandescent bulbs needed to last as long as one compact fluorescent.

(f) If your family replaced the light bulbs currently in use in the rooms investigated, how much could be saved over 10 000 h of use? Design a table to display your data.

Table 1	Table of Equivalent Bulbs
Incandescent	**Compact Fluorescent**
60 W	15 W
75 W	20 W
100 W	25 W
50/100/150-W trilight	13/23/34-W trilight

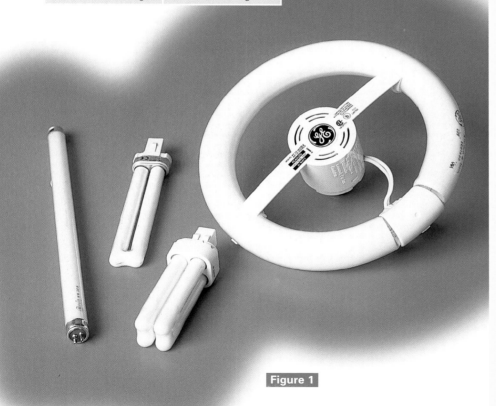

Figure 1

Caution: Appliances on Standby Use Energy

Many electrical devices are always using electricity because they are in a standby mode when not being used. Take a tour of your home and make a list of the devices in this category.

Various types of electrical devices take a significant amount of time to become operational when turned on. For example, the device that produces beams of electrons in the picture tube of a TV set takes several seconds to warm up before it will produce a picture. Therefore, a small electrical current must always be flowing to maintain the tube's operating temperature. Other devices, like microwave ovens and VCRs, have built-in clocks that are necessary for them to operate. Still others, such as cordless telephones, answering machines, and security systems,

Did You Know

You can save energy by

- thawing, or partially thawing, frozen foods in the refrigerator before cooking;
- using a microwave oven instead of a conventional oven whenever possible; it cooks up to 75% faster and saves up to 70% of the energy;
- cleaning dust and dirt from light fixtures;
- using timer switches to turn your lights on periodically, instead of leaving lights on continuously, when you're away from home for long periods of time;
- drying laundry on a drying rack or a clothesline whenever possible, instead of using the dryer;
- blocking the sunlight by closing drapes or blinds to reduce the need for air-conditioning.

must be ready to function when they receive a signal from outside your home. Computers and alarm clocks use low-voltage batteries to retain their memories when household electrical energy fails.

(g) Why are these devices on standby when not in use?

(h) What problems would occur if they were shut down completely when turned off?

(i) What common electrical devices require a standby mode independent of household electrical energy? Why?

(j) Use **Table 2** to calculate the cost per week, per month, and per year of providing the standby electrical energy for the devices listed.

Table 2

	Electrical Device	Standby Power Required (W)
appliances	microwave oven	4
communications	answering machine	3
	cordless telephone	3
entertainment	cable TV box	12
	colour TV	6
	compact audio system	11
	digital satellite TV	15
	VCR	5
miscellaneous	baby monitor	1
	cordless tools/vacuum	2
	doorbell	2
	electric toothbrush	2
	garage door opener	2
	security system	18

 **Try This** Saving

How much money can you save by reducing the amount of electricity you use? Here's how to find out first-hand.

- Record how much electrical energy you use in your home in a normal week by reading the electric meter at the beginning and end of one week.
- For the next week, do everything you can to reduce the amount of electricity you use. You could switch off appliances when they're not needed, use more energy-efficient methods of cooking, reduce the amount of air-conditioning, and read instead of watching TV.
- At the end of the second week, check your meter again.
- Calculate the amount of electrical energy used. How much did you reduce your consumption? At the present cost of electricity, how much money did your family save?

Exploring

1. Visit the web site of a lighting research centre (3A) (e.g., the Lighting Research Center at Rensselaer Polytechnic Institute in Troy, New York) to investigate different forms of lighting and the projects under way to make lighting more efficient. Make a visual presentation to illustrate more efficient lighting in a particular room.

Engineer in Business

When Robert Williams was still a student in high school, he was fascinated by new stuff, especially electronic devices. He fixed stereos and TV sets and helped the teacher to set up demonstrations for physics class. Young Williams wanted to work with existing electrical and electronic equipment and also explore new applications and different approaches. He enrolled in the Cooperative Electrical Engineering program at the University of Waterloo, where "the problem-solving skills taught in the program were very valuable and could be applied in many different situations." Williams enjoyed working with other students and professors to come up with unique solutions to problems, meanwhile earning a minor in Management Science. This helped him to realize that products need to be both well-financed and promoted to be successful.

The best technological developments in the world can be failures if the products are not promoted properly.

Rather than work for a large company after graduation, he chose to join a series of small software and communication companies just getting established. All flourished during his time with them.

Williams' current employer, Nexsys-Commtech, is a young company developing wireless electronic meter-reading systems that gather information on electrical, water, and natural gas consumption in homes and businesses. The information is sent via radio waves to a receiving antenna at a central site within a city. There, the customers' use of electricity, water, and gas can be analyzed, and plans developed to improve the utilities' distribution.

Williams says that the technology cycle today is so short that concepts must be brought on line very quickly. Engineers must have the problem-solving skills that allow them to translate ideas into reality to meet customers' needs. But, they must also be able to sell their ideas and raise the support to make it all happen.

Exploring

1. What is a "cooperative program"? What are its advantages over a regular college course?

2. Williams added courses in Management Science to his Engineering Degree. What benefit might these courses have given him?

3. In what other situations might electronic meter reading be useful?

Optimizing the Use of Electrical Energy

Most people agree that we need to conserve energy in as many ways as possible. The vast majority of Canadians obtain the electrical energy they use in their homes from the local or provincial energy suppliers. At present there are no practical alternatives for providing the general population with electrical energy. So if we want to reduce the amount of electrical energy we use, there are two basic approaches we can take. The first approach considers the formula used to calculate electrical energy use. It suggests that the energy used by an appliance depends on two factors, the power rating of the appliance, and the time that it operates. If we can reduce either of these two factors, we can use less energy. A second approach, of course, is to find another, more energy-efficient way to do the work being done by the appliance (**Figure 1**).

Figure 1

Reduce Energy Consumption

Consider the power rating of the appliance. If the power rating is reduced, so is the energy used. Of course, you can't change the power rating of the appliances you already own. However, when it is time to replace them, your family can choose a model with a reduced power rating by comparing the Energuide labels. The efficiency of some appliances has been improving over the years.

The second factor that can change the amount of energy used is the amount of time that you use appliances. For example, is there a way to reduce the time you operate the clothes dryer? How did people dry clothes before electric clothes dryers became common?

> ### Try This Reduce the Use
>
> In a small group, prepare a plan that would help a family to reduce the cost of operating a dryer. In the same group, look at **Table 1** on page 373 and decide how many of the appliances you could use for less time. Prepare a practical plan for reducing the use of at least five appliances.

Figure 2

Alternative Approaches to Energy Sufficiency

The second approach is to complete the task done by a particular appliance by using another form of energy altogether. Some groups and individuals have made a deliberate decision not to use electrical energy at all. They use a combination of energy sources that were available before electrical energy was commonly used. Others use a combination of different sources of energy, including electrical energy, and use each energy source at the appropriate time to be as efficient as possible. There are many more alternatives available than you might think. Thinking about what people use when they go camping provides some examples.

Future Possibilities for Energy Sufficiency

In this unit you have learned about many different kinds of energy sources, both large and small. You have learned how to design and construct electric circuits, and calculate the energy and power used in many different kinds of electrical loads. These valuable tools will provide you with the background to make informed decisions about conserving energy resources and selecting the most appropriate energy sources to operate your home in the future.

Many different kinds of small-scale sources of electrical energy are being developed (**Figure 2**). What are known as "hybrid" systems are being tested and used to generate some of the electrical energy we need in our homes. It may some day be economically feasible to use various combinations of photoelectric cells, wind energy, biomass-operated fuel cells, hydroelectric generators, batteries, and portable fossil-fuel generators to provide most or all of our energy needs.

Understanding Concepts

1. Explain how the electrical energy formula can be used to identify two ways to conserve electrical energy.

2. Describe what you could do, using some other form of energy, to replace each of the major appliances we normally use in our homes. What are the major disadvantages of not using electricity for these appliances?

3. What is meant by the term "hybrid" when applied to the source of electrical energy for a house or community?

Exploring

4. Research different renewable

(3A) hybrid systems for producing electrical energy. What would be the most practical hybrid combination of renewable sources of electrical energy for year-round use in your part of Ontario?

5. Research outdoor equipment that could be used to replace electrically operated appliances in your home.

Challenge

What key messages do you want to incorporate in your questions so others will understand the importance of conserving energy?

Achieving Self-Sufficiency

The number and variety of small-scale devices for generating electricity are increasing rapidly. You have looked at a number of web sites of companies that produce devices that generate electricity. Some people are already using these renewable energy sources to make their homes completely self-sufficient in electrical energy. The present methods of generating electricity by using nonrenewable fossil or nuclear fuels cannot be sustained in the long term. The emphasis must be on the development of sustainable methods for producing electricity using renewable resources.

Alternative Sources of Electrical Energy

In Chapter 11, you learned about the major alternative sources of renewable energy. Some alternatives, such as tidal, hydroelectric, and geothermal energy sources, are somewhat limited by location. Two of the most promising are wind energy and photoelectric cells. Others, such as fuel cells, are beginning to develop as a versatile energy source that can operate using a variety of energy inputs. Fast-growing hybrid trees, such as hybrid poplar trees, are also being considered as a renewable biomass alternative in some areas as a means of replacing the fossil fuels presently used to produce electrical energy.

Lifestyle Considerations

When considering self-sufficiency, we have to make decisions concerning our lifestyle and work. We have to consider what appliances, tools and other devices we consider to be essential, and which are conveniences, and what our energy needs really are.

In this activity, you will research and devise a self-contained electrical energy generation plan using renewable energy sources to meet the energy requirements of a dwelling, farm, or community in Ontario. Be creative. Who knows, maybe you will actually do this yourself in the not-too-distant future.

Procedure

1. Choose a dwelling. It could be a cottage, a house, a condominium complex, or even a houseboat or yacht. If you choose a farm, remember to include all the outbuildings and the needs of livestock. You might be really ambitious and choose to design the generating plant for a manufacturing facility or a small, self-sufficient residential community.

2. Your teacher will provide you with an outline that specifies the detailed requirements for your electrical energy generation plan.

Figure 1

3 Define the problem. This is more than half of developing the solution. You need to consider such factors as location, weather conditions throughout the year, long-term, worst-case weather conditions, and the construction of the building or buildings and their orientation to the Sun.

(a) List the factors to consider as you develop your self-sufficiency plan.

4 Earlier in this chapter, you analyzed the use of electrical energy in your home.

(a) List all the electrical appliances and devices that will be required.

5 Determine the total energy needs for times of peak energy usage throughout the year. You also need to determine the maximum amount of electrical power needed at any one time. This will indicate how many kilowatts of power your system will have to be able to supply when your worst-case number of electrical appliances and devices are operating at the same time.

✎ (a) Calculate and record the maximum usage.

6 Consider which sources of electricity you will use. When you are reviewing the information about each of the alternative energy sources, carefully consider the effect of the weather at all times of the year. You must be able to meet your critical energy needs, both day and night, all the time. You should also consider the impact of your plan on the environment.

(a) List possible sources of electricity and note their advantages and disadvantages.

7 There will be times when some of the energy sources will not be functioning, or will break down, or will require maintenance. Consider what kind of energy backup systems will be needed. If you need to use batteries, how many will be required? Will batteries alone provide enough energy, or will you need another energy source for peak power requirements? How will they be connected to the different kinds of generators, and how can you operate the appliances themselves if you are using batteries?

✎ (a) Record your answers and decisions.

8 Plan how the electricity will be supplied to your dwelling and how it will be distributed within it.

(a) Draw a floor plan of the building(s) and indicate where the wall outlets and light fixtures are to be placed. Identify what the current rating of the circuit breakers in the distribution panel should be for each circuit.

(b) Prepare a written description of the way your generating system functions, and how you have prepared for each of the various problems that may occur during its operation.

Making Connections

1. What are some of the benefits and drawbacks of being self-sufficient for energy?

Exploring

2. Are there any communities or homes in your (3A) area that are self-sufficient? Find out as much as you can about them, including their sources of energy.

3. It is more efficient to use the heat from (3B) burning wood or coal directly, rather than to use electricity generated from burning fuels to heat your home. Research how much energy is lost in the conversion to electricity, and the transmission to your home. What other issues need to be considered?

Reflecting

4. How practical would it be to convert your present home to self-sufficiency in electrical energy? Write a brief article about the advantages and drawbacks.

Challenge

Make a chart comparing the advantages and disadvantages of the alternative sources of energy mentioned.

Need More Energy?

The township is located in a scenic, yet quiet, rural area. Many tourists come to the main town that is on the edge of one of the largest lakes in the region. The tourists and the town's residents enjoy the sandy beaches near the town, and the nearby rivers. The resorts in the area are designed to appeal to families, providing such activities as swimming, fishing, hiking, canoeing, and tennis. There are several hundred cottages surrounding the edge of the lake, and the other lakes nearby.

A business corporation that builds theme parks has purchased a large piece of land, including several hundred metres of waterfront on the lake right next to the town limits. They plan to build thrilling roller coasters, wild rides, and a number of video arcades, restaurants, and bars, as well as facilities for water sports and boat rentals. However, there is a catch. The theme park will use large amounts of electrical energy, so the high voltage transmission lines currently delivering electricity to the township will have to be upgraded. The new transmission lines will require taller support towers and a wider clearance corridor below them. The old transmission line corridor passes through farmland in the valley and along the edge of the town, close to homes in a new subdivision and an elementary school. Because the new transmission line corridor must be wider, broad strips of land must be added alongside the old corridor as it passes through the region, including the sites of several farms and homes.

High voltage transmission lines emit electromagnetic radiation that increases in strength as the voltage increases. Some studies have linked exposure to electromagnetic radiation with increased cancer rates, particularly leukemia, brain, and breast cancer. However, other similar studies have not found such a link.

The town is split over the issues: do they want the new theme park? Do they want the new transmission lines? How will landowners and homeowners, some of whom have lived on the same land for generations, be compensated for no longer being able to live on their property?

There are those who feel that things are just fine the way they are and visitors are happy and enjoy coming for a quiet vacation. They argue that the family atmosphere will be ruined by a large, noisy theme park. Others are concerned about the possible health effects of the new transmission lines. Still others feel that the theme park is the shot in the arm needed to revitalize a tired, outdated resort. They believe that adequate compensation can be offered to people who have to move out of their homes and farms.

The public meeting has been organized to give everyone the opportunity to air their views on the project.

Issue Do we need the theme park? ⑧B

As a class, carry out a brief brainstorming session to list positive and negative aspects of the project, and identify the possible positions on the issue.

What do you think?

Form small groups. (Your teacher may allocate students to the groups.) As a group, select a role from the list below and prepare a presentation expressing that individual's views on the project. Your presentation should describe who you are, your position on the project, and supporting reasons for your position.

Individuals voicing support or concerns about the project are:

- a 16-year-old who has a part-time job during the tourist season
- a person who enjoys recreational fishing
- a representative of the electricity utility company
- an unemployed sawmill operator with a family
- an environmental activist
- a farmer whose farmhouse and outbuildings will be within the expanded power corridor
- an entrepreneur who rents cottages to summer visitors

- the chairperson of the local cottagers' association
- a construction worker who commutes to a nearby city to do most of his work
- the owner of a local hotel and restaurant
- the principal of the elementary school
- a local real estate agent
- the government representative for the Ministry of Natural Resources
- the owner of the local lumberyard and hardware store
- a retired couple who have a newly built home near the proposed resort
- the 14-year-old son of unemployed parents
- the supervisor of the local Employment Office
- a spokesperson for the housing subdivision next to the present transmission line

After all the presentations are heard, vote on whether or not to proceed with the construction of the theme park.

Chapter 12 Review

Key Expectations

Throughout the chapter, you have had opportunities to do the following things:

- Determine quantitatively the percent efficiency of an electrical device that converts electrical energy to other forms of energy. (12.4, 12.7, 12.8)

- Describe the relationship among electrical energy, power, and time interval, and determine electrical energy using these physical properties. (12.5, 12.6, 12.7)

- Use safety procedures when conducting investigations. (12.1, 12.6)

- Investigate the use and cost of electricity, and organize, record, analyze, and communicate results. (12.1, 12.6)

- Formulate and research questions related to safe and efficient use of electricity, and communicate results. (12.2, 12.3, 12.8, 12.9, 12.10)

- Describe practical applications of current electricity. (12.3, 12.4, 12.7, 12.9, 12.10)

- Devise a self-contained system to generate energy, using renewable energy sources, to meet the energy requirements of a dwelling, farm, or community in Ontario. (12.9)

- Describe and explain household wiring and its typical components. (12.2)

- Explain how some common household electrical appliances operate. (12.2, 12.3, 12.7)

- Explore careers requiring an understanding of electricity. (Career Profile)

KEY TERMS

circuit breaker	input energy
distribution panel	live wire
electric meter	main breaker switch
fuse	neutral wire
ground fault circuit interrupter (GFCI)	output energy
	percent efficiency
ground terminal	polarized plug
grounding pin	

Understanding Concepts

1. Make a concept map to summarize the material that you have studied in this chapter. Start with the words "energy conservation."

2. List three safety precautions related to the use of electricity that should be observed in the home and explain why.

3. List the three basic ways to conserve the use of electrical energy. Identify an example of each of the three ways.

Reflecting

- "Electricity can be used thoughtfully with due regard for safety, economy, and the environment." Reflect on this idea. How does it connect with what you've done in this chapter? (To review, check the sections indicated above.)

- Revise your answers to the questions raised in Getting Started. How has your thinking changed?

- What new questions do you have? How will you answer them?

---

4. Describe the two ways that the neutral wire is normally grounded when it is connected to the distribution panel inside a building.

5. Why should a burnt-out fuse never be replaced by one with a higher current rating, or with a piece of aluminum foil?

6. What is the typical range for the current rating of the main breaker in houses?

7. (a) What is the purpose of the main breaker switch?
   (b) What is the purpose of the distribution panel?

8. (a) What is the primary purpose of the circuit breaker or fuse?
   (b) What might happen if the circuit breaker did not operate and an unsafe amount of current continued to flow through the circuit?

9. (a) Why would having too many outlets on one circuit in a house cause problems?
   (b) If you were planning the wiring of a house, how would you try to minimize such problems?

Applying Skills

10. (a) Calculate the percent efficiency of an electric motor that uses 32 000 J of energy to produce 26 000 J of useful energy.

    (b) Calculate the percent efficiency of a fluorescent light bulb that produces 4 500 J of useful energy from 22 500 J of electrical energy.

11. (a) Calculate the energy used by a waffle maker that has a power rating of 0.750 kW, and operates for 3.6 h.

    (b) How much energy is used by a 1-kW heavy-duty dough mixer that operates for 6.50 h per day for one month?

    (c) Calculate how much energy a VCR that has a power rating of 0.1 kW uses in one week if it operates 9 h per day, every day.

12. Calculate the cost of operating each of the devices in question 11, if electrical energy costs $0.08 per kW·h.

13. (a) Draw a sketch of an electrical outlet, identifying the live terminal, the neutral terminal, and the ground terminal.

    (b) Describe the function of each terminal.

14. Describe two methods, a practical one and a theoretical one, to check if you could plug a number of appliances into a power bar, and have them all operate continuously.

15. Design and describe a procedure for selecting the most efficient appliances for a particular use.

16. **Figure 1** shows the dials of an electric meter.

January 1

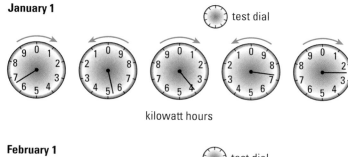

kilowatt hours

February 1

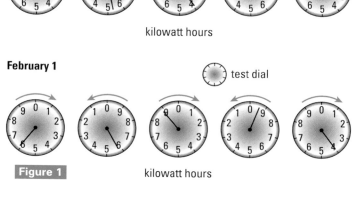

kilowatt hours

Figure 1

(a) Record the meter readings and then calculate the energy used between the readings.

(b) If electrical energy costs $0.08 per kW·h, calculate the cost of the energy used.

Making Connections

17. What can you do to make wall sockets safer to protect young children?

18. In recently built homes, which devices are most commonly used to control the individual circuits in the house, circuit breakers or fuses? Explain why.

19. If you were inspecting a house that your family was thinking of buying, make a list of the parts of the electrical system of the house, and any existing appliances, that you would check as part of your inspection. Explain the importance of each item on your list.

20. Why are some countries more concerned than others about the efficient operation of appliances?

21. Why are long-life energy-efficient fluorescent light bulbs more economical to use than normal incandescent bulbs if the fluorescent bulbs cost so much?

22. Every year people are electrocuted as they move ladders around buildings while doing such activities as painting or working on a roof. Explain why it happens. What could be done to help prevent such tragedies?

23. In recently built houses, the outlet sockets in kitchens are wired differently from those in the other rooms of the house. Each of the two sockets in every wall outlet is the only socket on that particular circuit. Explain why this is necessary in the kitchen, but not in other rooms.

24. Describe how you would use the Energuide labels attached to large appliances to decide which dishwasher you would buy.

25. (a) Which of your appliances would save the greatest amount of money if your family used it less often?

    (b) If most of the families in Canada were to achieve similar energy savings, what economic and environmental benefits would occur?

Challenge

Conserving Energy

When we use the television or radio, we may take for granted that there will always be a supply of energy available at the flick of a switch. As global energy consumption increases, however, consumers need to understand the importance of energy conservation. Public awareness programs can inform people about the effects of our present consumer habits and about ways of conserving energy.

The challenges below develop public awareness of energy consumption and the related environmental, social, and economic consequences. Your goal is to communicate the information in a way that will encourage personal action. Each challenge requires an electric circuit board to be built as a method of communicating the information.

1 An Electric Quiz on Energy Conservation

Your community is holding an information night in a local school to raise awareness of environmental concerns. You have been asked to produce an interactive display that will inform the public of issues related to the consumption of electrical energy and its effects.

Design and make an electric quiz board that asks questions about the importance of energy conservation and provides an immediate way of checking answers. Describe how you designed your quiz in a brief report.

Your quiz board should:
- be self-operating
- ask 12 questions that will raise awareness about electrical energy consumption and the conservation of energy
- engage the interest of people walking by
- be made of material that will be resistant to static electric shocks when a user is approaching or touching the board
- be supported by a personal action plan that users can fill out when they have completed the quiz

Your report should:
- include a schematic diagram of the circuit you designed
- explain your rationale for choosing the questions you asked

2 A Display of Renewable and Nonrenewable Resources

Your company has been hired to inform consumers about the viability of using renewable sources of energy to meet their household energy needs. You have been asked to create a display that will compare a renewable with a nonrenewable source of energy and that shows how each is transformed to electricity for household use.

Design your display using an electric circuit board to display the information. Users should be able to press a button or switch and view the sequence of necessary energy conversions. Describe the design and rationale of your display in a brief report.

Your board should:
- be self-operating
- clearly outline the sequence of energy transformations
- be made of material that will be resistant to static electric shocks when a user is approaching or touching the board
- be supported by written material that helps consumers understand the social, environmental, and economic costs of each transformation and guides them to make an informed choice

Your report should:
- include a schematic diagram of the circuit you designed
- explain your rationale for choosing the alternative source of energy you displayed

3 A Consumer Awareness Display

We often use energy inefficiently in the home. Your company has been hired to encourage the public to conserve energy.

Design a display using an electric circuit board that informs the general public of the consumption rates of common household appliances and devices.

Your product should:
- be self-operating
- be supported by written material outlining the operating costs of the devices displayed
- be made of material that will be resistant to static electric shocks when a user is approaching or touching the board
- be supported by a personal action plan to help people identify how they can save money by reducing the amount of energy they consume

Your report should:
- include a schematic diagram of the circuit you designed
- explain your rationale for choosing the electrical devices you highlighted

Unit 3 Review

Understanding Concepts

1. In your notebook, write the letters (a) to (p), then indicate the word(s) needed to complete each statement below.

   (a) A positively charged object has a(n) _____?_____ of electrons.

   (b) The law of electric charges states that unlike charges _____?_____.

   (c) A negative ion has a(n) _____?_____ of electrons.

   (d) When a positively charged object touches an uncharged object, the uncharged object becomes _____?_____ charged.

   (e) In a dry cell the _____?_____ electrode is used up by the chemical reaction.

   (f) When cells are connected in _____?_____ the voltage of the battery increases.

   (g) To increase the amount of time a battery of a particular voltage will last, the cells are connected in _____?_____.

   (h) When additional loads are connected in a parallel circuit, the total current _____?_____.

   (i) The amount of electrical _____?_____ used depends on the time the appliance is operating.

   (j) As the voltage drop across a resistor increases, the current _____?_____.

   (k) As the number of loads in a parallel circuit is increased, the effective resistance _____?_____.

   (l) Nuclear fuel and fossil fuels are both _____?_____ sources of energy.

   (m) The primary purpose of a circuit breaker is to prevent _____?_____.

   (n) The third, rounded pin on the plug of an appliance connects to the _____?_____ wire when it is plugged into a socket.

   (o) The unit of electrical energy measured by the electric meter is the _____?_____.

   (p) The efficiency of a device compares the _____?_____ to the input energy.

2. For each of questions (a) to (p), indicate whether the statement is TRUE or FALSE. If you think the statement is FALSE, rewrite it to make it true.

   (a) Neutral objects are attracted to charged objects.

   (b) The amount of electrical charge on an electron is the same as that on a neutron.

   (c) A metal-leaf electroscope charged by induction receives the same charge as that on the charging object.

   (d) Lightning never strikes in the same place twice.

   (e) A primary cell converts electrical energy into chemical energy.

   (f) A battery is a very large secondary cell.

   (g) To connect cells in series, the positive terminal of one cell is connected to the negative terminal of the next cell.

   (h) When the filament of a bulb in a series circuit burns out, all the other bulbs remain lit.

   (i) The amount of electrical energy in a battery depends on the electrical load connected to it.

   (j) A 1000-W electric heater uses more electrical energy than ten 100-W light bulbs in the same amount of time.

   (k) Photovoltaic cells produce electrical energy less efficiently than fuel cells.

   (l) Most of the electrical energy in Ontario is produced by nonrenewable energy sources.

   (m) The primary purpose of the circuit breaker and the fuse is the same.

   (n) The neutral wire is connected to the circuit breaker in the distribution panel.

   (o) The GFCI device measures the current flowing in the ground wire.

   (p) The electric meter measures the electrical energy being used in a house.

3. Describe the similarities and/or differences between each pair of terms listed below:

   (a) electron; proton

   (b) positive charge; negative charge

   (c) insulator; conductor

   (d) charging by contact; charging by induction

   (e) open circuit; closed circuit

   (f) series circuit; parallel circuit

   (g) dry cell; wet cell

(h) primary cell; secondary cell

(i) ammeter; voltmeter

(j) electrical energy; electrical power

(k) joule; kilowatt hour

(l) renewable energy source; nonrenewable energy source

(m) sustainability; energy conservation

(n) neutral wire; "live" wire

(o) circuit breaker; fuse

(p) input energy; useful output energy

For Questions 4 to 11, choose the best answer and write the full statement in your notebook.

4. A piece of fur is positively charged when:
   (a) it has an excess of electrons.
   (b) it has a deficiency of electrons.
   (c) the nuclei of its atoms are positively charged.
   (d) the electrons of its atoms are positively charged.

5. When a metal leaf electroscope is charged negatively by contact
   (a) electrons move from the metal ball to the charged rod.
   (b) electrons move from the charged rod to the metal ball.
   (c) protons move from the metal ball to the charged rod.
   (d) protons move from the charged rod to the metal ball.

6. The kind of circuit in which an electrical load may be disconnected without affecting other loads is called a(n):
   (a) series circuit
   (b) closed circuit
   (c) open circuit
   (d) parallel circuit

7. Two light bulbs, A and B, are connected in a series circuit to a dry cell. The switch is closed. If bulb B is removed from its socket, the brightness of bulb A will:
   (a) decrease
   (b) increase
   (c) become zero
   (d) remain the same

8. Which energy source can be converted to electrical energy most efficiently?
   (a) falling water

(b) nuclear fuels
(c) fossil fuels
(d) wind

9. Which form of energy can electrical energy be converted into most efficiently?
   (a) chemical energy
   (b) thermal energy
   (c) mechanical energy
   (d) sound energy

10. A circuit breaker is connected:
    (a) in series with the live wire
    (b) in parallel with the live wire
    (c) in series with the neutral wire
    (d) in parallel with the neutral wire

11. The electrical power of an appliance depends on:
    (a) the resistance of the load, the voltage drop across it, and the current through it
    (b) the resistance of the load, and the time it operates
    (c) the voltage drop across the load, and the time it operates
    (d) the voltage drop across the load, and the current through it

12. Why do two different substances build up static charges when rubbed together?

13. If you touched the end of a plastic comb with a negatively charged rod, and then did the same to the end of a metal rod that was stuck into the ground, what would happen to the charge in each case? Explain why.

14. If metal substances conduct electric charges away, why do science teachers use metal balls to hold electric charge when they do static electricity demonstrations? What is special about the construction of the apparatus?

15. How many times greater is the electrical potential energy of each electric charge at the negative terminal of a 24-V battery, compared to that of electric charges at the negative terminal of a 1.5-V dry cell?

16. What happens to the total current in a series circuit when one of the electrical loads is removed from the circuit? Explain why.

17. Calculate the voltage drop across the following electrical loads:
    (a) A food blender that has a resistance of 65 Ω and has a current of 1.85 A flowing through it.
    (b) A current of 0.15 A flows through a resistance of 800 Ω.
    (c) A bulb that has 0.75 A flowing through it. The resistance of the bulb is 16 Ω.

18. A current of 12 A flows through a red-hot stove element when the voltage drop across the element is 240 V.
    (a) Calculate the resistance of the element under these conditions.
    (b) When it had cooled, a student removed it from the stove and connected a 6-V battery to its terminals. She measured the current, and noted that it was greater than she had expected, based on Ohm's law. Explain why. Using Ohm's law, calculate the current the student expected to observe.

19. (a) Calculate the energy released from a battery in an electric drill. The drill was switched on for 3 min. The voltage drop was 9 V, and the current was 1.4 A. (Answer in joules.)
    (b) Calculate the energy released from a portable hedge trimmer using a 12-V battery. The current flowing in the motor was 2.8 A, and it operated for 20 min. (Answer in watt hours.)

20. Calculate the power rating of a portable electric sander that operates at a voltage of 12 V. A current of 2.5 A flows through the motor operating the sander.

21. Calculate the power rating of a water heater that operates at 240 V, and a current of 15 A flows through the heating element.

22. Could the efficiency of an electrical device ever be greater than 100 %? Explain your answer.

23. Calculate the efficiency of an electric motor that produces 15 000 J of useful energy while using an input energy of 22 500 J.

24. If electricity costs $0.07/kW·h, how much does it cost to operate each of the following appliances for the time indicated?
    (a) a 1200-W dishwasher for 1.2 h
    (b) a 240-W colour television set for 13 h
    (c) a 3500-W water heater for 6 h per day for 28 days

25. How much current flows in the filaments of a trilight reading lamp when each of the following settings is used? The voltage drop across the filaments is 120 V.
    (a) 100 W
    (b) 200 W
    (c) 300 W

Applying Skills

26. (a) Describe how to charge a metal-leaf electroscope with a positive charge by induction.
    (b) Draw a series of diagrams to show the movement of electrons as the electroscope in part (a) is charged.

27. Design and carry out an experiment to show the difference between an insulator and a conductor using a rod made of an insulator, one made of a conductor, an ebonite rod, some fur, and a pith-ball electroscope.

28. **Figure 1** shows three circuit diagrams.
    (a) In which circuit will the ammeter reading be (i) the greatest, and (ii) the least? Explain why in each case.
    (b) Calculate the current in the circuit in **Figure 1a**.

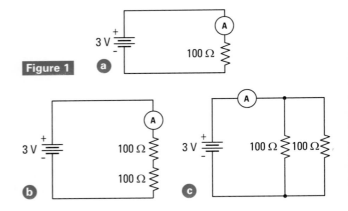

Figure 1

29. You are given a 6-V battery and a box full of identical bulbs, each having a resistance of 24 Ω. Design and draw a circuit diagram, using the required number of bulbs, such that the total current flowing in the circuit through the bulbs is 1 A. Show all calculations.

30. Draw a labelled diagram showing how a typical circuit is connected from the distribution panel to the circuit breaker and then to three wall outlets.

31. Determine the values of voltage indicated by the meter needle positions A and B shown in **Figure 2**.

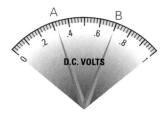

Figure 2

32. Determine the values of current indicated by the meter needle positions A and B shown in **Figure 3**.

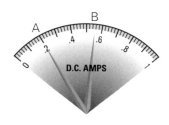

Figure 3

Making Connections

33. List three ways to control static electricity.

34. (a) Where are insulators and conductors used together on the huge electrical transmission towers that distribute electrical energy? Explain why.

(b) Why do problems occur on these transmission towers during a freezing rain storm?

35. Probably someone you know has had problems with allergies caused by airborne particles. What suggestions could you make to help them, now that you have learned about charging by induction?

36. Estimate how long the wires in the complete circuit are that control the light on the top of the tail of a jumbo jet, considering that the pilot must be able to control it from the cockpit.

37. Why are vehicles operated by fuel cells considered to be less polluting than those operated by diesel electric generators? How would you convince someone to use such a vehicle?

38. Why are portable generators so important in rural areas?

39. If the demand for electricity exceeded the amount being generated, what might happen? How would Ontario Hydro solve the problem?

40. Why would transmission line losses be greater in Ontario than in most European countries?

41. Why is the efficiency of the energy conversion process so important when considering different ways to produce electrical energy from both renewable and nonrenewable energy sources?

42. Identify some ways by which electrical energy could be conserved in your home and at school.

43. How can GFCIs be used to upgrade the safety of bathrooms and other hazardous areas in a house?

44. Why is a GFCI considered to be superior to a circuit breaker as a safety device?

45. What kinds of lights are used in most office buildings and schools? Explain why.

46. How can an Energuide label help a person who is buying an appliance?

47. Suggest three ways in which you could save energy by using the appliances in your home more wisely.

48. (a) Why should only low power appliances (radios, lamps, and stereos) be connected to ordinary extension cords?

(b) A certain extension cord has a maximum current rating of 8 A. What is the maximum power rating of a device that can be connected to the cord if the operating voltage is 120 V?

Unit 4

Space

Unit 4 Overview

Have you ever wondered what it would be like to leave Earth and travel far away into the vastness of outer space? How would conditions change as you left Earth? What objects would you discover? Would you find life somewhere else?

13. Sky-watching and the Solar System

Observation of the stars has affected many human beliefs and activities, including navigation, calendar-making, and the creation of myths.

In this chapter, you will be able to:

- recognize natural objects in the sky

- describe and compare the properties and motions of the components of the solar system and explain how space probes have contributed to our knowledge

- plan ways to answer your own questions about the motion of objects visible in the sky

- conduct investigations into the motions of objects in the sky, and gather, record, and communicate data found during the investigations

14. The Nature of the Universe

The universe is always changing. The many objects that can be found in the universe are in continual motion.

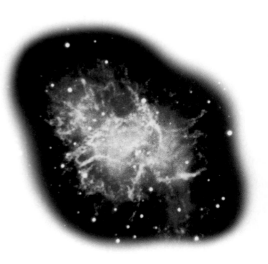

In this chapter, you will be able to:

- safely observe the Sun to investigate your own predictions about its motions

- describe and compare properties of stars, including the Sun

- recognize and compare various components of the universe, such as star clusters and galaxies

- research answers to questions about the motion and properties of objects in the universe

15. The History of the Universe

Scientists use models and simulations to visualize and explain the dynamic processes that form the universe.

In this chapter, you will be able to:

- understand how indirect evidence can provide insight to develop models and theories

- describe the life cycles of stars of various masses

- outline the current theory used to explain the formation of the solar system

- illustrate, using models and simulations, the concept of an expanding universe

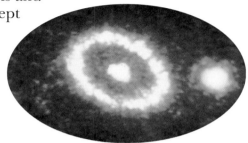

- examine and evaluate evidence of the origin and evolution of the universe

- describe how computers are used to enhance our understanding of the universe

16. Space Research and Exploration

Space exploration and the related technology contribute to our understanding of Earth and the universe and provide many useful applications for life on Earth.

In this chapter, you will be able to:

- interpret data provided by satellites and describe how it contributes to our understanding of Earth and the universe

- simulate and compare the effects of free fall and gravity

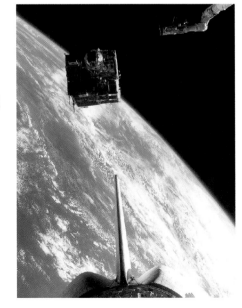

- design and plan an experiment that could be conducted in free-fall conditions

- describe Canada's participation in space research and international space programs

- describe and evaluate the spinoffs of space technologies

Challenge

In this unit, you will be able to...
demonstrate your learning by completing a Challenge.

Applying Ideas and Skills of Astronomy and Space Technology

As you learn about what you can see in the sky and how humans explore space beyond Earth, think about how you would accomplish these challenges.

1   **Planetarium Shows**

Design three shows for a "planetarium" to teach others about astronomy.

2   **A Space Research Colony**

Design a space colony suitable for permanent human habitation.

3   **A Space Technology Information Package**

Organize information about space exploration and the study of astronomy, emphasizing their influence on our lives.

To start your own Challenge see page 514.

Record your ideas for the Challenge when you see

Challenge

13 Sky-watching and the Solar System

1 When you first look at the night sky, you probably just see random points of light. Look more closely and more often, and you will begin to see patterns that look familiar. When you look at the stars shown in the photograph, what patterns do you see? Do those patterns change if you hold the book sideways or upside down?

2 You know that the Sun and planets are important objects in the solar system. What are other objects in the solar system, and how do we know about them? What technology are we using to investigate them further? What is the object shown in the photograph, and what are some of its properties?

3 It takes skill and practice to be able to tell the difference between a planet and a star in the night sky. What are the differences? How would you find this information?

4 How would you try to locate a specific star in the sky? Once you have found it, how would you describe to someone else where to look in the night sky to see this object? What information do you think you would need to communicate in order to solve the problem? How has a knowledge of the position of the stars helped travellers in the past? ▼

Reflecting

Think about the questions in **1,2,3,4**. What ideas do you already have? What other questions do you have about sky-watching and the solar system? Think about your answers and questions as you read the chapter.

Try This Modelling the Solar System

You may have already learned something about the solar system and other parts of the universe. To find out what you and your classmates remember, try the following class demonstration.

1. Choose some students to represent the Sun and other objects in the solar system (or beyond, if you wish).

2. Make signs to identify which object each student represents.

3. Simulate the motions of the objects they represent. Each "actor" should explain the motion being made.

4. As a class, discuss how to improve any weaknesses of the demonstration, then try it again.

5. In your notebook, draw a diagram showing the objects and their motion.

6. Also in your notebook, list questions you would like answered in order to improve the demonstration.

What Can We See in the Sky?

What are some of the objects you can see when you look up into the sky? Of course you can see the Sun and the Moon, and on a dark, clear night you can see many stars. You are also likely to see airplanes and satellites. But have you ever seen Mars and other planets, or moons that travel around planets such as Jupiter? When people begin viewing the sky, and keep records of what they see, they rediscover what many sky-watchers have discovered in the past: there are patterns and rhythms in the slow dance of the night sky.

By studying stars, planets, and other objects in the sky, you will learn where Earth is located in the universe. The **universe** is everything that exists, including all matter and energy everywhere. The study of what is beyond Earth is called **astronomy**.

You don't need a telescope, a camera, or any other special equipment to study the sky. You can start simply by looking at the sky at certain times, learning how to tell stars from planets, and recognizing patterns of stars. This skill can be developed further by looking through binoculars or a telescope.

Star Constellations

When you look up at the sky on a clear night without the aid of binoculars or a telescope, you see countless stars spread unevenly across the night sky. A long time ago, sky-watchers noticed that certain patterns of stars seemed to suggest the shapes of animals, mythical heroes or gods, and other objects. People gave different names to the patterns they saw (**Figure 1**). Groups of stars that seem to form shapes or patterns are called **constellations**. The stars of a constellation may not actually be close to each other. Some stars may be much farther from Earth than others; they just *appear* to be close together.

When you look at a constellation, you may not see the shapes that

Figure 1
The Big Dipper is part of the constellation Ursa Major, also called the Great Bear. It is probably the best-known group of stars in the Northern Hemisphere.

people in ancient times saw. What looked like a bird to them may look like a cartoon character to you. Today, rather than showing diagrams of animals or people, many books simply show shapes to represent the constellations (**Figure 2**).

Constellations have been used for thousands of years as calendars, timekeepers and direction finders by people travelling in unknown territory—both on land and at sea. Different constellations were, and still are, used in different parts of the world.

Try This   Looking for Patterns of Stars

Use a star map that represents the stars visible during the winter in Canada. The larger dots represent the brighter stars. On your copy, draw lines between stars that seem to make patterns. The patterns can be animal shapes, geometric figures, cartoon characters, or anything that suggests itself to you.

Objects in the Solar System

You probably already know that the **solar system** consists of our Sun and all the objects that travel around it, including nine known planets and the moons around some of those planets (**Figure 3**). It is easy to see the Sun and Earth's Moon in the sky, although you should never look at the Sun directly. On a clear night, you may also be able to see some of the planets in the solar system. Everything in the solar system is much, much closer to us than the stars and other objects in the universe.

Planets and moons are **nonluminous**: they do not emit their own light. We can see them in the sky only when light from the Sun reflects off them toward Earth. Can you see in **Figure 3** that we would not easily be able to see

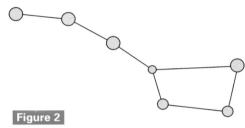

Figure 2

A common way of drawing the Big Dipper

Mercury in the sky whenever its path takes it close to the Sun? The Sun is very bright, so objects close to it get hidden in the daytime glare.

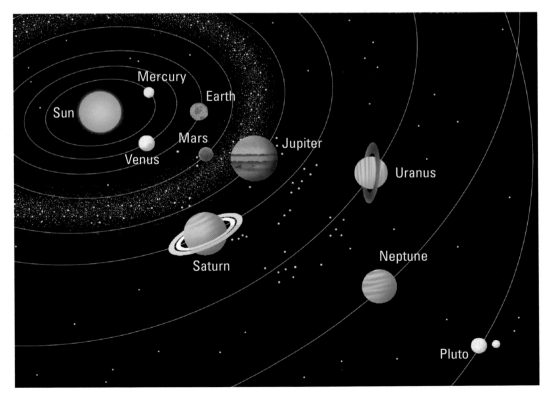

Figure 3

This drawing shows the main objects that make up our solar system. However, it does not show true sizes or distances. For example, the Sun is about 100 times bigger in diameter than Earth.

Observing Stars and Planets

A **star** is matter that emits huge amounts of energy. A **planet** is matter, generally spherical, that revolves around a star. When you look at the night sky, how can you distinguish a planet from a star (**Figure 4**)? Only five of the planets can actually be seen with the unaided eye: Venus, Mars, Jupiter, Saturn, and Mercury. Of these five, planets that are closest to Earth appear through a telescope as somewhat larger than stars. (Stars are very far away, so they look small.) Furthermore, planets appear to have steady light, whereas stars appear to twinkle. **Table 1** summarizes these and other features of stars and planets.

Table 1   **Comparing Planets and Stars**

Feature	Planet	Star
location	in the solar system	far beyond the solar system
distance from Earth	fairly near	very far
real size	smaller than most stars	usually larger than planets
reason we see object	reflects light from the Sun	emits its own light
surface temperature	usually cool or very cold	very hot
what object is made of	usually rocks or gases	gases under high pressure and temperature
observable feature	does not appear to twinkle	appears to twinkle
long-term observable feature	very slowly wanders through constellations	appears to move through sky as part of a constellation

Figure 4

The brightest star in the night sky is Sirius in the constellation Canis Major. It is brighter than the planet Saturn, but not nearly as bright as Venus or Jupiter.

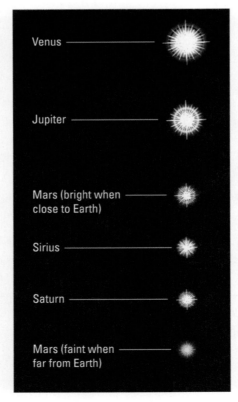

Venus ————

Jupiter ————

Mars (bright when close to Earth) ————

Sirius ————

Saturn ————

Mars (faint when far from Earth) ————

Understanding Concepts

1. The universe is everything that
   9D exists including all matter and energy. Create a concept map of the objects in the universe starting with the word "universe." Keep your map in a safe place so you can use it later.

2. Choose a planet and draw a sketch to show how light travels to allow us to see it.

3. Give at least one reason why
   (a) it is very difficult to see Neptune and Pluto with the unaided eye;
   (b) it is not as common to see Mercury in the sky as it is Jupiter.

4. Prepare an explanation of the differences between a star and a planet for someone who knows little about astronomy.

Making Connections

5. Describe how an astronomer's method of gathering information differs from that of other scientists. Give reasons why.

6. Until fairly recently, people used constellations to help them find their way when they travelled. Why do you think this was possible? Describe a situation where it can still be helpful today.

Reflecting

7. Astronomy is an important area of study. Do you agree or disagree? Give your reasons.

Challenge

Which objects in the universe will you represent in your planetarium shows?

Sunrise and Sunset

Scientists conduct investigations in a variety of ways. Sometimes they construct experiments to measure and collect their own data. At other times they take previously published data, analyze it, and extract new information. In both cases they ask questions, develop hypotheses, and use many of the skills of planning and conducting an investigation. You can apply the same skills they use as you perform your own research.

Question

How do the times of sunrise and sunset vary in your area throughout the year?

Hypothesis

The day-to-day differences in sunrise and sunset times are the same throughout the year.

Materials

- local newspapers, books, Internet access, or other sources of data on sunrise/sunset times in your area

Procedure

1 Find out the times at which the Sun rises and sets in your area. You could either collect 20 to 50 sets of data for consecutive days, or data for one day of each week for a year.

6D ✎ (a) Record your data in a table.

7B ✎ (b) Plot your data as a graph.

Analysis and Communication

2 Analyze your results by answering the following questions:

(a) Describe the shape of the curve in your graph.

(b) Answer the initial question.

3 Discuss your results as a class. If different students worked on different times of the year, combine and compare your results. Write a conclusion about how the sunrise and sunset times change throughout the year.

(a) Do your results support the hypothesis?

(b) If necessary, write a new hypothesis.

4 Use a diagram to explain how the results of your graph relate to Earth's motion.

Understanding Concepts

1. How would this investigation performed by a student in Australia compare with your investigation? Explain why.

2. Speculate how the number of hours of sunlight might affect the demand for electricity across various regions of Canada.

Reflecting

3. Imagine you are an early astronomer. Other astronomers think the Sun orbits Earth. They say their model explains the Sun's daily motion across the sky. But you think sunrise and sunset are caused by Earth's motion. Create a model to convince other astronomers.

Challenge

When working on your challenge you will need to use published information. How will you have confidence in the accuracy of your data?

The Effects of Planetary Motion

Effect of Earth's Rotation

If you watch the sky at night for at least an hour, you will notice that the positions of the stars and planets slowly change. Like the Sun, most stars appear to rise in the east, travel across the sky, and set in the west. You can understand these changes by thinking about Earth's motion.

One type of motion of Earth is called **rotation**, which is the spinning of an object around its axis. One rotation of Earth takes 24 h. This motion causes most stars (as well as the Sun, Moon, and planets) to appear to rise in the east and set in the west.

Earth's **axis** is an imaginary straight line joining the North Pole and the South Pole. If the axis were continued northward, out into space, it would pass almost through Polaris, the North Star. **Figure 1a** shows why people who live in Canada are able to see Polaris all year long. As Earth rotates, the stars near Polaris seem to travel in circles around Polaris. **Figure 1b** shows a way of demonstrating this.

Effect of Earth's Revolution

Another type of motion of Earth is **revolution**, which is the movement of one object travelling around another. It takes Earth one year to travel, or revolve, in a circle around the Sun. This motion enables us to see different stars and constellations during different seasons, as shown in **Figure 2**. Combined with the angle of Earth's axis, Earth's revolution also causes the seasons.

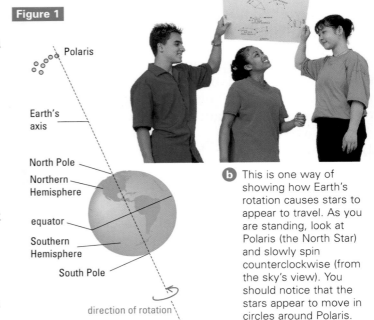

Figure 1

Polaris
Earth's axis
North Pole
Northern Hemisphere
equator
Southern Hemisphere
South Pole
direction of rotation

a Polaris is visible from Canada all year long. People south of the equator cannot see Polaris or any of the stars near it. Are there stars that Canadians never see?

b This is one way of showing how Earth's rotation causes stars to appear to travel. As you are standing, look at Polaris (the North Star) and slowly spin counterclockwise (from the sky's view). You should notice that the stars appear to move in circles around Polaris.

Did You Know ?

Has anyone ever asked you what "sign of the zodiac" you were born under? Your answer may be one of 12 signs, such as Aries, Taurus, Leo, or Scorpio. These signs are named after the zodiac constellations.

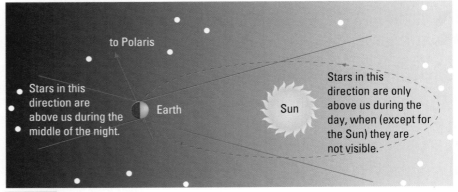

to Polaris

Stars in this direction are above us during the middle of the night.

Earth

Sun

Stars in this direction are only above us during the day, when (except for the Sun) they are not visible.

Figure 2

For observers in the Northern Hemisphere, the only constellations we can see in all seasons are those that appear close to Polaris. We can see the other constellations only when they are not in the sky at the same time as the Sun.

Effect of Planets' Motions

In ancient times, when there were no city lights to spoil people's views and no televisions to occupy their evenings, people were more aware of the night sky than most people are today. Even then, astronomers noticed that the bright objects in the sky moved, and that those in constellations moved together. They saw that five particularly bright points of light moved differently when compared with the other stars. They called these objects "wandering stars," or planets, from the Greek word *planetes*, meaning "wanderers." We now know that the reason for the changing positions is that the planets are travelling around the Sun and they are much closer to us than the stars in the background.

The planets appear to move through certain constellations, many of which were named after animals. Since the Greek word for "animal sign" is *zoidion*, the constellations were called the **zodiac constellations**. Thus, another way to distinguish a planet from a star is to observe its motion over weeks or months.

Try This   You Are Earth

Make your classroom into a planetarium. Place diagrams of the Big Dipper, the Little Dipper, and Cassiopeia on the ceiling. Place these four constellations on the walls:
- Aquila (summer) on the east wall
- Pegasus (autumn) on the north wall
- Orion (winter) on the west wall
- Leo (spring) on the south wall

Choose one person to represent the Sun in the middle of the room. Pretending you are Earth, stand east of the Sun and slowly spin around counterclockwise. When your back is toward the Sun, it is night on Earth. Aquila and the constellations above the Sun are visible. As you spin around to daytime, the other constellations may be located near the horizon, but you can't see them because the sunlight is too bright. Try this at locations north, west, and south of the Sun.

Understanding Concepts

1. Where do stars "rise" in the night sky?
2. Describe and compare Earth's rotation and revolution.
3. Explain why a constellation
   (a) appears to change position from hour to hour during the night;
   (b) is at a different location at the same time on different nights.
4. In **Figure 3**, the stars appear to revolve around a single spot. Where was the camera pointed?

Figure 3
Night sky observed in the Northern Hemisphere

Making Connections

5. What would be the difficulties of using stars to indicate direction while travelling? Would one star be more useful than any of the others?

Challenge

How can the model suggested in this section help you design a model of a planetarium?

Recognizing Constellations

In this activity, you will develop skill in recognizing some constellations like Orion (**Figure 1**). As you do, think about what sailors experienced hundreds of years ago when they relied on stars to find their way across the northern oceans. Why was the North Star the most important star in the sky to them?

Materials
- copy of star map
- ruler

Procedure

Part 1: Three Year-Round Constellations

1 **Figure 2** shows the shapes of three star
(5C) patterns that you can see in the Canadian sky all year. Near the middle of the star map, locate the Big Dipper in the constellation Ursa Major.

(a) Use a ruler to draw lines joining the stars that make up the Big Dipper.

2 Near the upper-centre of your star map, locate the Little Dipper and the star at the tip of the handle. This important star is called Polaris, or the North Star. Anybody in the Northern Hemisphere looking toward Polaris is facing toward the North Pole.

(a) Draw straight lines joining the stars that make up the Little Dipper. Label Polaris.

3 Find the two stars of the Big Dipper that are farthest from the handle.

(a) From these stars, use a ruler to draw a dashed line to Polaris. This line shows how to use pointer stars to locate other stars or constellations. (This is important when you try to view the night sky because the Little Dipper is not nearly as bright as the Big Dipper.)

4 From the same pointer stars you used in step 3, continue past Polaris until you are

Figure 1
To some people in ancient times, this constellation looked like a hunter whom they called Orion.

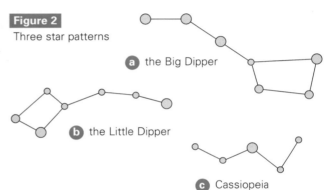

Figure 2
Three star patterns

a the Big Dipper

b the Little Dipper

c Cassiopeia

close to Cassiopeia, a constellation in the shape of a spread-out *W* or *M*.

(a) Draw straight lines joining the stars that make up Cassiopeia.

5 Identify the three constellations you have found.

(a) Ask your teacher to check your star map now. Label the constellations.

Part 2: Seasonal Constellations

6 **Figure 3** shows constellations that you can see in the night sky over Canada during the winter months. The easiest one for you to see is the constellation Orion, with three bright stars that line up to make this imaginary hunter's belt. Locate Orion on your star map. It is near the part of the map marked winter. This means that Orion can be seen most easily in

December, but it can also be seen in November and January.

(a) Draw straight lines joining the stars that make up Orion. Label Orion as well as the bright stars in Orion: Rigel and Betelgeuse.

7 Use the stars of Orion's "belt" as pointers to locate the brightest star in the night sky, Sirius.

(a) Draw a dashed line on your map to show this method. Also draw Canis Major, the constellation in which Sirius is found.

8 Locate the constellation Boötes.

(a) Draw the lines joining the stars of this constellation. (The bright star Arcturus is part of this constellation.)

(b) Describe how you could use the stars of the handle of the Big Dipper as pointers to find Arcturus. Draw a dashed line on your map to illustrate this method.

9 Leo is located about midway between Canis Major and Boötes, and contains the star Regulus.

(a) Draw the constellation Leo on your map.

(b) Describe how you could use two stars of the Big Dipper as pointers to locate Regulus. Show this with a dashed line on your map.

10 Find the three bright stars Deneb, Vega, and Altair.

(a) Join them with dashed lines to show the Summer Triangle. Draw the constellations to which Deneb, Vega, and Altair belong: Cygnus, Lyra, and Aquila.

11 Locate the constellations Andromeda and Pegasus. You can find these constellations by starting from Polaris and going past Cassiopeia.

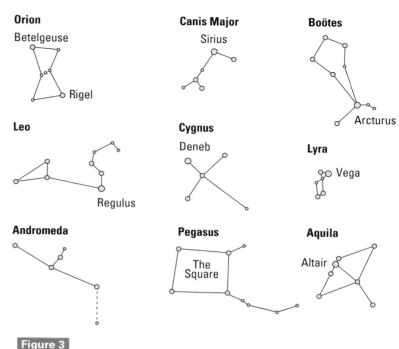

Figure 3

You can see these nine constellations only during certain seasons. Where are they at other times of the year?

(a) Draw and label Andromeda and Pegasus. Also label the Square of Pegasus.

Understanding Concepts

1. Why was Polaris important to early sea navigation?

2. Is Polaris still important to observers in the Northern Hemisphere?

3. Would people in Australia be able to see Polaris? Explain your answer with a diagram.

4. What are "pointer stars" and why are they useful?

Making Connections

5. If you were looking for stars while surrounded by bright city lights, which stars would you try to find first? Why?

Exploring

6. To know their exact location at sea, sailors often use two instruments: the sextant and the chronometer. Find out when these instruments were invented, how they were used, and how effective they were in determining location.

Measuring Angles in the Sky

When we are looking at things that are far away, it is hard to describe their location. For example, if you're looking at two airplanes coming toward each other at an air show, it's hard to judge how far apart they are. Sometimes it looks as though they're about to collide, even though they may be quite far apart, as shown in **Figure 1**. They seem close together because the angle between them is small. In this activity, you will use two ways to describe the location of distant objects using angles.

Figure 1

From one point of view, these aircraft look close together. From another point of view, they are clearly quite far apart.

a From the side they seem about to collide.

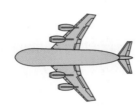

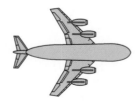

b From below there is clearly a horizontal distance between their wingtips.

Materials
- thin cardboard
- small protractor
- ruler
- pencil
- scissors
- string
- rubber stopper
- tape
- drinking straw

Procedure

1 Set up a data table based on **Table 1**. Two examples are given for reference. You will choose and record your own observations.

Table 1

Viewing location	Description of angle measured	Angle using astrolabe	Angle using hand method
edge of playing field	angle between the horizontal and the top of the nearest tree	26°	25°
height of roof	angle between the bottom and top of a roof	28° − 21° = 7°	8°

2 Using the illustration for reference, design and build an astrolabe. Start by drawing a large protractor on thin cardboard, then cut it out.

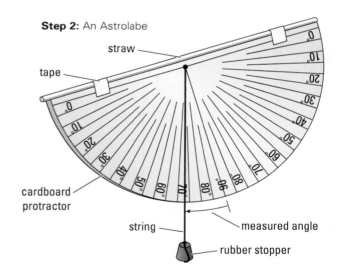

Step 2: An Astrolabe

straw

tape

cardboard protractor

string

measured angle

rubber stopper

3 Use the astrolabe to measure the angle between the horizontal (eye level) and three or four objects in the classroom or outdoors. Use the diagram in step 3 to help you see how to measure the angles.

✎ (a) Record the data in the first three columns of your table.

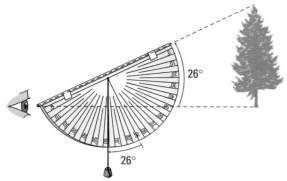

Step 3: The angle on the astrolabe gets bigger as you look higher in the sky. In this case, the angle on the astrolabe equals the angle between the horizontal and the top of the tree.

4 Practise using the astrolabe to measure the angle between two parts of the same object. Remember that the horizontal angle is 90° minus the reading on the astrolabe. For example, you can find the angle between the bottom and top of a roof, as shown. Do this for three or four situations.

(a) Record the data in the first three columns of your table.

Step 4: Use subtraction to find the angle between objects.

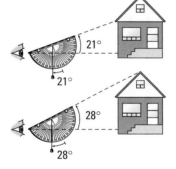

5 Learn how to use your outstretched arm and different parts of your hand to measure angles. The illustration below shows how to do this. Then repeat the measurements you made in steps 3 and 4.

(a) Record the data in column 4 of your table.

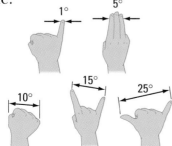

6 Use the hand method to measure the horizontal angle between two positions. One example is shown.

(a) Record the data in your table.

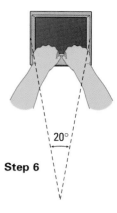

Step 6

7 Compare the astrolabe method and the hand method.

(a) Which method, do you think provided better angle measurements? Explain why.

(b) Which method works better for measuring horizontal angles? Why?

8 Compare the answers you found with those other students found for the same angles.

Understanding Concepts

1. What is meant by the statement "The Big Dipper is 25° long"?

Making Connections

2. A traveller in pioneer times wanted to cross a lake, in a direction 20° east of north. How might he or she have kept on course?

Exploring

3. How many "fists" are required to measure
   (a) a right angle?
   (b) from the eastern horizon to the western horizon going across the sky?
   (c) from the east to the west going around the horizon?
   (d) a complete circle?

4. Use the hand method to measure the angle from the horizontal and a point directly above your head. Compare your angle with 90°.

5. Predict what happens to the angle between two objects as you get farther away from the objects. Design an activity to check your prediction.

Different Views of the Sky

In your science course, you are responsible for learning concepts, developing skills, and applying your knowledge and skills. Many cultures also developed concepts, skills, and applications as they investigated the sky and developed their own ideas of the universe.

A Variety of Ideas

Evidence found in caves and tombs and on pottery shows that people studied astronomy thousands of years ago. This makes astronomy one of the oldest of the sciences (**Figure 1**).

Figure 1

This ancient Aztec carving shows that Aztec astronomers studied the sky and were familiar with many constellations.

People all over the world have used legends (traditional stories) to explain events and objects in nature. Canadian First Nations peoples have many legends that describe the Sun, stars, Moon, planets, and special events such as eclipses and meteor showers.

- The Menomini of the Great Lakes region tell this legend about meteors: "When a star falls from the sky it leaves a fiery tail. It doesn't die; rather its shade goes back to shine again."
- The Kwakiutl of the West Coast region believed that a lunar eclipse occurred when the sky monster tried to swallow the Moon. By dancing around a smoky fire, they hoped to force the monster to sneeze and cough up the Moon.
- The Tsimshian, also of the West Coast, believed that each night the Sun went to sleep in his house but allowed the light from his face to shine out of the smoke hole in the roof. The stars were sparks that flew out of the Sun's mouth. The full Moon was the Sun's brother who rose in the east when the Sun fell asleep.

(a) What evidence supported each belief?

(b) Identify the concept behind each legend. What does the story say to the listeners?

Developing Skills

About 7000 years ago, people began settling on farms and in towns. They began using the regular cycles of the Sun and Moon to plan a calendar. The calendar helped them decide when to plant and harvest crops, and when to hold their feasts. Astronomers developed great skill in mapping the night sky.

- More than 3000 years ago the Chinese developed a very accurate calendar of 365 days by using their observations of the Sun and stars. They recorded comets, meteors, and what they called "guest stars." (We now know that these are exploding stars, which are visible for a few weeks or months.) Both the Chinese and the Greeks created star maps more than 2000 years ago, showing the position and approximate brightness of more than 800 stars. They did this using only their eyes and some instruments to measure angles.

- The Maya are a native people who live in Central America. Over 1000 years ago they had developed ways of accurately keeping track of the movements of the planets. Many of their temples were used as observatories. One of the few remaining ancient Mayan books shows that the Maya were able to predict the appearance of Venus very accurately: their predictions were off by only two hours in 500 years.

(c) What skills must Chinese and Greek astronomers have acquired, before they could construct their calendars?

(d) List ways that a calendar may have helped people in ancient times.

(e) List several ways that a calendar helps you organize your life.

(f) There were excellent astronomers among the Chinese, Greeks, and other peoples in ancient times, yet their star maps contained only about 800 stars. What prevented these ancient astronomers from cataloguing more stars?

Evidence of Applications

Permanent structures built thousands of years ago verify that people applied their knowledge of astronomy to help them predict events.

- In Egypt, a pharaoh's temple was designed with a long, narrow entranceway through which the Sun's rays shone directly onto a statue of the pharaoh on two special days of the year: one in February and the other in October.

- A stone structure called Stonehenge was built in England about 5000 years ago (**Figure 2**). Historians believe its purpose was to indicate the longest day of the year and to predict the occurrence of eclipses.

(g) If the Egyptian statue was in sunlight for only two days of the year, the Sun must be in that position only on those days. How might ancient astronomers have discovered this fact?

Figure 2

The huge stones of Stonehenge are arranged in a well-organized pattern. After many years of studying the pattern, scientists think that the people who built Stonehenge used the pattern to predict repetitive events.

Making Connections

1. In Canada the Sun is highest in the sky on the first day of summer and lowest on the first day of winter. Use this information to design a simple calendar.

Exploring

2. What kind of instrument might ancient Chinese (3A) astronomers have used to measure angles? Propose a design, then research the original design. Compare the two.

3. Research ancient astronomical achievements until the end of the Roman Empire. Place these on a time line.

4. The reason we have a day-and-night cycle is that Earth rotates on its axis once every 24 h. Create a legend to give a different reason for this cycle.

5. Research information about Canadian First Nations' understanding of calendars and astronomy.

6. Find out why the Maya were so interested in Venus that they developed an accurate way to predict its appearance.

Reflecting

7. Compare and contrast how ancient peoples viewed the sky with your own modern view.

A Seasonal Star Map

A skill that is useful in everyday life is reading a map. As you use a star map to find directions and star locations, you will improve your skill at reading other types of maps. Like a street map, a star map is a tool that helps you to identify what is around you. In this activity you will build your own seasonal star map, similar to those used since early times for navigation. You will learn how to use it in the classroom before you try it outdoors.

Part 1: Building a Seasonal Star Map

Materials
- copy of seasonal star map
- cardboard
- scissors
- glue stick
- copy of star map "window frame"
- piece of acetate as large as the window frame
- paper fastener
- phosphorescent paints (optional)

Be careful when using scissors. Also, if you use any phosphorescent paints, be sure to follow the instructions on the container.

Procedure

1. Using **Figure 1** as a reference, cut out the seasonal star map and a piece of cardboard the same size. Glue the map onto the cardboard. Make a hole in the centre for the paper fastener.

2. Add all the dashed lines needed to show the use of pointer stars, as you did on the star map in Activity 13.4.

3. If possible, use one colour of phosphorescent paint to highlight the constellations on the star map and another for the labels on the window frame and the star map.

(This will help you see the map in the dark for a few hours after the painted parts have been in bright light.)

4. Cut out the window frame, including the actual window from the star map window frame. Glue the window frame to the piece of acetate. Cut the outer part of the acetate, but do *not* cut the window out of the acetate. In the centre of the circle, make a hole for the paper fastener.

5. Use the paper fastener to fasten the acetate to the star map. Be sure the window frame can rotate freely.

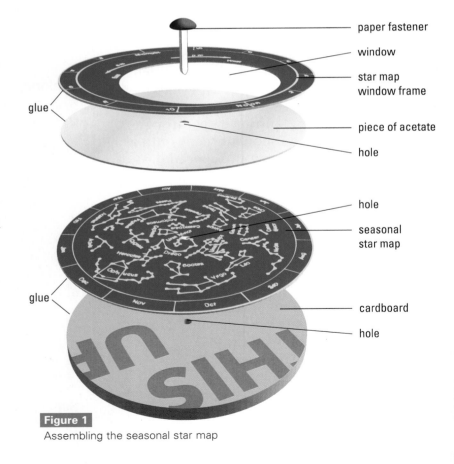

Figure 1
Assembling the seasonal star map

Part 2: Using the Seasonal Star Map

Materials

- the seasonal star map assembled in Part 1

Procedure

6 While in your classroom, pretend you are outdoors. Decide which directions in the room are east, west, north, and south.

🖉 (a) Record which wall in the room is the south wall.

7 To learn how to use the map at midnight on December 15, rotate the window frame so "midnight" is in the middle of December. Hold the map over your head, map facing downward, with the middle of the window directly above your eyes and "midnight" pointing north. What you see on your star map is what you would be able to see in the night sky.

(a) Predict which constellations you would be able to see near each horizon: north, south, east, and west.

(b) Predict which constellations you would be able to see high above your head.

8 Compare your predictions with those of other students.

(a) Based on your discussions with other students and your teacher, make changes to your predictions.

9 Repeat steps 7 and 8 for midnight on May 15.

🖉 (a) Record your predictions.

10 If you look at the sky before midnight, the stars you expect to see will be rising in the east. If you look at the sky after midnight, the stars you expect to see will be setting in the west. To prove this to yourself, set your star map at 8:00 p.m. on December 15.

(a) Describe what you discover.

Understanding Concepts

1. As you rotate the window frame, which constellations can be seen no matter what the month is?

2. Name the constellations that are visible nearer the horizon at midnight in (a) July and (b) March.

3. How would you adjust the window frame if you were using the seasonal star map on different dates of the same month, such as November 1, 15, and 30?

4. Name the constellations that have at least one very bright star in them and would be easier to identify in the sky than other constellations.

5. Pretend that a star is located where the front wall of your classroom meets the ceiling. Determine its angle above the horizontal from where you are located.

Making Connections

6. In societies where calendars are uncommon, the seasonal positions of the constellations are used to indicate when various festivals or activities should take place. If a crop has to be planted in October, what constellations should a North American farmer look for?

Exploring

7. Use resources to find out which planets are visible in the sky during the current month. Describe how you would try to find a planet in the sky if you went observing now.

8. A planisphere or star finder is a device designed to show the stars and constellations of the current night sky at any time, or season, during the year. Check all your local newspapers for an astronomy column with star maps, or the Canadian publication *SkyNews*. Compare your star finder with the star maps you found.

◯Challenge

How would a seasonal star map be a useful tool in the planetarium challenge? How would a tool like this be of use in your space colony?

Observing the Night Sky

When you gaze at the night sky, you are doing what millions of people have done for thousands of years. However, when you observe carefully, taking measurements and keeping notes, you are following in the footsteps of a much smaller group, but one almost as ancient.

Ever since people developed techniques for writing and map-drawing, dedicated astronomers have kept records of changes in the night sky.

In this activity you will use the seasonal star map. Of course, the sky will look very different from the star map because there are many more stars in the sky than on your map. However, with some practice, you can become just as skilled as the ancient astronomers in observing the stars.

Part 1: Getting Ready to Stargaze

Choose an evening when you can see the stars clearly. Follow the Procedure steps outlined below.

Materials
- copies of observation sheets

Procedure

1 You will need an observation sheet. Use the following headings: Date; Time of Observation; Constellation Seen; Diagram of Pattern; Angle above the Horizon; Direction; Questions I would Like Answered.

2 Organize the materials you will need for the trip. Refer to the list of materials suggested for Part 2.

3 As a group, and with the assistance of an adult, plan the trip. Choose a safe location, free from hazardous terrain. Consider when and where you will meet and who will bring the materials. Also consider the weather forecast and the safety of the group.

Part 2: Stargazing

Materials
- appropriate outdoor clothing
- compass
- seasonal star map
- flashlight
- red cellophane (to cover the flashlight)
- telescope or pair of binoculars (if available)
- astrolabe (if available)
- compass (if available)

✋ When you go stargazing, get prior permission from your parent or guardian, and always go in a familiar group.

Procedure

4 On a clear, preferably moonless, night after you are able to see the stars, go to a dark location away from bright lights. Allow your eyes at least 10 min to become used to the darkness. Organize the materials so you can begin observing.

5 Use the compass to find north. Set the map to the current date and time. Hold the star map up over your head, with "midnight" pointing north. Use your covered flashlight to look at the map (**Figure 1**). Compare what is in the sky with what is on the map.

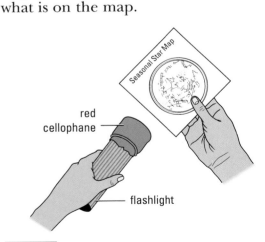

Figure 1

When using a flashlight to view a star map, it is best to cover the light with red cellophane. If you didn't cover the light, how might it affect your eyes?

6 Find the Big Dipper on the map. Then locate the Big Dipper in the sky.

✎ (a) Record your observations. When you draw any diagram, include objects near the horizon for reference. When you measure angles, include the angle above the horizon as well as angles between the ends of the constellation. For example, at 9:00 p.m. on a certain night, the belt of Orion may be two "fists" above the horizon. Record your reading in degrees.

7 Use the two pointer stars from the Big Dipper to locate the North Star, Polaris. Find the Little Dipper, first on your map and then in the sky.

✎ (a) Record your observations.

8 Find Cassiopeia on the side of Polaris opposite the Big Dipper.

✎ (a) Record what you observe.

9 Look for other constellations, bright stars, and planets. When you observe a planet, record its position relative to nearby constellations. (Remember that planets do not appear to twinkle the way stars do.)

✎ (a) Record your observations. Include any questions you have in your table of observations. For instance, you might observe a pair of equally bright stars one night, but the next night they might not be equally bright. You could ask, "Which star has changed its brightness?" You may have other questions about the planets, the moons of planets, meteors, satellites, or other objects in the sky.

10 If possible, use a telescope or a pair of binoculars to look at a specific star.

✎ (a) Record what you see.

Understanding Concepts

1. In your opinion, what are the most important constellations for beginners to recognize? Why do you think so?

2. Why is star-watching difficult in cities and towns?

Exploring

3. Compare the angle between the northern horizon and Polaris with the latitude of your location (found on an atlas or a globe). What do you conclude?

4. Plan an investigation to determine how the
(4B) positions of constellations and planets change over a period of several weeks. Include safety precautions. Have your teacher approve your plan, then carry it out. Keep a record of what you discover.

5. Find a computer software program that simulates the motion of the constellations and planets in the sky. Observe what happens as the program advances by some period of time, such as two weeks. Explain what you observe.

Reflecting

6. The night sky has always inspired people to create stories, music, and images. Choose a piece of sky-inspired art. What is your personal response to it?

7. Astronomy is an observational science not an experimental science. Do you agree or disagree with this statement? Give supporting reasons for your viewpoint.

Light Pollution

When you look up at the night sky from a city, you can see far fewer stars than if you are in a rural location. In fact, you will probably only ever see the brightest stars and planets. Why is this? It is because of the amount of artificial light around us. Every light source sends some of its light directly where we need it, and some where we don't. The light that is not used is wasted. In particular, light that shines upward reflects off hard surfaces and clouds, and is scattered by molecules and dust particles in the atmosphere. It may end up making the night sky up to 100 times brighter above a city than it is in an unpopulated, unlit rural area. This unused light is called **light pollution**.

Why Is Light Pollution a Problem?

- Astronomers and skywatchers are frustrated in their attempts to study the night sky by the amount of artificial light entering their eyes or telescopes. Just as it is difficult to see a candle flame in a brightly lit room, it is almost impossible to view distant stars through an atmosphere glowing with light from nearby towns and cities.
- It is a waste of energy. If 30% of the light output from outdoor lighting goes up into the sky, instead of down toward the street, we are using 30% more energy than we need to. This translates into a waste of money.
- Unnecessary light that travels horizontally can glare into the eyes of motorists, making it difficult for drivers to see obstacles, and thus causing a road safety hazard.
- Glare from poorly shielded lights may shine into the windows of homes, where it can be an unpleasant nuisance.
- Environmentalists are concerned about the number of migratory birds that are killed each year. The birds are dazzled by city lights and then fly into lighted windows.

How Can Light Pollution Be Reduced?

People with many different interests are trying to get light pollution reduced. There are several ways to achieve this.
- Persuade owners and operators of large office buildings to turn off most of their lights at night.
- Install better shielding around streetlights so the light is directed downward, to the street, not up into the sky.
- Design lighting systems that produce uniform illumination, reducing the depth of shadows (**Figure 1**). This would allow us to use lower light levels.
- Inform private users of electricity about the issues, and educate them to use only what light they need, where they need it. For example, lighting an outdoor car lot at night, if there is nobody there guarding it, is just helping thieves to see better!

Figure 1

10° - cut glare

10° to max

good uniformity

ⓐ "sharp cut-off" luminaire

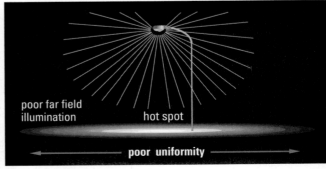

poor far field illumination

hot spot

poor uniformity

ⓑ "standard cobra" luminaire

SKILLS HANDBOOK: **8B** Exploring an Issue

Saturn

- Saturn, at about five-sixths the diameter of Jupiter, is the second largest planet in the solar system. The least dense of all the planets, it is possible that it has no solid core.
- Saturn's atmosphere is cloudy and, because of its quick rotation, windy. Saturn is farther from the Sun than Jupiter, so its average temperature is lower: about −180°C.
- For hundreds of years, people thought that Saturn was the only planet with rings. Detailed images sent by *Pioneer 11* and the two *Voyager* space probes showed that there are actually over 1000 separate rings. Astronomers are not certain whether the rings formed at the same time as the planet or are the crumbled remains of one of Saturn's many moons or some other object that came too close to the planet.

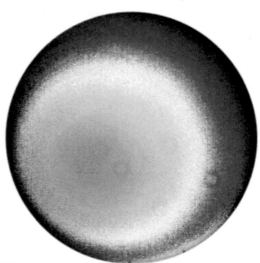

Uranus

- Although Uranus's diameter is almost four times that of Earth, it is so far away that it looks like a faint star. It was actually thought to be a star until its motion was discovered in 1781.
- Astronomers gathered considerable data about Uranus when *Voyager 2* passed near it in 1986.
- Uranus is unusual because its axis of rotation is in nearly the same plane as its orbit. This means that Uranus rotates on its side; the orange patch in the computer-enhanced photograph is a polar hood over the south pole.
- The atmosphere of Uranus is made up mostly of hydrogen, with some helium and methane. It has winds that blow up to about 500 km/h.

Neptune

- The story of the discovery of Neptune, the second farthest planet from the Sun, is one of great scientific achievement. Neptune is so far from Earth that it is barely visible, even through powerful telescopes. After 19th-century scientists established that Uranus was a planet and not a star, they studied its orbit and discovered that the orbit was not a smooth circular path. They hypothesized that some other object must be "tugging" on Uranus, causing its uneven orbit. Using detailed calculations, they predicted where this hidden object must be, searched, and discovered the "missing" planet in 1846.
- In 1989, computer-enhanced images from *Voyager 2* revealed that Neptune has bright blue and white clouds, and a dark region—the Great Dark Spot—that appears to be the centre of a storm. *Voyager 2* also uncovered the existence of at least eight moons and some thin rings orbiting Neptune.

Pluto

- Pluto is an unusual planet because it is not a gas giant and it does not seem to be terrestrial. It was discovered in 1930 after a painstaking search.
- Pluto is so far away that it takes 248 years to orbit the Sun. Although astronomers haven't yet observed a complete orbit, they have seen enough to detect that Pluto's orbit is elliptical and not quite centred on the Sun. Pluto actually passed within the orbit of Neptune, making it the eighth planet from the Sun, from January 1979 until February 1999. Pluto's unusual orbit has led some astronomers to suggest that it may have been a moon of Neptune at one time.
- Images taken by the Hubble Space Telescope have given us our best information yet about tiny, cold, distant Pluto and its moon, Charon.

Understanding Concepts

1. Why are the four closest planets to the Sun called the "terrestrial planets"?

2. Describe two features that make Earth unique among the planets, and two that make it similar to other planets.

3. Why is Jupiter easy to see in the night sky (when viewing conditions are right)?

4. There may be other groups of planets similar to our solar system, but they are very difficult to detect. Why?

5. How would the tilt of Uranus affect its seasons?

Making Connections

6. What are some features of a roving robot you would design for exploration on Mars?

7. List five or six ways that humans have had an impact on Earth and mention how each has had positive and negative results on life on Earth.

Exploring

8. Research the orbits of Mercury and Earth around the Sun. Draw a diagram to illustrate them and use it to explain why Mercury is so difficult to see from Earth.

Reflecting

9. List the steps that were followed in discovering Neptune. How do these steps relate to the process of scientific discovery?

Prize-Winning Astronomer

Mary Lou Whitehorne is a stargazer who, within eight years, captured Canada's highest award for an amateur astronomer—the Chant Medal. How did she do it, and why?

As a Girl Guide in Bedford, Nova Scotia, Whitehorne was interested in the sky but, with no expert to talk to her and no local library, her interest waned. After high school, she graduated in medical laboratory technology and pathology, but left her medical career to raise a family with her husband, Lloyd.

She went back to school, at age 31, to study at the Astronomy Department of St. Mary's University. She undertook a three-year study of a rare type of star known as a B-emission star. "B-stars" vary in brightness, so she decided to investigate the light they emit. She spent many hours examining their spectra through a telescope to investigate their atomic composition. "It is challenging raising two kids while observing the stars every clear night past midnight," she says, but she did it. She published two scientific papers, winning the 1993 Chant medal for her research efforts.

She has since completed ground school and flight training and has been awarded her Private Pilot Licence. She has also helped to establish a hands-on astronomy program for schools in Nova Scotia. In her spare time, she opened a resource centre for the Canadian Space Agency in the Atlantic region.

It's Terry Dickinson's fault. In 1985, I saw a hokey little star chart in his newspaper column. In it, he said that you could see four moons of Jupiter all aligned on one side—with binoculars. That was all it took and I was hooked.

Exploring ③A

1. Find out if there are any introductory astronomy courses or programs in your area. Attend a stargazing party, if you can, and learn about the sky from an expert.

2. A beginner's telescope is usually priced at $350 or less. Most astronomers will tell you that it is a big mistake to buy this type of instrument to explore the sky. Why?

3. Search the Internet for astronomical societies or amateur observing groups and write a brief summary of their activities.

Other Objects in the Solar System

About 60 million years ago, the dinosaurs that had roamed on Earth for millions of years died out in a fairly short time, along with many other species. What could have caused this extinction? Scientists have found evidence that a fast-moving object from outer space crashed into Earth, sending material flying into the atmosphere. This material reduced the amount of sunlight reaching Earth's surface, causing the climate to change and numerous life forms to die out.

As scientists study more about these objects from space, they hope to find clues about how life may have formed on Earth.

Did You Know ?

As of 1997, astronomers had observed 15 moons orbiting Uranus. Then Canadian astronomers discovered two more moons, bringing the count to 17.

Planetary Moons

Large natural objects that revolve around planets are called **satellites**, or moons. Several planets have more than one moon, but the chunks of rock that make up the rings of the gas giants are far too small to be considered moons.

Probably the most famous satellite of any planet is Earth's Moon, which has a diameter about one-quarter that of Earth. Six visits to the surface of the Moon by humans from 1969 to 1972 provided much new information (**Figure 1**). The Moon has no atmosphere, and its surface is filled with hills and valleys as well as craters caused by the impact of large and small objects from space.

Figure 1

The *Apollo* astronauts brought moon rocks to Earth for detailed study and collected data on the moon's soil, surface conditions, and moonquakes.

The moons of the other planets were discovered after the invention of the telescope. In 1610 Galileo Galilei looked at Jupiter through his telescope and became the first person to see four of Jupiter's moons.

Although humans have not yet been to other satellite moons besides our own, space probes have investigated several at close range, including the two small moons of Mars and many of the largest moons of Jupiter and Saturn.

What has surprised astronomers most about moons is the great differences in their sizes and surfaces, as **Figure 2** shows.

Table 1 lists the number of known moons revolving around the planets of the solar system. Only the numbers of moons revolving around the four terrestrial planets are known for certain. Future space probes may reveal more moons in our solar system.

Table 1

Planetary Moon Count (1998)

Planet	Number of known moons
Mercury	0
Venus	0
Earth	1
Mars	2
Jupiter	16
Saturn	18
Uranus	17
Neptune	8
Pluto	1

Studying the many planetary moons in the solar system helps us understand more about the origin and evolution of the solar system. But we may find other uses for these moons in the future. The moons contain huge amounts of useful minerals that humans may mine one day and use for construction on Earth or on other planets. Who owns the moons and the minerals they contain? Who will hold mining rights? Do humans have the right to affect the environments of other places, when we have shown little regard for Earth's environment? These controversial questions will stimulate much debate in the years to come.

Figure 2

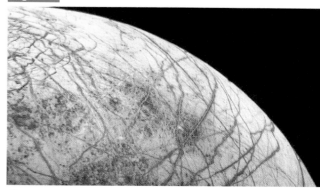

a Jupiter's moon Europa has an icy surface with very few craters and is nearly as big as Earth's Moon.

Asteroids

Refer to the model of the solar system on page 401. There is a large gap in the solar system, between the orbits of Mars and Jupiter. In that space is a ring around the Sun, made up of thousands of small rocky objects called **asteroids**. This ring of asteroids is called the **asteroid belt** and has been explored by several probes. The word asteroid comes from the Greek word *astron* meaning "starlike," but a better name would be "minor planet." Scientists think that the asteroids might have formed into a planet if the gravitational force of the large planet, Jupiter, had not been so strong.

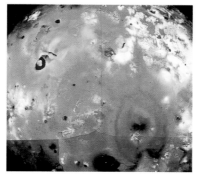

b Io, the closest moon to Jupiter, is the only moon in our solar system known to have active volcanoes, more violent than any on Earth.

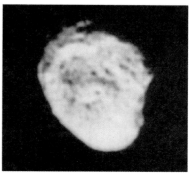

c Saturn's moon Hyperion is only 360 km across and has an irregular shape, possibly as a result of repeated collisions with large space rocks.

There are also asteroids sharing Jupiter's orbit, while others travel in paths that may take them closer to the Sun or Earth. In 1937 an asteroid named Hermes came within 800 000 km of Earth, only about twice the distance from Earth to the Moon. **Figure 3** shows the orbits of some asteroids and their names.

Like the planetary moons, asteroids are rich in minerals, which humans may someday mine. The largest asteroid is only about 1000 km in diameter, so it has low gravity. This means that a spacecraft carrying mined minerals would require much less energy to blast off from an asteroid than from Earth, Mars, or even our own Moon.

Meteors and Meteorites

A **meteoroid** is a lump of rock or metal that is trapped by Earth's gravity and pulled down through Earth's atmosphere. As it falls, it rubs against the molecules of the air. This rubbing, called friction, causes the meteoroid to

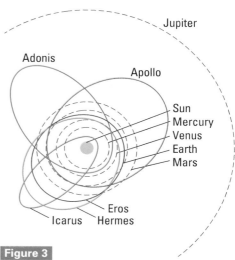

Figure 3
Orbits of several asteroids (Note: not drawn to scale.)

become hot and vaporize, and the air to glow. This produces a bright streak of light across the sky, called a **meteor**, that you can see at night.

Where do meteoroids come from? The largest are probably asteroids in orbits that cross Earth's orbit. The millions of tiny meteoroids that produce spectacular displays called meteor showers probably come from the debris left behind by comets. Most meteoroids vaporize completely before they reach Earth's surface. If the object is large enough to hit the ground before totally vaporizing, it is called a **meteorite**. If a large meteorite hits Earth's surface, it can produce a crater. The best-preserved meteorite crater in North America is the Barringer Crater near Winslow, Arizona (**Figure 4**), but there are also many in Canada.

Comets

One evening an amateur astronomer from Japan, Yuji Hyakutake, was viewing the night sky through a powerful pair of binoculars (20 × 50). In an area of the sky he had examined before, he noticed an object he had never seen. Like any thorough amateur astronomer, he recorded what he saw, then he reported it to professional astronomers. To everyone's excitement, Hyakutake had discovered a **comet**, which is a chunk of frozen matter that travels in a very long orbit around the Sun (**Figure 5**). Because Hyakutake had reported finding the comet, it was named after him.

One of the most interesting and exciting things you could see in the night sky is a comet with a tail that is millions of kilometres long. A few comets have been so bright that they were visible even in daylight.

During most of its elliptical orbit around the Sun, a comet remains far out in the solar system. When it approaches the Sun, however, the comet is warmed by solar radiation and the frozen substances become gases. As the comet travels closer to the Sun, these gases are pushed outward by solar wind and radiation, forming a bright, glowing tail that may be over 10^8 km long. The glowing tail may be seen for several months as the comet travels near the Sun.

Many comets have regular periods of revolution around the Sun. This allows us to predict when they will be close enough to the Sun to

 Try This Meteor Showers

Have you heard of shooting stars or falling stars? They are not stars at all: they are meteors. If you view the sky on a clear night, you might see a meteor. View the sky during a meteor shower and your chances of seeing one become very high. The three most active meteor showers are the Perseid shower (August 12), the Geminid shower (December 14), and the Quadrantid shower (January 3 or 4).

Figure 4

Evidence gathered from local rocks suggests that the Barringer Crater was formed 20 000 to 30 000 years ago by a large meteorite. The crater is 1.2 km in diameter and 120 m deep. Why are most craters this shape?

Figure 5

Comet Hyakutake, discovered in 1996, was the brightest comet seen in 20 years. It passed within 1.5×10^7 km of Earth—only a tenth of the distance from Earth to the Sun.

be seen. For example, a regular visitor is Halley's comet, last seen in 1986. It has a period of 76 years.

Exploring the Minor Bodies

Think of a chocolate bar that has many mixed ingredients covered with chocolate. In order to figure out how the bar was made, you would start by finding out the ingredients. Scientists go through a similar process when they try to figure out what formed Earth and other planets. Scientists think that the ingredients of the planets and moons were objects similar to asteroids and comets—the minor bodies of the solar system. Unlike Earth, many of the minor bodies have changed little since the birth of the solar system. This is why scientists send probes to the minor bodies to learn more about them, and perhaps learn more about the origin of the solar system.

A probe called *Deep Space 1*, launched in 1998, is the first of a new type of probe, sent to explore minor bodies. This light-weight probe was sent to study an asteroid about 1.9×10^8 km away. The probe is unique because its main source of energy uses ions (charged particles) of xenon and mercury gases. These ions escape at extremely high speeds from the rear of the vehicle, causing a forward thrust on the vehicle. The amount of fuel needed is only about 10% of what a conventional rocket uses. Another feature of this probe is that, when it gets closer to the asteroid it is chasing, it will use an on-board computer to decide on its final approach. This form of artificial intelligence is being used more and more for space exploration (**Figure 6**).

Figure 6

Deep Space 1

Understanding Concepts

1. **(a)** What is an asteroid?
   (b) Where is the asteroid belt?
2. **(a)** Explain the difference between a meteoroid and a meteorite.
   (b) Why are meteorites less common than meteors?
3. Using a labelled diagram, describe what causes the glowing tails of comets.
4. Both comets and planets orbit the Sun. How do their orbits differ?
5. When will Halley's comet next be close enough to the Sun to be seen?

Making Connections

6. Describe what might happen if a giant meteorite crashed into Earth's surface (a) on land (b) on water. On a map of the world, mark the spots where a meterorite would have the least impact on human life.

Exploring

7. The names of many astronomical objects may sound romantic or old-fashioned to us: many of them were named thousands of years ago. However, some recently discovered objects are given exotic names, too.
   (a) What are the planets of our solar system named after?
   (b) Find out what the features on the surface of Venus are named after.
8. One theory suggests that the extinction of dinosaurs may have been caused by the collision of a small asteroid or comet. Investigate this theory. Do you agree or disagree?
9. Of all the planetary moons in our solar system, which one would you be most interested in visiting on a scientific expedition? Research and describe the special features of this moon, and explain what you would try to discover about it while you were there.
10. Chart the names and dates of meteor showers, as well as the point in the sky from which the meteor appears to orginate.

Reflecting

11. Space exploration is costly. Do you think sending probes to explore the minor bodies is justified? Give reasons to support your opinion.

Chapter 13 Review

Key Expectations

Throughout the chapter, you have had opportunities to do the following things:

- Describe and compare constellations. (13.1, 13.3, 13.4, 13.7, 13.8)

- Investigate the motions and characteristics of objects visible in the sky and organize, record, and communicate your results. (13.2, 13.4, 13.5, 13.7, 13.8, 13.11, 13.13)

- Analyze data and use them to predict future observations, such as when to see a seasonal constellation or the return of a comet. (13.2, 13.7, 13.8, 13.13)

- Plan ways to model answers to questions about the motion of objects in the sky, and communicate results. (13.2, 13.8, 13.11, 13.13)

- Describe and compare the properties and motions of the objects in the solar system. (13.1, 13.2, 13.7, 13.8, 13.11, 13.13, 13.14, 13.15)

- Formulate and research questions related to sky-watching and the solar system, and communicate results. (all sections)

- Describe and explain how data from space probes contribute to our knowledge of the solar system. (13.12, 13.15)

- Describe different ways in which various cultures have understood the universe (13.6)

- Evaluate the impact of light pollution on the work of astronomers. (13.9)

- Identify careers related to the exploration of space. (Career Profile)

KEY TERMS

asteroid	orbital period
asteroid belt	outer planet
astronomical unit	planet
astronomy	revolution
axis	rotation
comet	satellite
constellation	solar system
gas giant	star
inner planet	Sun
light pollution	terrestrial planet
meteorite	universe
meteoroid	zodiac constellation
nonluminous	
orbit	

Reflecting

- "Observation of the stars has affected many human beliefs and activities, including navigation, calendar-making, and the creation of myths." Reflect on this idea. How does it connect with what you've done in this chapter? (To review, check the sections indicated above.)

- Revise your answers to the questions raised in Getting Started. How has your thinking changed?

- What new questions do you have? How will you answer them?

Understanding Concepts

1. Make a concept map to summarize the material that you have studied in this chapter. Start with the word "Sun."

2. Each of the following descriptions fits one of the planets in the solar system. Name the planet described by each sentence.
   - (a) It was discovered in 1846 after careful observations.
   - (b) It has more mass than all the other planets combined.
   - (c) It has surface temperatures ranging from −180°C to 400°C.
   - (d) It has an atmosphere containing oxygen.
   - (e) It is neither a gas giant nor a terrestrial planet.
   - (f) It has over 1000 rings around it.
   - (g) It appears reddish in colour.
   - (h) It has a very warm surface caused by its thick atmosphere.
   - (i) It rotates on its side.

3. Why does a meteor appear as a streak of light in the sky?

4. (a) Describe Earth's position among the nine planets of the solar system.

   (b) Describe several features of Earth that make it unique in the solar system.

5. Choose one planet and name three similarities and three differences between it and Earth.

6. Why are some comets seen much more often than others?

7. (a) State one constellation that you can see best during each of the four seasons.

   (b) Why can you see some constellations only during certain seasons?

8. Mercury is much closer to the Sun than Earth is, yet at night its surface temperature can fall much lower than the lowest surface temperature on Earth. Why do you think this happens?

9. (a) Why do you think craters caused by meteorites are rare on Earth?

   (b) Why do you think there are many meteorite craters on a planet such as Mercury?

10. Would the asteroid belt be dangerous to travel through in a spacecraft? Explain why or why not.

11. Describe a demonstration you could perform with other students to illustrate both revolution and rotation. Try your ideas.

12. Look back to Table 1 on page 418. On which planets would you weigh more than you do on Earth? On which would you weigh less?

Applying Skills

13. If you wanted to see the planets, where would you look in the night sky? Explain your answer.

14. (a) Use your hands to measure the angular sizes of at least five different objects outdoors. Create a table for your measurements, and draw conclusions about the relative sizes of the objects.

    (b) If the objects you investigated had been constellations, would you have been able to draw similar conclusions about their relative sizes?

15. Why do the planets all have different surface temperatures? Interpret the information in Table 1 on page 418 to find reasons for the differences.

16. Describe how you would use "pointer stars" to locate each of the following stars or constellations in the night sky:

    (a) Polaris   (d) Arcturus
    (b) Cassiopeia   (e) Regulus
    (c) Sirius

17. Many books are available on how to build your own telescope. Obtain one or two such references and write plans to build such an instrument. If possible, begin your project soon so you can use the telescope to view the planets.

Making Connections

18. What are some benefits that scientists would achieve by sending space probes to objects in outer space?

19. Before humans are sent to explore other planets, robots are sent to study areas of the planet.

    (a) Describe two advantages of using robots rather than humans for such exploration.

    (b) Describe two advantages of sending humans rather than robots.

20. The table below shows a total of 63 known planetary moons in the solar system in 1998. If you checked other reference books published in earlier years, you would find the number of moons listed as follows:

Year of publication	Number of known moons
1969	28
1984	44
1987	53
1991	60
1998	63

    (a) Why has the number of known moons changed so greatly in such a short period of time?

    (b) Do you think the number of moons will change as much in the next 25 years or so? Explain why or why not.

21. (a) Describe some of the problems that humans might face as they try to set up a settlement on Mars.

    (b) Suggest ways to overcome each of the major problems you listed in (a). Explain your answer in each case.

14 The Nature of the Universe

Getting Started

1 Darkness in the middle of the day is an eerie sensation. Before people understood solar eclipses, this phenomenon was associated with bad luck, a message from the gods, or even a warning of the end of the world. Although we now know that a solar eclipse is caused by the Moon passing between Earth and the Sun, it is still an exciting and fascinating event. Observing a solar eclipse also helps astronomers learn about properties of the Sun, and stars in general. What do you notice when you look at this image of a solar eclipse? Do all the stars, including our Sun, have the same properties?

2 Every 10 or 11 years Earth's communication systems may be disrupted. This happens just after violent storms occur on the surface of the Sun. We know that the Sun gives us light and heat, but how else does it affect Earth? We can learn a lot about the Sun, and other stars, by observing effects of the Sun on Earth's atmosphere. Is there a change in the environmental impact of the Sun on Earth? What can you judge by looking at the photo shown here?

3 You have learned about the solar system, and stars and constellations. Are there other objects in space? What do you think the image here shows? How far away is it? How was it taken? What can you predict about the motion of what is shown in the image?

Try This Using Angles to Estimate Distances

An important concept you will explore in this chapter is how vast the distances are in the universe. You will discover how astronomers measure those distances by comparing the apparent position of an object viewed from two different places. In this activity you can model how they measure large distances.

• Choose a partner and obtain a metre stick. Stand as far as you can from one of the walls in your classroom. Cover one eye with one hand, then stretch out the other arm and hold your index finger up so it hides an object on a distant wall (**Figure 1**). Use the metre stick to carefully measure the distance from your finger to your eye.

• Holding your finger still, cover your other eye. Notice what object your finger now hides. Determine the angle between the first and second objects hidden. (Use the hand method from Activity 13.5.)

• How does the angle between the two objects depend on the distance between your finger and your eye? Plan steps to answer the question. Will you need to make any other measurements? Then carry out the steps, record your observations, and communicate your discoveries using a diagram, table, graph, and/or spreadsheet.

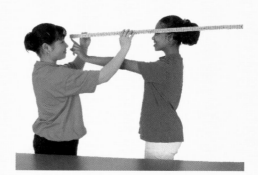

Figure 1

When your hand covers one eye and then the other, the position of the index finger shifts against the background.

Changing Ideas About the Universe

Movies about aliens from space or wars in far-away galaxies are more popular than ever. It seems that many people like to fantasize about what exists in the rest of the universe. Hollywood movies are an expression of our fascination with what is "out there."

People in ancient times didn't make science fiction movies, but they were still fascinated by the universe. Their beliefs and religious ceremonies reflected their ideas of what they thought the universe was like. These people left behind traditions, structures, and writings that give us clues about how they perceived the night sky.

In this section, you will learn how and why ideas about the universe have changed.

Ancient Ideas

Imagine that you are on a spinning merry-go-round. Your friends standing on the ground look as if they are moving in the opposite direction. Now imagine slowing the rate of spinning so that one rotation takes 24 h. At that speed, it would be difficult for you to judge whether it was the ride or the ground moving. Earth certainly seems to be stationary so, in ancient times, people thought everything else moved around it.

Everything in the sky appeared to be in motion. The Sun, Moon, stars, and planets all seemed to rise in the east and set in the west. One early idea about the stars was that they were attached to a large ball that revolved around Earth once every day. People thought that the stars were only slightly farther away than the Sun, Moon, and the planets they could see. In **Figure 1**, you can see one possible

arrangement of the universe. This idea is called the **Earth-centred universe**.

Improvements in Making Observations

The Earth-centred idea of the universe was popular for thousands of years. Then, around 500 years ago, some scientists began to question this idea. Scientific ideas were starting to change for two major reasons: the invention of the telescope and the fact that scientists began experimenting to learn more about nature, Earth, and the universe.

The telescope was invented in Europe in the early 1600s. The great Italian scientist Galileo Galilei improved upon the invention and was the first person to use it to view the night sky. His first telescopes made distant objects look seven to thirty times larger than normal. He saw that the planets were not merely points of light: they were circular, like the Moon. The stars, however, still appeared as points of light. From these observations, Galileo inferred that the stars must be much farther from Earth than the planets. The universe was clearly much bigger than was previously thought.

Before Galileo had discovered how different the sky looked through a telescope, other scientists had begun to question the Earth-centred view of the universe. For example, in the mid-1500s, Copernicus had presented mathematical evidence that the planets all orbit the Sun. Galileo's discoveries, especially his observation that Venus has phases like the Moon, supported the new theories. Galileo became

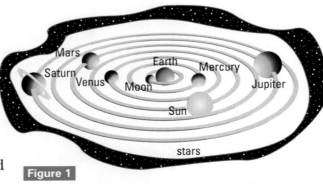

Figure 1

Many people in ancient times thought that Earth was the centre of the universe and that all the other objects revolved around it.

convinced that Earth and other planets travel around the Sun. Persuading other people to accept his theory was difficult and sometimes dangerous work. The predominant view in Europe at the time was that Earth, made by God, was the centre of the universe. Anyone who challenged this idea was considered "heretical" (against the Christian Church) and was even threatened with torture. These threats forced Galileo to deny his findings that Earth travels around the Sun. Eventually, however, despite religious persecution, the idea of the **Sun-centred solar system** replaced the Earth-centred view (**Figure 2**).

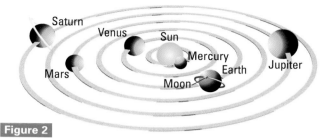

This is the Sun-centred view of the solar system of about 400 years ago. Compare this model with Figure 1. Which planets are not shown? Why not?

Today's Ideas

Nearly 400 years have passed since the invention of the telescope. In that time other inventions have led to many new discoveries about the universe. We now know that the planets orbit the Sun and that the Sun is just one of countless stars. Astronomers have observed that the Sun and other stars are also moving. Stars are gathered in large groups, surrounded by gas and dust. The group of stars that our Sun belongs to is called the Milky Way Galaxy. A **galaxy** is a huge collection of gas, dust, and hundreds of billions of stars and planets.

Modern telescopes show us that there is a vast region beyond the Milky Way Galaxy that appears to be almost empty. Far off in the distance there are countless more galaxies and other fascinating objects (**Figure 3**).

Our knowledge of the universe is always increasing as scientists develop new tools that let astronomers see farther into the distance. For instance, by detecting radiation that may have taken millions of years to reach Earth, they are able to learn about the early history of the universe.

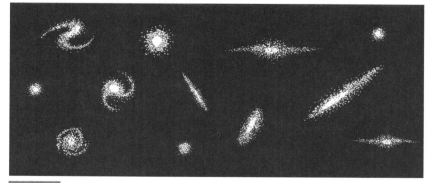

Figure 3
The universe contains huge groups of stars, called galaxies, separated by vast distances. The galaxies and everything in them are constantly in motion.

Understanding Concepts

1. **(a)** What does the expression "Earth-centred universe" mean?
   (b) What evidence helped scientists change their ideas about an Earth-centred universe?

2. How and why has our estimate of the size of the universe changed in the past few centuries?

3. **(a)** What is a galaxy?
   (b) What is our galaxy called?

Making Connections

4. Describe how "new" technology in Galileo's time helped to change an established idea. What societal effects were related to the change?

Exploring

5. Research how one of the
   (3A) following scientists contributed to the development of ideas about the universe: Nicholas Copernicus; Tycho Brahe; Johannes Kepler; Isaac Newton. Create a written report or an audio-visual presentation.

Reflecting

6. The Earth-centred view was replaced by the Sun-centred view of the solar system. How does this illustrate the process of science?

Who Owns the Solar System?

The human race has a long history of exploration, colonization, and exploitation. From Stone Age migrations across the Bering Strait to 20th-century forays into the jungles of South America, we have set out to see what we can find and what we can make use of. Most of Earth's surface has now been explored, charted, and investigated with a view to extracting valuable resources. More recently, humans have visited Earth's Moon and sent space probes to explore outer space objects such as comets, asteroids, and other planets and their moons. These explorations have shown scientists that there may be many valuable resources on several of these objects. Mars is rich in iron oxide, which could yield both iron and oxygen, and Jupiter's moon Ganymede appears to contain hydrogen peroxide. Other outer space objects are quite likely to contain other useful substances. An additional advantage of mining a low-gravity object is that it would be easier to lift extracted resources up, into outer space, than it would be from Earth. This has potential implications for the construction of future space stations.

Some people suggest that colonies and mines could indeed be built on such bodies as Mars, our Moon, or a few other moons or large asteroids. But who owns outer space objects? Should ownership rights be awarded on a "first-come, first-served" basis? If colonies are set up, who should live in them? Who would benefit from the existence of the colonies? Who owns any resources that are mined on other bodies? Would it be a case of the rich countries getting richer, and the poorer countries having no chance of participating? What about the environmental aspect? In many instances we have not shown that we can responsibly manage Earth's resources. Do we have any right to export our mismanagement? It has been suggested that this is a way to ease the pressure of a rapidly increasing population here on Earth. Is this a valid reason? If we were to find any form of extraterrestrial life, would our decisions be altered? These questions should be discussed before such expensive projects begin.

Issue A Moratorium on Extraterrestrial Mining

"We should learn to manage Earth's resources better, before exploiting extraterrestrial resources."

Imagine that you have been asked to participate in a panel discussion, either supporting or opposing the statement above. You might be a mining engineer, an investor, an unemployed miner, an astronomer, a lawyer, an environmentalist, a social worker, a space engineer, or anyone else who might have an opinion in the discussion. Choose your position and support it by developing a three- to five-minute presentation. Refer to the ideas below and add some more of your own. 8C

Point

- There are many untapped resources here on Earth. If we were to use and recycle them more efficiently, we would have no need of space mining.

- The cost of the technology to exploit space resources is far too high. The money would be better spent improving efficiencies on Earth.

- We have no right to take resources from outer space objects. They do not belong to us.

Counterpoint

- Using resources from outer space objects would reduce our dependence on Earth resources.

- By having resource extraction facilities set up in space, we would have bases for refuelling and manufacturing, which would make further space exploration much more economical.

Challenge

What issues presented in the panel discussion do you need to consider when participating in any endeavour involving space exploration?

Using Triangles to Measure Distances

We know that the Moon is closer to us than the Sun, and that the planets are closer than the stars. How do we know this? Because of their movements relative to each other: the stars form an almost unchanging backdrop for the wandering planets. But just how far away are these objects? How can we measure such vast distances? Try using only one eye to judge how far away something is. It is not as easy as using two eyes to judge distances. The reason is that the lines of sight from your eyes form an angle where they meet an object, and the size of the angle helps you judge distances. You may have noticed this when you did the Getting Started activity on page 437.

In this activity you will measure distances using an indirect method. You will measure angles between a baseline and the object, and draw a scale diagram to calculate the distance to the object. This indirect method of measuring the distance to an object using geometry is called **triangulation**. The more carefully you draw the triangles, the more accurate your answers will be.

Materials
- astrolabe designed in Activity 13.5 (with the string attached but the rubber stopper removed)
- metre stick
- paper
- small protractor
- centimetre ruler

Procedure

1 Your teacher will suggest an unknown distance that you will measure in the classroom. (For example, the distance could be from one wall to a door hinge on the opposite wall.) Use your eyes (and experience) to estimate the distance.

✎ (a) Record your estimate.

2 Mark off two points, Point 1 and Point 2, **6B** that are at least 5 m apart, and use the metre stick to measure the distance. This distance is called the triangle baseline.

✎ (a) Record the length of the baseline.

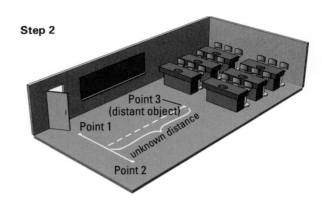

Step 2

Point 3 (distant object)
Point 1
unknown distance
Point 2

3 Have one student stand at Point 1 holding the astrolabe horizontally, so that its baseline lines up with the triangle baseline. Another student should use the string to be sure this is as straight as possible. While one student holds the astrolabe still, another student should move the string around until it lines up with the distant object, Point 3. Measure the angle formed, Angle 1.

✎ (a) Record the value of Angle 1.

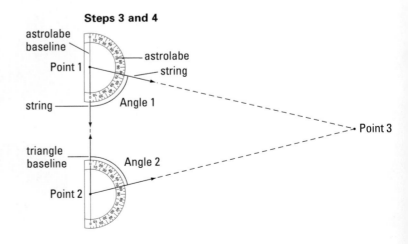

Steps 3 and 4

astrolabe baseline
Point 1
astrolabe string
string
Angle 1
triangle baseline
Angle 2
Point 2
Point 3

4 Repeat step 3 from Point 2.

✎ (a) Record the value of Angle 2.

5 On a piece of paper, use the centimetre ruler to draw a diagram of your experiment. Choose an appropriate scale so that the triangle baseline will be at least 5 cm long. Label Points 1 and 2. Use a small protractor to draw Angles 1 and 2.

6 From Points 1 and 2, draw straight lines that meet at Point 3. Finally, draw a straight line from Point 3 back to the triangle baseline. (Try to make this line meet the baseline at a 90° angle.) Measure the distance from the baseline to Point 3. Label all the points and distances on your diagram.

Steps 5 and 6

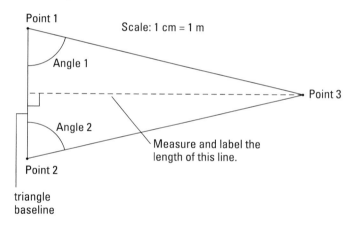

Point 1
Scale: 1 cm = 1 m
Angle 1
Point 3
Angle 2
Measure and label the length of this line.
Point 2
triangle baseline

7 Design your own investigation to measure large distances indirectly, either outdoors or in an appropriate area inside the school. After your teacher approves your design, carry out the investigation.

Understanding Concepts

1. Why is the method of finding distances in this investigation called an indirect method?

2. Do you think that measurements in this investigation would be more accurate with large baselines or smaller ones? Why?

3. Draw two diagrams to show the lines of sight from your eyes to an object that is near, then to one that is farther away. (Your two eyes form a baseline.) How do the angles in your diagrams relate to your ability to judge distances?

4 How could triangulation be used to find out the distance from Earth to a neighbouring star?

Exploring

5. The widely spaced eyes of hawks, eagles, and other birds of prey enable them to judge distances very well. Explain, with the aid of a diagram, why this is so.

6. Astronomers call their way of using triangulation to measure distances "parallax." Look up parallax in an astronomy reference book or other source and describe how it relates to the method you learned in this activity.

7. Use direct measurements to find out the true value of your group's measurements. Use the following equation to determine the percent error of each of your indirect measurements.

$$\% \text{ error} = \frac{\left| \begin{array}{c} \text{indirect} \\ \text{measurement} \end{array} - \begin{array}{c} \text{direct} \\ \text{measurement} \end{array} \right|}{\text{direct measurement}} \times 100\%$$

Reflecting

8. In Chapter 13 you learned how to use fists and fingers to measure angles. Could fists and fingers be used to measure distances indirectly as you did in this activity? Explain your answer.

9. How does the method of triangulation relate to what you discovered in the Getting Started activity on page 437?

Distances in Space

Using Scientific Notation

Distances in space are very large, so measurements written out in kilometres can become long. Scientific notation is a mathematical abbreviation for writing very large or small numbers. Using this notation, a number is written with a digit between 1 and 9 before the decimal, followed by a power of 10. That is why Canada's population of 32 million could be written as 3.2×10^7 instead of 32 000 000. The exponent, in this case 7, indicates the number of places you have to move the decimal point.

Using Long Baseline Measurements

A surveyor trying to measure the distance across a raging river can use the indirect method of triangulation. Astronomers can apply the same method to measure distances to objects in the solar system and beyond. To measure large distances with as much accuracy as possible, they use the largest baselines available.

How can scientists obtain long baselines to measure the huge distances to stars? One way to obtain a large baseline is to use the diameter of Earth, about 1.3×10^4 km (**Figure 1**). This method could be used to determine the distance to the Moon or a nearby planet.

What is the largest possible baseline available to observers on Earth? It is the diameter of Earth's orbit, a distance of about 3.0×10^8 km (**Figure 2**). This baseline has been used to calculate the distances to some of the stars nearest to our solar system.

Distances to the Stars

The distance from our Sun to the next nearest star that you can see without using a telescope or binoculars is about 4.1×10^{13} km. This star is called Alpha Centauri. Most other distances in space are much greater than this. To avoid using such huge numbers, scientists have developed units of distance other than kilometres or metres.

One common unit used to measure large distances is the light-year. One **light-year** is the distance that light travels in one year. Light

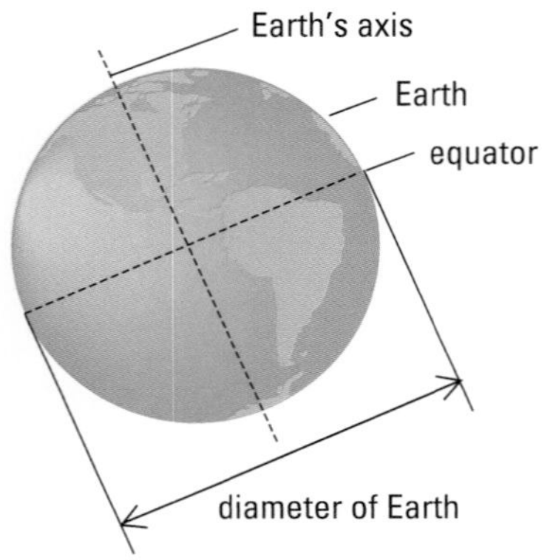

Figure 1

One example of a long baseline is the diameter of Earth at the equator. Because Earth rotates on its axis, it takes 12 h for an observer to move from one side of Earth to the other.

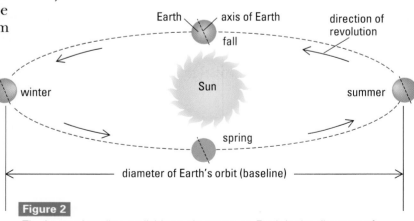

Figure 2

The largest baseline available to observers on Earth is the diameter of Earth's orbit. As it takes a year for one complete orbit, angles to the stars are measured six months apart.

travels at an enormous speed in space, about 300 000 km/s. So in one year it can travel about 9.46×10^{12} km*. Thus, the distance to Alpha Centauri, given above, is equal to 4.3 light-years. (You can prove this by dividing the distance to Alpha Centauri by the distance light travels per year.) **Table 1** lists examples of distances measured in light-years. Notice that all the stars named are different distances from Earth. This is true even for stars in the same constellation. For example, Betelgeuse and Rigel are both in the constellation Orion, but their distances from Earth are quite different.

One interesting fact about light-years is that they tell us how far back we are looking in time: the light from Alpha Centauri that we might see tonight actually left Alpha Centauri 4.3 years ago. Whatever we view on Alpha Centauri *now* has already happened! By looking at stars thousands or even billions of light-years away, astronomers can look back in time and see what the universe was like when it was much younger. Measurements in light-years tell us not only how far away an object is, but also how much time has passed since the light we see left that object.

Table 1	Some Distances of Objects from Earth
Star or object	**Approximate distance (light-years)**
Alpha Centauri	4.3
Sirius (brightest star in the sky)	8.8
Vega	26
Arcturus	36
Betelgeuse	700
Rigel	900
Deneb	1400
Most-distant known galaxy in the universe	15 000 000 000

Understanding Concepts

1. Why is the length of the baseline important when you use triangulation to measure distances?

2. A surveyor (**Figure 3**) measures off a baseline of 120 m along the shore of a river. He then measures the angle from each end of the baseline to a rock on the opposite shore. The two angles that he measures from the baseline to the rock are 65° and 50°. Draw a scale diagram to determine the width of the river.

Figure 3

A surveyor measures angles with an instrument called a transit theodolite. Why do surveyors need to measure exact angles and distances?

3. (a) Describe the longest baseline possible on Earth.

   (b) How much time passes between the measurement of angles using the baseline you suggested in (a)?

4. What is the difference between a year and a light-year?

5. Determine the distances listed in **Table 1** in (7C) kilometres and write the final answers using scientific notation.

Exploring

6. Design a way to use a triangle method to measure the height of a wall or tree indirectly. Try your method and draw a scale diagram to show your calculations.

*distance = speed × time
= (300 000 km/s) (1 year × 365 d/year × 24 h/d × 3600 s/h)
= 9 460 000 000 000 km or 9.46×10^{12} km

Scaling the Universe

About how much time do you think it would take to
- drive a car across Canada?
- fly around the world in an airplane?
- travel from Earth to the Moon in a spacecraft?
- travel from Earth to the Andromeda Galaxy in a spacecraft?

Although many people may think of travelling far away in space, they first must find out how long the distances are. This activity will give you an idea of how tiny our Earth is compared with the entire universe.

The distances between different objects in the universe are given different names. Distances between planets in the solar system are called interplanetary distances (*inter* is the Latin word meaning "between"). Distances between the stars are called interstellar distances (*stellar* refers to stars). And distances that separate the galaxies are intergalactic distances (*galactic* refers to galaxies).

As you have learned, large distances in space are often measured in light-years. (Recall that a light-year is the distance light travels in one year, at a speed of 300 000 km/s. Thus, a light-year is a distance of about 9.46×10^{12} km.) In this activity, you will use a model of the light-year to help you understand the vastness of the universe. To make your calculations easier, always round off your answers to easy numbers, such as 10, 100, 1000, and so on.

Materials
- this textbook
- ruler or metre stick with millimetre and centimetre divisions
- calculator

Did You Know ?

If the space shuttle could travel in a straight line for a year at its average cruising speed, it would travel 2.5×10^8 km. Light travels 9.46×10^{12} km in a year: more than 30 000 times farther!

Procedure

1 Count out 50 sheets of paper in your textbook. Measure the thickness of those 50 sheets together in millimetres. Use your measurement to calculate the approximate number of sheets of paper per millimetre (sheets/mm). You will need this number in the next two steps, so before going on, check with your teacher to be sure that your calculation is reasonable.

Step 1

(a) Record your measurement and calculation.

2 Repeat this two more times.

(a) Record your measurements.

(b) Calculate a mean value of your results. Remember to include the units.

3 Assuming that the thickness of one sheet of paper represents a distance of one light-year, determine the number of light-years in: 1 mm; 1 cm; 1 m; 1 km; 1000 km; 2000 km (the approximate distance from Montreal to Winnipeg as shown in **Figure 1**).

(a) Record your answers.

4 Copy the first three columns of **Table 1** into your notebook and complete the required information.

(a) Write the actual distance in kilometres using scientific notation.

Table 1 **Sample Distances**

Model distance	Number of sheets thick (or actual distance in light-years)	Actual distance (km)	Example of distance in the universe
0.1 mm	?	?	maximum distance of some comets from the Sun
0.4 mm	?	?	approximate distance from the Sun to the nearest star (Alpha Centauri)
thickness of two pennies (almost 3 mm)	?	?	approximate distance to the star Vega
approximate length of an adult's thumb (7 cm)	?	?	distance to the star Betelgeuse
height of a wall in a home (2.5 m)	?	?	distance from Earth to the centre of the Milky Way Galaxy
length of a science classroom (10 m)	?	?	diameter of the Milky Way Galaxy
length of two football fields (200 m)	?	?	distance to the Andromeda Galaxy
30 km	?	?	distance to the Coma cluster of galaxies
Montreal to Winnipeg (about 2000 km)	?	?	estimated size of the entire universe

Figure 1

Winnipeg   Montreal

Understanding Concepts

1. Do you think the size of Earth's orbit around the Sun could be shown in a model of the universe, using the scale in this activity? Explain your answer.

2. Using **Table 1**, state an example of (a) an intergalactic distance and (b) an interstellar distance.

3. Why are interplanetary distances usually not stated in light-years?

Making Connections

4. Based on what you have learned in this activity, do you think it would be possible for humans to ever travel to (a) Mars, (b) the moons of Jupiter, (c) Alpha Centauri, or (d) the Andromeda Galaxy? Give reasons for your answers.

Reflecting

5. In studying distances and sizes in astronomy, it is important to develop skills in designing models. How do you think the model of distances suggested in this activity could be improved?

Challenge

How would you help audiences at the planetarium understand the different distances related to the universe? How would you alter the explanation for a young audience?

Telescopes

We can learn about space beyond Earth by looking at images of planets, stars, and other objects in the universe. If you were to compare the images in this textbook with the images in astronomy books published even a few years ago, you would discover that we can now see objects it was not possible to see then. We can now see better and farther than ever before.

Using Telescopes

Telescopes may cost from a few hundred dollars to hundreds of millions of dollars. From Galileo's first astronomical telescope to the most sophisticated instruments now orbiting Earth, the main purpose of a telescope is to gather light. The light forms an image that can be seen or recorded using cameras or other devices. The design of optical telescopes has changed little over the years, but they are being constructed larger and more powerful than ever before. **Figure 1** shows different types and locations of telescopes. What might be their advantages and disadvantages?

Figure 1

a In a **refracting telescope**, light rays refract (bend) as they pass through a light-gathering lens, called the objective lens. Unfortunately, there is a limit to how big objective lenses can be built. When the lenses reach about 1 m in diameter, the glass becomes too heavy, sags under its own weight, and distorts the image. Galileo developed the first refracting telescope in the early 1600s. Through it he could see that the planets are spheres, not just points of light.

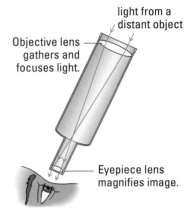

light from a distant object
Objective lens gathers and focuses light.
Eyepiece lens magnifies image.

b A **reflecting telescope**, first constructed by Isaac Newton in 1668, uses a concave mirror to gather light. Such a mirror can be supported from underneath, so it can be built much larger than the objective lens in a refracting telescope. Large reflecting telescopes were used in the early 1800s to investigate nebulas.

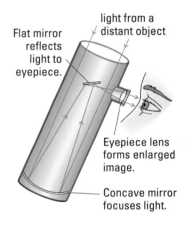

light from a distant object
Flat mirror reflects light to eyepiece.
Eyepiece lens forms enlarged image.
Concave mirror focuses light.

c The biggest reflecting telescopes are located in **observatories**, which are large buildings with opening domes. Many observatories, such as the Canada-France-Hawaii observatory in Hawaii, shown here, are built on the tops of mountains. There are no city lights and the atmosphere is thin and steady. The thin atmosphere is helpful because it absorbs and scatters less light than the denser atmosphere at lower altitudes. Clearer views led astronomers to the discovery that there are other galaxies, besides our own.

d An expensive but successful way of overcoming the problem of Earth's atmosphere is to place the telescope into orbit around Earth. The Hubble Space Telescope, shown here, was put into Earth orbit in 1990. Its reflecting telescope can obtain a much more detailed view of distant objects and see much farther away than ground-based telescopes. The Hubble Space Telescope is shown in Figure 2 on page 483, and one of its famous images is featured on page 480.

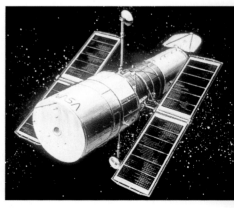

Looking through a telescope at objects in the sky can be rewarding and informative. However, astronomers use another way to obtain detailed permanent images of objects in the sky. They attach an instrument, such as a digital camera, to the telescope. As the telescope follows the object across the sky, gathering light for minutes or hours or even days, more details are captured. Because the camera gathers more light from the object over time, the resulting images show stars and galaxies too faint to be seen with our eyes alone (**Figure 2**).

Using Invisible Energies

What do you do when you want to receive radio signals sent out by a radio station? You simply tune your radio to receive them. What do astronomers do when they want to receive the radio waves sent out by some star or other object in the sky? They aim a radio receiver toward the object and try tuning the receiver until it receives waves from space.

Radio waves belong to a broad band of energies called the **electromagnetic spectrum**. This spectrum consists of radio waves, microwaves, infrared rays (heat), visible light, ultraviolet rays, X rays, and gamma rays. These types of energy are emitted (given off) by stars, galaxies, and other objects in the universe. These waves all travel at the speed of light in a vacuum, and they have energies that become greater as their wavelengths become smaller. Studying these types of energy helps astronomers understand more about the universe.

A device that receives radio waves from space is called a **radio telescope**. It is able to detect radio waves that are emitted by stars and galaxies. A radio telescope can work even on cloudy days because radio waves can pass through Earth's atmosphere, including clouds, very easily.

Radio telescopes often look like giant satellite dishes. The dish part, which may be made of wire mesh, reflects the radio waves to a collector held just in front of the dish. The radio telescope shown in **Figure 3** is the largest single radio telescope in the world, measuring more than 300 m in diameter. It was made by placing a concave wire mesh in a valley in the mountains. This radio telescope receives radio signals from many different parts of the universe, day and night, as Earth rotates.

Figure 2

These views of the same object show how the amount of detail increases as the time that the camera gathers light increases. In order from top to bottom, these images show light-gathering times of 1, 5, 30, and 45 min.

Figure 3

The Arecibo radio telescope is built into the mountains of Puerto Rico. What is the triangular structure just above the "dish"?

Figure 4

An array of radio telescopes produces the same results as a much larger single radio telescope.

Radio telescopes can also be made to work together in sets called arrays. **Figure 4** shows an array of radio telescopes in New Mexico, linked together to produce the same results as a much larger radio telescope. Radio signals collected over a period of time are combined, using a computer, into a map of objects in the sky.

Some parts of the electromagnetic spectrum are absorbed by Earth's atmosphere, so they cannot be detected from the surface of Earth. To overcome this problem, scientists use satellites that are put into orbit above the atmosphere. One example of such a satellite is the Infrared Astronomical Satellite (IRAS), which can detect objects in space that emit very tiny amounts of heat (**Figure 5**). Launched in 1983, the IRAS made some exciting discoveries, including evidence that planets may be forming around nearby stars.

Canada is involved in the planning for a new orbiting telescope scheduled for launch in 2007: the Next Generation Space Telescope. It will have much greater sensitivity to infrared radiation than existing telescopes, allowing astronomers to look even farther back in time.

Figure 5

The IRAS is sensitive enough to detect infrared (heat) radiation from a source on Pluto, over 40×10^8 km away, which gives off as little heat as a 20-W light bulb.

Understanding Concepts

1. What is the purpose of a telescope?

2. Describe the similarities and differences between refracting and reflecting telescopes. Include a diagram.

3. **(a)** Why is Earth's atmosphere a problem for astronomers? Include a diagram.

   (b) Where would you build an observatory to overcome this problem?

4. Large astronomical telescopes are usually used together with digital cameras or other electronic instruments, and astronomers seldom look through such telescopes directly. Suggest reasons for this.

Making Connections

5. Write a brief description of how improvements in technology have altered our view of the universe.

6. Predict how future technological developments might bring new discoveries.

Challenge

Research Canada's role in the development of the Next Generation Space Telescope. Include the results in your information package.

Space-Age Communicator

How do you get from Amherstburg, a small Ontario town, to the Jet Propulsion Laboratory in Pasadena, California? Science journalist Ivan Semeniuk knows; he's made the journey.

At the age of eight, he visited a planetarium and was hooked on astronomy. He eventually joined the Royal Astronomical Society of Canada and, as a camp counsellor in Haliburton, Ontario, enjoyed exploring the night sky far from the city lights, and also teaching others about his favourite pastime. "It is not surprising I loved communicating," says Semeniuk. "I like writing and I did well in English at school. In high school, my English teacher was probably disappointed when I decided to pursue science."

In 1986, he got his first break: the job of running the planetarium at the Ontario Science Centre. "It was a natural fit for me because the Centre is involved in the public communication of science," he says. "I got to do it all, from writing scripts to tracking down the latest astronomical images."

By 1994, the Internet was allowing Semeniuk to make his planetarium as current as the evening news. "You can download images and communicate with experts easily. When Comet Shoemaker-Levy slammed into the planet Jupiter, the Internet allowed me to have images of the first impact in the show before the last piece of the comet had even hit the planet," he recalls.

Semeniuk writes for *Sky News*, a Canadian astronomy magazine and reports on astronomy for Discovery Channel.

To develop his abilities further, Semeniuk decided to return to school to train as a science journalist. He learned to access and synthesize large amounts of information into stories that would interest readers and viewers.

Ivan Semeniuk continues to reach people through his writing and broadcasting, and through his work at the Ontario Science Centre.

> Science is like a rolling landscape: the scientist works a lifetime in a small area, just uncovering a bit of it. What I get to do is wander all over the whole terrain designing voyages for people to see the boundaries and the mysteries.

Exploring

1. What service does the Royal Astronomical Society of Canada provide for its members?

2. Speculate about the kinds of traits that one would require as a science journalist.

3. Comment on the contributions that a mission, such as *Pathfinder*, makes to the understanding of the world around us.

The Sun: An Important Star

By far the most important star to us is the one at the centre of the solar system: our Sun. It provides the energy needed by all the plants and animals on Earth, and its gravitational pull keeps us in our steady orbit. Learning about the Sun helps us understand about the nature of other stars. As you read about the features of the Sun, remember that we are continually learning more: about its origin, its chemistry, its radiation, perhaps even its future.

Because the Sun is the closest star to Earth, it is also the brightest object in the sky. In fact, it emits so much light energy that you cannot see the other stars until the Sun has set.

Where does the Sun's energy come from? Like all stars, the Sun produces energy through a process called **nuclear fusion**. Inside the Sun, the temperature and pressure are so high that substances fuse (join together) to form new substances. For example, hydrogen nuclei fuse to form helium nuclei. This process produces large amounts of heat, light, and other forms of energy (**Figure 1**) that travel out from the Sun through space. Every second, the Sun makes more energy than humans have used throughout our entire history (**Figure 2**).

Scientists have calculated that the Sun has been producing energy for about 5 billion years and is still 75% hydrogen. (The remaining 25% is helium, with small amounts of other gases.) Scientists estimate that it will continue producing energy for about another 5 billion years before it runs out of fuel.

A Close Look at the Sun

From Earth, astronomers can study the parts of the Sun near the equator, but not the areas near the Sun's poles. To overcome this problem, the space probe *Ulysses* was launched in 1990 to study the poles of the Sun. Another important probe, *SOHO*, has 12 instruments on board for observing the Sun. These probes, and others, have sent back data and photographs to help scientists study both the atmosphere and interior of the Sun.

Based on the observations made by these probes, and calculations made by astronomers, we can draw a model showing the various layers of the Sun (**Figure 3**).

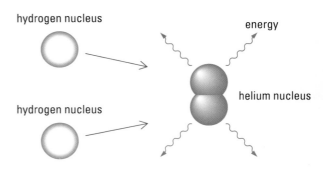

hydrogen nucleus

hydrogen nucleus

energy

helium nucleus

Figure 1

During nuclear fusion, substances fuse to form a new substance, releasing large amounts of energy. This is how the Sun produces energy, and it may someday become an important energy source on Earth.

Figure 2

Individuals are taking advantage of the Sun as a source of energy. Why isn't this being done on a larger scale?

The Sun's Effects on Earth

While astronomers can see solar flares as they happen, other effects only become apparent a few days afterwards. Solar flares emit charged particles, which travel much more slowly than light. When these particles reach Earth, they are focused, by Earth's magnetic field, at the north and south poles. The resulting electrical effects in the atmosphere interfere with the transmission of radio waves. This is why many communities in the far north of Canada sometimes lose radio communication for days at a time. The same charged particles produce the beautiful auroras seen over the North Pole (known as the Northern Lights or Aurora Borealis) and the South Pole (the Southern Lights or Aurora Australis). Photographs of auroras are shown on pages 394 and 436.

Did You Know

Sunspots are huge cooler areas in the Sun's photosphere. Even the smallest sunspots observed are larger than Earth.

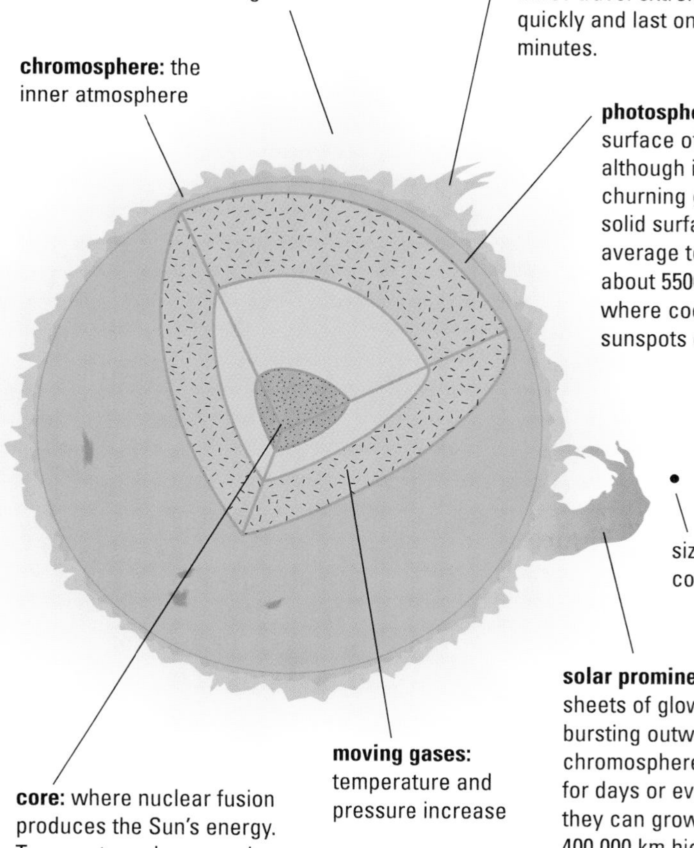

corona: the hot outer part of the Sun, where the gases reach temperatures of about 1 million degrees Celsius.

chromosphere: the inner atmosphere

solar flare: travels outward from the chromosphere through the corona. Solar flares travel extremely quickly and last only a few minutes.

photosphere: called the surface of the Sun, although it is made up of churning gases: not a solid surface at all. Its average temperature is about 5500°C. This is where cooler, darker sunspots occur.

size of Earth, for comparison

core: where nuclear fusion produces the Sun's energy. Temperatures here reach perhaps 15 million degrees Celsius, and the pressure is enormous.

moving gases: temperature and pressure increase

solar prominences: large sheets of glowing gases bursting outward from the chromosphere. They can last for days or even weeks, and they can grow as large as 400 000 km high, which is greater than the distance from Earth to the Moon.

Figure 3
The structure of the Sun

Understanding Concepts

1. Describe the differences between a solar flare and a solar prominence. Which affects us and how?

2. Describe the process that occurs in the Sun's core to produce so much energy.

Making Connections

3. You have read that the Sun can continue producing energy for about 5 billion more years. Is the possible "death" of the Sun at that time a problem we should worry about? Why or why not? Discuss this with your class.

Exploring

4. The Sun's diameter is about 110 times as big as Earth's diameter, yet the Sun could hold about 1.3 million Earths. To find out why there is such a difference, use your calculator to cube the number 110. Then relate what you discover to the difference between diameter (a length) and capacity (a volume or length cubed).

5. In 1990, the space probe *Ulysses* was launched from Earth. It was a joint European-American probe and was the first probe intended to look closely at the poles of the Sun. Research *Ulysses* and describe what you find out.

6. The Sun is an almost inexhaustible source of energy. Research the technologies that are being developed to store and use that energy.

Reflecting

7. Why do we consider the Sun to be the most important star?

Observing Our Closest Star

You know that Earth rotates on its axis once each day. Do other bodies in the universe, such as the Sun, also rotate? In this investigation you will discover a way of observing whether the Sun, our closest star, rotates.

Approximately every 11 years, scientists observe problems that occur with our telephone and other communication systems, as well as problems with our distribution of electricity. They found that these problems occur just after they observe violent storms on the Sun's surface. These storms appear to occur when there are many dark regions on the Sun, called **sunspots**, as **Figure 1** shows. As you observe the image of the Sun in this investigation, look for evidence of sunspots.

🛑 Never look directly at the Sun and never look through binoculars or any other instrument at the Sun. The energy in sunlight is strong enough to permanently damage your eyes.

Question
Does the Sun rotate?

Hypothesis

 **1** Write a hypothesis about whether the Sun rotates and, if it does, predict how you might detect this rotation.

Materials
- binoculars (focused for distant viewing)*
- tripod
- mounting bracket (or masking tape)
- piece of cardboard
- scissors
- screen (e.g., a regular sheet of smooth white paper attached to cardboard)
- clock or watch
- log book
- a sunny day!
- digital or regular camera (optional)

*Note: This investigation can be done using a telescope instead of binoculars. If you use an astronomical telescope, however, be aware that the image you obtain may be inverted.

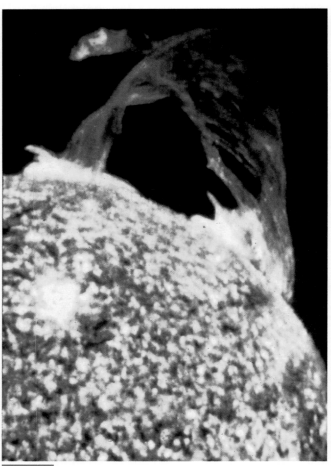

Figure 1
The numerous small, dark regions in this photograph of the Sun are called sunspots.

Procedure

2 **Figure 2** shows how to set up the binoculars for safe viewing of the Sun. Mount the binoculars on the tripod. (If a mounting bracket is not available, secure the binoculars with masking tape.) Use scissors to cut a hole in the cardboard the same size as the lens to allow light to pass through one lens. Cover the second lens with cardboard. Draw a circle about 12 cm in diameter on the screen.

3 Without looking at the Sun, aim the binoculars at the Sun. **Do not look through the binoculars.** Place the screen below the eyepiece of the binoculars. Move the screen until a clear image of the Sun fills the circle on the screen. Determine east, west, north, and south.

✎ (a) Record the starting time in your log book.

(b) Draw a diagram of the image you observe on the screen. Show parts that are lighter or darker. Indicate which way the Sun's image moves. Label the directions east, west, north, and south on your diagram. Include any other features you observe. Alternatively, photograph the image and make notes.

✎ (c) Record the stop time.

4 Repeat your observations a few more times, on other sunny days. It is a good idea to always draw the image of the Sun the same size.

✎ (a) Record what you see.

(b) Describe any changes you observe.

Analysis and Communication

5 Analyze your observations by answering the following questions:

(a) Describe how you figured out which edge of the Sun was north, and which was west.

(b) What is the advantage of using a tripod in this investigation?

(c) Describe the process you used in getting the Sun's image to be clear.

(d) How could the binoculars' shadow be used to aim them at the Sun?

(e) Describe the features of the Sun that you observed.

(f) What evidence would you need to answer the initial question?

6 Present your discoveries as a poster or, if you used a digital camera, as an audio-visual presentation.

Exploring

1. The predicted years of maximum sunspot activity are late 2001, 2012, and 2023, and the years of minimum activity are 2007 and 2018. Predict how what you saw would compare with what may be observed during the years of maximum and minimum sunspot activity.

Reflecting

2. Do you think that studying the Sun is important to a technological society such as ours? Why or why not?

Challenge

What considerations do you need to incorporate in your design for a space colony to prepare for potential sunspot activity?

Cardboard covers one lens and leaves the other exposed. The cardboard must be large enough to cast a shadow on the screen.

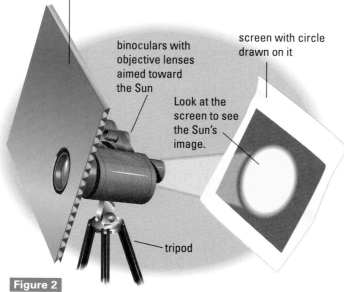

binoculars with objective lenses aimed toward the Sun

screen with circle drawn on it

Look at the screen to see the Sun's image.

tripod

Figure 2
A safe way to view an image of the Sun

The Brightness of Stars

What are some of the factors that affect the brightness of ordinary light bulbs? Does the brightness change as your distance from the bulb increases? Does the power of the bulb affect its brightness? Are there other factors? If you think about these questions, you may get clues about how the brightness of stars depends on certain factors.

Part 1: The Stars in Cassiopeia

Question

What affects the apparent brightness of a light source?

Hypothesis

1 Write a hypothesis predicting which of Cassiopeia's stars are brightest. (You could refer to star maps for information.)
(4A)

Materials

- observation sheet
- flashlight covered with red cellophane
- pencil
- binoculars (optional)

Procedure

Do not go out alone to a dark area without permission from your parent or guardian. Make sure to dress warmly for night observations.

2 On a clear night, position yourself in an area well away from lights, with a clear view of the sky. After about 10 min, your eyes will be adapted to the dark. Make sure your flashlight, observation sheet, and pencil are ready.

3 Locate the constellation Cassiopeia.

(a) Sketch the stars that you can see in the constellation. Name as many of the stars as possible. Add any other relevant observations.

4 Rank the stars you see in Cassiopeia in order of their brightness to your eyes. Use 1 for the brightest, 2 for the second brightest, and so on.

Analysis and Communication

5 Compare your rankings of star brightness with those of your classmates. Discuss as a class what problems you think there may be with your method of ranking star brightness. Explain how you think your method could be improved.
(9C)

6 Your teacher will provide you with a list of the brightness of the stars in Cassiopeia.

(a) How do your rankings compare with the rankings on the list?

7 Compare any two stars that you observed.

Part 2: The Brightness of Light Sources

Question

What affects the apparent brightness of a light source?

Hypothesis

8 Write a hypothesis about the factors that might affect the apparent brightness of a light source. If the light source were a star, would those factors be the same? Explain your answer.

Materials

- solar cell
- galvanometer
- dolly or cart with wheels
- ray box
- metre stick or tape measure
- graph paper

Procedure

9 Copy **Table 1** into your notebook.

Table 1

Bulb number	Distance from bulb (m)	Reading on galvanometer
1	?	?
?	?	?

10 Set up the equipment as shown. Place the ray box at one end of the room. Cover all the windows to prevent stray light from affecting the results.

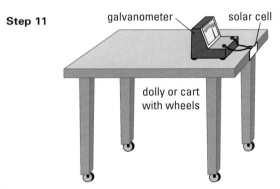

Step 11
galvanometer
solar cell
dolly or cart with wheels

11 Place the galvanometer apparatus 1 m from the ray box bulb or at some other distance suggested by your teacher.

✎ (a) Record the electric current, with the units indicated by the galvanometer needle.

12 Repeat step 11 at distances of 2 m, 3 m, and 4 m from the ray box. (Your teacher may suggest different distances.)

✎ (a) Record your readings.

13 Repeat steps 10, 11, and 12 using a second bulb of different power.

✎ (a) Record your readings.

14 Create a graph that has the electric
(7B) current on the vertical axis and the distance from the ray box on the horizontal axis.

✎ (a) Plot the data for each bulb from your table of observations. Label the two lines on the graph.

Analysis and Communication

15 Examine the graph you made in Part 2.

(a) Describe the relationship you see between the galvanometer reading and the distance from the light source.

(b) Describe the relationship between the observed brightness of a light source and the power (or size) of the source.

16 Suppose that two stars, A and B, give off the same amount of light. From Earth, star A appears to be 100 times brighter than star B.

(a) What would you conclude about the distances of these two stars from Earth?

17 Use what you have learned about the brightness of light sources to explain the apparent brightness of the stars in Cassiopeia.

Exploring

1. Use your graph to predict what the reading would be if the galvanometer was moved to distances other than those you have already used. Extend the investigation to check your answer.

Reflecting

2. Is it possible to predict the relative distances to any two stars in the sky, using only what you have learned in this investigation? Explain.

Characteristics of Stars

We know that the Sun has planets orbiting it, and life exists on at least one of those planets. There are billions of other stars in the universe and there is indirect evidence that some of these are also orbited by planets. It would be exciting to discover that life exists on these planets. To estimate the chances of finding life elsewhere, astronomers begin by studying the characteristics of stars. These characteristics include colour, temperature, size, and brightness.

Figure 1

Star colours vary from blue (the hottest) to red (the coolest). Star sizes vary from supergiants to dwarfs. The Sun is bigger than about 95% of the stars.

The Colour, Temperature, and Size of Stars

You can tell that the element of your electric stove is hot because it glows. As it gets hotter it changes colour: red, then orange. Kitchen stoves rarely get hot enough to display other colours of the spectrum, but you have probably heard about metals being heated until they are white-hot, to make them soft enough to be bent and moulded.

The colour of a metal gives us some information about the amount of energy it has: it indicates the temperature of the metal. In the same way, scientists have discovered that the colours of stars tell us something about their temperature. A relatively cool star glows red; a very hot one glows bluish-white or even blue. **Table 1** lists the approximate temperature ranges of different colours of stars and gives examples of stars we can see in the sky. **Figure 1** shows the relative sizes of various stars.

Table 1

Colour	Temperature Range (°C)	Example(s)
blue	25 000–50 000	Zeta Orionis
bluish-white	11 000–25 000	Rigel, Spica
white	7500–11 000	Vega, Sirius
yellowish-white	6000–7500	Polaris, Procyon
yellow	5000–6000	Sun, Alpha Centauri
orange	3500–5000	Arcturus, Aldebaran
red	2000–3500	Betelgeuse, Antares

Figure 2

A spectroscope splits light energy into a spectrum of colours.

Spectroscopes and the Electromagnetic Spectrum

Scientists use special devices to look closely at the light given off by the Sun and other stars. One of the most useful instruments that astronomers use is the **spectroscope** (**Figure 2**), a device that splits light into a pattern of colours so we can see them as separate lines of colour.

This pattern of colours is called the **visible spectrum**. You may have seen one as a rainbow or when looking through a prism. The visible spectrum is a small part of the electromagnetic spectrum.

Scientists have found that when a chemical element is heated or energized, it gives off energy that shows a unique spectrum when viewed through a spectroscope. Each element tested has its own spectrum (**Figure 3**).

Scientists have also used the spectroscope, attached to a telescope, to look at stars. Much of what we know about stars today has resulted from using the spectroscope. The spectrum of a star can tell us which chemical elements make up the star, how much of each element the star contains, the temperature of the star, and in which direction the star is moving relative to Earth.

Figure 3
Each element that is heated or energized produces its own spectrum that can be seen when you look through a spectroscope.

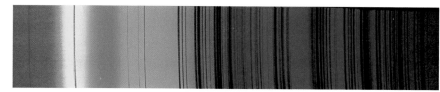

a light from the Sun

b sodium light

c hydrogen light

The Brightness of Stars

You have learned how stars can be classified according to their colour, temperature, and size. Stars can also be classified by their age, distance from Earth, or brightness.

Almost 2200 years ago, the Greek astronomer Hipparchus developed the idea of classifying stars by their brightness. He divided stars into six categories. The brightest stars were called first-magnitude stars, and the faintest stars were called sixth-magnitude stars. Astronomers still use this classification system. Since more advanced sky-watching tools have improved our ability to see fainter stars, the magnitude scale set up by Hipparchus has been revised. Astronomers now use the word "magnitude" in two ways.

Apparent magnitude refers to the brightness of a star as it appears to us. This is the magnitude recorded by Hipparchus, or by you, if you are looking at the sky at night. In fact, two stars that have the same apparent magnitude can actually be giving off very different amounts of light. One star may simply be much closer to Earth than the other star. The term **absolute magnitude** refers to the *actual* amount of light given off by a star at a standard distance. Astronomers calculate the absolute magnitude of stars by determining how bright stars would appear if they were all the same distance from Earth.

Did You Know ❓

Astronomers have discovered some very unusual stars and star systems. Binary stars are pairs of stars that revolve around each other.

A simple example will demonstrate how astronomers compare magnitudes. Imagine that you are looking at two lights, one a flashlight located close to you, the other a bright floodlight that is far away (**Figure 4a**). Both lights appear to have the same brightness; this means that they have the same apparent magnitude. What can you do to compare their absolute magnitudes? Simply move both light sources so they are the same distance away from you (**Figure 4b**). Then you observe that the floodlight has a much brighter absolute magnitude than the flashlight.

Since we cannot move the stars around to test their absolute magnitude, astronomers have had to develop an indirect method of measuring their magnitude. They measure a star's apparent magnitude and its distance from Earth and calculate the absolute magnitude.

The Sun has the brightest apparent magnitude of any star because it is the closest star to Earth. But the Sun's absolute magnitude is only the absolute magnitude of an average star. Some stars, if they were as close to Earth as our Sun, would be nearly one million times brighter than the Sun. Others would be only one-millionth as bright as the Sun.

Figure 4

To the observer in **a**, the flashlight and the bright floodlight have the same apparent magnitude. In **b**, when the two light sources are the same distance from the observer, the floodlight is brighter. It has a much brighter absolute magnitude.

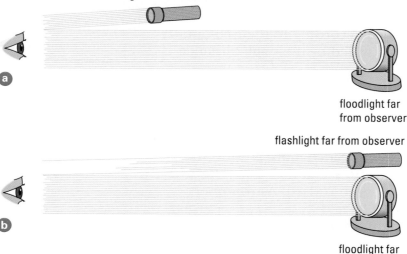

flashlight close to observer

a

floodlight far from observer

flashlight far from observer

b

floodlight far from observer

Understanding Concepts

1. How is the colour of a star related to its temperature?

2. Explain why a cooler star could actually appear brighter than a hotter star.

3. Astronomers use two systems of magnitude to measure the brightness of a star. Which system of magnitude would be more useful in comparing how two stars appear in the sky? Explain why.

4. One night, you observe two stars that have the same apparent magnitude. Could these two stars be giving off different amounts of light? Explain.

5. What effect would pollution in the atmosphere have on a star's
   (a) apparent magnitude?
   (b) absolute magnitude?

6. (a) What instrument does an astronomer use to determine the spectrum of a star?
   (b) Why is using this instrument better than using only a telescope to view the spectrum?

Exploring

7. Who were Hertzsprung and
   (3A) Russell? What is the Hertzsprung-Russell diagram? Research how they developed their diagram, and how it helps astronomers determine the absolute magnitude of stars.

Challenge

Which characteristics of stars can you incorporate in your planetarium display?

Galaxies and Star Clusters

If you look at a map of your province, you see cities, towns, and villages. These are places where people live close together. Between these places are large rural (country) regions where people are scattered quite far apart. Similarly, if you look at a model of the universe, you see different-sized groups of stars with different characteristics.

Galaxies

In Section 14.1, you learned that a galaxy is a huge collection of gas, dust, and hundreds of billions of stars. These stars are attracted to each other by the force of gravity, and they are constantly in motion. Astronomers can see galaxies as far away as the power of their telescopes will permit.

We are part of the Milky Way Galaxy. You might be able to see the Milky Way, in the summer or the winter, looking like a trail of milk spilled across the night sky (**Figure 1**). Astronomers estimate that there are at least 400 billion stars in the Milky Way Galaxy. The Milky Way is roughly disk-shaped, with our Sun located near the outer part of the disk (**Figure 2**). The thicker inner region of the disk is called the central bulge of the galaxy, where the stars are so numerous that they appear very close together even though they are separated by large distances. Most of the stars outside the bulge are arranged in long ribbons, called arms, which curve around the bulge. The entire Milky Way Galaxy is rotating around the bulge.

The Milky Way Galaxy is called a spiral galaxy because of its circular, spiral shape. **Figure 3** shows photographs, taken from the edge-on rather than face-on, of three other spiral galaxies.

Did You Know

Gravity is a force exerted by all objects but is only really noticeable if at least one of them is very large. A dropped cup falls to Earth because it is within Earth's gravitational field and is pulled "down." Meteors crash into the Moon because the Moon's gravity attracts them. The planets orbit the Sun because it has a huge gravitational field. Any large objects close to each other in space will attract each other because of the force of gravity.

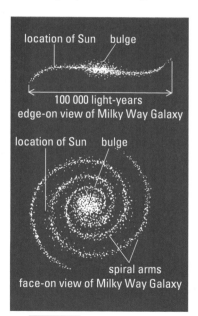

location of Sun   bulge

100 000 light-years
edge-on view of Milky Way Galaxy

location of Sun   bulge

spiral arms
face-on view of Milky Way Galaxy

Figure 2
The spiral arms of our galaxy contain great concentrations of stars. Our Sun is one of the many stars in the less concentrated regions between the spiral arms. If our Sun were in the central bulge of the galaxy, what do you think our night sky would look like?

a This giant spiral galaxy resembles the Milky Way Galaxy.

Figure 3

b This spiral galaxy is coloured to show young giant stars (blue), which are hotter than older stars.

c An example of a barred-spiral galaxy, coloured to show the central bar. The spiral arms come out from the ends of the bar-shaped area that contains the galaxy's central bulge.

There are other shapes of galaxies besides the spiral shape and barred-spiral shape. Some are elliptical galaxies and others that have no distinct shape are called irregular galaxies (**Figure 4**).

Figure 4
Two views of the Large Magellanic Cloud: an irregular galaxy about 160 000 light-years away

Unusual Galaxies

As astronomers see farther into the distance, they find answers to questions about the universe, but what they discover makes them ask more questions. Below are some observations that astronomers continue to research.

- Some galaxies appear to be in the process of colliding and recombining, tearing stars away from each other. Sometimes small galaxies are swallowed by larger ones.
- Some violent galaxies emit far more energy than average galaxies.
- Strangest of all are **quasars**, objects that look like faint stars but emit up to 100 times more energy than our entire galaxy (**Figure 5**). The word "quasar" was taken from the expression "quasi-stellar radio source," which means a starlike object that emits radio waves.

Did You Know ?

There are billions of galaxies in the universe. Some galaxies that gather together in groups are called galaxy clusters or superclusters. Our Milky Way Galaxy belongs to a group called the Local Group and to the supercluster called the Virgo Cluster.

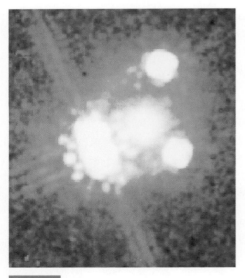

Figure 5
Quasars produce huge amounts of energy, yet appear only as faint points of light.

Try This **Model a Spiral Galaxy**

If you look at a photograph of people swimming under water, you can tell which way they are swimming by how their hair streams out behind them. In a similar way, astronomers can look at images of a galaxy and draw conclusions about its movement from the shape and position of its arms. You can make your own model of a spiral galaxy.

1. Plan a method for modelling a spiral galaxy, using a clear beaker or glass container, water, a stir stick, and a drop of food colouring.

2. Write your own hypothesis for what you predict you will observe.

3. Show your plan to your teacher. After getting approval, carry it out. Be sure to record your observations carefully. Include a diagram.

4. Explain your observations and how they apply to a spiral galaxy.
(8A)

Star Clusters

Groups of stars that are fairly close and travel together are called **star clusters**. These clusters may have as few as 10 stars or as many as a million: too few to be called a galaxy. Smaller star clusters are found in the main parts of the Milky Way Galaxy. Larger star clusters are found just outside the main parts of the galaxy. If galaxies are like cities and towns, then the small star clusters are like neighbourhoods and the large star clusters are like suburbs.

One of the more interesting sights to observe in the sky is the star cluster called the Pleiades, in the constellation Taurus (**Figure 6**). With the unaided eye you can see up to six or seven stars, and many more with binoculars or a telescope.

Figure 6

a The Pleiades is a star cluster.

b To locate the Pleiades, begin by finding the winter constellation Taurus. It is near the constellation Orion.

Capella

GEMINI

Procyon

TAURUS · Pleiades

Aldebaran

Betelgeuse

ORION

Sirius

Rigel

CANIS MAJOR

Understanding Concepts

1. How are galaxies classified? Draw and label an example of each.

2. Arrange the following in order of size, starting with the largest: star cluster; galaxy; universe; star; planet.

3. A certain star is located 60 light-years away from us. Which galaxy do you think this star is in?

Exploring

4. Calculate the speed at which the Sun is travelling around the central bulge of the Milky Way Galaxy. Use the equation: speed = distance/time. Assume that the Sun travels in a circular path of radius 1.38×10^{17} km, taking 225 million years for each trip around the central bulge. (If you change the units of time to hours, your answer will be in km/h.)

5. Make a model of the Milky Way Galaxy. You could use a sheet of polystyrene, toothpicks, and modelling clay, or any other materials approved by your teacher. Draw diagrams of the top and side views.

Challenge

Much of the pioneering work in investigating globular star clusters was done by a Canadian astronomer, Helen Sawyer Hogg. Research about her life and contributions.

Chapter 14 Review

Key Expectations

Throughout the chapter, you have had opportunities to do the following things:

- Compare modern scientific views of the universe with the beliefs of various cultures. (14.1)

- Recognize, describe, and compare the major components of the universe. (14.1, 14.7, 14.9, 14.10, 14.11)

- Describe the Sun and its effects on Earth. (14.7, 14.8, 14.10)

- Investigate distances and properties of celestial objects, and organize, record, analyze, and communicate results. (14.3, 14.5, 14.8, 14.9)

- Formulate and research questions related to the nature of the universe, and communicate results. (all sections)

- Determine how astronomers compare distances and sizes in the universe. (14.3, 14.4, 14.5, 14.9, 14.10)

- Describe and explain how data provided by ground-based and satellite-based astronomy contribute to our knowledge of the Sun and other objects. (14.1, 14.4, 14.6, 14.8, 14.10)

- Describe and evaluate the possible impact of future human exploration and exploitation of planets. (14.2)

- Explore careers related to the exploration of space. (Career Profile)

- Organize and record information in an appropriate manner. (all sections)

- Communicate using appropriate language and formats. (all sections)

KEY TERMS

absolute magnitude	radio telescope
apparent magnitude	reflecting telescope
chromosphere	refracting telescope
corona	solar flare
Earth-centred universe	solar prominence
electromagnetic	spectroscope
spectrum	star cluster
galaxy	Sun-centred solar
light-year	system
nuclear fusion	sunspot
observatory	triangulation
photosphere	visible spectrum
quasar	

Reflecting

- "The universe is always changing. The objects in it are in continual motion." Reflect on this idea. How does it connect with what you've done in this chapter? (To review, check the sections indicated above.)

- Revise your answers to the questions raised in Getting Started. How has your thinking changed?

- What new questions do you have? How will you answer them?

Understanding Concepts

1. Make a concept map to summarize the material that you have studied in this chapter. Start with the word "universe."

2. Explain why the distances to planets are given in kilometres, whereas the distances to stars are given in light-years.

3. (a) What is the process that produces energy inside the Sun?

   (b) How fast does light from the Sun travel through space?

4. (a) Which theory do we accept today: the Sun-centred theory of the solar system or the Earth-centred theory of the universe?

   (b) How do these two theories differ?

5. Copy these distances into your notebook. Beside each number, indicate whether the distance would be described as intergalactic, interstellar, interplanetary, or interstudent.

   (a) 0.001 km   (c) 100 million km

   (b) 10 billion   (d) 100 million
      billion km   million km

6. Name and use illustrations to describe three types of telescope.

7. How do the wavelengths of radio waves compare with the wavelengths of visible light? Use your answer to explain some of the characteristics of radio telescopes.

8. Explain the difference between the absolute and apparent magnitudes of stars, using our Sun as an example.

9. The Milky Way is a strip of stars that we can see most easily during winter and summer. Describe how our view of the Milky Way would change if our solar system were located in the bulge of the Milky Way Galaxy.

10. Compare galaxies and star clusters.

11. Using examples, explain why humans could not travel to the stars at the speeds reached by today's spacecraft.

12. Make a list of characteristics used to describe stars. Beside each characteristic, indicate how the Sun compares with other stars.

13. What features of quasars make them unique?

14. Describe effects that particles from solar flares have on Earth.

15. (a) List models that have been used in this chapter to help you understand concepts.

    (b) Describe other models that you think might help explain other concepts.

16. Between 400 and 500 years ago, ideas about the universe began to change. What main factors caused that change?

17. A store has several telescopes on display.

    (a) How can you judge which telescopes are reflecting and which are refracting?

    (b) If you were buying a telescope, which type would you prefer? Why?

18. Some cultures in ancient times were able to catalogue stars even better than most people today, although they did not know about most of the stars in the universe. Explain why.

19. Write at least five questions about the universe that you would like to have answered.

20. Why is it difficult for scientists to measure interstellar and intergalactic distances? How do they solve these problems?

Applying Skills

21. An astronomer uses the diameter of Earth's orbit as a baseline to estimate the radius of Saturn's orbit. As shown in **Figure 1**, the angles to Saturn, taken six months apart, are both 84°. Use a scale diagram to find the distance from Saturn to the Sun. (When astronomers use triangulation to measure such large distances, they take into consideration the movement of the distant object.)

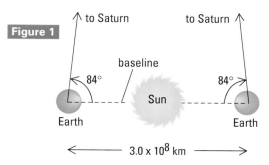

Figure 1

22. What advice would you give to someone who is just starting to learn how to draw scale diagrams?

23. What advice would you give to someone who wants to use binoculars to obtain an image of the Sun on a screen?

24. What safety precautions must you take when observing the Sun? Why are these precautions so important?

25. The Milky Way Galaxy is approximately 90 000 light-years across. Using correct scientific notation, express this distance in both astronomical units and kilometres. (1 light-year = 9.46×10^{12} km; 1 a.u. = 1.5×10^{8} km)

26. When using triangulation to determine distances indirectly, what can you do to improve the accuracy of the measurements?

Making Connections

27. Look back at Table 1 in Section 14.10. Describe how you think conditions on Earth would differ from conditions today if the Sun were (a) as hot as a blue star, and (b) as cool as a red star.

28. The thin atmosphere at the tops of mountains may be a great advantage for telescopes, but it is a great disadvantage for people who work there. People often become light-headed or even ill when they first arrive at a high elevation. Research altitude sickness, and write a brief report describing the causes and effects of this problem.

The History of the Universe

1 This is the Cat's Eyes Nebula. Using technology that allows them to see events that happened billions of years ago, scientists have collected evidence that nebulas are one of the stages in the lives of stars. What is this amazing technology? Are the stars we see at different stages of their existence? What do we know about the "birth" of stars? Are stars still being formed? Do stars die?

2 The nebula shown here, called the Lagoon Nebula, is a nursery for new stars. Scientists have proposed that planets sometimes form at the same time as stars, so our solar system is probably not the only planetary system in the universe. What is their evidence? Why is it so difficult to prove that there are planets travelling around other stars? How old is the solar system? How did the planets form? Will they last forever?

3 Canadian astronomers join other astronomers in the search for answers to questions about the universe. Did the universe have a beginning? If so, when was it? What evidence do we have that supports theories of the origin, growth, and future of the universe? If astronomers are unable to find direct evidence to support their hypotheses, how could they use indirect evidence? ➤

Reflecting

Think about the questions in **1**, **2**, **3**. What ideas do you already have? What other questions do you have about the history of the universe? Think about your answers and questions as you read the chapter.

Try This Using Indirect Evidence

Many scientists, from space scientists to chemists, are sometimes unable to make direct measurements or collect direct evidence for their investigations. Instead they use many kinds of indirect evidence. You have already used an indirect method to measure distances that you could not measure directly. Now you can discover features of something you can't see by gathering indirect evidence.

1. At a station set up by your teacher, a flat, horizontal board hides something from your direct view. Observe carefully as one student at a time rolls a small steel ball along the surface of the board.

2. Continue observing until you are satisfied that you can describe what is hidden.

3. Draw a diagram of what you think is hidden beneath the board.

4. What is "indirect evidence"?

The Life of a Star

We say that stars have a "life" because they evolve from clouds of gas and dust and follow a predictable series of stages: they begin (are "born"), develop, and end ("die"). Each life may take billions of years or more, a length of time that is difficult for humans to imagine.

Using modern instruments and looking at stars billions of light-years away, astronomers have been able to examine millions of stars at different stages in their lives. Scientists have put together many pieces of information, like the pieces of a puzzle, to develop a theory of the life of a star.

One of the first pieces of information came in 1678 with Isaac Newton's discovery of **gravity**. We now know that gravity is a force that pulls objects toward each other. The more mass an object has, the more gravity it exerts, so the Sun has stronger gravity than Earth. But the force gets smaller as the distance between objects increases.

All stars begin their lives in **nebulas**, which are huge clouds of dust and gases, mainly hydrogen and helium (**Figure 1**). The dust and gases swirl around, breaking into clumps and contracting because of gravitational forces. As the clumps bump into each other and get bigger, their gravity gets stronger, and they are able to attract more particles and pack more tightly together. Eventually, the clumps are dense and hot enough for nuclear fusion to start. They have become new stars.

Just as no two clouds on Earth are identical, no two stars are identical. Although scientists understand the stages of stars fairly well (**Figure 2**), our knowledge of the details is far from complete. That is part of the excitement of studying astronomy. All these ideas are theories that are undergoing confirmation, development, or change based on further observations and evidence.

Figure 1

The Trifid Nebula has two separate regions. The pink region has a high concentration of hydrogen, and the bluish-violet region is reflecting light from nearby stars. The dark patches are caused by dust clouds located between the nebula and Earth.

Try This Graphing Gravity

What factors affect the force of gravity between two objects? Make a prediction. The equation for two objects interacting is $F = (Gm_1 m_2)/d^2$, where F is the force in newtons (N), G is a constant (6.67×10^{-11} Nm^2/kg^2), m_1 is the mass of object 1, m_2 is the mass of object 2, and d is the distance between the centres of the objects.

1. Find the masses and radii of the planets in our solar system and calculate the force of gravity on a 1.0-kg rock at the surface of each one.

2. Determine the force of gravity on the rock when its distance from the centre of Earth is 2, 4, 6, and 8 Earth radii. Plot a graph to show your discoveries.

3. How accurate was your prediction?

Figure 2 **The Life of Different Types of Stars**

Type of Star	Small or Medium Star (mass the same as the Sun's or less; the most common type of star)	Large Star (mass about 10 times the mass of the Sun; rare)	Extremely Large Star (mass about 30 times the mass of the Sun; very rare)
Birth	Forms from a small- or medium-sized nebula.	Forms from a large nebula.	Forms from an extremely large nebula.
Early Life	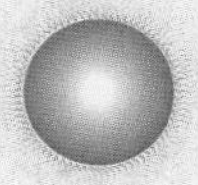 Gradually turns into a hot, dense clump that begins producing energy.	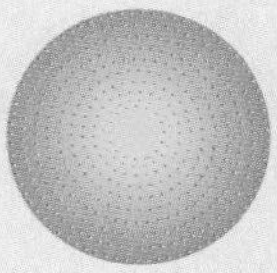 In a fairly short time turns into a hot, dense clump that produces large amounts of energy.	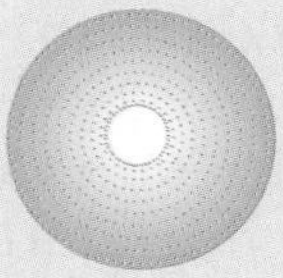 In a very short time turns into a hot, dense clump that produces very large amounts of energy.
Major Part of Life	 Uses nuclear fusion to produce energy for about 10 billion years if the mass is the same as the Sun's, or 100 billion years or more if the mass is less than the Sun's.	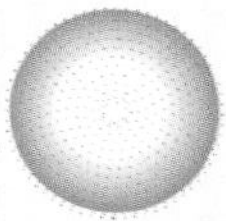 Uses nuclear fusion to produce energy for only a few million years. It is perhaps 5000 times as bright as our Sun.	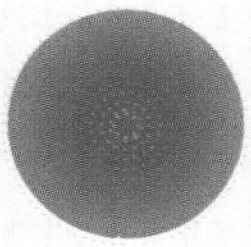 Uses nuclear fusion to produce energy for only about one million years. It is extremely bright.
Old Age	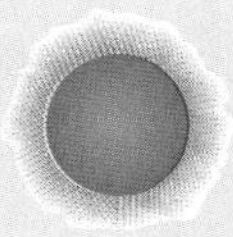 Uses up hydrogen and other fuels, and swells up into a large, cool red giant.	Uses up hydrogen and other fuels, and swells to become a red supergiant.	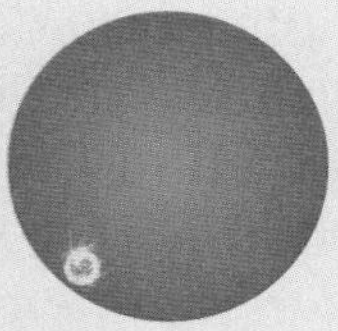 Uses up hydrogen and other fuels, and swells to become a red supergiant.
Death	Outer layers of gas drift away, and the core shrinks to become a small, hot, dense white dwarf star.	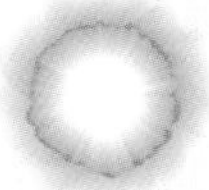 Core collapses inward, sending the outer layers exploding as a supernova.	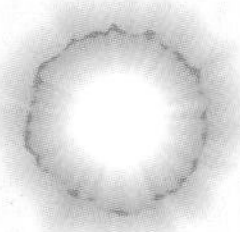 Core collapses, sending the outside layers exploding in a very large supernova.
Remains	White dwarf star eventually cools and fades.	 Core material packs together as a neutron star. Gases drift off as a nebula to be recycled.	Core material packs together as a black hole. Gases drift off as a nebula to be recycled.

Note: These drawings are not to scale.

Giants and Dwarfs

You can see in **Figure 2** that when a star nears the end of its life it runs out of hydrogen and other fuels needed to produce energy. When this happens, the pressure holding the star together becomes reduced, so the star swells up while at the same time cooling down. Thus, in "old age" the star becomes larger and red. Stars the size of the Sun or smaller become **red giants**, while stars with masses 10 times (or more) larger than the Sun's become **red supergiants**.

A star the size of the Sun or smaller is said to "die" when the nuclear reactions die down, the core shrinks, and the outer layers of the star drift away. The remaining material becomes a **white dwarf**, a small star with a higher temperature than red or yellow stars.

Supernovas

A **supernova** is an enormous explosion that occurs at the end of a large star's life. By this stage, the star has used up the fuels needed to keep producing energy by the process of nuclear fusion. With reduced pressure on the core, the core collapses inward to become either a neutron star or a black hole (described below). At the same time, shock waves cause the outer layers to explode outward in a rapidly expanding nebula of dust and gases.

Supernovas are rare events. A famous one was observed in the year 1054 and was recorded in both China and India. It was so bright that people could see it even in daylight. It was visible for about 21 months. The gases spreading out from this supernova now form the Crab Nebula, which can be seen in the constellation Taurus (**Figure 3**).

Since the telescope was first invented about 400 years ago, only one supernova has ever been seen with the unaided eye. This event, now called Supernova 1987A (**Figure 4**), was discovered by a Canadian named Ian Shelton working at an observatory in Chile, South America.

Neutron Stars

When a star about 10 times the mass of the Sun dies, the resulting core is called a **neutron star**, an extremely dense star composed of neutrons. The neutrons are so tightly packed,

Figure 3
The supernova that created the Crab Nebula was observed in 1054. Other nebulas seen in the sky provide evidence that supernovas have occurred many times.

Did You Know

When our Sun eventually swells into a red giant star, its outer layers will grow to be about 100 times its present size swallowing up Mercury, Venus, Earth, and maybe even Mars.

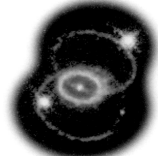

Figure 4
Supernova 1987A, photographed three years after the supernova was first seen by Ian Shelton. Astronomers are observing this supernova with great care. What they learn from their observations may help them improve their theories about the lives of stars.

 **Modelling a Black Hole**

In a large, open area, securely tie a one-holed rubber stopper onto a 1-m length of string. Whirl the stopper around a finger (without hitting anybody!), allowing the string to wind up on your finger until the stopper crashes into the finger.

1. In this model, what acts as the black hole? What acts as the material near it? What force represents the force of gravity?

2. What happens to the motion of the rubber stopper as it gets closer and closer to the finger? Relate this to what happens to material near a black hole.

with no space between them, that a cupful of a neutron star would have a mass of millions of kilograms. A **pulsar** is one type of neutron star. Pulsars emit pulses of very high energy radio waves. Pulsars are very small, only about 20 km in diameter, but very dense, with about the same mass as a normal star. Pulsars rotate while emitting energy in the form of light and radio waves, like the rotating light of a lighthouse—the effect is the energy reaches us as pulses. Today more than 400 pulsars have been identified.

Black Holes

When a star about 30 times the mass of the Sun dies, the resulting core is called a **black hole**, which is a small, very dense object with a force of gravity so strong that nothing can escape from it. Even light cannot be radiated away from its surface. That is why it is called a black hole (**Figure 5**).

Because light cannot escape the surface of black holes, they can exist undetected. Yet a Canadian astronomer, Tom Bolton, was able to confirm the existence of black holes. What was his evidence? He observed a star that appeared to be circling an invisible partner and noticed that this partner was emitting X rays as it drew matter toward itself from the star. These X rays confirmed Bolton's suspicions that the "invisible partner" was a black hole. The word "hole" is misleading because it sounds as though there is nothing there; it actually has a huge amount of matter packed into a sphere only a few kilometres across. It is so dense that a handful of it would have a mass of perhaps 10 billion cars! This explains its strong gravitational pull. Perhaps "dark body" would be a better name for this extremely rare type of object.

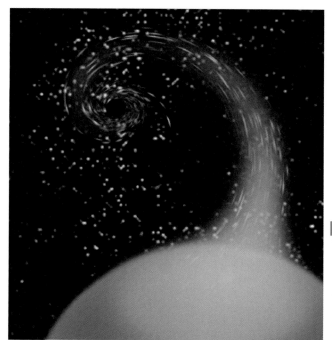

Figure 5

This is an artist's view of what happens near a black hole. As it twirls around, the hole attracts materials from a nearby star toward it. Why is this an example of indirect evidence?

Understanding Concepts

1. Describe how a star forms.
2. Describe the differences between the life of a low-mass star and that of a star 10 times the Sun's mass.
3. What is a supernova?

Exploring

4. Design and make a set of three-dimensional models to illustrate the life of one of the types of stars described in **Figure 2**.
5. Design a model of a black hole. Your model could use a circular magnet, metal washers, and thread. If your teacher approves your design, try it. In what ways is it accurate?

Reflecting

6. Describe why observing evidence of a black hole is an example of indirect evidence.
7. Think of circumstances in your life where you draw conclusions using indirect evidence. In each situation, what do you have to know already, before you believe the indirect evidence?

Challenge

How does the use of models help to communicate the information in the challenge you have chosen? What kind of models are most effective?

The Origin of the Planets

You know how difficult it is to see the planets in our own solar system. Now imagine looking for planets travelling around stars other than our own: other **planetary systems**. Planets are smaller than stars and do not emit light, so the task is not easy. Astronomers are continually using newer and better technology and making use of both direct and indirect evidence to search for other planetary systems in the universe. Learning about the formation of planets at various stages will help us better understand the development of the solar system and our own planet.

Evidence of Planetary Systems

What evidence do astronomers have of other planetary systems? Shortly after its launch into orbit in 1983, the Infrared Astronomical Satellite or IRAS made some exciting discoveries. It detected a large cloud of tiny particles in orbit around the star Vega. This was the first direct evidence that solid matter exists around a star other than our own Sun. Astronomers thought that the clouds of particles could be an early stage of the development of planets. Four months later, IRAS discovered solid material orbiting another star.

More recently, the Hubble Space Telescope has obtained images of planetary systems being formed. Astronomers also have indirect evidence of other planetary systems. The pull of gravity between a star and a nearby planet causes a slight wobble of the star's motion. This wobble can be detected by observing the star's spectrum.

Formation of the Solar System

Figure 1 shows the steps that may have occurred in the formation of the Sun, planets, and other parts of our solar system. In its early stages, our solar system was part of a nebula consisting mainly of hydrogen and helium. Minute particles of matter, such as iron, rock, and ice, made up about 1% of the nebula. This solid matter formed as a result of neighbouring supernova explosions from nearby stars.

Scientists think that our Sun formed about 4.6 billion years ago. At the same time, smaller clumps of matter began to contract in the outer regions of the solar nebula. Attracted by the force of gravity, the clumps formed planets. This theory explains the formation of the gas giants

Figure 1

Astronomers think there were three main stages in the formation of the solar system.

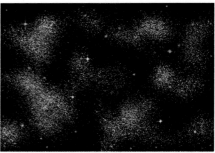

a Gravity caused components of a rotating nebula to unite. Because of its rotation, the nebula flattened out as it contracted.

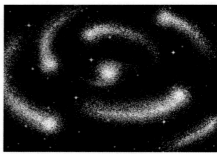

b As the process continued, a bulge formed toward the centre. This bulge later became the Sun. Some of the dusty disk of cooler material further from the centre gathered together to form smaller chunks.

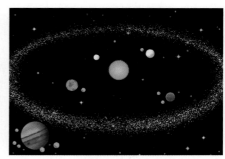

c These smaller chunks away from the bulge gradually grew into larger chunks, which eventually formed into planets.

Did You Know ?

The belt of asteroids located between Mars and Jupiter likely would have formed into a planet if Jupiter's strong force of gravity hadn't prevented the particles from gathering together.

(Jupiter, Saturn, Uranus, and Neptune). The gas giants are composed mainly of hydrogen and helium gases: the same gases that make up much of the Sun. These planets probably formed in a way similar to the Sun—through the clouds of gases contracting due to gravity. But what about the terrestrial planets? What is the current theory about their formation?

The Terrestrial Planets and Minor Bodies

Mercury, Venus, Earth, and Mars are composed largely of rock and iron, with little hydrogen and helium. In the early stages of solar system formation, the gases in the inner regions of the solar system were probably too hot to condense. The young Sun produced enough solar wind (bursts of charged particles) to blast most of the hydrogen and helium out of the inner regions of the solar system. Only the chunks of heavier matter—iron and rock—were left behind (**Figure 2**).

The terrestrial planets formed from those solid chunks. As chunks of solid matter circled the Sun, they eventually collided with one another and joined together. During millions of years, some of the chunks grew in size until their mass was large enough to have strong gravity. The more massive chunks pulled in all the matter in the space around them. They eventually became the planets.

Over a long time (perhaps 10 million years), most of the matter in the solar system became concentrated into the terrestrial planets and gas giants. The remaining matter makes up the asteroids, meteoroids, and comets— the minor bodies. Studying these minor bodies provides information about the early stages of our solar system, before most of the matter formed into the planets.

Pluto's origin is not certain as it has characteristics of a minor body, but is too far from the Sun to be explained by the theory of formation for the terrestrial planets. It may be more like a giant comet nucleus. The research is continuing.

Figure 2
The history of the terrestrial planets is

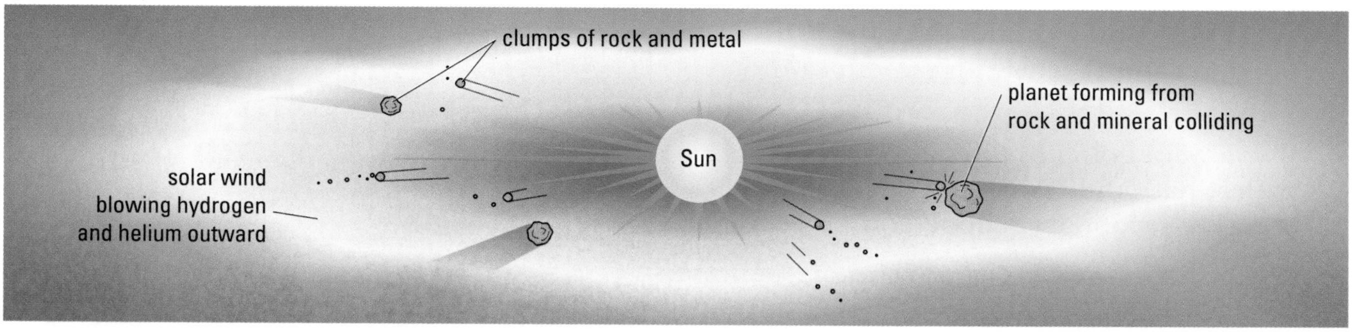

Understanding Concepts

1. What force is responsible for bringing together the particles found in space?

2. Describe, in order, the stages of the formation of the solar system.

3. **(a)** Which of the planets in the solar system are thought to have formed in a similar manner to the Sun?

   (b) Why is it unlikely that the other planets formed this way?

4. **(a)** What are the minor bodies of the solar system?

   (b) What do astronomers hope to learn from the minor bodies?

5. Assume that 10 years from now astronomers discover planets travelling around hundreds of different stars. Are the planets they discover more likely to be gas giants or terrestrial planets? Explain.

Exploring (3A)

6. Research the Kuiper Belt. Why is its discovery a good example of a theory being supported by observations? Why do astronomers think Pluto and Neptune's moon Triton may have come from the Kuiper Belt?

7. Research and report on the Oort cloud. Explain why comets provide an opportunity to study what the solar system may have been like before the planets formed.

8. Use an Internet resource to research recent discoveries of planetary systems. Make a visual display of your discoveries.

clumps of rock and metal

planet forming from rock and mineral colliding

Sun

solar wind blowing hydrogen and helium outward

Space Artist

Paul Fjeld gives a thumbs up sign as he steps out of the high performance jet in which he has just pulled 6 *g*'s. His call sign is "Fingers," and he flies in jet fighters whenever he can as part of his research: Paul Fjeld is a Canadian artist interested in space.

Between 1971 and 1973 the *Montreal Star* newspaper sent a young artist to cover the last three launches of *Apollo*. He convinced NASA to include him in their official Art Program; soon Fjeld got to go where even the press could not, including the *Apollo* simulator. His greatest thrill was working in Mission Control during the 1975 *Apollo/Soyuz* mission where he was made welcome and encouraged to "document what we were doing in a different way—for history."

Fjeld's first painting for NASA was of a crippled Skylab module with torn thermal shield and damaged solar arrays. Working with engineers from the manufacturer, he had to assess what the damage might look like and prepare a visual image for the press. The Canadian Space Agency has commissioned Fjeld to create large images of the *International Space Station*, the Canadarm, the SPDM (Figure 2, p. 499), and the various satellites. His work has appeared in *National Geographic* and on the cover of *Aviation Week*.

His favorite spacecraft is the *Apollo Lunar Module* (seen in the painting with Armstrong and Aldrin). The skin of the insect-like robot is a paper-thin wrapping of aluminum. It is a great subject to paint against the blackness of space. Actor Tom Hanks hired Fjeld to recreate that lunar moment on a huge set for the TV series *From the Earth to the Moon*.

What's next for Paul Fjeld? Another painting? A script for a movie? Whatever it is, it will almost certainly involve his passion: space.

> "I want to paint a picture as if you are actually there."

Explore

1. What skills and attitudes does the artist need to capture what the camera cannot capture?

2. If you were going to follow a similar career path, what would you study at secondary and post-secondary school, and why?

3. When might humans again land on the Moon? Provide a rationale for your answer.

A Model of the Expanding Universe

In this activity, you will use a model to illustrate the concept of an expanding universe. As you use the model, think of ways it might be improved.

Materials
- 1 round balloon
- black, fine point, felt tip pen
- metric tape measure

Procedure

1 Copy **Table 1** into your notebook.

Table 1

	Measured distances (cm)			Calculated distances (cm)	
	A to B	B to C	C to D	A to C	B to D
"Orange" stage	?	?	?	?	?
"Basketball" stage	?	?	?	?	?

2 Blow up the balloon to the size of an orange. While your partner keeps the balloon at that size by pinching the opening, use the felt tip pen to mark the balloon with four dots, each separated by about 1 cm. Measure the exact distances between the dots with the metric tape. Label the dots in order, A, B, C, D.

✎ (a) Record the measured distances between the dots in your data table.

3 Now inflate the balloon until it is about the size of a basketball. Measure the distances between the dots.

✎ (a) Record the measured distances in your data table.

4 Calculate the distances A to C and B to D for both the first and second stages.

✎ (a) Record these calculated distances in your data table.

5 Look at how much the distances A to B, B to C, and C to D increased when you inflated the balloon to the second stage.

(a) Compare those increases.

6 Suppose you continued to blow up the balloon.

(a) How do you think the change in the distances A to B, B to C, and C to D would compare with the change from the first stage to the second stage?

7 Imagine that you are standing on dot A while the balloon is expanding.

(a) Which dot would appear to be moving away from you most quickly?

(b) Which dot would appear to be moving away from you most slowly?

(c) Would any dot appear to be moving toward you?

8 Imagine that the dots on your expanding balloon are galaxies of stars in an expanding universe.

(a) What difference would you expect to find in how quickly nearby and distant galaxies are moving away from Earth?

9 Now imagine that you are on a planet in galaxy B looking at a planet in galaxy C.

(a) Do you think the spreading of the galaxies would appear to be any different from these two locations? Explain.

(b) If galaxy B is the Milky Way Galaxy, is it correct to say that galaxy B is the centre of an expanding universe? Explain.

Reflecting

1. Although the balloon model was useful to illustrate the concept of an expanding universe, it also had a major limitation. Identify the limitation and suggest a way to overcome it.

Evidence of an Expanding Universe

A balloon is a useful model in helping us imagine how the universe expands: as the balloon expands, closely spaced dots on the surface spread apart slowly; dots that are far apart spread apart more quickly. However, as a model, an expanding balloon has limitations. For example, galaxies are not found on a skin or membrane; instead, they are scattered throughout the universe.

How can we tell whether a galaxy is moving away from us? After all, the galaxies are very far away.

Before looking at evidence of an expanding universe, let's review the properties of light. Light is a form of energy that travels as a wave. Each colour has a wavelength, which is the length of one wave. The visible spectrum (**Figure 1**) consists of different colours, ranging from the longest wavelengths (red) to the shortest wavelengths (violet).

You already know that different stars have characteristic colours. The same is true for galaxies: each galaxy emits its own spectrum or range of colours in a pattern that astronomers recognize. As early as 1912, astronomers observed evidence that the galaxies are moving away from Earth and from one another. This evidence came from looking at the light spectra (patterns of colours) given off by nearby galaxies. The spectrum of a moving object in space is different from the spectrum of an object that is not moving. An example will help explain this.

Imagine a duck bobbing up and down on the surface of a pond (**Figure 2**). Suppose the duck represents a source of energy. The ripples that spread out from the duck represent waves of energy emitted by the source. The distance from the top of one ripple to the top of the next ripple represents one wavelength.

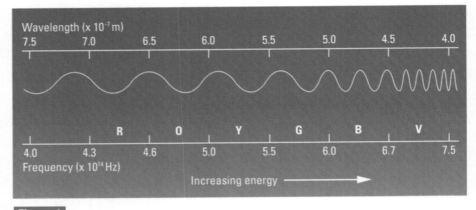

Figure 1
The visible spectrum

Figure 2

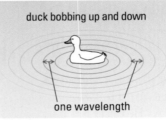

a This model shows an energy source (the duck) sending out waves (the water ripples) with a constant wavelength (the distance from the top of one ripple to the top of the next).

b Side view of the waves.

Figure 3

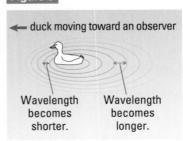

a The wavelengths become shorter in one direction and longer in the other direction.

b A side view of the waves created by a bobbing duck moving to the left.

Now, imagine yourself standing on the shore, watching the duck. If the duck swims toward you, the ripples in front of it are squeezed together and therefore have shorter wavelengths (**Figure 3**). Meanwhile, the waves behind the duck are spread farther apart and so have longer wavelengths. Even if you could not see the duck itself, you could observe from the changing wavelengths which way the duck was moving. Shorter wavelengths indicate that the duck is moving toward you; longer wavelengths indicate that the duck is moving away from you. Of course you have to know what the wavelengths would be if the energy source were stationary, relative to you.

When astronomers look at a distant source of light, such as a galaxy, through a spectroscope, they recognize the *pattern* of the spectrum but observe that the *colours* of its lines are not the same as they would expect. The lines are moved toward the red end of the spectrum (**Figure 4**) meaning that they have a longer wavelength. Searching for an explanation, astronomers concluded that the light has a longer wavelength than normal because the galaxy is moving away from us. This movement, or shift, into the red end of the spectrum is called **red shift**.

The light energy from all distant galaxies shows red shift. However, there are different amounts of red shift. The greatest shifts occur for galaxies that are farthest away from us. From this information, we can infer that all galaxies are moving away from us and away from each other. But what is causing this movement? Why is the universe expanding? Astronomers are investigating the answers to these questions.

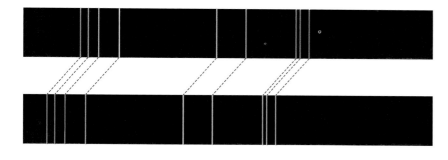

Figure 4
The top diagram represents the nine lines of a spectrum emitted by a stationary galaxy. The second diagram represents the same nine lines of the same spectrum emitted by the galaxy if it were moving away from us rapidly. The wavelengths of all the lines in the spectrum have shifted (moved) toward the red end. The bottom white light spectrum is shown for reference.

Understanding Concepts

1. Copy **Figure 5** into your notebook. It is a pattern of ripples. Label the ripples that have shorter wavelengths and those that have longer wavelengths. Indicate the direction of movement of the "object" causing the waves. Explain your answer.

Figure 5

2. What does "red shift" mean?

3. What evidence do astronomers have that the universe is expanding?

4. If astronomers were to observe a "violet shift" for a certain star, what could they infer? Explain why.

5. If the duck in **Figure 3a** were travelling from the top of the diagram toward the bottom, what would you observe? Draw a diagram and explain why.

Exploring

6. How could you use water in a flat container and a bobbing finger to illustrate changes in wavelength as the source of the waves moves?

7. The scientific name for the cause of red shift is the "Doppler effect." Research this effect and relate it to what you have learned about red shift. (The Doppler effect relates to all types of waves, including sound waves. For now, focus on light waves.)

The Origin of the Universe

Do you enjoy solving mysteries or putting the pieces of a puzzle together? If so, perhaps a career in astronomy is ideal for you! One of the main goals of many astronomers is to solve the mystery of the origin of the universe. These astronomers search for an answer to the question, "How did the universe begin?"

If stars go through cycles, beginning and ending as nebulas, does the entire universe also go through some type of cycle? Scientists do not know the answer to this question, but they have gathered enough information to formulate a theory. This theory is based on evidence that the size of the universe is not constant but is expanding. This is the accepted theory, but it may be modified as more evidence is collected and analyzed.

The study of the origin and changes of the universe is called **cosmology**.

The Big Bang Theory

If the galaxies are moving apart, then presumably they used to be closer together. Still earlier, they must have been even closer together. Continue to move backward in time in your imagination and you will reach what we call "time zero." Scientists estimate time zero as being between 10 and 15 billion years ago. At that time, all the matter of the entire universe was packed together into one small, extremely dense, hot mass under enormous pressure. The event that occurred when the universe emerged from this state of enormous density and temperature is known as the Big Bang (**Figure 1**). Scientists use the **Big Bang theory** to describe the beginning of the universe that we know.

Scientists have more than just the red shift of distant galaxies as evidence of the Big Bang. They have also collected data that suggest that, even today, the entire universe is "glowing" from that initial explosive event, just as a piece of firewood continues to glow even after the fire has been put out. In 1965, for example, two scientists made an

Big Bang

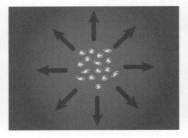
galaxies forming

the present:
galaxies dispersing

Figure 1

These diagrams illustrate the Big Bang theory of the formation of the universe. The density of the material that made up the universe shortly after the Big Bang was immense. One cubic centimetre of the material is estimated to have had a mass of one billion kilograms!

accidental discovery using a radio telescope. They detected unexpected radiation coming from all directions in space. After checking that there were no defects in their equipment, they concluded that this faint, unknown energy is all that remains of the enormous amount of energy that exploded outwards at the time of the Big Bang. The intense radiation from the tremendous explosion has faded to only a faint "whisper" after travelling for thousands of millions of years through space.

The Big Bang theory can be used to explain our observation that the universe is expanding. But we still don't know everything about how the universe began. For example, while scientists can describe what they think occurred a fraction of a second after the Big Bang, they do not know what the universe was like (or even if it existed) just before the Big Bang explosion. Changing our theories about the universe is a continuing scientific process as we try to explain the new discoveries that astronomers make year after year (**Figure 2**).

However, the Big Bang theory is not the only way that the origin of the universe has been explained. In the past, many cultures have held different beliefs of their own. These beliefs, however, are not the same as scientific theories—they cannot be tested, explored, and either supported or disproved by evidence. But each is a kind of model that helps people understand how the world works.

Understanding Concepts

1. How does the Big Bang theory explain an expanding universe?

2. How does the development of the Big Bang theory illustrate the scientific process?

3. How could you use all the students in your class to act out a model of the expanding universe shortly after the Big Bang?

Exploring

4. Older theories about the universe include the oscillating theory and the steady state theory. Find out about these theories, then write an article for a science magazine describing one of them. Draw illustrations representing the theories to go with your article.

5. People in all parts of the world have had, and still do have, other ideas about the origin of the universe. Research and compare at least two contrasting cultural views. Present your findings in a creative way.

Reflecting

6. How does a scientific theory, such as the Big Bang theory, differ from a belief?

Challenge

The Big Bang theory is a difficult idea to conceptualize. What model or diagrams will you use to explain it to the general public?

Figure 2
The Antennae galaxies are a pair of galaxies whose collision has resulted in a firestorm of star birth activity. It is proving to be an excellent "laboratory" for studying the formation of stars and star clusters.

The Hubble Deep Field

Figure 1 shows one of the most amazing images ever taken of outer space. It is called the *Hubble Deep Field*. When you look at it, you are peering back in time up to 8 billion years, into a part of the sky that appears no bigger than a grain of sand resting on a fingernail of your outstretched hand! Although this picture only shows you a tiny piece of the sky, it contains a lot of information. The only stars in the photograph are the objects that appear to have spikes. (This feature results from the wave nature of light.) All the other objects are galaxies.

Figure 1
This image was captured by the Hubble Space Telescope when it was aimed toward a part of the sky above the Northern Hemisphere.

(a) What can you infer about the number of galaxies in the universe? (In your answer, be sure to consider the concept of the "grain of sand" mentioned above.)

(b) Which types of galaxies can you see?

The image in Figure 1 was obtained by the Hubble Space Telescope over a period of about 100 h. It is similar to a time-lapse photo that might be taken here on Earth.

(c) Why is time-lapse imaging important when viewing faraway galaxies?

The "optical" portion of the Hubble Space Telescope includes a very large mirror, specially engineered with great precision. Unfortunately, there was an error in the manufacturing that was not detected until the telescope was launched: the curvature of the mirror was out by the thickness of a human hair, so it could not focus as well as had been hoped. Fortunately, astronauts were able to correct the problem in 1993 by adding a secondary mirror during a series of space walks. It was a delicate but successful operation. Other instruments have since been added to Hubble, improving its performance still further.

(d) Why are more instruments added to the Hubble Space Telescope, instead of being put into orbit separately? What might be the advantages of each alternative?

Advanced technology is needed to keep a telescope, travelling at almost 30 000 km/h while orbiting Earth, aimed steadily at one tiny spot for such a long time.

(e) Why wouldn't the telescope naturally stay pointed in the same direction? Make some predictions.

This image sees deeper into space than any previous image. That is why it is called the "Hubble Deep Field." The farthest objects, which are galaxies, are up to 8 billion light-years away. Galaxies 8 billion years ago had less structured shapes than galaxies now appear to have. This feature helps support the Big Bang theory.

(f) Explain how the less structured shape of early galaxies supports the Big Bang theory.

Astronomers observe evidence of several galaxies colliding with each other shortly after the Big Bang.

(g) Propose a reason why galaxies in the past were more likely to collide than they are now.

Studying such images in more detail will help scientists better predict what might happen to the universe in the future.

(h) If you were a scientist studying images like this, what would you look for? Why?

Understanding Concepts

1. How was the universe different 8 billion years ago than it is now?

2. Explain how this image allows scientists to "peer back in time."

Exploring

3. If you were responsible for aiming the Hubble Space Telescope to take a new Deep Field image, in which direction would you aim it? Give reasons.

4. Research the problems with Hubble's mirror and how they were corrected. Draw a labelled diagram to explain the repairs.

5. Research the costs involved with the Hubble
3A Space Telescope. Do you think it is a good use for the money? Explain reasons for your opinion in a brief letter to a science magazine.

6. Find out what other Deep Field images have been taken. View the images on a NASA web site.

How Astronomers Use Computers

When you imagine an astronomer at work, you probably think of someone gazing up at the sky or perhaps peering through a telescope. Actually, today's astronomers spend most of their time in front of computers. The computers analyze the vast amounts of data that are collected by various instruments and reorganize the information into forms that help the astronomer pick out what is important (**Figure 1**). Without computers, most research in astronomy would be slowed down because we would not be able to analyze information as fast as we are collecting it. Computer technology has made a large contribution to the extensive knowledge and understanding we now have of Earth and its place in the universe.

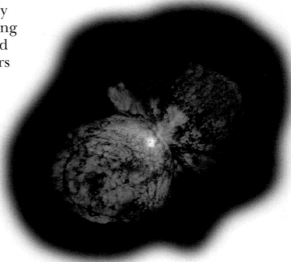

For example, an astronomer can study the glowing image of an exploding star on a computer screen. The computer created the image from data gathered by a series of radio telescopes. The information is stored in the computer, which creates a radio-wave "picture" of the exploding star—what you would see if you had eyes that could see radio waves. Images like this give astronomers a new view of the universe. The astronomer may use the computer to display the image in different ways to show certain features more clearly and to take measurements and make comparisons with previous data.

Figure 1

This is a false-colour image generated by computer using the image from a telescope. It shows Eta Carinae—a dying star.

Computers are also essential in the radio search for extraterrestrial life. Only by using computers can scientists listen to all parts of space, at many different radio frequencies, for long periods of time. Scientists hope to detect radio signals that unknown beings might have sent from other parts of the universe.

The Hubble Space Telescope (**Figure 2**) is constantly sending computerized images of space to Earth. Astronomers can display these images on their computer screens, explore certain features, and make calculations. They can also store the data, keeping it on hand for easy retrieval, should it be needed again.

As well as helping astronomers analyze, display, and store data, computers are used to control the operation of astronomical instruments. Imagine the problem of focusing on a galaxy when your telescope is moving along with Earth as it rotates. In the past, astronomers had to constantly turn their telescopes in the opposite direction to Earth's rotation. Now, computers do all the calculations and control the movement of the instruments.

Simulations are another important use of computers. Scientists may apply known equations to simulate conditions deep within the Sun or test models of the formation of the universe. Simulations can help

Did You Know ?

Earth may be the only planet in our solar system with the right conditions to support life. But does life exist only on this one planet in the entire universe? Scientists look at many factors when trying to determine if a planet can support life. In one ongoing investigation, a radio telescope scans the sky 24 h a day. It is linked to computers that sift through all the incoming radio signals in a search for signs of intelligent civilizations. So far, no signals have been observed, but research continues. This kind of search is called SETI, which stands for Search for ExtraTerrestrial Intelligence.

This view of the Hubble Space Telescope shows the solar panels that change light energy into electrical energy. Earth is seen in the background.

people learn about the universe and our own fragile and beautiful planet as it travels in one tiny corner of a galaxy that contains about 400 billion stars.

Computers record large amounts of data, analyze them, create pictures, do mathematical calculations, and operate complicated astronomical instruments. They help to bring the past into the present. But to some astronomers, the most exciting thing computers do is electronically link scientists all over the world. Through their computer networks, astronomers can share ideas, data, calculations, computer images, and the latest scientific information.

Understanding Concepts

1. Why is the concept of an astronomer continuously peering through a telescope not valid today?

2. In a chart, compare three uses astronomers make of computers today with how information used to be managed.

3. Suggest how a computer could help an astronomer analyze images of a star cluster.

Making Connections

4. How can computers help in time-lapse imaging of components of the sky? How do these images enhance our understanding of the history of the universe?

Exploring

5. List concepts in the study of astronomy that you think you could understand better if you could watch computer simulations of the concepts. If a simulation program is available, try it out and evaluate its benefits.

Challenge

How can you use your computer as a tool to help you in the challenge you have chosen?

Chapter 15 Review

Key Expectations

Throughout the chapter, you have had opportunities to do the following things:

- Outline the current theory explaining the origin, evolution, and fate of the Sun and other stars. (15.1)

- Describe these components of the universe: nebulas; supernovas; neutron stars; black holes. (15.1)

- Outline the current theory explaining the formation of the solar system. (15.2)

- Describe and evaluate scientific evidence of origin and evolution of the universe. (15.1, 15.4, 15.5, 15.6)

- Plan ways to model answers to questions about the history of the universe, and communicate results. (15.3)

- Formulate and research questions related to the history of the universe, and communicate results. (all sections)

- Describe how computers are used to enhance our understanding of the universe. (15.7)

- Identify contributions of Canadian scientists to the exploration of the universe. (15.1)

- Explore careers related to the exploration of space. (Career Profile)

KEY TERMS

Big Bang theory	pulsar
black hole	red giant
cosmology	red shift
gravity	red supergiant
nebula	supernova
neutron star	white dwarf
planetary system	

Reflecting

- "Scientists use models and simulations to visualize and explain the dynamic processes that form the universe." Reflect on this idea. How does it connect with what you've done in this chapter? (To review, check the sections indicated above.)
- Revise your answers to the questions raised in Getting Started. How has your thinking changed?
- What new questions do you have? How will you answer them?

Understanding Concepts

1. Make a concept map to summarize the material that you have studied in this chapter. Start with the words "Big Bang."

2. (a) Arrange the following stages of the life of a star in the order in which they occur: black hole; supernova; nebula; red supergiant.
   (b) Is there more than one possible answer in (a)? Explain.
   (c) Which type of star would go through these stages?

3. (a) What is a nebula?
   (b) How are nebulas important to the formation of stars and planets?

4. What process causes stars to reach "old age"?

5. What force leads to the formation of planets?

6. Why are white dwarf stars much more common than black holes in the universe?

7. Describe the relationship between a red giant and a white dwarf.

8. The following is a list of stages in the lives of stars: white dwarf, supernova, red giant, neutron star, black hole.
   (a) Which of these stages, if any, will the Sun pass through in the future?
   (b) Explain why the Sun will not pass through the other stages.

9. (a) Since the invention of the telescope, how many supernovas have been observed with the unaided eye?
   (b) How do astronomers benefit from studying supernovas?

10. Describe the Big Bang theory.

11. Why is it difficult to observe planets in orbit around distant stars?

12. Describe the current theory of the formation of the planets of our solar system. Illustrate your answer with drawings.

13. (a) What is red shift?

(b) How does red shift provide evidence for the Big Bang?

14. What evidence besides red shift supports the Big Bang theory?

15. As scientists continue to observe the spectra of stars and galaxies, what do you think they would conclude if they observed the following?

(a) Red shift continued.

(b) Red shift was no longer observed for any star or galaxy.

(c) A shift was observed toward the violet end of the spectrum.

16. State, with reasons, whether you agree with this statement: "Nebulas provide a good example of recycling in the universe."

Applying Skills

17. If you look at photographs of the Moon, Mercury, Mars, and the moons of other planets, you will see evidence of collisions.

(a) What is this evidence?

(b) How does this evidence support the theory of the formation of the solar system?

(c) What evidence of collisions on Earth supports the theory?

18. Describe examples in this chapter of indirect evidence used by astronomers.

19. Make a physical model or a computer model to illustrate one of the concepts covered in this chapter, such as the life of a star, or the expanding universe.

20. The discovery of red shift supports the theory of an expanding universe. However, it is indirect evidence, not proof, that galaxies are moving away from each other. What other interpretations could there be of red shift?

21. Write at least five questions about the universe that you would like more complete answers to.

22. Choose one of the questions from your list in 21. Describe how you would go about finding the answer(s) to that question.

23. Research the most recent ideas about black holes and prepare a presentation.

24. The light from two distant galaxies, A and B, is examined through a spectroscope. The two line spectra have a similar pattern of lines, but the lines of Galaxy A's spectrum have slightly longer wavelengths than those of Galaxy B.

(a) Draw what their two spectra might look like.

(b) What might you conclude about the motions of the two galaxies, relative to Earth?

25. The star probe *Stardust* was launched in 1999 to collect dust from the tail of comet Wild-2. Research the purpose and current status of this mission.

Making Connections

26. Design a visual image that could be attached to a space probe leaving the solar system. The image should communicate information about ourselves and our planet to intelligent aliens discovering the probe. Remember, they would not be able to read our written communications! Research plaques that have been attached to probes that have already been launched.

27. Assume you have unlimited funds to carry out investigations to test one of the theories presented in this chapter. Which theory would you test? How would you test it? Predict your results.

28. Some radio telescopes are used to try to detect intelligent life that may exist on planets revolving around other stars in the universe. What other methods could you suggest for trying to find evidence of intelligent life beyond Earth (extraterrestrial intelligence)? Explain your answer.

29. Do you expect computers will be more or less important in the study of astronomy in the future? Why?

1 This is what the *International Space Station (ISS)* will look like when it is completed. The *ISS* will be the biggest technological project ever built. It involves the cooperation of many different countries, including Canada. What is Canada's role in the *ISS*? What would it be like to work up in space? *ISS* is a kind of satellite. What is the purpose of other satellites?

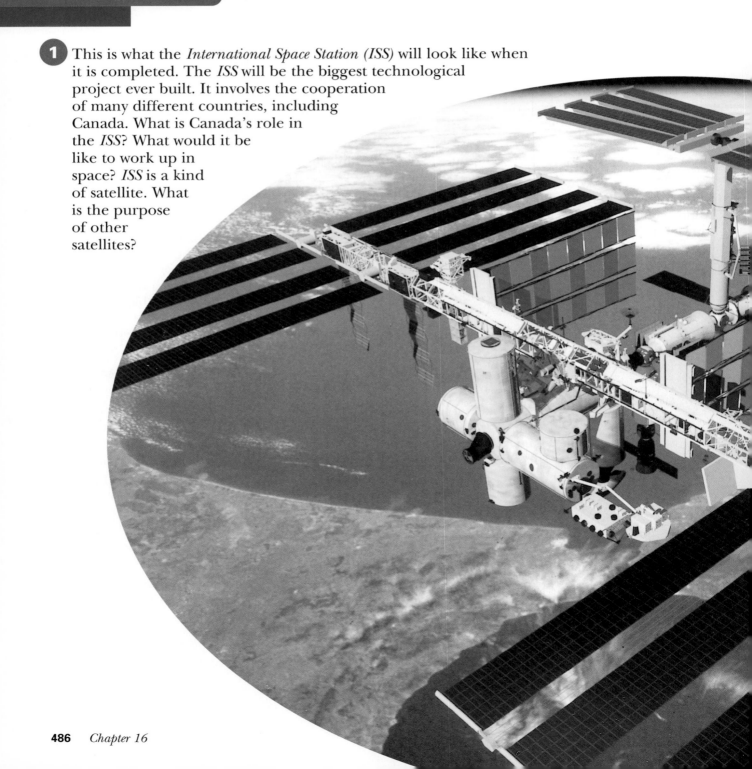

2 Although Canada's population is relatively small, our contribution to space technology is impressive. How does Canada contribute to space research, telecommunications, and international space programs? What are the positive and negative effects of this contribution on Canadians and on the rest of the world?

3 Astronauts travelling in a spacecraft in orbit around Earth look as if they are floating around in the craft. Why does this happen? How does this type of motion affect the human body and contents of the craft?

Reflecting

Think about the questions in **1**, **2**, **3**. What ideas do you already have? What other questions do you have about satellites and humans in space? Think about your answers and questions as you read the chapter.

Try This A Packing List

Make a packing list of all your personal requirements for a three-month working tour on the Space Station. Your essential needs are air, warmth, water, and food.

1. Consider a typical day in your life here on Earth. Make a list of the activities you perform and what you need to carry out those activities.

2. Decide which of your daily needs on Earth you could live without for three months in space. Alter your list accordingly.

3. Consider what you would require in space that you would not require on Earth. Again, alter your needs list accordingly.

4. Assume that you will be in space for three months. Complete your needs list for the trip.

5. Compare your list with that of other students or groups. If necessary, make changes to your list.

Getting into Space

A basic human characteristic is curiosity—our drive to explore. People have crossed the oceans, climbed Earth's highest mountains, and walked on the Moon. The biggest single frontier left to explore is **outer space**, which is everything outside of Earth's atmosphere. The atmosphere—the air and everything it holds, such as water vapour—extends upward from Earth's surface, gradually becoming less dense until it finally becomes almost nothing at an altitude of about 150 km. This is where "outer space" begins.

Comparing Aircraft and Spacecraft

Can an aircraft travel in outer space? Can a spacecraft travel in Earth's atmosphere with no engines operating, as it can in outer space? The answer to both questions is no. To find out why, we need to consider the characteristics of each type of craft.

An **aircraft** is a vehicle that travels through the air; examples include jet airplanes, propeller airplanes, and helicopters. The engines of these vehicles must operate continuously to keep them above the ground. The engines use oxygen in the air to burn the fuel they need to travel. Most aircraft do not travel higher than about 20 km above Earth's surface. If they go too high, there is not enough oxygen to support the burning of the fuel. Also, air is needed to support the aircraft. Whether it is a helicopter or a plane, air keeps the aircraft aloft. The rotors or wings are curved so that the air pressure below them is greater than that above (**Figure 1**), so they are forced upward. Without the air, an aircraft would plummet toward Earth.

A **spacecraft** is a vehicle designed to travel in the near vacuum of space, usually 200 km or more above Earth's surface (**Figure 2**). Once the spacecraft rises above the atmosphere and is travelling extremely fast parallel to Earth's surface (which tends to make it leave Earth), its forward speed balances the pull of gravity (which tends to make things fall toward Earth). When these two opposing pulls are perfectly balanced, the spacecraft no longer needs to operate its engine full-time in order to travel around Earth. The spacecraft must be above the atmosphere, otherwise air resistance (the friction of the air molecules brushing past the craft) would slow it down.

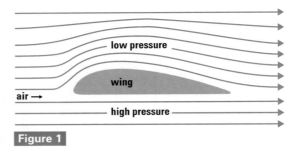

low pressure

wing

air →

high pressure

Figure 1

 **Events in Space Exploration**

Make an illustrated time line of the major events in space exploration. Use reference books and CD-ROM resources to help you design interesting illustrations. Early events include Konstantin Tsiolkovsky's explanation of rocketry in 1903 and Robert Goddard's first liquid-fuelled rocket launch in 1926.

Although many spacecraft travel in orbits around Earth, some go to the Moon or to other parts of the solar system. A spacecraft must carry its own fuel and source of oxygen, and it must travel fast enough so it doesn't fall back to Earth.

Unpiloted and Piloted Spacecraft

Spacecraft can be piloted (with people on board) or unpiloted (also called robotic spacecraft). Satellites are the most common type of spacecraft. As you've learned, a satellite is an object that travels in an orbit around another object; for example, the Moon is a natural satellite of the Earth. Another important type of spacecraft, called a **space probe**, is one with many instruments on board that is sent to discover more about moons, planets, comets, the Sun, and other parts of the solar system.

The first piloted spacecraft was sent into space in 1961 by the former Soviet Union (Russia and its allies). The cosmonaut aboard was Yuri Gagarin. Piloted craft tend to gain more public attention than unpiloted spacecraft, but the added expense of sending humans into space is a disadvantage. Piloted spacecraft always have to be brought back to Earth, whereas unpiloted craft often do not have to return.

One type of piloted spacecraft is the space station. The first one, *Salyut 1*, was placed into orbit in 1971. A newer station is the *International Space Station (ISS)*, shown on page 486. It is designed to stay in space for many years, allowing people to live and work in space. **Figure 3** shows an example of another piloted spacecraft: the space shuttle built by the United States. One task of the shuttle is to take materials and astronauts to and from the *ISS*.

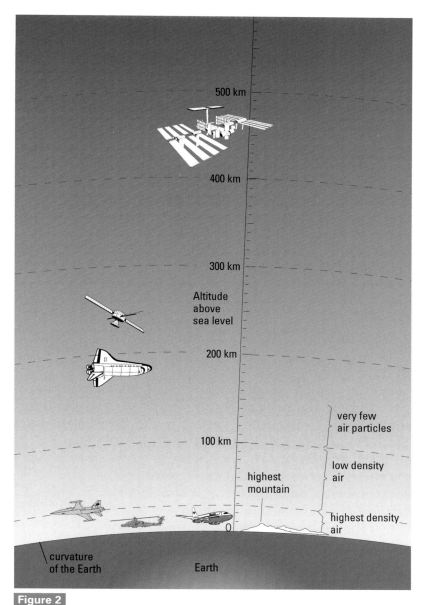

Figure 2

Vehicles in Earth's atmosphere and in outer space beyond

Figure 3

This U.S. space shuttle is about to land. What clues reveal that it is a piloted spacecraft rather than an unpiloted one?

Payloads and Launchers

If you were to throw a ball to a friend, your arm could be called the launcher, and the ball, the payload. In space exploration, a **payload** is any satellite, piloted spacecraft, or cargo launched into space. The **launcher** is the device that carries a payload into space. Its main part is a rocket engine.

What happens if you try to throw a ball straight up? It soon falls back down, pulled by gravity. Large bodies, such as Earth, exert a strong force of gravity on nearby objects. A vehicle can be launched into space only by counteracting Earth's strong force of gravity. The huge force needed to counteract gravity is provided by a powerful rocket engine.

A toy balloon is a simple model that helps explain how a rocket engine works. When the air in an inflated balloon is released, the balloon flies quickly and uncontrollably around the room. As the air under pressure inside the balloon escapes from the open neck, the balloon is forced in the opposite direction. The force that causes an object to move is called **thrust**. In the case of the balloon, the thrust is exerted by the pressure of the gas coming out of the balloon. If the air is escaping to the left, the thrust on the balloon is to the right (**Figure 4**).

Like a balloon, a rocket engine is enclosed except at the nozzle end. Unlike a balloon, a rocket engine operates through the action of two chemicals. **Figure 5** shows that exhaust gases travelling rapidly in one direction cause thrust on the rocket in the opposite direction.

direction of thrust

direction of air

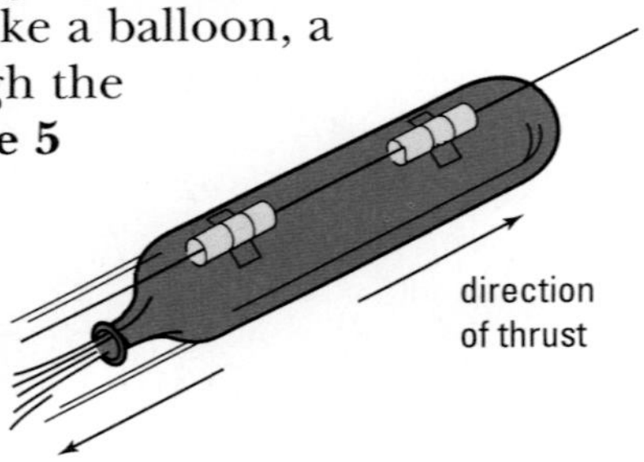

Figure 4

A balloon can demonstrate thrust. To guide the balloon, feed fishing line through two pieces of a drinking straw taped to a blown-up balloon. When the balloon is released, air leaves the open neck rapidly in one direction, creating a thrust on the balloon in the opposite direction.

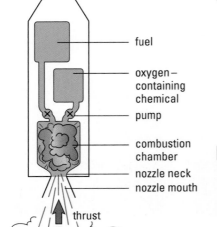

fuel

oxygen– containing chemical

pump

combustion chamber

nozzle neck

nozzle mouth

thrust

exhaust gases

Figure 5

This represents the basic design of a liquid-fuelled rocket engine. The fuel and an oxygen-containing chemical are pumped into the combustion chamber where the fuel burns rapidly, producing gases under high pressure. The gas molecules speed up as they escape through the nozzle neck; then they speed up even faster when they get to the nozzle mouth. These escaping gas molecules exert an upward thrust on the rocket engine.

Understanding Concepts

1. Name the force that must be overcome in order to launch a vehicle into space.

2. Classify each of the following as a payload or a launcher:

   (a) a weather satellite

   (b) a rocket

   (c) an astronaut

3. Describe thrust and how it causes a rocket to move.

Making Connections

4. Why do humans want to explore space? Give reasons to support your views.

5. Suppose you needed raw materials to build a space colony on another planet. What would be some of the advantages of obtaining those materials from the Moon rather than from Earth?

Exploring

6. How could you make a flying balloon into a payload-carrying device? What might an appropriate payload be? With your teacher's approval, try "launching" a rocket balloon (**Figure 4**) with and without the payload. Account for your observations.

16.2 Investigation

SKILLS MENU
- Questioning
- Hypothesizing
- Planning
- Conducting
- Recording
- Analyzing
- Communicating

Launching Water Rockets

A toy water rocket is a safe and easy way to study how rockets work. Although it is powered only by water and air, it works on the same principles as a rocket powered by chemical fuels. Your challenge is to design modifications to make the water rocket go as high as possible.

Materials

- toy water-rocket kit
- instruction manual for the rocket
- clean water
- safety goggles
- astrolabe

Question

1 Ask a question that your experiment will
(2A) answer.

Hypothesis

2 Write a hypothesis and predict how each controlled variable will affect the dependent variable.

Experimental Design

3 Examine the water rocket (**Figure 1**) and read the instruction manual. Design a controlled experiment in which you vary one factor—"controlled variable"—at a time. The height reached by the rocket is the "dependent variable."

(a) Make a list of the steps you would take to safely launch the water rocket, including safety precautions, measurement of variables, and recording your data. Ask your teacher to approve the steps.

Figure 1
After adding some water to a water rocket, you can use the pump to increase the air pressure inside the rocket. What factors do you think control how high the rocket will fly? Which of these could become controlled variables?

Procedure

4 Discuss with your group how you will determine the height the rocket reaches. (Think of triangulation, measuring angles, and drawing scale diagrams.)

(a) Record the steps you will take to measure the height. Ask your teacher to approve these steps.

5 After your design has been approved, carry out your investigation outdoors.

(a) Record your observations, measurements, and calculations.

✋ Operate the rocket outdoors only. Do not aim it at anybody. Stand clear of the line of flight. Wear safety goggles.

Analysis and Communication

6 Analyze your observations by answering the following questions:

(a) How did the height of the rocket change as you altered each controlled variable?

(b) How is thrust obtained in a water rocket?

(c) Explain the factors that affect the thrust.

(d) Write up your investigation in a compete lab report.

(e) Do your results support your hypothesis? Explain.

Making Connections

1. Compare the operations of a water rocket and a rocket fuelled by chemicals. Include both similarities and differences.

Exploring

2. If possible, use simulation software to research rocket launching.

3. Find information about a model rocketry club in your area. Invite a member of the club to share ideas about launching rockets and, if possible, demonstrate a model rocket launch.

Earth-Orbit Satellites

Have you ever watched a "live" event on television as it is happening in a country halfway around the world? Such live coverage on TV is made possible by satellites that continually orbit Earth relaying signals from place to place.

Satellites in Low Earth Orbit

In 1962, Canada became the third country in the world to send a satellite into space. *Alouette 1* was used for scientific research: to study particles in Earth's upper atmosphere by circling above it and collecting data.

Alouette 1 was an example of a satellite in **low Earth orbit**, an orbit at an altitude of between 200 km and 1000 km. At speeds of about 28 000 km/h, such satellites take only about 1.5 h to travel once around Earth (**Figure 1**).

Besides carrying out scientific research, the main function of low Earth-orbit satellites is **remote sensing**: making observations from a distance using imaging devices. We are familiar with images made from light: photographs. They can be recorded using traditional film cameras, but satellites use digital cameras so that the images can be sent back to Earth. Other types of images are created from infrared (heat) waves, radio waves, and other invisible waves of the electromagnetic spectrum. Each type of radiation registers different information and is used to watch for different phenomena (events). For example, radio waves can pass through cloud, so they can be used to create images of events that would otherwise be hidden by cloud cover. Satellites detect these different kinds of waves by using special cameras or antennas. The information gathered by the satellites is beamed to Earth using radio waves (**Figure 2**), which travel at the speed of light. On Earth, the signals are analyzed by computers and converted into visible images. Certain features of the image may have characteristics that make them particularly visible. For example, vegetation appears red in an image constructed from infrared radiation. This makes the image specially useful to foresters monitoring the regrowth of trees in a clear-cut area.

Weather forecasting relies heavily on remote sensing, using images of cloud patterns around the world. With experience, weather forecasters can interpret these images to help predict what the weather will be like locally. They may even save lives by providing information on approaching storms, such as hurricanes (**Figure 3**).

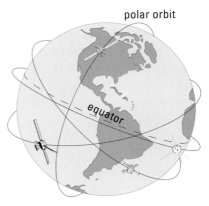

Figure 1

Examples of low Earth orbits may be around the equator, around the poles, or at any angle between the equator and the poles.

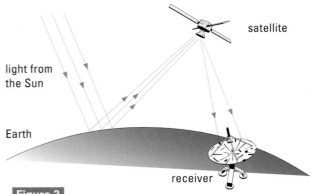

Figure 2

In daylight, visible light from the Sun reflects off Earth's surface or clouds. Digital cameras aboard a satellite capture the reflected light; then antennas on the satellite beam the digital information, using invisible radio waves, to a receiver on Earth.

Did You Know

The cameras aboard some spy satellites are so sensitive that they are able to read letters and numbers about the size of a person's hand from hundreds of kilometres away.

Figure 3

A satellite image showing a hurricane

Remote sensing is also used by the military. For example, some nations have spy satellites that obtain images of other nations' activities on Earth's surface.

Scientists also use images from satellites to study water resources, crop management, forests, insect damage, animal migration, pollution, and fault lines on Earth's surface.

Remote sensing satellites do not detect only naturally occurring radiation. They can also emit their own radiation and then pick up the reflections (**Figure 4**). Computers analyze the reflected signals and convert them into images. The types of radiation most often used are radio waves and microwaves. A Canadian satellite system called RADARSAT, featured on page 496, is used to construct radio-wave images of Earth's surface.

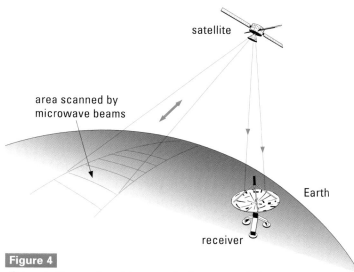

satellite

area scanned by microwave beams

Earth

receiver

Figure 4

This satellite sends out beams of microwaves that can pass through clouds and reflect off Earth's surface. The satellite receives the reflected waves and converts the information to digital format and beams it back to a receiver on Earth.

Satellites in low Earth orbit can also be directed out toward space. To astronomers looking out into space from Earth's surface, the atmosphere acts like a fog, reducing their ability to see the stars, galaxies, and other parts of the universe clearly. Furthermore, there are parts of the electromagnetic spectrum that never get through Earth's atmosphere. Astronomers have known for a long time that they could improve their understanding of the universe if they could observe it from above the atmosphere. Telescopes aboard orbiting satellites, such as the Hubble Space Telescope, now enable them to make observations from space.

Did You Know

Satellite phones and pagers can be used anywhere in the world if they are part of a communications network called Iridium. The system consists of 66 satellites in low Earth orbit (780 km).

Satellites in Higher Orbits

Our lives would be very different if we did not have telephones, televisions, and radios, and if we could not transmit computer data almost instantaneously. These products of the telecommunications industry help us communicate over small and large distances. (The prefix "tele" is from the Greek word *tele*, meaning "far.") People who live and work in remote areas are especially reliant on satellites to keep in contact with the rest of the world.

Satellites in low Earth orbits have a major disadvantage if used for TV communication. Imagine a television satellite dish trying to receive signals from such a satellite. The dish would have to track the satellite as it moved across the sky. Then the satellite would go below the horizon, and all signals from the satellite would be lost. In order for a satellite dish to receive signals constantly, the satellite would have to remain in the same location above Earth's surface.

The orbit of such a satellite is called a **geosynchronous orbit**. "Geo" means Earth and "synchronous" means taking place at the same rate. Since it takes Earth 24 h to turn once on its own axis, a satellite in geosynchronous orbit must travel at the correct speed so that it takes 24 h to orbit Earth. This speed must be carefully calculated, taking altitude into account, so that the satellite neither hurtles off into space nor crashes to Earth. From the ground, it appears that the satellite is not moving at all. The easiest place to control such an orbit is directly above Earth's equator, at an altitude of 36 000 km above sea level, which is much higher than low-orbit satellites. To keep its position at this altitude, the satellite must travel at a speed of 11 060 km/h (**Figure 5**).

Try This Dish Hunt

What are the characteristics of the dishes that receive signals from television communication satellites? To answer this question, look at several satellite dishes in your area, answering these questions:

1. What are the size, shape, and design of the dishes?

2. Which way are they pointed?

3. Are they all pointed in the same direction?

4. Hypothesize why the dishes have the characteristics you observed. In your hypothesis, refer to shapes of curved reflectors, focal length, focal point, and wavelengths of waves.

Did You Know

When Earth passes through a comet's tail, tiny particles of the tail can seriously damage a satellite, tearing its solar panels or even knocking it out of orbit.

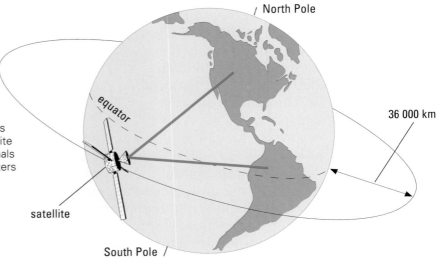

Figure 5

A geosynchronous orbit lies above the equator. A satellite located there receives signals from Earth-based transmitters and sends signals back to Earth-based receivers.

Canada has become a leader in telecommunications. The first satellite for domestic communications put into geosynchronous orbit was Canada's *Anik 1*, launched in 1972. There have been many versions of the *Anik* series, improving the area of coverage and the quality of radio, telephone, and TV signals. A recent technological advance includes a satellite that sends and receives signals from mobile telephones in cars, trucks, ships, and airplanes.

Some satellites not used for TV communication travel in orbits somewhat lower than geosynchronous orbits. For example, 24 satellites used for search and rescue operations travel in 12-hour orbits, 20 000 km above Earth's surface. They form the **Global Positioning System** (GPS). A portable GPS receiver is a small device that receives signals from three or more of the orbiting satellites. It then uses accurate clocks and preprogrammed information to calculate and display its own position. This position could then, if necessary, be relayed by radio to other people, such as a rescue team searching for a downed plane or a lost hiker.

Space Junk

As you have now read, we receive many benefits from orbiting satellites. Unfortunately there are also some disadvantages. One is the huge expense. Another is the accumulation of useless artificial objects orbiting Earth, commonly referred to as *space junk*. Satellites, for instance, are not designed to return to Earth once they become obsolete or malfunction. Another concern is that chunks of satellites and space stations could fall to Earth. There are other pieces of space junk, ranging in size from microscopic to huge pieces: blown-out bolts; bits and pieces from accidental explosions; discarded rocket engines; and even real garbage from piloted craft. These pieces are spread far apart in their orbits, so are difficult to clean up. Countries involved in the space program are discussing what to do about the problem of space junk.

Did You Know

In January 1978, the Russian nuclear satellite *Cosmos 954* crashed into the Thelon Game Sanctuary in the Northwest Territories, narrowly missing some explorers camping in the wilderness. Picking up every piece took months and, although it was somebody else's garbage, Canadian taxpayers had to pay most of the bill.

Understanding Concepts

1. What are the features of a low Earth orbit?

2. What is a geosynchronous orbit, and what type of satellite must use this orbit? Explain why.

3. What is the advantage of having an observatory in space?

4. Using the data given in the text and the equation

    speed = distance/time,

    calculate the speeds (in kilometres per hour) of

    (a) a geosynchronous satellite;

    (b) a satellite with an orbit 200 km above Earth's surface.

    Assume that Earth's radius is 6400 km. Compare your answers with the values given in the text.

Making Connections

5. Who should be responsible for the cleanup of pieces of spacecraft that crash to Earth? Give reasons. Are your answers affected knowing that some of these falling vehicles contain nuclear power plants?

Exploring

6. What else would you like to know about satellites? Make a list of questions, then try to find answers.

7. Search the Internet for satellite images of Ontario. What types of electromagnetic radiation are used to compile each image? What features do they reveal?

Challenge

Identify Canada's role and use of satellites to include in your information package.

Make a three-dimensional scale model of Earth and space beyond it to show the location of various satellites in orbit for your planetarium show.

RADARSAT

In the spring of 1997, people in Manitoba were faced with one of the worst floods in their history. Soon after the flood began, satellite images helped emergency crews plan disaster control and relief (**Figure 1**). These images were provided by RADARSAT, a highly successful satellite system designed, built, and operated by Canadians.

Helping in emergencies is only one of the many uses of RADARSAT, as you will see.

The word **radar** is short for "radio detection and ranging." This means that a radar device emits bursts of radio waves and picks up their reflections to detect objects (detection) and find out how far away they are (ranging). Radio waves travel at the speed of light and can pass through clouds, so they can be used in all types of weather and at night.

(a) What are advantages of radio waves over visible light?

(b) Brainstorm some other uses of radar.

As its name suggests, RADARSAT is a satellite system that employs radar. It looks at features on land and on oceans using radio waves. Bats use a similar system of emitting high-pitched sounds and interpreting the reflections off obstacles or food sources.

In addition to floods, RADARSAT helps in various large-scale emergencies, such as earthquakes, mudslides, ice storms, ice jams, and oil spills on the ocean.

(c) How might RADARSAT help in your local area?

Many industries benefit from RADARSAT images. Many resources, such as oil, natural gas, water, and minerals are found underground. Often the surface features on the ground help scientists predict where these resources are. These surface features are much easier to find using satellite images. Images also let us monitor crop conditions, forests, soil humidity, river flows, fish stocks, and shipping conditions.

Figure 1

This radar image of part of Manitoba has been computer enhanced, so different features show up as different colours. The red line shows the normal course of the river. The blue area shows the area under floodwaters in 1997.

(d) Why might it be better to search for underground resources by using RADARSAT images than by ground surveying?

In order to provide a healthy planet for future generations, humans must learn to protect our environment. Satellites help us monitor the environment and make wise decisions about our actions.

(e) RADARSAT is an expensive system, but it provides great benefits. Is it worth the cost? How do we decide?

Understanding Concepts

1. In a list, summarize the benefits of using RADARSAT.

Exploring

2. Compare the properties of radio waves with the properties of other parts of the electromagnetic spectrum. (For more information about the electromagnetic spectrum refer to Section 14.6, page 449, or use a separate resource.)

3. Research other animals, besides bats, that use a ranging/detection system similar to RADARSAT. Compare the systems.

Tracking Satellites and Other Objects

Have you ever looked up into the sky on a clear evening just after sunset and seen a bright object slowly drifting across the sky? It could have been a satellite, a space shuttle, or perhaps even the *International Space Station*. In this activity, you will perform research to find out when satellites and other objects might be seen, then you can try to observe them and track their paths.

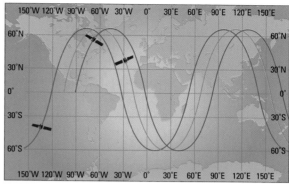

Figure 1

Example of a world map showing paths of satellites

Materials
- access to the Internet or appropriate software
- copy of seasonal star map (or a planisphere)
- observation sheets
- binoculars (optional)

Procedure

1 Research which large satellites (or other human-made objects) will be passing over your region in the next few weeks. Interpret the maps showing the satellites' paths (**Figure 1**).

 (a) Describe what you have learned about predicting where and when to see a human-made object in the sky.

2 On your star map, trace the paths of two or more objects that may soon be visible on a clear evening. Label when the objects might be seen at various parts of the path. (Be sure to correct to local time if your resources use standard time.)

3 Choose a time to view the sky. Remember that just after sunset the sky will be dark but the satellite will still be in bright sunlight, and also that it will be difficult to see a satellite near the Moon.

Always obtain your parent or guardian's permission before going out to watch the night sky. Dress appropriately.

4 On a clear evening, try to observe the satellites you have researched. Binoculars might help you view the satellite once you have located it.

 (a) Record whether or not you observed the satellites. Give reasons.

Understanding Concepts

1. Why can we see human-made objects in the night sky?

2. If you observe two satellites in the sky and they take different times to travel all the way across the sky, what can you judge about the motion or path of the satellites? Explain.

3. Do you think you would be able to see geosynchronous satellites in the sky? Explain.

4. Write a short report for the student newspaper, describing what satellites or other artificial objects can be seen at what times. Include appropriate safety precautions.

Exploring

5. Find out why people in Canada are unable to see the Hubble Space Telescope in the sky, yet people in the southern United States can. Share what you discover with your class.

Reflecting

6. Describe features of the software or Internet site you used that (a) you liked and (b) you think could be improved.

Challenge

Many people are unaware that satellites are often visible overhead. How would you try to increase that awareness in your planetarium show?

The International Space Station

Long before humans are able to colonize places like the Moon or Mars, we need to learn more about how the human body can survive in space. Much of the research on living in space can be carried out in space stations orbiting Earth. The first space station was *Salyut*, followed by *Skylab*. The first continuously occupied station, *Mir*, was constructed and controlled by the former Soviet Union and visited by astronauts and cosmonauts from many other countries. For many years, studies on surviving in space, and the function of the space station itself, provided data for scientists planning the next big space project: the *International Space Station*.

The **International Space Station (ISS)** is the biggest technological project ever built by humans. It involves the cooperation of space agencies from Brazil, Canada, Europe, Japan, Russia, and the U.S.A. When it is complete, it will have four research modules, a service module, a habitation module, remote robotic controls, a cargo block, a docking station for shuttle craft, and huge solar panels, all connected to a central truss over 100 m long. Forty-five launches of space shuttles and Russian boosters will carry up the more than 100 pieces—totalling 4.5×10^5 kg—for construction about 450 km above Earth's surface.

On board the station, six astronauts will live for three months at a time, performing numerous science experiments related to plants, animals, humans, materials research, crystal growth, chemical reactions, the environment, and other areas. Many of these experiments depend on the constant free fall (microgravity) conditions of the space station. The astronauts will also retrieve and repair satellites. This research could lead to better medicines, better crops, and better liquid fuels.

An artist's impression of the *ISS* is shown on page 486.

Did You Know ?

The Canadian astronaut program, part of the CSA, holds recruiting drives every few years, searching for applicants with the right aptitude and qualifications. These qualifications include scientific studies in medicine, physics, or engineering.

Canada's Contribution to the *International Space Station*

Canada's space exploration activities fall under the control of the Canadian Space Agency (CSA). The National Research Council of Canada (NRC) is also involved in the development of technology for use in space. Canada, with its reputation for building sophisticated robots (**Figure 1**) as well as visual systems used on the ground and in space, is an important contributor to the *ISS*. **Figure 2** describes the robotic and visual systems designed, built, and maintained by Canadians.

Try This **Manual Dexterity**

While working outside the *ISS*, astronauts have to wear bulky space suits to protect them from the cold, the vacuum of space, and the Sun's radiation. At the same time, they have to perform delicate procedures with extremely sensitive equipment. To get an idea of the challenge faced by astronauts, try the following tasks while wearing bulky or heavy gloves:
- tighten a nut on a bolt
- operate a VCR
- use tweezers to pick up a feather

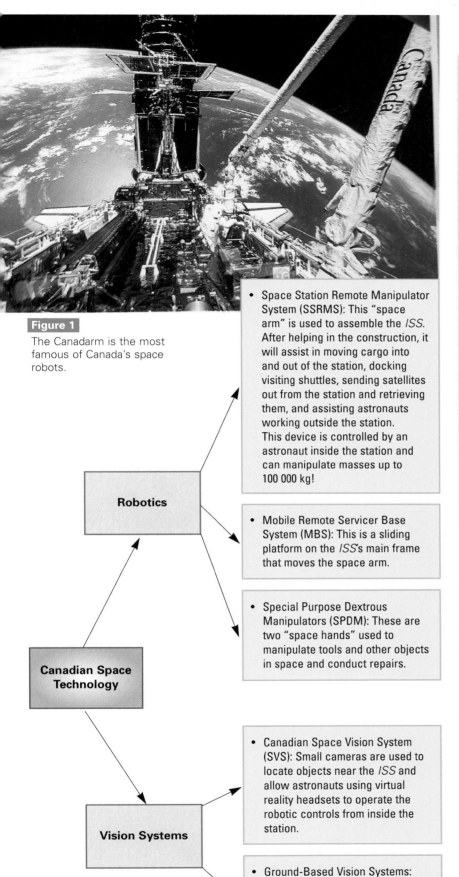

Figure 1
The Canadarm is the most famous of Canada's space robots.

Canadian Space Technology

Robotics

- Space Station Remote Manipulator System (SSRMS): This "space arm" is used to assemble the *ISS*. After helping in the construction, it will assist in moving cargo into and out of the station, docking visiting shuttles, sending satellites out from the station and retrieving them, and assisting astronauts working outside the station. This device is controlled by an astronaut inside the station and can manipulate masses up to 100 000 kg!

- Mobile Remote Servicer Base System (MBS): This is a sliding platform on the *ISS*'s main frame that moves the space arm.

- Special Purpose Dextrous Manipulators (SPDM): These are two "space hands" used to manipulate tools and other objects in space and conduct repairs.

Vision Systems

- Canadian Space Vision System (SVS): Small cameras are used to locate objects near the *ISS* and allow astronauts using virtual reality headsets to operate the robotic controls from inside the station.

- Ground-Based Vision Systems: These virtual reality systems are used for testing robotic devices and training astronauts in preparation for their mission.

Figure 2

Understanding Concepts

1. Describe uses of space stations.
2. What areas of expertise do Canadians bring to the *International Space Station*?

Making Connections

3. Assume you have to design a robot to turn the pages of a book. What features would you use in your design? Start by breaking down the task into its component parts.

Exploring ③A

4. To learn why Canada has a good reputation in the robotics industry, find out more about the Canadarm used on space shuttles. Identify the role of the National Research Council of Canada in its development. Make a visual presentation of what you discover.

5. Find out about robotic uses in space other than those mentioned in this section. Make a list of the uses you discover. In a group, create a mural that illustrates past accomplishments of and future expectations for robots in space.

6. The Russian space station, *Mir*, experienced many technical difficulties in the 1990s, sometimes threatening the lives of the crew. Research these problems and how they were solved.

Challenge

Which experiments on the *ISS* will be necessary to know how to survive on a space colony? How would you simplify an understanding of the significance of Canada's role in space technology for a younger audience?

Humans in Space

Dr. Valeri Polyakov has spent a total of 22 months in space. He also holds the record for the most consecutive time in space: 14 months! He was a cosmonaut aboard the *Mir* space station, the first spacecraft designed to keep people in orbit for extended periods of time. The first module was launched in 1986, and several other components have been added since. It has housed between two and six cosmonauts and astronauts at a time, including Canadian Chris Hadfield, all of whom are transported to and from *Mir* by space shuttles. Several of the crew have remained in *Mir* for over a year.

Imagine that you are one of Canada's astronauts about to be taken in the space shuttle to the *International Space Station*. You are a payload specialist: an expert on the scientific research you will be doing. Another job you have is to replace the fuel tank and replace a damaged part of a Canadian communications satellite.

As the shuttle blasts off from the launch pad, the sound is tremendous, and you are pushed back in your seat with a force that makes you feel three times as heavy as normal. Within two minutes, you feel a jolt as the solid rocket boosters separate from the shuttle. About six minutes later, you feel another jolt as the main fuel tank separates. The force pushing you back eases off, and you feel as if you are floating; only the seat belt keeps you in your seat.

Floating in Space: A Result of Free Fall

Often people and objects in orbiting spacecraft appear to be floating (**Figure 1**). In an effort to explain this effect, the terms "zero gravity" or "microgravity" are sometimes used. These expressions are misleading: there is plenty of gravity acting on the spacecraft and everything in it. The *ISS*, for example, operates at 450 km above the ground. At that altitude, the pull of Earth's gravity is almost 90% of what it would be on the ground, which is far from zero! In fact, without the force of gravity pulling on it, the *ISS* could not orbit Earth at all. For a spacecraft to follow Earth's curvature, the downward motion caused by gravity is essential. The floating occurs because the spacecraft and its contents are

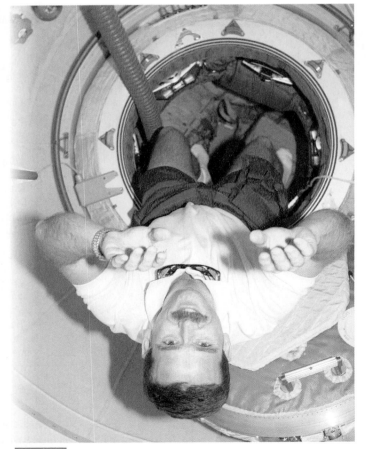

Figure 1

Canadian astronaut Chris Hadfield appears to be floating. What is really happening to create this effect?

continuously falling toward Earth at the same time as they are speeding forward almost fast enough to shoot out into space. This continuous falling is called **free fall**. The result of continuous free fall is that the spacecraft and everything in it remain balanced in orbit (**Figure 2**).

Sometimes the expression "weightless" is used to describe the effect of constant free fall. **Weight** is the force of gravity acting on an object. When an Earthbound object feels heavy, it is because your muscles must exert a large upward force to overcome the downward pull of gravity, whether it is yourself or some other object you are moving. Although the force of gravity pulls you down, your awareness of gravity comes mainly from the upward forces you must exert to oppose it. In an orbiting spacecraft, the craft and everything in it are continuously falling at the same rate, so the astronaut does not have to exert an upward force to hold onto anything in the spacecraft. Thus, astronauts feel as if there is no gravity; they feel weightless. However, as there is gravity, "weightless" is a poor expression to use.

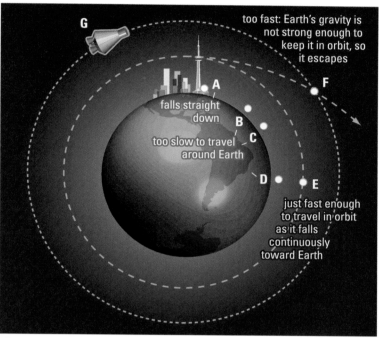

Figure 2

This diagram illustrates constant free fall. Imagine six steel balls that leave a tall tower. Ball A falls straight down while balls B to F get shot outward. Only ball E has the right speed to travel in an orbit around Earth. It follows a curved path just like the spacecraft G, in which astronauts and objects fall together toward Earth as they travel forward at great speed.

Living on the *ISS*

After the shuttle docks with the *ISS*, you enter the station and are greeted by other crew members who have been there for three months. With them, you eat your first space meal, part of which is prepared by adding water to a dry, crumbly mixture in a plastic pouch (**Figure 3**). The other part of the meal consists of fresh vegetables grown on the station as an experiment to help absorb carbon dioxide from the air and provide oxygen for the crew. Water, both hot and cold, is available as a byproduct of the fuel cells that produce some of the energy aboard the station.

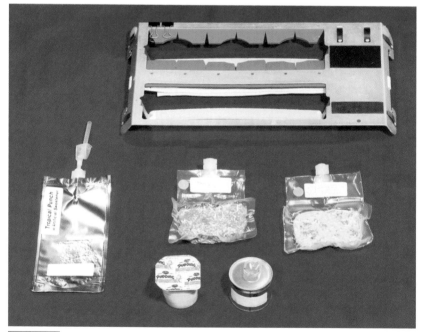

Figure 3

Food in sealed plastic bags is heated in a food warmer.

You do your share of work to help keep the station clean. Sanitation is important in space because microorganisms can grow rapidly in conditions of continuous free fall and an enclosed system like a spacecraft. Allowed to grow unchecked, these microorganisms could cause infections in the astronauts, so the eating equipment, dining area, toilet, and sleeping facilities must be cleaned regularly. Garbage and worn clothing are sealed in airtight bags to be returned to Earth later.

After doing some work at a computer terminal, you talk to your family on Earth and spend a long time exercising on the specially designed treadmills, cycles, and rowing equipment. You must exercise every day to keep your muscles and bones strong and your blood circulating properly. Then, after another meal and relaxation, you fasten yourself into the sleeping hammock tied to one wall.

You spend the next few days working at computers and practising the use of the manipulator arms. You are not allowed to leave the station to carry out your repair tasks until you have been in space at least three days, adapting to continuous free fall.

On the fourth day, you prepare for your first space walk by spending a few hours in a sealed area, breathing pure oxygen, since this is what you will be breathing in your space-walking suit. Breathing pure oxygen means that the air pressure inside your suit can be much lower than in the *ISS*, which allows your joints to be much more manoeuvrable (**Figure 4**) than they would be if your suit pressure was high. Protected by the suit from the vacuum of space and from harmful ultraviolet radiation from the Sun, you go outside the station and attach yourself to the Mobile Remote Servicer Base System (MBS). Using the manipulator system, you grab the satellite that has moved close to the station and pull it close to you. You then replace the satellite's used fuel tank with a fresh one and use the small manipulator arms to replace the damaged part. After testing the satellite, you use the large manipulator arm to push the satellite safely away from the *ISS*.

Try This Artificial Gravity

How can you make "up" into "down"? This model of artificial gravity will show you.

This activity should be done outdoors.

You will need a plastic bucket with a strong handle, a small rubber duck, and water. Half fill the bucket with water. Place the duck in the water. Swing the bucket back and forth until you are ready to swing it in a complete, vertical circle at a fairly high speed. Try not to spill the water as you bring the bucket to a stop.

1. Does the water fall out of the bucket when it is swung quickly in a vertical loop? Explain.

2. From the duck's point of view at the top of the loop, how important is Earth's gravity and which way is down?

3. In what way is this activity a model for "artificial gravity"?

4. How could you apply what you have learned in this activity to a space vehicle that takes humans to and from Mars?

Earth-based controllers then start up the satellite's engines to get it back into its proper orbit.

After three busy months on the *ISS*, you board the shuttle for your trip back to Earth. When the shuttle lands, you may need assistance as you step out and try to walk under the influence of gravity. You feel very heavy at first, but soon you get used to being back on Earth.

Understanding Concepts

1. In your own words, describe why astronauts appear to be floating in the *ISS* or a space shuttle in orbit around Earth, even though gravity is pulling on them.

2. Neither weightlessness nor microgravity are accurate ways to describe the conditions experienced by astronauts. Explain why. What is a more accurate term?

3. Draw a diagram to show how an astronaut who is working outside the *ISS* can use a backpack maneuvering unit (the EMU) to move from left to right.

4. Compare a day of living on *ISS* with a day on Earth. What are the similarities and differences?

Exploring

5. Design an amusement park ride that produces artificial gravity.

6. Research one of the following Canadian astronauts' qualifications, background, interests, and contributions to the space program: Roberta Bondar; Marc Garneau; Chris Hadfield; Steve MacLean; Julie Payette; Robert Thirsk; Bjarni Tryggvason; Dave Williams.

7. Compose an application letter to join the CSA's astronaut-in-training program. Indicate why you want to become an astronaut, what you can add to the space program, and other details.

Challenge

Humans have designed special features and products to survive in spacecraft. What features and products do you need to consider for the space colony challenge?

Figure 4

Space-walking astronauts move around with the help of a backpack called an Extravehicular Maneuvering Unit (EMU). The unit contains compressed nitrogen gas, which can be released through a nozzle. How does the EMU resemble the balloon rocket model or the rocket engine model in Section 16.1, Figures 4 and 5?

Gravity and Free Fall

Have you ever been on an amusement park ride that lets you fall straight downward? If so, you were in free fall. Astronauts on space shuttles or the *International Space Station* experience free fall continuously.

Part 1: Weight and "Weightlessness"

Question
What do the astronauts feel and why?

Hypothesis

1 Write a hypothesis about the relationship between mass and weight of an astronaut in orbit.

Materials
- spring scale that measures weight in newtons (N)
- a hanging mass that registers about halfway on the spring scale
- pillow

Procedure

2 Squat down, then jump up vertically as high off the floor as you can.

  (a) Describe what you feel in your leg muscles
    (i) as your legs are pushing off the floor;
    (ii) when you are in the air;
    (iii) as you are landing back on the floor.

3 Hang the mass on the hook at the base of the spring scale. Determine the apparent weight of the mass (as registered on the spring scale) when

Step 3:
Record the reading—quickly!

    (i)   the mass is resting on the floor;
    (ii) the mass is jerked gently upward;
    (iii) the mass is suspended in midair;
    (iv) the mass and spring scale are allowed to fall freely toward a pillow on the floor.

  (a) Record the four readings you observed.

4 Repeat step 3 twice.

  (a) Average your readings for the three trials.

Analysis and Communication

5 Analyze your observations by answering the following questions:

  (a) When, in step 2, did
    (i)   your weight feel normal?
    (ii) you feel heavier than normal?
    (iii) you feel lighter than normal?

  (b) What are the differences between mass and weight?

6 Relate your measurements in steps 3 and 4 to the sensations you felt in step 2.

Part 2: What Affects Free Fall?

Question
What factors affect how fast something falls from a person's raised hand to the floor?

Hypothesis

7 Write a hypothesis to answer the question.

Experimental Design

8 Design an experiment to test your hypothesis.
(4B)

  (a) Identify the factors you will investigate and describe how you think each will affect the motion of the falling objects.

  (b) List the materials you will need.

  (c) Write out your procedure. Include any useful diagrams.

Procedure

9 After your teacher has approved your procedural steps and materials list, carry out the steps.

✎ (a) Record what you observe.

Analysis and Communication

10 What can you conclude from your observations?

11 A 500-g ball and a 200-g ball are allowed to fall freely from the same height to the floor. Predict how their motions compare.

12 A piece of paper and a marble are allowed to fall toward the floor from the same height. Compare the motions of the two objects if
   (i) the paper is flat;
   (ii) the paper is crumpled into a ball.

Part 3: Falling While Moving in a Path

Question
What effect does horizontal motion have on the "falling time" of an object?

Hypothesis

13 Write a hypothesis and predict what will happen when you launch one coin horizontally, compared with what will happen when you allow a coin to fall freely downward.

Materials
- coin-launching apparatus (shown in **Figure 1**)
- 2 coins

Experimental Design

14 **(4B)** Design an experiment that uses a coin-launching apparatus to test your hypothesis.

Figure 1

Fold the cardboard and place the coins on it as shown. When you flip the launcher sideways, can you tell which coin will fall straight downward and which will fly sideways? Which coin do you think will land first?

Procedure

15 Make the coin-launching apparatus.

16 After your teacher approves your design, conduct your investigation.

🖐 Do not aim the coins at anybody. Launch them only where they will land safely.

✎ (a) Record your observations.

Analysis and Communication

17 Does the sideways motion of a falling object affect its downward falling? Explain.

18 **(8A)** Write a lab report describing and giving reasons for the different types of motion you observed.

Understanding Concepts

1. As the *ISS* travels in its path around Earth, it is always undergoing free fall. Astronauts inside the station are also experiencing free fall. How does this relate to what you observe about astronauts' motion in the station?

2. Imagine you are in an elevator that is falling freely. You hold out a pen and let it drop. How will its motion compare with yours?

3. Under what circumstances would an astronaut truly be weightless?

Exploring

4. When astronauts train, they experience free fall by going up in a special airplane that falls toward Earth for several seconds at a time. Sometimes the astronauts suffer from motion sickness on this airplane.

   (a) Research information about this airplane and share what you discover with the rest of your class.

   (b) Find out how the actors in the movie *Apollo 13* experienced free fall to make their flight look authentic.

Challenge

What is "artificial gravity" and how might it be applied for long-term space flights? What considerations related to gravity do you have to consider for Challenge 2?

Spinoffs from the Space Industry

How do the hard plastics used for in-line skates, safety helmets, and many other products, relate to space technology? Hard plastics were first designed by the space program to fulfill a specific need, then applied to the products we use (**Figure 1**). This is just one of many examples of a **spinoff**—an extra benefit from technology originally developed for another purpose.

How important will space science be in the 21st century? People who oppose spending on space missions argue that the money would be better spent cleaning up the environment and reducing poverty on Earth. People who promote space exploration argue that the money spent on space missions is only a small portion (under 2%) of any nation's budget. Furthermore, they point out, society benefits from space exploration (**Table 1**).

Procedure

1 Choose a recently developed product, and research how its development began with the space program (see **Table 1** for examples).

2 Design a poster or give an oral presentation tracking your product's research and development, evaluating the product, and giving your opinion on whether to invest in the spinoff.

Table 1	Types and Examples of Spinoffs
Area of Research	**Examples**
microelectronics	digital watches, home computers, pacemakers, hand-held calculators
new materials	nylon strips used to fasten clothing and objects, nonstick coating, flame-resistant materials (**Figure 2**)
metal alloys	dental braces
hard plastics	safety helmets, in-line skates
robotics	mining, industry, and offshore oil exploration where the conditions are too dangerous or the tasks too precise or repetitive for humans
vehicle controllers	controller for those with disabilities
safety devices	smoke detectors
recycling processes	water recycling
energy storage	solar cells, chemical batteries
food	freeze-dried convenience foods
pharmaceutics	antinausea medication to overcome the effects of motion sickness or the side effects of other drugs
pump therapy	method to provide insulin continuously to diabetes patients
scanning	medical scanning using imaging techniques developed for satellites
space vision technology	satellite data applied to improving the efficiency of agricultural spraying
lasers	improved laser surgery

Figure 1

Figure 2

Understanding Concepts

1. What is meant by the term "spinoff"?

2. Which spinoffs listed in this section are most likely linked to Canada's contribution to space science and technology?

Making Connections

3. List some space spinoffs that are now part of your daily life, both at school and at home.

4. What can we learn in space that would benefit the elderly?

Aerospace Engineer

Amyn Samji is not your typical engineer, but he was the one on NASA's graveyard shift monitoring Canada's famous robot, the Canadarm. He had to help remedy the problem of icicle growth on the shuttle.

"It's a long way from Overlea Secondary School to sitting at a console at NASA's Johnson Space Centre," he quietly muses.

Amyn Samji was born in Tanzania, Africa and came to Toronto with his family when he was eight years old. At first he struggled with mathematics, but things changed in high school. "I began to understand relationships and patterns." He joined the mathematics club and a special geometry class.

At the University of Waterloo, Samji took a co-op course in Applied Mathematics and Engineering, followed by a Masters degree in Aerospace Science at the University of Toronto Institute for Aerospace Studies.

Samji's research on robots began when he joined Spar Aerospace, developing computer models for the assembly of *International Space Station* (ISS). Spar engineers were developing the Space Station Remote Manipulator System (SSRMS). Samji developed computer simulations to show how astronauts would exchange parts of the SSRMS.

Later, at NASA's Johnson Space Center, Samji assumed flight support duties, monitoring the Canadarm during missions and solving operation problems. "At 4:00 a.m. on one shift the SSRMS break slip (alarm) went off during a thruster firing. NASA was trying to shake loose an icicle on the orbiter. It was tense. We solved it. It was O.K.," he smiles.

Samji then became the manager of the Mobile Servicing System (MSS) program, Canada's proud contribution to the *International Space Station*.

> Becoming a team player, learning to solve problems, and working hard are important to success in the space business.

Exploring

1. If you were designing a space robot, how would you deal with the harsh conditions it would encounter in outer space?

2. Identify several important characteristics that make a successful project manager.

3. Research post-secondary institutions in your area providing programs that could lead to a career in the space field.

Space Medicine

Human health is an important issue. The study of human health in space—space medicine—is not only fascinating, it is also a great benefit because it helps us learn more about human health on Earth.

Three of Canada's astronauts, Roberta Bondar, Robert Thirsk, and Dave Williams, are medical doctors working in the field of space medicine. Dr. Williams is director of NASA's Space and Life Sciences program.

Space Sickness

An immediate and annoying problem for a human body in orbit is space sickness, a type of motion sickness. During constant free fall, signals from the eyes, skin, joints, muscles, and the balance organs get rearranged. The conflicting clues lead to signs and symptoms of space sickness, such as dizziness, nausea, and vomiting. **Figure 1** explains how the balance organ, located in the inner ear, functions.

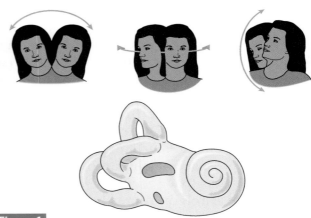

Figure 1

The drawing shows the structure of the inner ear of a human. The three fluid-filled tubes, called the semicircular canals, are responsible for balance. The fluid moves when the head moves: side to side, spinning around, or up and down.

Try This **Your Sense of Balance**

Here are two simple activities you can use to test your sense of balance. How does your sense of balance compare with that of other students?

1. Stand on one leg with your eyes open. Then close both eyes and try to maintain your balance.

2. Stand on one leg facing a striped sheet or blanket held by two students. Try to maintain your balance as the blanket is moved sideways, as shown in **Figure 2**.

Have two spotters nearby in case you lose your balance.

Figure 2

The moving stripes may fool your brain into thinking that you are moving, so your muscles compensate, and you fall over!

Other Effects of Space Flight

Imagine you have had a full-length cast on one leg for three months. The broken bones have healed, and your doctor removes the cast. But your leg still doesn't work normally: although the bones have healed, the muscles have weakened and shrunk from lack of use. It will probably take weeks or even months for you to regain the normal use of your muscles.

Similar problems occur when astronauts are in constant free fall while orbiting Earth. On Earth, your *awareness* of gravity comes mainly from the upward forces needed to oppose it. But in free fall, astronauts are not aware of this gravitational pull. They *feel* weightless, just like the hanging mass that fell with the scale in Investigation 16.8. The lack of forces against the body causes the muscles to become smaller and the bones to lose calcium and become brittle. It seems as though the normal aging process is vastly speeded up. This process can be slowed with a vigorous exercise program during the space flight, but it is not yet known whether this process can be stopped entirely. The data collected during space missions is used in research into the normal aging process. Scientists would like to reduce the deterioration that occurs during both processes.

Another possibly dangerous side effect of space travel involves body fluids. Because of constant free fall, about two extra litres of blood remain in the upper half of the body, swelling the heart and the blood vessels and making the astronauts' faces look puffy and their legs look thinner. More importantly, this condition alters the fluid balance of the body and, in turn, affects the kidneys and causes excess urination. Much is yet to be learned about long-term effects, and research is ongoing.

Astronauts are also exposed to more cosmic radiation on a single trip than they would experience in several years on Earth. We do not yet know whether this exposure is harmful because any effects might not show up for several years. Thus, scientists continue to check the health of space travellers carefully. Space medicine is still a young science, and much remains to be learned about the effects of space travel on the human body.

Understanding Concepts

1. Describe the effects of constant free fall on the human body. Why do these effects occur?

2. Astronauts must spend a long time exercising each day. Why?

Making Connections

3. How do you think an exercise program in space might be applied to design an exercise program for seniors on Earth?

4. Choose one of the health problems described. How does the research related to it affect the lives of other people not involved in space flight?

Exploring

5. Research the invention of the antigravity suit by Canadian, Dr. W. R. Franks in 1940. Write your own short story about it.

6. Make up an exercise program for astronauts either while aboard the *ISS* or when they return to Earth.

Challenge

Check the NASA web site to obtain more information about NASA's Space and Life Science program. Which contributions have Canadians made to this program?

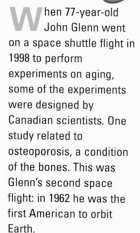

Did You Know ?

When 77-year-old John Glenn went on a space shuttle flight in 1998 to perform experiments on aging, some of the experiments were designed by Canadian scientists. One study related to osteoporosis, a condition of the bones. This was Glenn's second space flight: in 1962 he was the first American to orbit Earth.

16.11 Investigation

SKILLS MENU
- Questioning
- Hypothesizing
- Planning
- Conducting
- Recording
- Analyzing
- Communicating

Experimenting in Free-Fall Conditions

Did you know that you are slightly taller when you get up in the morning than when you go to bed? In the daytime, when you are upright, the force of gravity pulls on your spinal column, causing it to crush together. When you are lying down, this crushing is reduced, so the bones spread apart a bit.

This same thing happens to astronauts in constant free fall: they become several centimetres taller. What is more, their bones lose mass, and their muscles weaken because they don't have to work against gravity. Many of their experiments involve finding out what happens to human bodies in space and why. Knowing the answers will help future space exploration as well as the treatment of aging problems on Earth.

In this investigation, you will design an experiment suitable for astronauts to perform on the *ISS* or for the Canadian Space Agency (CSA) to send on a space probe mission.

Question

1 Choose one area of research from **Table 1** that interests you. Plan how you will find out more about that area.

2 Find the information, then choose one topic on which you will base an experiment that will prove useful either to future space flight or to life here on Earth.

(a) Write out the question your experiment will attempt to answer.

Hypothesis

3 Write a hypothesis to answer the question.

4 Predict what results you would expect from your experiment.

(a) Write out your prediction, including the controlled and dependent variables.

Experimental Design

5 Design a controlled experiment to answer your question. List the materials needed, the procedural steps, and any safety considerations. Also mention how you would analyze the results.

(a) Write out your experimental design and have it approved by your teacher.

Procedure

6 Conduct your experiment.

Analysis and Communication

7 Create a proposal that would persuade a space agency to conduct your experiment, then submit your proposal.

8 How does performing experiments in space differ from performing experiments on Earth?

Table 1	Canada's Specialties in Space Science
Area of Research	**Examples of Topics Researched**
Space Life Sciences	bone and muscle loss, neurobiology, early development, radiation, heart and circulation system research
Science in Constant Free Fall	crystals, including protein crystals, metals research, fluids research, materials for semi-conductors and fibre optics
Atmospheric Sciences	global warming, Arctic-region research, ozone depletion, middle/upper atmosphere science
Space Astronomy	research using various parts of the electromagnetic spectrum
Solar-Terrestrial Relations	Earth's ionosphere and magnetosphere, the ionosphere on Mars, solar wind, space plasma, aurora research

Making Connections

1. How do you think your proposed experiment could benefit people or things on Earth?

Exploring

2. Write a short letter, including a relevant question, to a Canadian astronaut on the *ISS*.

Our Future in Space

If space stations are successful, the next stages of space travel may be exploration of Earth's Moon, and the planet Mars and its small moons, Phobos and Deimos. These objects could be mined and the materials used to build structures in space. Eventually, human colonies may be established on the Moon and on Mars.

Many problems must be overcome before we can travel to, and live on, Mars. One of the biggest problems is the reaction of the human body to constant free fall. To reduce the effects on long space trips, a feature of any piloted trip to Mars will probably be artificial gravity, created by rotating the spacecraft (**Figure 1**). Once on Mars, the astronauts will find a gravitational force only 38% of that on Earth. This may also cause problems for the human body over long periods of time.

Visitors to Mars will require water, food, oxygen, and warm shelter. Oxygen and small amounts of water could be extracted from the low-density atmosphere, which consists mostly of carbon dioxide. Water could also be extracted from the permafrost on Mars and from the recycling of plant and human wastes. Large greenhouses could be used to grow fruits and vegetables. Even though Mars is farther from the Sun than Earth is, the Sun's radiation is far more harmful there because the atmosphere on Mars has no ozone to offer protection. As a result, living quarters would have to be buried beneath the Martian soil to protect humans from harmful solar radiation.

Some scientists predict that by the time humans have settled on Mars, we will have propulsion systems powerful enough to allow spacecraft to reach extremely high speeds. Perhaps such a craft, if unpiloted, would be sent to explore neighbouring stars in our galaxy.

Figure 1

Artist's concept of a spacecraft travelling to Mars

Issue Space Exploration

People in favour of space exploration give many good reasons why it will benefit the human race. People opposing space exploration feel strongly that there are drawbacks that outweigh the benefits. Working in a group, choose either a supporting or an opposing position on space exploration. (9C)

- Brainstorm a list of reasons that support your group's viewpoint. Consider economics, ethics, environmental issues, politics, and available technology as you brainstorm.
- Discuss these as a group and carry out further research where necessary.
- Prepare a 5-min presentation, putting forth your position.

Challenge

What situations should you consider for sustained survival when designing a space colony?

Key Expectations

Throughout the chapter, you have had opportunities to do the following things:

- Describe and explain the effects of continuous free fall (microgravity) on organisms and other contents of orbiting spacecraft. (16.8, 16.9, 16.11, 16.12)

- Investigate questions related to sending satellites and humans into space, and organize, record, analyze, and communicate results. (16.2, 16.4, 16.6, 16.7, 16.9, 16.10, 16.12)

- Formulate and research questions related to space exploration, and communicate results. (all sections)

- Describe, evaluate, and communicate the impact of research and achievements in space on other fields of endeavour. (16.3, 16.4, 16.5, 16.9, 16.10, 16.11, 16.12)

- Identify the purpose and accomplishments of space initiatives such as satellites, space probes, and the *International Space Station*. (16.1, 16.3, 16.4, 16.5, 16.6, 16.7, 16.13)

- Investigate and provide examples of ways in which Canada participates in space research and international space programs. (16.3, 16.5, 16.7, 16.11)

- Explore careers related to space exploration. (Career Profile)

KEY TERMS

aircraft	outer space
free fall (microgravity)	radar
geosynchronous orbit	remote sensing
Global Positioning System	space probe
International Space Station	spacecraft
launcher	spinoff
low Earth orbit	thrust
payload	weight

Reflecting

- "Space exploration and the related technology contribute to our understanding of Earth and the universe and provide many useful applications for life on Earth." Reflect on this idea. How does it connect with what you've done in this chapter? (To review, check the sections indicated above.)

- Revise your answers to the questions raised in Getting Started. How has your thinking changed?

- What new questions do you have? How will you answer them?

Understanding Concepts

1. Make a concept map to summarize the material that you have studied in this chapter. Start with the words "space exploration."

2. What are the purposes of space stations?

3. During blast-off, an astronaut feels heavier than usual; later, lighter than usual. Explain why.

4. What is "artificial gravity"? How might it be created on a spacecraft?

5. List benefits of studying space medicine.

6. Classify each of the following vehicles as Earth-based, atmosphere-based, low Earth orbit, or geosynchronous orbit: the *International Space Station*; a space shuttle; a helicopter; a sailboat; a TV satellite; a remote-sensing satellite.

7. State the purpose or function of: remote sensing satellites, RADARSAT, GPS satellites, and an orbiting observatory.

8. Compare an aircraft engine and a rocket engine, explaining why an aircraft cannot fly in space.

9. Describe situations in which you have experienced free fall.

10. Why are robotics and vision systems important on the *International Space Station*?

11. What would happen to a spacecraft in orbit around Earth if its speed becomes too slow? Explain.

12. Can a human be a payload? Explain your answer.

13. Which electromagnetic waves, infrared or radar, would be better for viewing the heights of mountains and craters on the moons of planets? Why?

14. Why are remote sensing satellites that use radar not used to predict daily weather patterns?

15. (a) Why is the expression "zero gravity" misleading for an astronaut inside a spacecraft in an Earth orbit?

    (b) Under what condition(s) would the expression "zero gravity" not be misleading?

16. During the last few days, identify ways you have benefited from space exploration. (Include both direct benefits and spinoffs.)

17. If you were standing on your head, your blood distribution would change. Relate this change to what an astronaut experiences in orbit around Earth.

18. At the same instant, one car drives horizontally off a cliff at a high speed and another falls straight down.

    (a) Compare the times the cars take to land.

    (b) Relate your answer to space vehicles that travel around Earth.

19. A sports event takes place in Australia, and at almost the same time you are able to watch that event on television. Describe how this form of telecommunication works.

20. Space probes are sent to explore asteroids and comets. How does this exploration help in the study of the origin of the solar system?

21. Why are we more protected from the Sun's harmful radiation on Earth than astronauts are in space?

22. The air pressure in an *ISS* suit is less than atmospheric pressure. Find out why, and what advantages and disadvantages this brings.

Applying Skills

23. If a satellite takes 90 min to travel once around the world, for about how long will it be in view, each orbit? Show your calculations.

24. Describe what conditions would allow you to observe a satellite or the *ISS* travelling across the sky. Draw a diagram to illustrate your answer.

25. Find the results of a Canadian research project, involving students growing seeds that had been in space.

26. Select a career in the field of astronomy or space science, and find out what aptitudes and qualifications you would need to enter this career. If possible, interview someone in this career and prepare an audio report on the person's professional life.

27. Write a short science fiction story about space exploration using as many of the key terms as you can.

28. Describe what steps you would take to discover what fuels are used to launch a space shuttle.

29. Describe how to access information on the Internet about RADARSAT.

30. Assume you are an astronaut on the Moon carrying out an investigation to determine the factors affecting how fast objects fall. Describe in detail the steps you would take. Show that you understand what a controlled experiment is.

31. Draw a graph to predict how the speed of a balloon (vertical axis) depends on the air pressure inside the balloon.

Making Connections

32. Create a chart of Canadian contributions to the study of the universe, giving the Canadians' names and outlining their work.

33. List two or three careers in the space industry that require a background in (a) engineering and (b) technical studies.

34. Choose five interesting space spinoffs and for each, name at least one possible career associated with it.

35. To find out more about how the human body reacts in space, investigate Dr. Dave Williams' research on flight STS-90.

36. Defend or oppose one of the following statements:

    (a) Space exploration benefits the human race.

    (b) Money spent on space exploration would be better spent cleaning up the global environment.

Challenge

Bringing Your Ideas Together

Scientists and technologists share ideas in many ways. They may design simulations to understand concepts and to predict future events. They may develop technologies that improve our lives or that can be used to make further scientific investigation possible. They may conduct research and record and communicate what they discover. You can choose one of these approaches in the challenges described below:

1 Planetarium Shows

A planetarium is a structure in which people can learn more about astronomy. The night sky is simulated using a projection of points of light representing stars and other objects onto the inside of a dome. By moving the projector, the appearance of the night sky at any time and any date can be represented. Planetariums usually offer a variety of shows designed for different purposes and audiences.

Your job is to design and prepare three shows that could be used in a planetarium. One show should be designed to teach grade 6 students about the solar system. The other two shows should be designed for the general public. You must include a script for each show and be prepared to present one of them.

Your simulations should include:
- models or other representations of year-round constellations, seasonal constellations, planets, comets, and other natural objects discussed in this unit
- representations of human-made objects such as satellites and space probes
- a model that demonstrates the motions of as many of the objects as possible
- design features that take into consideration the size, construction, and layout of a large planetarium

2 A Space Research Colony

You are part of a company that has designed a colony that would conduct research in outer space. The company is preparing to display its plans at the International Space Conference, in the hope of attracting investors for the project.

Design and create a display depicting a sustainable space research colony suitable for permanent human habitation.

Your display should include:
- a representation of the destination of your choice in outer space
- a demonstration of how the colony will be adapted to the environment of the destination chosen
- a plan that outlines how colonists will achieve self-sufficiency in both the short and long terms
- evidence of how knowledge gained from existing space explorations and space technology will be used
- proposals for the type of research that would be best suited for the destination you chose
- models and diagrams of relevant components

3   A Space Technology Information Package

Vast amounts of money are spent on technology related to space exploration. The merits of this expenditure are frequently debated in public. Your company has been commissioned by the Canadian Space Agency to develop an information package about the usefulness of space technology. You are to demonstrate how the use of space technology contributes to our understanding of the universe and how applications of that technology benefit us in our everyday lives. The package is designed for distribution to the general public.

Design an information package that outlines the diverse tools used both in space exploration and the study of astronomy and the influence they have had on our lives.

Your information package should include:
- an organizational tool for accessing the information
- examples of the uses of space technology
- links to Canada's role in the development of space technology
- illustrations of what we have learned about the universe using the technology
- examples of applications of the technology that positively affect our daily lives

Assessment

Your completed challenge will be assessed according to how well you:

Process
- understand the specific challenge
- develop a plan
- choose and safely use appropriate tools, equipment, and materials when necessary
- conduct the plan applying technical skills and procedures when necessary
- analyze the results

Communication
- prepare an appropriate presentation of the task
- use correct terms, symbols, and SI units
- incorporate information technology

Product
- meet established criteria
- show understanding of concepts, principles, laws, and theories
- show effective use of materials
- address the identified situation/problem

Unit 4 Review

Understanding Concepts

1. In your notebook, write the letters (a) to (k), then indicate the word(s) needed to complete each statement below.

   (a) The ___?___ is everything that exists, including all the matter and energy everywhere.

   (b) The four planets closest to the Sun can be called the ___?___ planets or the ___?___ planets.

   (c) ___?___ was the first scientist to use a telescope to obtain evidence that stars were much farther away than planets.

   (d) The ___?___ is a broad band of energies that can travel in a vacuum.

   (e) The process used by stars to produce energy is called ___?___ .

   (f) Stars start and end their lives as clouds of dust and gas called ___?___ .

   (g) ___?___ is the force of attraction between all objects that have mass.

   (h) A payload is to a ___?___ as an arrow is to a ___?___ .

   (i) A satellite in ___?___ orbit appears to remain stationary when viewed from Earth.

   (j) ___?___ is the force of gravity acting on an object. In the metric system its unit is the ___?___ .

   (k) A ___?___ is a benefit that comes from space science and technology research.

2. Indicate whether each of statements (a) to (o) is TRUE or FALSE. If you think the statement is FALSE, rewrite it to make it true.

   (a) Our Sun is the only star that has planets orbiting around it.

   (b) An object that is 30 astronomical units from the Sun could be either a gas giant or a comet.

   (c) The light-year is a unit of time.

   (d) A comet's tail is visible only when the comet's path is close to the Sun.

   (e) When using triangles to measure distances to objects in the sky, the longest baseline possible to observers on Earth is Earth's diameter at the equator.

   (f) The speed of electromagnetic waves in a vacuum is 3×10^5 km/s.

   (g) Auroras are caused by light pollution.

   (h) Most stars are bigger than the Sun.

   (i) The Sun is in the galaxy called the Milky Way.

   (j) In the lives of stars, a red supergiant results from stars that have a smaller mass than the Sun.

   (k) Both stars and planets form from nebulas.

   (l) Evidence of an expanding universe comes from red shift of the spectra of stars and galaxies.

   (m) Space probes have been sent to explore stars nearest to the solar system in our galaxy.

   (n) RADARSAT uses waves that belong to the invisible part of the electromagnetic spectrum.

   (o) The problems of increased body length and harmful radiation disappear after astronauts have been in orbit for a week or more.

3. Describe the similarities and/or differences between each pair of objects listed below:

   (a) star constellation; star cluster

   (b) Earth's rotation; Earth's revolution

   (c) asteroid; comet

   (d) galaxy; star cluster

   (e) interplanetary distance; intergalactic distance

   (f) solar flare; solar prominence

   (g) apparent magnitude; absolute magnitude

   (h) neutron star; black hole

   (i) aircraft; spacecraft

   (j) vacuum; atmosphere

   (k) natural satellite; artificial satellite

   (l) space shuttle; space station

   (m) geosynchronous orbit; low Earth orbit

For Questions 4 to 10, choose the best answer and write the full statement in your notebook.

4. Pointer stars:

   (a) can be used to locate planets in the night sky

(b) are found only in the year-round constellations

(c) can be used to locate constellations and other stars

(d) appear to be pointed

(e) all of the above

5. Compared with the terrestrial planets, the gas giants tend to:

(a) be hotter, rotate faster, and have a lower density

(b) be colder, rotate faster, and have a lower density

(c) be colder, rotate slower, and have a lower density

(d) be colder, rotate slower, and have a higher density

(e) be hotter, rotate faster, and have a higher density

6. Which does not belong in this list?

(a) radio waves

(b) visible light

(c) X rays

(d) sound waves

(e) infrared radiation

7. The order of stars from coolest to hottest is

(a) blue, yellow, red

(b) red, blue, yellow

(c) yellow, red, blue

(d) blue, red, yellow

(e) red, yellow, blue

8. A possible order of events in the evolution of stars is

(a) nebula, energy produced through fusion, core collapse

(b) core collapse, energy produced through fusion, nebula

(c) nebula, core collapse, energy produced through fusion

(d) energy produced through fusion, core collapse, nebula

(e) energy produced through fusion, nebula, core collapse

9. The force caused by expanding gases leaving a rocket engine is called

(a) chemical energy

(b) gravity

(c) thrust

(d) friction

(e) air resistance

10. The best explanation of why orbiting astronauts appear to be floating is that they are experiencing

(a) microgravity

(b) weightlessness

(c) no gravity

(d) continuous free fall

(e) none of the above

11. Which planets are the most likely targets for exploring by robotic rovers in the near future? Give reasons for your choices.

12. Some planets cannot be explored by robotic rovers. What ways can be used to study these planets at close range?

13. Which planets tend to have the greatest number of moons? Explain why this situation occurred as the solar system developed.

14. (a) What are sunspots?

(b) How can they be safely observed?

(c) What evidence do they provide of the Sun's rotation?

15. Stars move at great speeds. Why did people long ago believe that stars were fixed in place?

16. Can two stars with the same luminosity have different apparent magnitudes? Explain.

17. How are the temperatures and colours of stars related? Give examples.

18. Which types of stars tend to become supernovas? What happens after that stage?

19. The inner planets and gas giants formed at about the same time. Which stages of their formation were similar, and which were different? Explain why there was a difference.

20. According to the present theory, what are the main stages of the evolution of the universe?

21. What happens to low Earth orbit satellites if their orbits become too low? Why?

22. (a) Describe the conditions needed for you to feel you are weightless for a short period of time.

(b) Relate your answer in (a) to what astronauts experience in orbiting spacecraft.

23. (a) As you observe the night sky in the Northern Hemisphere, which star appears not to move?

(b) Why does it appear not to move?

(c) How long do constellations take to appear to travel once around that star?

24. You are conducting an investigation to compare the motions of planets and constellations seen in the night sky.

(a) How would the motions compare if you made observations three hours apart?

(b) How would the motions compare if you made observations three weeks apart?

(c) Explain the difference between observations (a) and (b) above.

25. What is the Hubble Deep Field? Why is it important?

26. If you see two stars through a telescope and one looks bluish, the other yellowish, what can you say about the difference between the stars?

27. What is the relationship between the speed of a satellite and the height of its orbit?

Applying Skills

28. In observing the night sky, it is important to judge the differences between stars and planets. Describe the main ways you have learned to distinguish stars and planets.

29. A student records these observations in a log book: "The constellation was 3 fists to the right of south, and 4 fists up from the horizon." Describe what these observations mean.

30. A certain comet, visible tonight, has a period of 185 years.

(a) Draw a diagram showing the basic orbit of the comet around the Sun. Include Earth in your diagram as well as the comet's tail at a few locations.

(b) Predict when the comet will again be visible.

31. Write these measurements out in long form

(a) 3.4×10^6 km

(b) 7.9×10^{12} kg

32. Describe a safe way to make observations of the Sun when it is high in the sky.

33. The figure below shows a device in which a rolling ball moves faster and faster, finally getting gobbled into the centre. How is this a model of a black hole? In what ways is the model not realistic?

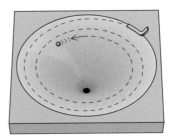

34. Describe briefly how each of the following devices helps in the exploration of the universe:

(a) an orbiting observatory

(b) a ground-based observatory

(c) a radio telescope

35. Describe how any or all of the following have helped you understand ideas in this unit:

(a) modelling

(b) performing student-designed investigations

(c) drawing scale diagrams

(d) graphing

(e) operating computer simulations

(f) researching information on the Internet

36. **Table 1** indicates the speed, orbital radius (from the centre of Earth), and period of revolution of four types of artificial satellites and the Moon.

(a) Determine the information needed to complete the table.

(b) What patterns do you observe in your completed table?

(c) For the artificial satellites, plot a line graph of period of revolution (vertical axis)

Table 1

Satellite	Orbital Radius	Period of Revolution	Average Speed
remote sensing	6.67×10^3 km	1.5 h	?
Iridium	7.18×10^3 km	1.7 h	?
GPS	2.67×10^4 km	12 h	?
geosynchronous	4.24×10^4 km	24 h	?
the Moon	3.84×10^5 km	656 h	?

against the orbital radius. Use the line on the graph to determine (i) the period of a satellite with an orbital radius of 3.5×10^4 km and (ii) the orbital radius of a planet with a period of 10 h.

(d) For each satellite, including the Moon, calculate the ratio of radius^3 to period^2. What do you discover?

37. Make a list of three important questions you would like answered to help you understand more about (a) astronomy and (b) space exploration.

Making Connections

38. What were some skills achieved by astronomers in ancient times?

39. What is the evidence that astronomy is a very old science?

40. (a) What are observatories used for?

(b) Does Canada operate or help operate any observatories outside the country? If so, where?

41. Professional astronomers rely on amateur astronomers to find objects such as comets in the sky. Why?

42. Give examples in which Canadians have participated in discoveries and/or research in astronomy.

43. Many devices you use in your daily life are digital, just like devices used in astronomy and space exploration. Give examples of digital devices and describe why they are called "digital."

44. How has Canada's space technology helped in each of these endeavours?

(a) robotics

(b) resource management

(c) navigation

(d) telecommunications

45. In what ways will future space probes help increase our understanding of the universe?

46. List three general areas that involve careers in space technology. Within each area, name one specific career, some of its aspects, and the education required for it.

47. Explain the advantages of using robotics rather than piloted probes to explore the planets and other bodies in the solar system.

48. Explain this statement: "Visual systems and robotics often operate together."

49. Choose an area of scientific research carried out by astronauts aboard the *International Space Station*. Describe what you know about that research.

50. You are the leader of a team of scientists, engineers, and technologists who are developing spinoffs of the space industry. Make a list of five problems you would like the team to work on.

51. Robotic rovers designed to roam across the irregular surface of Mars will have to be "smart"; in other words, they will have to make their own decisions about whether it is safe to move forward. Explain why such rovers could not be adequately controlled by Earth-based controllers.

52. Research the latest developments in communications systems that make use of geosynchronous satellites. How are they improvements on older systems?

53. Find out about one of the many traditional religions or cultures that base their calendar on the movements of the Sun and Moon. Create a display showing how the dates of various celebrations are fixed.

54. Do spacecraft have to be aerodynamic in shape? Explain.

55. Many satellites are powered by electricity converted by big shiny panels from solar energy. Investigate the advantages and disadvantages of this source of energy, and suggest what design challenges the spacecraft engineers face and how they might overcome these challenges.

56. Research into the specialized areas of science (such as fluid physics, or crystal growth) that make use of the free fall conditions in an orbiting laboratory.

57. What probes are in outer space this year? Research their purpose, and find out what kinds of data they are sending to Earth. How are these data changing our ideas about space?

58. Investigate the "What's New" page of the Canadian Space Agency's web site to check on the progress of the *ISS*. Prepare periodic verbal reports for your class, focussing on Canada's contribution.

Skills Handbook

Table of Contents

⒜ Safety Conventions and Symbols

Safety Conventions in *Nelson Science 9*

When you perform the investigations in *Nelson Science 9*, you will find them challenging, interesting, and safe. However, you should be aware that accidents can happen. In this text, chemicals, equipment, and procedures that are hazardous are highlighted in red and are preceded either by the appropriate WHMIS symbol (illustrated below) or by one of the following:

 general caution    electrical hazard

You should always read cautions carefully and make sure you understand what they mean before you proceed. If you are in doubt about anything, be sure to ask someone who knows (e.g., your teacher, a classmate, or a parent).

Hazardous Household Product Symbols (HHPS)

You are probably familiar with the warning symbols in **Figure 1**. They appear on a number of products that are common in most households. These warning symbols were developed to indicate exactly why and to what degree a product is dangerous.

Workplace Hazardous Materials Information System (WHMIS) Symbols

The Workplace Hazardous Materials Information System (WHMIS) symbols in **Figure 2** were developed to standardize the labelling of dangerous materials used in all workplaces, including schools. Become familiar with these warning symbols and pay careful attention to them when they appear in *Nelson Science 9* and on any products or materials that you handle.

Figure 1

Figure 2

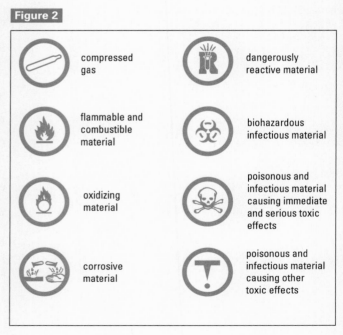

ⓐ Safety in the Laboratory

The Importance of Safety

Certain safety hazards exist in any laboratory. You should know about them and about the precautions you must take to reduce the risk of an accident.

Why is safety so important? Think about the safety measures you already take in your daily life. Your school laboratory, like your kitchen or the squash court, need not be a dangerous place. In any situation, you avoid accidents when you understand how to use materials and equipment and follow proper procedures. What do you use to take cookies out of the oven? What safety precautions do you take when you play squash? What are the common safety procedures related to swimming and boating?

Safety in the laboratory combines common sense with the foresight to consider the worst-case scenario. The activities in this textbook have been tested and are safe, as long as they are done with proper care. While your teacher will give you specific information about safety rules for your classroom and for conducting investigations, you should always consider setting safety rules on your own.

Preventing Accidents

Most accidents that occur in the lab are caused by carelessness. Knowing the most common causes of accidents can help you prevent them. These include:
- applying too much pressure to glass equipment (including microscope slides and cover slips)
- handling hot equipment without proper precautions
- measuring and/or mixing chemicals incorrectly
- working in a messy or disorganized space
- paying too little attention to instructions and working distractedly
- failing to tie back long hair or loose clothing

Setting Safety Rules

Before You Start

1. Learn the location and proper use of the safety equipment available to you, such as safety goggles, protective aprons, heat-resistant gloves, eye wash station, broken glass container, first-aid kit, fire extinguishers, and fire blankets. Find out the location of the nearest fire alarm.

2. Inform your teacher of any allergies, medical conditions, or other physical impairments you may have. Do not wear contact lenses when conducting investigations—if a foreign substance became trapped beneath the lens, it would be difficult to remove it.

3. Read the procedure of an investigation carefully before you start. Clear the laboratory bench of all materials except those you will use in the investigation. If there is anything you do not understand, ask your teacher to explain. If you are designing your own experiment, obtain your teacher's approval before carrying out the experiment.

4. Wear safety goggles and protective clothing (a lab apron or a lab coat), and tie back long hair. Remove loose jewellery. Wear closed shoes in the laboratory, not open sandals.

5. Secure any potentially dangerous or fragile equipment that could be hazardous if tipped over. Use the appropriate stands, clamps, and holders.

6. Never work alone in the laboratory.

Working with Chemicals

7. Do not taste, touch, or smell any material unless you are asked to do so by your teacher. Do not chew gum, eat, or drink in the laboratory.

8. Be aware of where the MSDS (Material Safety Data Sheet) manual is kept. Know any relevant MSDS information for the chemicals you are using.

9. Label all containers. When taking something from a bottle or other container, double-check the label to be sure you are taking exactly what you need.

10. If any part of your body comes in contact with a chemical or specimen, wash the area immediately and thoroughly with water. If your eyes are affected, do not touch them but wash them immediately and continuously with cool water for at least 15 min and inform your teacher.

11. Handle all chemicals carefully. When you are instructed to smell a chemical in the laboratory, take a few deep breaths before waving the vapour toward your nose. This way, you can smell the material without inhaling too much into your lungs. Only this technique should be used to smell chemicals in the laboratory. Never put your nose close to a chemical.

12. Place test tubes in a rack before pouring liquids into them. If you must hold a test tube, tilt it away from you before pouring in a liquid.

13. Clean up any spilled materials immediately, following instructions given by your teacher.

14. Do not return unused chemicals to the original containers, and do not pour them down the drain. Dispose of chemicals as instructed by your teacher.

Heating

15. Whenever possible, use electric hot plates for heating materials. Use a flame only if instructed to do so. If a Bunsen burner is used in your science classroom, make sure you follow the procedures listed below.
    - Obtain instructions from your teacher on the proper method of lighting and adjusting the Bunsen burner (**Figure 1**).
    - Do not heat a flammable material (for example, alcohol) over a Bunsen burner. Make sure there are no flammable materials nearby.
    - Do not leave a lighted Bunsen burner unattended.
    - Always turn off the gas at the valve, not at the base, of the Bunsen burner.

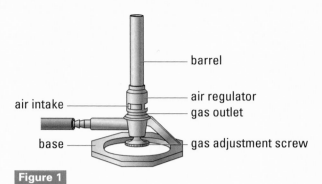

Figure 1
The Bunsen burner

barrel

air regulator

air intake

gas outlet

base

gas adjustment screw

16. When heating liquids in glass containers, make sure you use clean Pyrex or Kimax. Do not use broken or cracked glassware. If the liquid is to be heated to boiling, use boiling chips to prevent "bumping." Always keep the open end pointed away from yourself and others. Never allow a container to boil dry.

17. When heating a test tube over a flame, use a test-tube holder. Hold the test tube at an angle, with the opening facing away from you and others. Heat the upper half of the liquid first, then move it gently in the flame, to distribute the heat evenly.

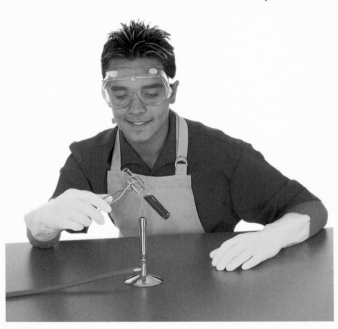

18. Be careful when handling hot objects and objects that might be hot. Hot plates can take up to 60 min to cool completely. Test that they are cool enough to move by touching first with a damp paper towel. If you hear sizzling or see steam, wait a little longer! If you burn yourself, immediately apply cold water or ice, and inform your teacher.

Other Hazards

19. Keep water and wet hands away from electrical cords, plugs, and sockets. Always unplug electrical cords by pulling on the plug, not the cord. Report any frayed cords or damaged outlets to your teacher. Make sure electrical cords are not placed where someone could trip over them.

20. Be sure to have your teacher inspect any electric circuits before you turn on the electrical power.

21. Place broken and waste glass in the specially marked containers. Wear heavy gloves while picking up the pieces.

22. Follow your teacher's instructions when disposing of waste materials.

23. Report to your teacher all accidents (no matter how minor), broken equipment, damaged or defective facilities, and suspicious-looking chemicals.

24. Wash your hands thoroughly, using soap and warm water, after working in the science laboratory. This practice is especially important when you handle chemicals, biological specimens, and microorganisms.

 **Student Safety Contract**

Many teachers have students sign Student Safety Contracts. They serve to show just how important safety really is. Create a Student Safety Contract that you and your teacher will sign that lists potential safety hazards and safety rules, and the reasons for these rules.

②A Controlled Experiments

Science is about observing things, asking questions, proposing solutions, and testing those solutions. One way of doing this is through controlled experiments. A controlled experiment is a test in which one variable (something that can change or vary in your experiment) is purposely and steadily changed to find out what (if any) effect occurs.

For instance, you may observe that plants grow upward and ask yourself, "Why?" Various solutions might suggest themselves. To test those solutions, you can design a controlled experiment. Perhaps you think that the reason plants grow upward is because they grow toward the light. This is your hypothesis. From your hypothesis, you can make a prediction that you can test through experimentation, for instance: "As the angle of light shining on plants steadily increases, the angle of the plants' growth should steadily increase."

To design a controlled experiment to test this hypothesis, you need to control, or keep constant, as many of the variables, or possible causes of the effect, as possible. You might conduct an experiment on five identical plants, ensuring that all growing conditions—water, nutrients, temperature, kind of light, etc.—are the same. You would then change one variable, the angle of light shone on the plant, and measure the results.

You can use a control group as well (**Figure 1**). In the plant experiment, you could have a plant that has sunlight shining on it from directly above. If the hypothesis is correct, this plant will not grow at an angle. It is the control for the experiment, and the other plants are the test cases. The control is set up exactly like the test cases, but no variable is changed.

Unfortunately, scientists can never be sure that they have controlled all possible causes of the effects. Because of this, they can never be absolutely sure that the conclusions they make are true. However, the more the results match the prediction, the more confident the scientists can be about their hypotheses.

The Process of Scientific Inquiry in Controlled Experiments

Scientists use an inquiry process to find answers to their questions. This process is also referred to as the Scientific Method. There are common components for an inquiry that allow scientists to duplicate experiments so results can be validated and communicated. These components are outlined in the flow chart in **Figure 2**.

It is important that you follow this process and use the related skills whenever you are asked to design and conduct an experiment, if you expect to find reliable answers to the questions you pose. You can use the flow chart as an overall checklist. You can also refer to the more detailed sections in this Skills Handbook that deal with each of the specific skills necessary for each part of the process.

Figure 1

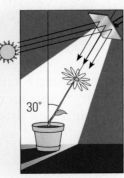

1 Asking a Question

All inquiry begins with curiosity. Our experience and observations of the world often lead us to ask questions. Ask a question that interests you or express an idea that can be tested. State the purpose of your experiment.

2 Making a Hypothesis

Research the subject of your question. Review the literature and find out as much as you can about previous information and discoveries surrounding your question.

Develop an educated guess that answers your initial question. This is your hypothesis. Make a prediction based on your hypothesis and state it as a cause-effect relationship.

3 Designing the Experiment

Identify all your variables.

Decide what materials and apparatus you will need to perform your experiment.

Write a procedure that explains how you will conduct your experiment. Be sure to take safety into account.

Draw a labelled diagram that illustrates your procedure and the materials and apparatus you will use.

Create a rough draft of tables for recording your data.

4 Conducting the Experiment

Follow the steps in the procedure carefully and thoroughly.

Use all equipment and materials safely, accurately, and with precision.

Record the variable(s) you are measuring, manipulating, and controlling.

Remember to repeat your experiment at least three times. If you are collecting quantitative data, take an average. This increases the accuracy and reliability of your results.

Make careful notes of everything that you observe during the experiment.

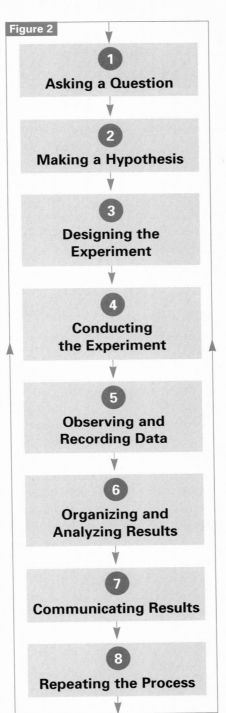

Figure 2

1. Asking a Question
2. Making a Hypothesis
3. Designing the Experiment
4. Conducting the Experiment
5. Observing and Recording Data
6. Organizing and Analyzing Results
7. Communicating Results
8. Repeating the Process

5 Observing and Recording Data

Describe the qualitative (anecdotal comments) and quantitative (numerical data) results clearly and accurately.

6 Organizing and Analyzing Results

When appropriate, create graphs to make better sense of quantitative data from your tables.

Study your qualitative observations and quantitative observations.

Identify patterns and trends.

Make a conclusion. Decide whether your results support, partially support, or refute your hypothesis.

Develop an explanation for your conclusion.

Apply your findings to your life today. Think about who will want to know about your discovery, how it will affect our lives, who it will benefit, and whether it could harm our world if used in a certain way.

Reflect on your experiment. Explore any sources of experimental error in your process. Think about any changes you would make if you were to conduct this experiment in the future.

7 Communicating Results

It is important that scientists share their findings. In order to have their investigations repeated and validated, it is common for scientists to publish details of their research, results, and conclusions.

8 Repeating the Process

Most scientists must complete an experiment many times before making important discoveries (e.g., a cure for cancer). If your experiment did not answer the question you initially asked, it must be revised and repeated until the question is answered or the problem is solved.

Try This Getting It Right

Research two famous Canadian scientists, Dr. F. G. Banting and C. H. Best, then answer the following questions:

1. How many days, months, or years did it take Banting and Best to discover insulin?

2. Did they find the solution to their problem by doing their experiment once? If your answer was no, find out how many times they had to repeat their experiment before getting it right.

②B Correlational Studies

Is it wise for a pregnant woman to consume alcohol, or smoke, or drink coffee? In attempting to answer questions like these, it is often difficult to know all the variables, let alone control them. A correlational study is an alternative to a controlled experiment. In a correlational study, a scientist examines whether a variable is affecting another variable without purposely changing any of the variables. Instead, variables are allowed to change naturally.

You are probably more familiar with correlational studies than with experiments. These studies are frequently summarized in newspapers, under headlines like "Chemical X Suspected in Heart Disease." Next time you are reading a newspaper, try to decide which articles would be classified as correlational studies.

Correlational studies are often chosen to test hypotheses that may be unsafe, impossible, or unethical to test with a controlled experiment. For instance, you might observe that some students appear to be hard of hearing and that they also spend a lot of time listening to loud music. This might lead you to hypothesize that listening to loud music damages hearing. It would be unethical to make students listen to loud music and then measure hearing damage. However, you could ask many students what the volume settings are on their radios and tape and CD players, determine how loud those settings are with a sound meter, and then test the students' hearing. Graphing one variable against the other will suggest whether there is a relationship between them.

It is more difficult to isolate cause and effect in correlational studies. Any two variables can be compared. It is important for the scientist to assess whether a reasonable link is possible. As an extreme example, one could check the annual tea production in China against the frequency of taxi accidents in Hamilton and discover that the years of highest tea production correspond to the years of the greatest number of taxi accidents. Could you expect to predict the frequency of accidents in the future by the amount of tea grown in China? This kind of correlation is likely to be a coincidence.

The Process of Scientific Inquiry in Correlational Studies

The flow chart in **Figure 1** outlines the components that are important in designing a correlational study. By following this format, investigators can do science without doing experiments or fieldwork. Instead, they can use data banks, CD-ROMs, the Internet, interviews, and surveys to find relationships between two or more variables. They can also, of course, make their own observations and measurements. You can use this flow chart as a checklist to make sure you use all the steps necessary in completing a reliable correlational study.

Look at the flow charts outlining Controlled Experiment and Correlational Studies. Can you identify major similarities and differences between a controlled experiment and a correlational study?

 What Kind of Investigation?

With a partner, look through some scientific journals and find five articles. Decide whether each article is a controlled experiment or a correlational study. Give reasons for your answers. Organize your information in a chart.

① Making an Observation

Choose a topic that interests or puzzles you.

② Asking a Question

Ask a question. State the purpose of the study.

③ Developing a Hypothesis

Research your subject. Review the literature and find out as much as you can about previous information, predictions, and discoveries surrounding your question.

Develop an educated guess that answers your initial question. This is your hypothesis.

Research alternative hypotheses and revise your original hypothesis.

④ Gathering Data

Your data can be collected by measuring, interviewing, or taking a survey. You can also use data collected by others (e.g., governments, banks, insurance companies, and scientific organizations).

Choose two variables for further study. (You are looking for one variable that may have a natural effect on the other.) Develop a hypothesis about them.

Develop or find ways to measure the two variables.

Plan to make as many measurements as it takes to get a continuous and wide collection of measurements of the variables.

Try to make measurements and observations in the same way each time.

⑤ Organizing and Recording Results

Plan appropriate tables to hold your observations.

⑥ Analyzing Results and Conclusions

Using the data in your tables, develop graphs in which the two variables you have chosen are plotted against each other.

Describe all results, including observations you made that you could not measure.

Describe the strengths and weaknesses of the methods you used.

Conclude whether the results support your original hypothesis. If not, suggest a modified hypothesis.

Make suggestions for future work on this topic.

⑦ Reporting on the Study

Tell others about your work and your conclusions. Consider any new questions that might require further study.

Figure 1

3A Research Skills

There is an incredible amount of information available in our society, from a number of different sources, including the newspaper, the Internet, and the neighbourhood or school library. Your generation has the ability to access more information than any other generation in history. Information is simply information, however. Before you can make effective use of it, you must know how to gather information efficiently. As well, you must know how to assess its credibility.

Where do you obtain your information? Think about all the information sources that you interact with in the course of a typical day. Now think about where you might go to access other sources of information. Compare your list of available resources with the list shown in **Table 1**. Certainly, there is no shortage of resources.

General Research Tips

- Before you begin, list the most important words associated with your research, so you can search for appropriate topics.
- Brainstorm a list of possible resources. Rank the list, starting with the most useful source.
- Use a variety of resources.
- Ask yourself: "Do I understand what this resource is telling me?"
- Check when the resource was published. Is it up-to-date?
- Keep organized notes or files while doing your research.
- Keep a complete list of the resources you used, so you can make a bibliography when writing your report. (See Reporting Your Work on page 564, for proper referencing formats.)
- Review your notes. After your research, you may want to alter your original problem or hypothesis.

Table 1

Information Consultants (people who can help you locate and interpret information)		Reference Materials (sources of packaged information)	
teachers	business people	encyclopedias	bibliographies
nurses	scientists	magazines/journals	newspapers
public servants	librarians	videotapes	slides
volunteers	veterinarians	data bases	almanacs
lawyers	senior citizens	yearbooks	maps
parents	doctors	charts	radio
farmers	politicians	films	dictionaries
members of the media		biographies	textbooks
		pamphlets	television
		filmstrips	records

Places (sources beyond the walls of your school)		Electronic Sources	
public libraries	shopping malls	world wide web (www)	
parks	colleges	CD-ROMs	
research laboratories	government offices	on-line search engines	
historic sites	zoos	on-line periodicals	
universities	volunteer agencies	computer programs	
businesses	museums		
farms	hospitals		
art galleries			

Some Specific Research Tips

Using a Library

School libraries use an international system, known as the Dewey Decimal Classification System, to organize books into the major subject areas shown in **Table 2**.

Table 2

Catalogue #	Subject Area
000	Generalities
100	Philosophy & Psychology
200	Religion
300	Social Science
400	Language
500	Natural Science & Mathematics
600	Technology (Applied Sciences)
700	The Arts
800	Literature
900	Geography/History

Using On-line Sources

"Think of an [Internet] search engine as a dog whistle. Blow it in a kennel and you'll just attract dogs. Blow it in a zoo, and you'll get a few dogs, plus many other creatures with good high frequency hearing: maybe some lions or tigers, hyenas, coyotes, timber wolves, perhaps a moose... The point is this—the Internet is a zoo." from Jeffery K. Pemberton's column in *Online User* magazine, May/June 1996.

The Internet is a large and fairly unstructured network. Many Internet navigation programs access and make use of several search engines. Search engines are on-line tools that find web sites (or hits) based on key words you enter into the search. When researching on-line, consider these additional research tips.

- If you have a specific web site address, go to the direct source first.
- If you need to use a search engine, find two good search engines, bookmark them, and use them. While the "Search" buttons in navigation programs are useful, they will not always give you the best search engine for your purposes.
- Learn about the features of the search engine. The better search engines make use of Boolean logic, operations that are used to combine key words when searching the Internet. In these cases, adding the word "AND" can let you narrow your search by combining two key words. Adding the word "OR" can let you expand your search by joining together two key words. Lastly, adding the word "NOT" enables you to disregard a key word. Be sure to refine your search as you continue. For example, entering the word "music" may give you 1 678 243 hits but adding "AND guitar" will reduce your search results to 18 860 hits. Including "NOT classical" will further reduce your number of hits to 2340.
- Familiarize yourself with the advanced search option within each search engine.
- Look over the first few pages of hits before you start exploring each web site. Be patient and go with your intuition.

Try This Canadian Scientists

With a partner, use the General Research Tips as a guide to research the contributions and life of one female Canadian scientist and one male Canadian scientist. You may want to choose your scientists from the following list:

Roger Daley (Meteorologist)
Birute Galdikas (Expert on Orangutans)
Gerhard Herzberg (Physicist)
Doreen Kimura (Behavioural Psychologist)

Julia Levy (Microbiologist)
John Polanyi (Chemist)
Endel Tulving (Expert on Human Memory)
Irene Uchida (Expert on Down Syndrome)

③B Critical Thinking

Communicating the results of your own work is important, but what about the work of others? Understanding and evaluating the work of others is an important part of communication. Think about how many messages, opinions, and pieces of information you hear and see every day. When you do research, you may access information via the Internet, textbooks, magazines, chat lines, television, radio, and through many other forms of communication. Is all of this information "correct"? Are all of these information sources "reliable"? How do we know what to believe and what not to believe?

Every day you see and hear extraordinary claims about objects and events (**Figure 1**). Often science, or the appearance of science, is used as a way of convincing us that the claims are true. Sometimes this method of reporting is used to get you to buy something or just to catch your interest. Even serious stories on scientific work are sometimes difficult to interpret, especially when they are reported in a way that makes the scientific work sound important, official, and somewhat mysterious.

To analyze information, you have to use your mind effectively and critically. When you encounter a "scientific" report in the media, analyze the report carefully, and see if you can identify the following:
- the type of investigation that is being reported (experiment or correlational study)
- the dependent and independent variables in any reported investigation (refer to Designing a Procedure, page 536)
- the strengths and weaknesses in the design of the investigation.

Try This — Scientific Journalism

Analyze a scientific report or story from one of the following magazines by performing the tasks listed above—

Scientific American   *Popular Mechanics*
Discover Magazine   *Maclean's*

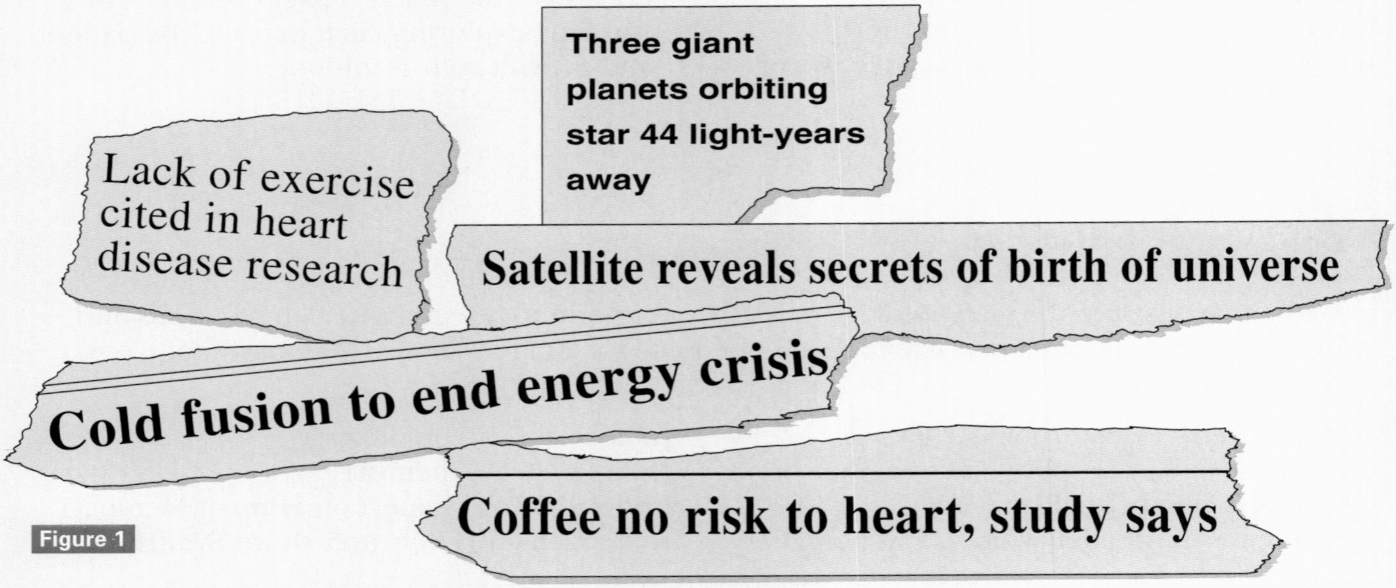

Lack of exercise cited in heart disease research

Three giant planets orbiting star 44 light-years away

Satellite reveals secrets of birth of universe

Cold fusion to end energy crisis

Coffee no risk to heart, study says

Figure 1

PERCS

A lot of research has been done to help people develop critical thinking skills. One very good, practical framework is known as PERCS. It was founded at Central Park East Secondary School (CPESS) in New York City, NY. In the interest of preparing themselves for their world, the students at CPESS use a series of questions to help them think critically about information and arguments. You might think of this framework as a box of useful "tools" that will allow you to effectively build an educated and critical opinion concerning an issue.

The PERCS Checklist

P = Perspective
From whose viewpoint are we seeing or reading or hearing?
From what angle or perspective?

E = Evidence
How do we know what we know?
What's the evidence and how reliable is it?

R = Relevance
So what?
What does it matter?
What does it all mean?
Who cares?

C = Connections
How are things, events, or people connected to each other?
What is the cause and what is the effect?
How do they "fit" together?

S = Supposition
What if….?
Could things be otherwise?
What are or were the alternatives?
Suppose things were different.

 **PERCS**

Try exercising your critical thinking skills. Reread the article that you selected in the last Try This or read the article below. Using the questions listed in the PERCS checklist, critically think about the article you chose.

Study Concludes Children of Smokers More Likely to Be Criminals

According to a study carried out by North American and European researchers this year, the child of a mother who smoked during pregnancy is more likely to commit a crime than the child of a mother who is a non-smoker.

Researchers looked at the statistics for violent and nonviolent crimes committed by men born around 1960 and investigated the smoking habits of their mothers during pregnancy. They found that the sons of women who smoked the most had a one in four probability of getting into trouble with the law.

The discoveries provide evidence to support the implementation of education campaigns aimed at persuading pregnant mothers not to smoke.

Suggestions for further study include an investigation into whether smoking affects fetal brain development, which might in turn affect antisocial behaviour. Alternatively, the observed factors might be coincidental, both possibly caused by genes carried by both mothers and sons.

4A Asking Questions and Hypothesizing

Asking Questions

Have you ever noticed that balloons stick to walls if you rub them on your head? This observation might lead you to wonder about a number of things: Does a balloon stick better if you rub it more times? Does the length or texture of a person's hair affect how well a balloon sticks? Does it matter how inflated the balloon is or what colour it is?

Each of these questions is the basis for a sound scientific investigation. With a little rewording, it will become clear how these questions can be tested. Not every question that you will want an answer for can be tested: some are too general, or too vague. Learning to ask questions that can be tested takes time and is a fundamental skill in scientific inquiry.

The first balloon-related question could be stated more usefully: "What is the effect of increasing the number of times that a balloon is rubbed on a person's head on the length of time that the balloon stays stuck to a wall?" You can probably already begin to plan an experiment that would answer this question. A testable question is often about cause-effect relationships. These questions often take the form: "What *causes* the change in variables?" and "What are the *effects* on a variable if we change another variable?" As you know, a variable is something that can change or vary in an investigation.

Scientists call the cause variable the independent variable. This is the one thing in the experiment that you purposely change. For instance, increasing the number of times that a balloon is rubbed on a person's head is

---

Try This · Can You Test It?

Decide whether each of the questions in **Table 1** would be testable or impossible to test. Then make up some testable questions of your own.

Table 1

Question	Testable		Not Testable
	Independent (Cause) variable	Dependent (Effect) variable	
What is the effect of increasing the number of times that a balloon is rubbed on a person's head on the length of time that the balloon stays stuck to a wall?	balloon rubbed on a person's head	the length of time the balloon stays stuck to a wall	?
Does listening to loud music affect a teenager's ability to hear?	?	?	?
Does a rock think?	?	?	?
Do tennis balls bounce differently when wet?	?	?	?
Is there extra-terrestrial life in our universe?	?	?	?

something you control: you are changing the independent variable. The effect variable is called the dependent variable. This is what you measure in your experiment (e.g., time, distance) and it "depends" on the variable you purposely change (independent variable). The amount of time the balloon stays on the wall is the dependent variable.

A scientific question that asks what happens to a dependent variable when we change the independent variable is a question you can test.

Hypothesizing

A suggested answer or reason why one variable affects another in a certain way is called a hypothesis. Often you have some idea about this even as you ask your question. Having noticed, for instance, that a balloon sticks to the wall after you rub it in your hair, you might predict that rubbing it more will make it stick longer. You might be wrong, of course, but your prediction is probably based on past observations, on logic, and on bits of scientific theory you may remember. If you're really interested, you may even do some research based on what you already know. If you then pull everything you know together and express it, you would have a hypothesis. For instance, you might predict that the balloon sticks to the wall because it is attracted by static electricity, and that rubbing the balloon more produces a greater static electric charge on the balloon.

Predictions and hypotheses go hand in hand (**Table 2**). The hypothesis is how you can explain a prediction. The prediction is what you test through your experiment. And, if the experiment confirms the prediction, you can have more confidence that your hypothesis is correct.

Table 2 **Sample Hypotheses and Predictions**

Hypothesis (possible reason for cause-effect relationship)	Prediction	
	Possible cause (independent variable)	Possible effect (dependent variable)
Candy contains sugar that is used for energy by germs in the mouth, and these germs produce an acid that decays the teeth.	As the amount of candy that a person eats increases…	…the number of tooth cavities increases.
A larger sail traps more air, which then provides a greater force to a boat.	As the size of the sail increases…	…the top speed of the sailboat increases.
Salt helps oxygen in the air combine with iron in the metal of a bicycle to form rust.	As the amount of salt on a road increases…	…the amount of rusting of the metal parts of a bicycle increases.

Try This **Hypothesize**

Now it's your turn. With a partner, think of a few hypotheses and predictions that could be tested. Use a chart similar to **Table 3** to record your thoughts.

Table 3

Hypotheses and Predictions of Cause-Effect Relationships

Hypothesis (possible reason for cause-effect relationship, an educated guess)	Possible cause (independent variable)	Possible effect (dependent variable)
?	?	?
?	?	?

4B Designing a Procedure

Now that you are an expert at asking testable questions and hypothesizing about their possible answers, it is time to learn to design an experiment that will test your hypothesis.

Remember that an independent variable is the possible cause and a dependent variable is what the effect will be. Another important variable in an experiment is called the controlled variable. One way to control an experiment is to control—keep constant—all known possible causes of the result except one. As is shown on page 526, for example, there are many factors that must remain the same in order for the experiment on plant growth to be reliable. What would the independent variable be in this experiment? What would the dependent variable be in this experiment? What must be controlled in this experiment to make the results reliable?

To design a procedure that will test your hypothesis, you must identify your variables and design a control. You must decide how you will change your variable, and what you will observe and/or measure with each change. You must also decide how regularly you will make your observations and how you will record them. This may require you to create some tables for recording your data. Obviously, you must also decide what materials and equipment you will use. It is useful to create a labelled diagram that illustrates the materials and equipment you will need, and the procedure you are going to use.

Writing a Procedure

This is an essential component of an experimental design. Anyone who is interested in learning about your experiment needs to be able to understand how it was performed, so that it can be duplicated exactly. Therefore, it is important that you be able to write an experimental procedure clearly, concisely, and accurately.

When writing a procedure you should use
- numbered steps
- passive voice (avoid using pronouns)
- past tense

For example, the first two steps of a procedure could look like this:

Procedure

1. The experiment was set up as shown in the diagram.

2. The temperature of the water was measured every 5 min.

Don't forget to consider all possible safety issues!

 **Write a Procedure**

Think about how you would conduct an experiment that would give the answer to the question, *Do plants grow toward sunlight?* Write a procedure for this plant experiment, using the format explained above.

5A Laboratory Equipment

You can do many science investigations using everyday materials and equipment. In your science classroom, there are other pieces of equipment. Some of these are illustrated in **Figure 1**.

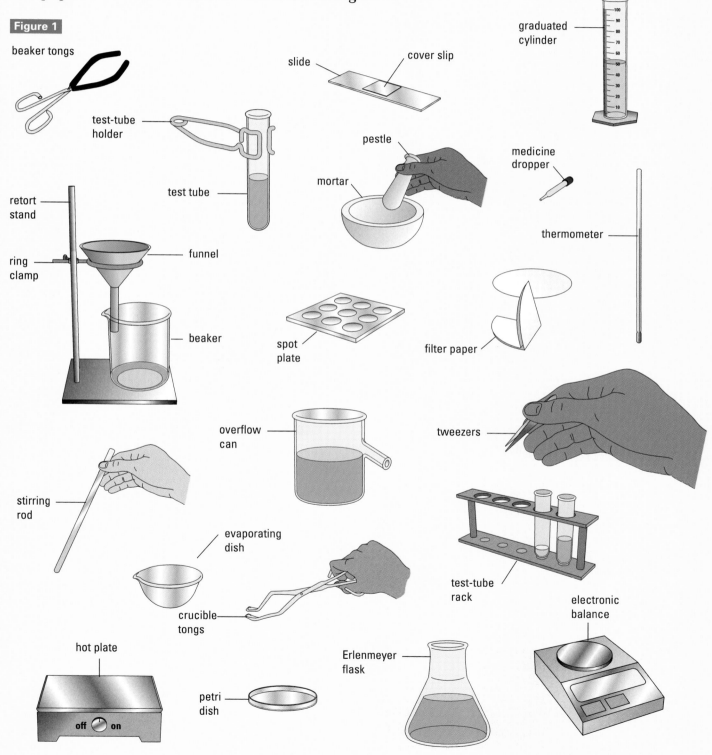

beaker tongs

slide

cover slip

graduated cylinder

test-tube holder

pestle

medicine dropper

test tube

mortar

retort stand

thermometer

ring clamp

funnel

beaker

spot plate

filter paper

overflow can

tweezers

stirring rod

evaporating dish

test-tube rack

crucible tongs

electronic balance

hot plate

off   on

petri dish

Erlenmeyer flask

⑤B Using the Microscope

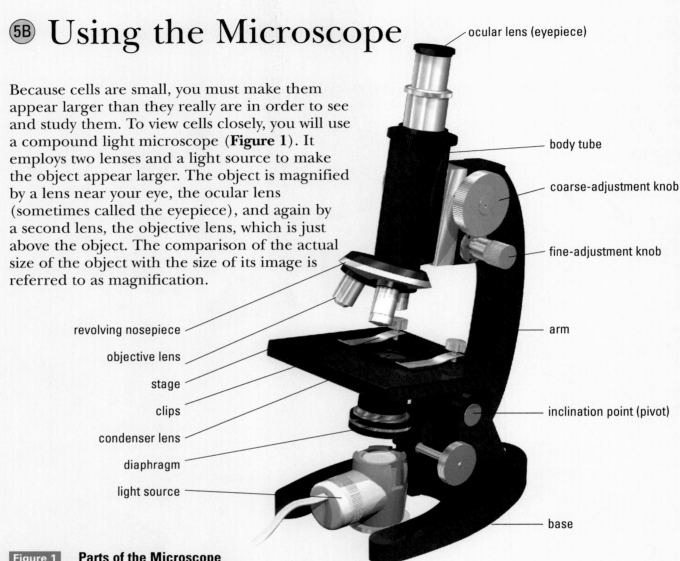

Because cells are small, you must make them appear larger than they really are in order to see and study them. To view cells closely, you will use a compound light microscope (**Figure 1**). It employs two lenses and a light source to make the object appear larger. The object is magnified by a lens near your eye, the ocular lens (sometimes called the eyepiece), and again by a second lens, the objective lens, which is just above the object. The comparison of the actual size of the object with the size of its image is referred to as magnification.

Figure 1   **Parts of the Microscope**

Structure	Function
stage	Supports the microscope slide. A central opening in the stage allows light to pass through the slide.
clips	Found on the stage and used to hold the slide in position.
diaphragm	Regulates the amount of light reaching the object being viewed.
objective lenses	Magnifies the object. Usually three complex lenses are located on the nosepiece immediately above the object or specimen. The smallest of these, the low-power objective lens, has the lowest magnification, usually four times (4X). The medium-power lens magnifies by 10X, and the long, high-power lens by 40X.
revolving nosepiece	Rotates, allowing the objective lens to be changed. Each lens clicks into place.
body tube	Contains ocular lens, supports objective lenses.
ocular lens	Magnifies the object, usually by 10X. Also known as the eyepiece, this is the part you look through to view the object.
coarse-adjustment knob	Moves the body tube up or down so you can get the object or specimen into focus. It is used with the low-power objective only.
fine-adjustment knob	Moves the tube to get the object or specimen into sharp focus. It is used with medium- and high-power magnification. The fine-adjustment knob is used only after the object or specimen has been located and focused under low-power magnification using the coarse adjustment.
condenser lens	Directs light to the object or specimen.

Basic Microscope Skills

The skills outlined below are presented as sets of instructions. This will enable you to practise these skills before you are asked to use them in the investigations in *Nelson Science 9*.

Materials
- newspaper that contains lower-case letter "f," or similar small object
- scissors
- microscope slide
- cover slip
- dropper
- water
- compound microscope
- thread
- compass or petri dish
- pencil
- transparent ruler

Preparing a Dry Mount

This method of preparing a microscope slide is called a dry mount, because no water is used.

1. Find a small, flat object, such as a lower-case letter "f" cut from a newspaper.

2. Place the object in the centre of a microscope slide.

3. Hold a cover slip between your thumb and forefinger. Place the edge of the cover slip to one side of the object. Gently lower the cover slip onto the slide so that it covers the object.

Step 3

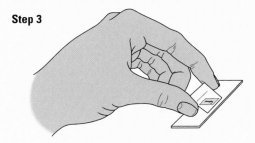

Preparing a Wet Mount

This method of preparing a microscope slide is called a wet mount, because water is used.

1. Find a small, flat object.

2. Place the object in the centre of a microscope slide.

3. Place two drops of water on the object.

Step 3

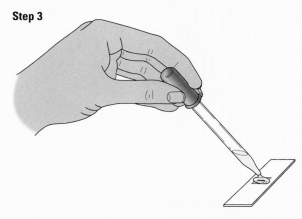

4. Holding the cover slip with your thumb and forefinger, touch the edge of the surface of the slide at a 45° angle. Gently lower the cover slip, allowing the air to escape.

Step 4

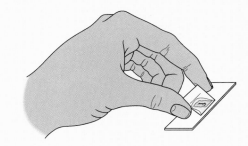

Positioning Objects Under the Microscope

1. Make sure the low-power objective lens is in place on your microscope. Then put either the dry or wet mount slide in the centre of the microscope stage. Use the stage clips to hold the slide in position. Turn on the light source.

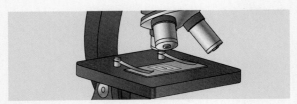

Step 1

2. View the microscope stage from the side. Using the coarse-adjustment knob, bring the low-power objective lens and the object as close as possible to one another. Do not allow the lens to touch the cover slip.

Step 2

3. View the object through the eyepiece. Slowly move the coarse-adjustment knob so the objective lens moves away from the slide, to bring the image into focus. Note that the object is facing the "wrong" way and is upside down.

4. Using a compass or a petri dish, draw a circle in your notebook to represent the area you are looking at through the microscope. This area is called the field of view. Look through the microscope and draw what you see. Make the object fill the same amount of area in your diagram as it does in the microscope.

5. While you are looking through the microscope, slowly move the slide away from your body. Note that the object appears to move toward you. Now move the slide to the left. Note that the object appears to move to the right.

6. Rotate the nosepiece to the medium-power objective lens. Use the fine-adjustment knob to bring the letter into focus. Note that the object becomes larger.

🖐 Never use the coarse-adjustment knob with the medium- or high-power objective lenses.

Step 6

7. Adjust the object so that it is directly in the centre of the field of view. Rotate the nosepiece to the high-power objective lens. Use the fine-adjustment knob to focus the image. Note that you see less of the object than you did under medium-power magnification. Also note that the object seems closer to you.

Investigating Depth of Field

The depth of field is the amount of an image that is in sharp focus when it is viewed under a microscope.

1. Cut two pieces of thread of different colours.

2. Make a temporary dry mount by placing one thread over the other in the form of an X in the centre of a microscope slide. Cover the threads with a cover slip.

Step 2

3. Place the slide on the microsocope stage and turn on the light.

4. Position the low-power objective lens close to, but not touching, the slide.

5. View the crossed threads through the ocular lens. Slowly rotate the coarse-adjustment knob until the threads come into focus.

6. Rotate the nosepiece to the medium-power objective lens. Focus on the upper thread by using the fine-adjustment knob. You will probably notice that you cannot focus on the lower thread at the same time. The depth of the object that is in focus at any one time represents the depth of field.

7. Repeat step 6 for the high-power objective lens. The stronger the magnification, the shallower the depth of field.

Determining the Field of View

The field of view is the circle of light seen through the microscope. It is the area of the slide that you can observe.

1. With the low-power objective lens in place, put a transparent ruler on the stage. Position the millimetre marks on the ruler immediately below the objective lens.

2. Using the coarse-adjustment knob, focus on the marks on the ruler.

3. Move the ruler so that one of the millimetre markings is just at the edge of the field of view. Note the diameter of the field of view in millimetres, under the low-power objective lens.

Step 3

4. Using the same procedure, measure the field of view for the medium-power objective lens.

5. Most high-power lenses provide a field of view that is less than one millimetre in diameter, so it cannot be measured with a ruler. The following steps can be followed to calculate the field of view of the high-power lens.

Calculate the ratio of the magnification of the high-power objective lens to that of the low-power objective lens.

$$\text{Ratio} = \frac{\text{magnification of high-power lens}}{\text{magnification of low-power lens}}$$

Use the ratio to determine the field of diameter (diameter of the field of view) under high-power magnification.

$$\text{Field diameter (high power)} = \frac{\text{field diameter (low power)}}{\text{ratio}}$$

Estimating Size

1. Measure the field of view, in millimetres, as shown above.

2. Remove the ruler and replace it with the object under investigation.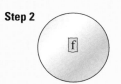

Step 2

3. Estimate the number of times the object could fit across the field of view.

4. Calculate the width of the object:

$$\text{width of object} = \frac{\text{width of field of view}}{\text{number of objects across field}}$$

Remember to include units.

Storage

When you complete an investigation using the micrscope, follow these steps:

1. Rotate the nosepiece to the low-power objective lens.

2. Remove the slide and cover slip (if applicable).

3. Clean the slide and cover slip and return them to their appropriate location.

4. Return the microscope to the storage area.

5C Using Star Maps

What Is a Star Map?

A star map shows the most easily seen stars in the sky, with many of the stars joined by lines into constellations. Each star map is designed for a range of latitudes, such as locations about 45° north of the equator. Thus, a star map designed for southern Canada cannot be used in Australia.

You can use a star map to help you recognize what you can see in the sky, and to observe the motions of objects as Earth goes through its cycles of rotation and revolution.

Maps for All Seasons

Different parts of the sky are visible during different times of the year. To show the different stars and constellations visible, different star maps have been designed for each of the seasons or even months.

Another way to take into consideration the changing skies is to use a seasonal star map (**Figure 1**) in which a "window" can be rotated to expose different parts of the map. Each visible section represents the portions of the sky that are visible at different times of the year. Activity 13.7 gives instructions on making this type of map.

Stargazing Trips

When you want to observe the night sky, consider these tips:

- Plan the trip in advance, taking into consideration the weather forecast, safety, transportation, location, what to wear, and what to bring.
- Choose a location far away, or at least screened away, from bright lights.
- Be prepared to record your observations.
- Before viewing, allow your eyes at least 10 min to adapt to the dark.
- Use a flashlight covered with red cellophane to view your star map.

Using a Star Map

To use a seasonal star map, follow these steps:
- Rotate the window to expose the part of the map closest to the current season or month.
- Hold the map, facing downward, above your head.
- Rotate the map so the top part (away from the window) is facing north. This means that Orion is facing south.
- Compare what you see in the sky with what is on the map.
- Also notice any planets or other objects besides stars.

Keeping Records

Use a table to record your observations. Possible titles for the columns are shown in **Table 1**. Be careful when recording dates because Dec. 15 becomes Dec. 16 after midnight.

Table 1

Date	Object seen	Description (including a diagram)	Location (including angles)	Questions I want answered
?	?	?	?	?
?	?	?	?	?

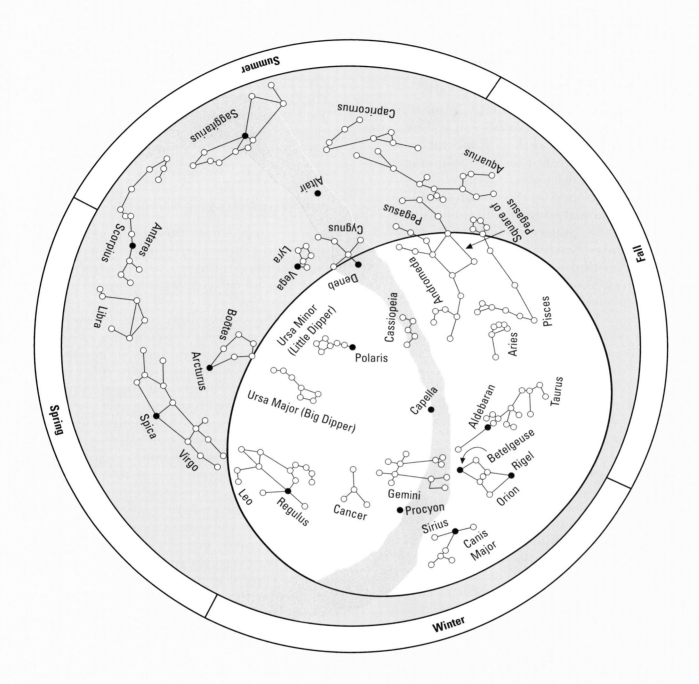

Figure 1

Only the unshaded part of the seasonal star map is visible during the winter months.

5D Drawing and Constructing Circuits

Sources of Electrical Energy

To provide the electrical energy in most of the circuits you use in this course you will be using combinations of dry cells or a special device called a power supply. Power supplies can be set to supply the voltage required.

The source used in the circuits you construct and test yourself will be a "direct current," or DC source of electrical energy. In DC circuits the current only flows in one direction around the circuit. We use DC circuits in this course because the operating voltages are much safer to use, generally below 28 V.

Wall outlets provide a different kind of electrical energy known as an "alternating current" or AC source. In AC circuits the electric current reverses its direction 60 times a second. AC appliances are specially designed for this energy source, and typically operate at 120 V or 240 V.

Drawing Circuit Diagrams

Before building a circuit, it is a good idea to draw a circuit diagram. This will remind you how components should be connected. There are some conventions to follow when drawing circuit diagrams: connecting wires are generally shown as straight lines or 90° angles, and symbols (shown in **Table 1**) are used to represent all components.

Safety Considerations

It is important to observe and use appropriate safety procedures, especially when you see ⚡ .
- Always ensure that your hands are dry, and that you are standing on a dry surface.
- Do not use faulty dry cells or batteries, do not connect different makes of dry cells in the same battery, and avoid connecting partially discharged dry cells to fully charged cells. Take care not to accidentally short-circuit dry cells or batteries.
- Do not use frayed or damaged connectors.
- Handle breakable components with care.
- Only operate a circuit after it has been approved by your teacher.

Constructing the Circuit

When constructing or modifying a circuit, always follow the instructions. If you are unsure of the procedure, ask for clarification. Check that all the components are in good working order.
- Check the connections carefully when linking connecting dry cells in series or in parallel. Incorrect connections could cause shorted circuits or explosions. Ask your teacher for clarification if you are unsure.
- When attaching connecting wires to meters, connect a red wire to the positive terminal and a black wire to the negative terminal of the meter. This will remind you to consider the polarity of the meter when connecting it in the circuit.
- Sometimes the ends of connecting wires do not have the correct attachments to connect to the device or meter. Use extra, approved attachment devices, such as alligator clips, but be careful to position the connectors so that they cannot touch one another.
- Open the switch before altering a meter connection or adding new wiring or components.
- If the circuit does not operate correctly, open the switch and check the circuit wiring and all connections to the terminals. If you still cannot find the problem, ask your teacher to inspect your circuit again.

Table 1 **Circuit Diagram Symbols**

	DC CIRCUITS		HOUSEHOLD CIRCUITS (additional symbols)	
Sources/Outlets		cell		wall outlet
		3-cell battery		range outlet
				single outlet
				double outlet (duplex)
				weatherproof outlet
				special-purpose outlet
Control Devices		switch		
		fuse		
		circuit breaker		
		switch and fuse		
		distribution panel		
	S	switch		
	S$_{WP}$	weatherproof switch		
		push button		
Electrical Loads		light bulb		ceiling light
		clock		wall light
		motor		lampholder with pull switch
		thermostat		recessed fixture
		resistor		television outlet
		variable resistor (rheostat)		fan
		fluorescent fixture		buzzer
		heating panel		bell
Meters		ammeter		
		voltmeter		
Connectors		conducting wire		
		wires joined		
		ground connection		

⑤E Using the Voltmeter and the Ammeter

As we cannot see electrons flowing in electric circuits, we have to rely on instruments that can detect and measure electricity. There are at least two you are likely to use: the voltmeter and the ammeter.

The Voltmeter

Figure 1

A voltmeter (**Figure 1**) measures the voltage difference between two different points in a circuit. The voltmeter can be connected across the terminals of a cell, to measure the voltage output of the cell, or across another component of a circuit, to measure the voltage drop across that component. In other words, the voltmeter is always connected in parallel with the component you want to investigate. The voltmeter could be digital (providing a digital readout) or analog (indicating voltage by the movement of a needle across a scale).

Reading an Analog Voltmeter

The needle on a voltmeter usually moves from left to right, with the zero voltage being on the left and the maximum voltage on the right of the scale. If the voltmeter scale has only one set of numbers, it is relatively easy to measure the voltage. Be sure that you know the voltage value represented by the smallest division on the scale.

The voltage in this circuit is 0.3 V.

Sample Problem:
If the voltmeter has several sets of numbers, identify which set of numbers matches the voltage range selected by the switch on the voltmeter.

If the switch on this voltmeter indicates that the voltmeter is measuring a maximum of 1 V, the voltage in this circuit would be 0.72 V. However, if the switch indicates a maximum of 5 V, the voltage in the circuit would be 3.6 V.

The two leads that connect a voltmeter to any part of the electric circuit must be attached so that the negative terminal of the voltmeter is connected to a more negative part of the circuit than the positive terminal. If the leads were attached incorrectly, the needle would try to move to the left, but would be unable to do so, and would not give a reading.

The Ammeter

An ammeter measures the amount of electric current flowing in a circuit. To measure the electric current we connect the ammeter directly into the circuit itself. In whatever part of the circuit we wish to measure the current, a wire is disconnected and the ammeter is connected, in series, to complete the circuit. A typical ammeter is shown in **Figure 2**. Reading digital and analog ammeters is very similar to reading digital and analog voltmeters. The unit of current is the ampere (A) or milliampere (mA).

Figure 2

6A Obtaining Qualitative Data

An observation is information that you get through your senses. You observe that a rose is red and has a sweet smell. You may also note that it has sharp thorns on its stem. You may count the petals on the flower and the leaves on the stem and measure the length of the stem.

When people describe the qualities of objects and events, the observations are qualitative. The colour of the rose, the odour of the flower, and the sharpness of the thorns are all qualitative observations.

Scientists have grouped qualitative observations into several categories, based on the kind of qualities of the object or event being described. The following is a list of categories that can be used to qualitatively describe objects or events:

State of Matter: One of three states—solid, liquid, or gas.

Colour: Objects can be described as being any colour or any shade of colour. Materials that have no colour should be described as colourless.

Smell: Also known as odour. There are many words to describe smells, including pungent, strong, spicy, sweet, and odourless.

Texture: The surfaces of objects can have a variety of textures, including smooth, rough, prickly, fine, and coarse.

Taste: Objects can taste sweet, sour, bitter, or salty. Other tastes are combinations of these basic tastes. Objects that have no taste can be described as tasteless.

Shininess: Also known as lustre. Objects with very smooth surfaces that reflect light easily, like mirrors, are said to be shiny or lustrous. Objects with dull surfaces are said to be non-lustrous.

Clarity: Some substances let so much light through that letters can be read through them. These substances are said to be clear or transparent. Other substances that allow light through, but not in a way that allows you to see through them, are translucent. Objects that do not let light through are opaque.

Other qualitative descriptions include form (the shape of a substance), hardness, brittleness (how easily the substance breaks), malleability (the ability of the object to be changed into another shape), and viscosity (a liquid's resistance to flow).

Another important characteristic that can be described qualitatively is the ability of substances to combine with each other.

Recording Data

Scientists must record rough notes and observations. Using a science journal or log book will help you organize and keep track of your rough work while performing investigations.

 **Observing**

With a partner, search the classroom or the outside environment. Find five objects and make as many qualitative observations as you can about each one.

⑥Ⓑ Obtaining Quantitative Data

Observations that are based on measurements or counting provide quantitative data, since they deal with quantities of things. The length of a rose's stem, the number of petals, and the number of leaves are quantitative observations.

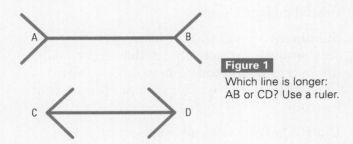

Figure 1

Which line is longer: AB or CD? Use a ruler.

In **Figure 1** you will find that AB and CD are the same length. Our senses can be fooled. That is one reason why quantitative observations are important in science and it is also the reason measurements must be made carefully.

Standards of Measurement

Units of measurement used to be based on local standards that the community had agreed to. Horse heights, for instance, are still measured in hands, based on the width of a hand, and measured from the ground to the horse's shoulder.

These standards may sound strange, but the unit of length that replaced most local standards, the metre, was based on an arc on Earth that ran from the equator, through Barcelona, Paris, and Dunkirk, to the North Pole. The length of this arc, divided by 10 000 000, equalled 1 m. That also sounds strange, but it established the first standard unit of length that the whole world could use.

Later, when more accurate measurements of Earth were possible, scientists abandoned this definition because it was not precise enough. The metre is now defined very accurately as the distance travelled by light in 1/299 792 458 s (3.33×10^{-9} s).

The metric system has been adopted by Canada. You should be familiar with, and use, the units from this international system (also called the SI system, from the French name: *Système International d'Unités*).

Base Units and Prefixes

There are seven base units, shown in **Table 1**.

Table 1	The Seven SI Base Units	
Quantity	**Unit**	**Symbol**
length	metre	m
mass	kilogram	kg*
time	second	s
electric current	ampere	A
temperature	kelvin	K
amount of substance	mole	mol
light intensity	candela	cd

*The kilogram is the only base unit that contains a prefix. The gram proved to be too small for practical purposes.

Larger and smaller units are created by multiplying or dividing the value of the base units by multiples of 10. For example, the prefix *deca* means multiplied by 10. Therefore, one decametre (1 dam) is equal to ten metres (10 m). The prefix *kilo* means multiplied by 1000, so one kilometre (1 km) is equal to one thousand metres (1000 m). Similarly, each unit can be divided into smaller units. The prefix milli, for example, means divided by 1000, so one millimetre (1 mm) is equal to 1/1000 of a metre.

To convert from one unit to another, you simply multiply by a conversion factor. For example, to convert 12.4 m to centimetres, you use the relationship 1 m = 100 cm.

12.4 m = ? cm

$$12.4 \text{ m} \times \frac{100 \text{ cm}}{1 \text{ m}} = 1240 \text{ cm}$$

To convert 6.3 g to kilograms, you use the relationship 1000 g = 1 kg.

6.3 g = ? kg

$$6.3 \text{ g} \times \frac{1 \text{ kg}}{1000 \text{ g}} = 0.0063 \text{ kg}$$

Any conversions of the same physical quantities can be done in this way. The conversion factor is chosen so that, using cancellation, it yields the desired unit.

Once you understand this method of conversion, you will find that you can simply move the decimal point. Move it to the right when the new unit is smaller and to the left when the new unit is larger. As you can see from **Table 2**, not all the units and prefixes are commonly used.

Table 2		**Metric Prefixes**	
Prefix	**Symbol**	**Factor by which the base unit is multiplied**	**Example**
giga	G	10^9 = 1 000 000 000	
mega	M	10^6 = 1 000 000	10^6 m = 1 Mm
kilo	k	10^3 = 1 000	10^3 m = 1 km
hecto	h	10^2 = 100	
deca	da	10^1 = 10	
		10^0 = 1	m
deci	d	10^{-1} = 0.1	
centi	c	10^{-2} = 0.01	10^{-2} m = 1 cm
milli	m	10^{-3} = 0.001	10^{-3} m = 1 mm
micro	μ	10^{-6} = 0.000 001	10^{-6} m = 1 μm

Table 3 shows the quantities that you should be familiar with.

Table 3	**Common Quantities and Units**	
Quantity	**Unit**	**Symbol**
length	kilometre	km
	metre	m
	centimetre	cm
	millimetre	mm
mass	tonne (1000 kg)	t
	kilogram	kg
	gram	g
area	hectare (10 000 m^2)	ha
	square metre	m^2
	square centimetre	cm^2
volume	cubic metre	m^3
	litre	L
	cubic centimetre	cm^3
	millilitre	mL
time	minute	min
	second	s
temperature	degrees Celsius	°C
	kelvin	K
force	newton	N
energy	kilojoule	kJ
	joule	J
pressure	kilopascal	kPa
	pascal	Pa

Problems in Measurement

Many people believe that all measurements are accurate and dependable. But there are many things that can go wrong when measuring. The instrument may be faulty. Another similar instrument may give different readings. There may be limitations that make the instrument unreliable. The person making the measurement may also make a mistake.

When measuring the temperature of a liquid, for instance, it is important to keep the bulb of the thermometer near the middle of the liquid. If the liquid is being heated and the thermometer is simply sitting in the container with its bulb at the bottom, you will be measuring the temperature of the bottom of the container, not the temperature of the liquid. There are similar concerns with most measurements. To be sure that you have measured correctly, repeat your measurement at least three times. If your measurements are close, calculate the average and use that number. To be more certain, repeat the measurements with a different instrument.

Try This Measuring Mass

Work as a group. Your teacher will give you a bottle of vitamin C tablets. Each tablet is advertised as having the same mass (measured in milligrams). Check to make sure this is true. Decide how your group will prove or disprove this. Be sure to record all data in a table. What is the average mass of one tablet? Compare your quantitative observations with those of other groups.

6C Scientific Drawing

Scientific drawings are done to record observations as accurately as possible. They are also used to communicate, which means they must be clear, well labelled, and easy to understand. Following are some tips that will help you produce useful scientific drawings.

Before You Begin

- Obtain some blank paper. Lines might obscure your drawing or make your labels confusing.
- Find a sharp, hard pencil (e.g., H or 2H). Avoid using pen, thick markers, or coloured pencils. Ink can't be erased—even the most accomplished artists change their drawings—and coloured pencils are soft, making lines too thick.
- Plan to draw large. Ensure that your drawing will be large enough that people can see details. For example, a third of a page might be appropriate for a diagram of a single cell or a unicellular organism. If you are drawing the entire field of view of a microscope, draw a circle with a reasonable diameter (e.g., 10 cm) to represent the field of view.
- Leave space for labels, preferably on the right side of the drawing.
- Observe and study your specimen carefully, noting details and proportions.

Drawing

- Simple, two-dimensional drawings are effective.
- Draw only what you see. Your textbook may act as a guide, but it may show structures that you cannot see in your specimen.
- Do not sketch. Draw firm, clear lines, including only relevant details that you can see clearly.
- Do not use shading or colouring in scientific drawings. A stipple (series of dots) shown in **Figure 1** may be used to indicate a darker area. Use double lines to indicate thick structures.

Label Your Drawing

- All drawings must be labelled fully in neat printing. Avoid printing labels directly on the drawing.
- Use a ruler. Label lines must be horizontal and ruled firmly from the structures being identified to the label (**Figure 2**).
- Label lines should never cross.
- If possible, list your labels in an even column down the right side.
- Title the drawing, using the name of the specimen and (if possible) the part of the specimen you have drawn. Underline the title.

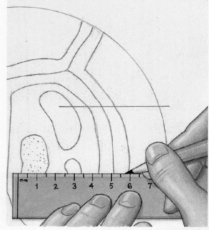

Figure 2

Figure 1

Scale Ratio

- To show the relation of the actual size to your drawing size, print the scale ratio of your drawing beside the title.

$$\text{scale ratio} = \frac{\text{size of drawing}}{\text{actual size of the specimen}}$$

For example, if you have drawn a nail (**Figure 3**) that is 5 cm long and the drawing is 15 cm long, then the scale ratio, which in this case is a magnification, is

$$\frac{15\ cm}{5\ cm} = 3X$$

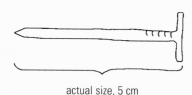

actual size, 5 cm

Figure 3

The magnification is always written with an "X" after it. In a fully labelled drawing, the total magnification of the drawing should be placed at the bottom right side of the diagram. If the ocular lens magnified a specimen 10X, the low-power objective (4X) was used, and the scale ratio was 3X, the total magnification of the diagram would be as follows:

$$\text{Total Magnification} = \text{Ocular Lens} \times \text{Low-Power Objective Lens} \times \text{Scale Ratio}$$
$$= 10 \times 4 \times 3$$
$$= 120X$$

The total magnification should be written on the bottom right-hand side of the diagram, as shown in **Figure 4**.

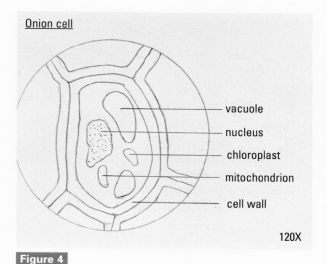

Onion cell

— vacuole
— nucleus
— chloroplast
— mitochondrion
— cell wall

120X

Figure 4

Checklist for Good Scientific Drawing

✔ Use blank paper and a sharp, hard pencil.
✔ Draw large. For example, the field of view for a microscope could be 10 cm in diameter.
✔ Do not shade or colour.
✔ Draw label lines that are straight and parallel and run outside the drawing. Use a ruler for this!
✔ Include labels, a title, and the total magnification.

Try This Draw It

Obtain a specimen from your classroom. It could be a piece of your hair, chalk dust, or something else you would be interested in looking at under a microscope. Prepare a dry mount slide and focus your specimen under the medium-power objective lens. Complete a scientific drawing of your specimen. Use the checklist to ensure that your diagram is accurate and complete. When your drawing is complete, exchange it with your friend's drawing. Note the strengths and weaknesses of the drawing, keeping in mind all the features of a good scientific drawing. Evaluate your friend's drawing using the checklist.

⑥D Creating Data Tables

Creating effective data tables in your investigations will help you record and analyze your data. Constructing a useful data table is one of the first steps in making sense of your experimental data. Take a look at **Tables 1** and **2**. What similarities exist? What strategies should you employ when constructing your data tables?

The following checklist will help you in constructing effective data tables in your investigations:

- List the dependent variable(s) (the effect) along the top of the table.
- List the independent variable (the cause) along the side of the table.
- Be sure that each data table has a descriptive, yet concise, title.
- Be sure to include the units of measurement along with each variable when appropriate.
- If you include the results of your calculations in a table, be sure to show at least one sample calculation in your data analysis.

Spreadsheets

A workbook, worksheet, or spreadsheet (**Figure 1**) is an electronic file that helps with the collection, manipulation, and analysis of data. There are many popular computer programs (e.g., Microsoft Excel, Lotus 1-2-3) that use these very powerful tools for data collection and analysis. In general, the programs treat each table unit as a cell or compartment with a specific address. The user can perform various mathematical, statistical, or graphical operations on the data.

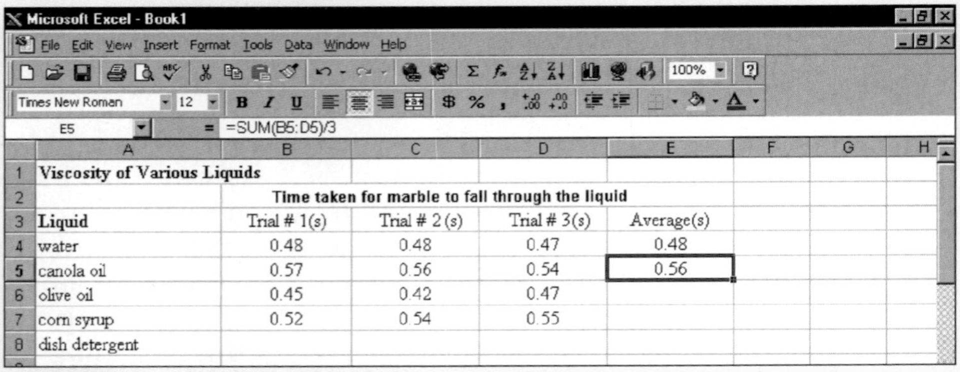

Table 1   **Data Table for Electricity Investigation**

Resistor Point	Connection Reading (V)	Voltmeter Reading (A)	Ammeter Ratio	V/I
1	A	?	?	?
1	B	?	?	?
1	C	?	?	?
1	D	?	?	?
2	A	?	?	?
2	B	?	?	?
2	C	?	?	?
2	D	?	?	?

Table 2   **Data Table for Reproduction Investigation**

Phase	Number of cells	Percentage of total in phase
prophase	?	?
metaphase	?	?
anaphase	?	?
telophase	?	?

Figure 1

(7A) Understanding Graphs

The Need to Graph

Scientists and science students often create huge amounts of data while doing experiments and studies—maybe hundreds, even thousands, of numbers for every variable. How can this mass of data be arranged so that it is easy to read and understand? That's right—in a graph. The sample table below doesn't have thousands of pieces of data, but it does have enough to become confusing. Can you make sense of the data in **Table 1** by simple inspection?

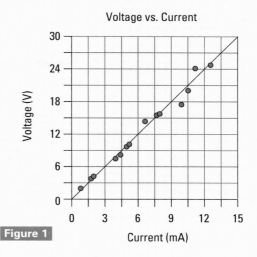

Voltage vs. Current

Figure 1

Table 1

Electric Current and Voltage in a Simple (one-resistor) Electric Circuit

Power Supply Setting	Current (mA)	Voltage Drop (V)
1	0.8	2.1
2	1.6	3.8
3	2.5	4.1
4	3.9	7.8
5	4.4	8.8
6	4.8	10.5
7	5.0	10.8
8	7.3	14.6
9	7.9	15.8
10	8.9	15.9
11	9.8	17.9
12	10.5	20.2
13	11.1	24.1
14	12.6	24.8

A graph is an easy way to see where a relationship or pattern exists. As well, it allows you to see more precisely what the relationship is, so it can be accurately described in words and by mathematics. **Figure 1** is a point-and-line graph that shows the data from **Table 1**.

The graph shows the relationship between the two variables as a fairly straight line. It could be described by saying that as the current through the resistor steadily increased, the voltage drop through the resistor also steadily increased. This simple relationship between current and voltage leads to a more advanced scientific concept known as electrical resistance. In more complex relationships such as this, the need for a graph is even stronger. Data are much easier to visualize and understand in the organized form of a graph than as numbers in a table.

Types of Graphs

There are many types of graphs that you can use when organizing your data. You need to identify which type of graph is best for your data *before* you start drawing it. Three of the most useful kinds are point-and-line graphs, bar graphs, and circle graphs (also called pie graphs).

Point-and-Line Graphs

When both variables are quantitative, use a point-and-line graph. The graph in **Figure 2** was created after an experiment that measured the number of worms on the surface of soil (quantitative) and the volume of rain that fell on the soil (quantitative).

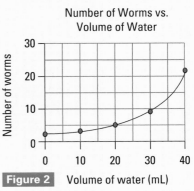

Number of Worms vs. Volume of Water

Figure 2 Volume of water (mL)

Bar Graphs

When at least one of the variables is qualitative, use a bar graph. For example, a study of the math marks of students (quantitative) who listened to different kinds of music (qualitative) while doing their math homework resulted in the graph in **Figure 3**. In this kind of graph, each bar stands for a different category, in this case a type of music. Notice also that the range on the vertical axis is chosen so that even the smallest bar is still visible.

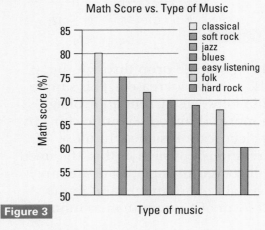

Figure 3

Circle Graphs

Circle graphs and bar graphs are used for similar types of data. If your quantitative variable can be changed to a percentage of a total quantity, then a circle graph is useful. For example, if you surveyed a class to find the students' favourite type of music, you could make a circle graph like **Figure 4**. In a circle graph, each piece stands for a different category (in this case, the kind of music preferred), and the size of the piece tells the percentage of the total that belongs in the category (in this case, the percentage of students who prefer a particular kind of music).

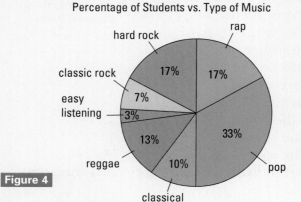

Figure 4

Reading a Graph

When data from an investigation are plotted on an appropriate graph, patterns and relationships become much easier to see and interpret—it is easier to tell if the data support your hypothesis. Looking at the data in a graph may also lead you to a new hypothesis.

What to Look for When Reading Graphs

Here are some guiding questions as you interpret the data on a graph:
- What variables are represented?
- What is the dependent variable? What is the independent variable?
- Are the variables quantitative or qualitative?
- If the data are quantitative, what are the units of measurement?
- Are two or more sample groups included?
- What do the highest and lowest values represent on the graph?
- What is the range (the difference between the highest and lowest values) of values for each axis?
- What patterns or trends exist between the variables?
- If there is a linear relationship, what might the slope (steepness) of the line tell us?
- Is this the best graph for the data?

Using Graphs for Predicting

If a graph shows a regular pattern, you can use it to make predictions. For example, you could use the graph in **Figure 5**, in which the mass of silver is plotted against volume, to predict the mass of 8 cm³ of silver. To do this, you would extrapolate the graph (extend it beyond the measured points) assuming the observed trend would continue.

Some common sense is needed in extrapolation. Sometimes a pattern extends only over a certain range. For example, if you extrapolated the graph of bacterial growth in **Figure 6**, you would predict that after 24 hours there would be 280 000 000 000 000 000 cells. These would have a total mass of about 40 kg. In another day and a half, the cells' mass would be greater than the mass of Earth. Something will happen to break the pattern!

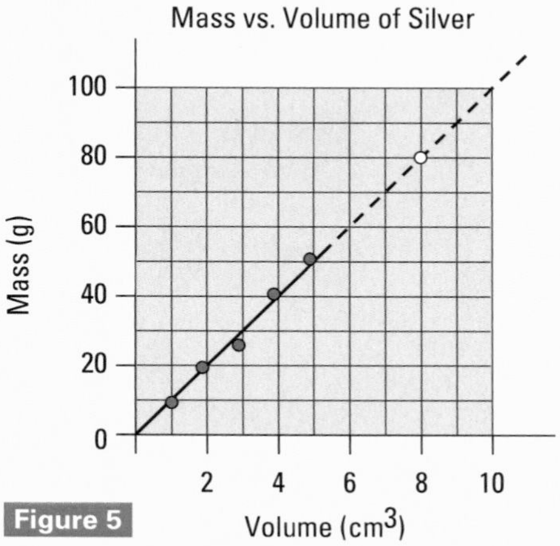

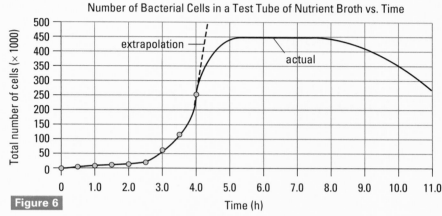

Figure 5 — Mass vs. Volume of Silver

Figure 6 — Number of Bacterial Cells in a Test Tube of Nutrient Broth vs. Time

---

Try This — Read a Graph

Until the 1970s, small amounts of lead were added to gasoline to improve the performance of car engines. Eventually, when people became aware of the hazards of lead, leaded gasoline was phased out. As you are probably aware, leaded gasoline is no longer used in Canada.

Lead pollutes the air. In one experiment in Ontario, the amount of airborne lead was measured near expressways and at several other locations. **Figure 7** shows the results of this study. Take a close look at the results and answer the following questions:

1. In what year was the level of airborne lead highest?

2. Was the amount of lead in the air increasing or decreasing during the time shown in the graph?

3. In any year, is the amount of airborne lead near expressways more than or less than the average of all sites?

4. (a) On the basis of the data, form a hypothesis to explain the presence of lead in the air.

   (b) Why do you think the amount of lead pollution changed over the measuring period?

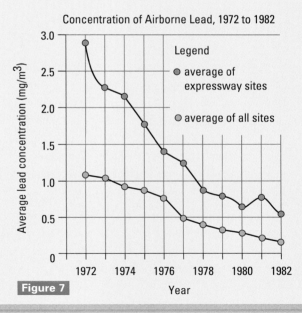

Figure 7 — Concentration of Airborne Lead, 1972 to 1982

Legend
● average of expressway sites
● average of all sites

⑦ⓑ Constructing Graphs

Making Point-and-Line Graphs

Point-and-line graphs are common in mathematics, economics, geography, science, technology, and many other subjects. This section will help you become more skilled at drawing them and at understanding point-and-line graphs produced by others.

As an example, the data in **Table 1** are used to produce a graph. When making a point-and-line graph, follow these steps:

1. Construct your graph on a grid. The horizontal edge on the bottom of this grid is called the x-axis and the vertical edge on the left is called the y-axis. Don't be too thrifty with graph paper—if you draw your graphs large, they will be easier to interpret.

Step 1: Draw the axes.

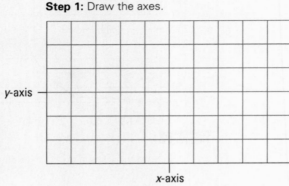

2. Decide which variable goes on which axis, and label each axis, including the units of measurement. It is common to plot the dependent variable (number of late students) on the y-axis and the independent variable (light level in candelas, cd) on the x-axis.

Step 2: Label each axis.

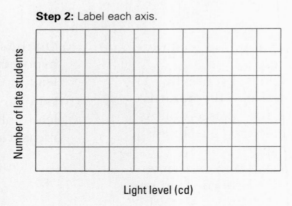

Table 1	
Light level (cd)	**Number of late students**
45	16
34	12
27	14
15	9
35	15
12	12
20	12
78	22
65	19
88	24
85	23
92	24
14	8
10	7
7	2
30	11
36	11
47	17
58	23
58	18
46	14

3. Determine the range of values for each variable. The range is the difference between the largest and smallest values. For the light levels in the table, the maximum is 92 cd, and the minimum is 7 cd, so the range is: 92 cd – 7 cd = 85 cd. For the number of late students, the range is 24 – 2 = 22.

4. Choose a scale for each axis. This will depend on how much space you have, and the range of values for each axis. Each line on the grid usually increases steadily in value by a convenient number, such as 1, 2, 5, 10, 20, 50, 100, etc. In the example, there are 10 lines for the x-axis and 6 for

Step 4: Choose a scale for each axis.

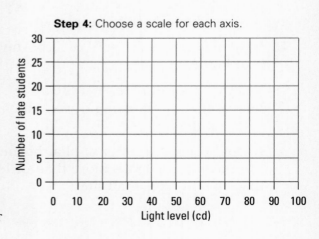

the *y*-axis. To calculate the increase in value for each line, divide the range by the number of lines:

$$\frac{85 \text{ cd}}{10 \text{ lines}} = 8.5 \text{ cd/line}$$

Then, round up to the nearest convenient number, which in this case is 10. The scale on the light level axis should increase by 10 cd every space. Repeat the calculation for the *y*-axis:

$$\frac{22 \text{ students}}{6} = 3.7 \text{ students/line, which is rounded up to 5.}$$

5. Plot the points. Start with the first pair of values from the data table, 45 cd and 16 late students. Place the point where an imaginary line starting at 45 on the *x*-axis meets an imaginary line starting at 16 on the *y*-axis.

6. After all the points are plotted, and if it is possible, draw a line through the points to show the relationship between the variables. It is unusual for all the points to lie exactly on a line. Small errors in each measurement tend to move the points slightly away from the perfect line. You must draw a line that comes closest to most of the points. This is called the line of best fit—a smooth line that passes through or between the points so that there are about the same number of points on each side of the line. The line of best fit may be a straight line or a curved line.

7. Title your graph (*y*-axis vs. *x*-axis).

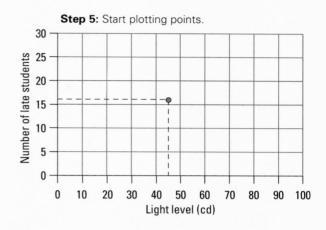

Step 5: Start plotting points.

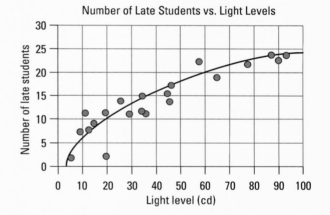

Step 6: Draw a line of best fit.

Try This A Point-and-Line Graph

Make labelled point-and-line graphs, with appropriate lines of best fit, for the following sets of data in **Table 2**:

Table 2

(a)

Air temperature (°C)	Snake heart rate (beats/min)
7	11
33	62
24	39
11	31
27	73
4	22
34	80
25	54

(b)

Wind speed (km/h)	Average speed of cars (km/h)
0	100
10	82
20	84
30	73
40	75
50	68
60	72
70	71

Making Bar Graphs

Bar graphs are useful when working with qualitative data and when a variable is divided into categories. In the following example science students did a study of the kind of music students listen to while doing mathematics problems, and got the results listed in **Table 3**. Follow these steps to plot a bar graph of this data.

Table 3

Type of music	Math score (%)
easy listening	69
hard rock	60
jazz	72
blues	70
classical	80
folk	68
soft rock	75

1. Draw and label the axes of your graph, including units. Some people prefer to have the bars based on the *x*-axis; others prefer to use the *y*-axis as the base. In the illustrations, the *x*-axis was chosen for the base.

Step 1: Draw the axes.

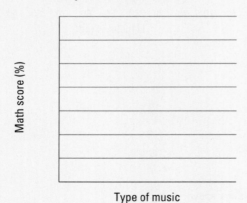

2. Develop a scale for the axis of the quantitative variable, just as you would for a point-and-line graph (pages 556–557). In this example, the *y*-axis increases by fives, starting below the lowest value. In the illustration, 50 was chosen as the starting point, so all the bars would be visible.

3. Decide how wide the bars will be, and how much space you will put between them.

This decision is based on:
- How much space you have. Measure the length of the axis on which the bars will be based, and divide that length by the number of bars. This will give you the maximum width of each bar.
- How you want the graph to look. Decide how much less than the maximum width your bars will be, based on the visual appeal of thick and thin bars.

4. Draw in bars. Start by marking the width of each bar on the base axis. Then, draw in the top of each bar, according to your data table, and the sides. You can shade the bars equally, or make each bar different from the others. It is important, however, to keep the graph simple and clear.

Step 4: Draw the bars.

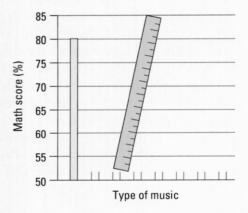

5. Identify each bar. There are several ways to do this. The best choice is the one that makes the graph easy to understand.

Step 5: The completed bar graph

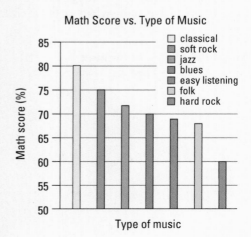

Making Circle (or Pie) Graphs

If your quantitative variable can be changed to a percentage of a total quantity, a circle graph is useful. A sample circle graph is worked out below, using the data in **Table 4**.

Table 4

Type of music	Number of students who prefer that type	Percentage of total (% and decimal)	Angle of piece of pie (degrees)
rap	5	17% = 0.17	61.2
pop rock	10	33% = 0.33	118.8
classical	3	10% = 0.10	36.0
reggae	4	13% = 0.13	46.8
easy listening	1	3% = 0.03	10.8
classic rock	2	7% = 0.07	25.2
hard rock	5	17% = 0.17	61.2
TOTAL	30	100% = 1.00	360.0

Follow these steps to construct a circle graph.

1. Convert the values of your quantitative variable into percentages, and then into decimal form. In the sample, each number of students who prefer a type of music was turned into a percentage of the total number of students.

$$\text{Percentage} = \frac{\text{number}}{\text{total}} \times 100\%$$

$$\text{Percentage for rap} = \frac{5}{30} \times 100\% = 17\% \text{ (decimal version} = 0.17)$$

2. Multiply the decimal version of each percentage by 360° (there are 360° in a circle) to get the angle of each "piece of the pie" within the circle.

Angle of piece of pie for rap = 0.17 × 360° = 61.2°

Try This Graphing

Conduct a survey of your class or a larger group, and make bar and circle graphs from the data you collect. You could use one of the following variables or one of your own: favourite sport, favourite school subject, eye colour, birth order (i.e., first-born, second-born, etc.).

3. Draw a circle using a compass. To make the graph easy to read (and make), the circle should be big. The more pieces there are, the bigger the circle should be.

4. Draw in each piece of pie, using a protractor.

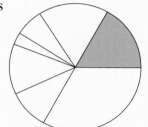

Step 4: Draw the pieces.

5. Shade each piece of pie using colours or patterns.

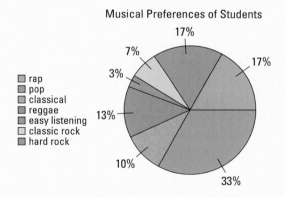

Step 5: Shade the pieces.

6. Label and title the graph. Put the percentages and the name of each category with its piece of pie (perhaps percentage inside and category outside the circle), or include them in a legend. Pick a title for your graph that describes the variables.

Step 6: The completed circle graph

Musical Preferences of Students

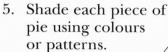

- rap
- pop
- classical
- reggae
- easy listening
- classic rock
- hard rock

Using Computers for Graphing

You should be aware that there are many useful computer programs that can help with the graphing process. For example, Microsoft Excel is a very powerful spreadsheet/graphing program that allows for the construction of point-and-line, bar, and circle graphs as well as many other types of graphs. In addition, such programs can make use of statistical analysis to compute the best straight line or line of best fit.

⑦C Using Math in Science

Amazingly, nature can often be analyzed, described, and predicted mathematically. For example, the initial growth of a population of bacteria follows an exponential function, and the mass and volume of a substance, when graphed, prove to be directly proportional to one another. Incredibly, mathematicians and scientists have recently developed mathematical representations of snail shells, ferns, and coastlines. While mathematics can sometimes appear abstract and daunting, as a scientist, your ability to understand some fundamental mathematical "tools" will improve your analytical ability.

The following key mathematical "tools" will help you work through the *Nelson Science 9* program and in your further studies.

Scientific Notation

In your study of Space, you use some very large numbers, whereas in your study of Matter, you work with some very small numbers. For example, the average distance from the Sun to Earth is 150 000 000 000 m and the average radius of a hydrogen atom is 0.000 000 000 05 m. Because we don't want to spend the better part of our day writing out zeros, it is convenient to use a mathematical abbreviation known as scientific notation. The scientific notations for these two values are 1.50×10^{11} m and 5.0×10^{-11} m, respectively. Can you see how this works?

Essentially,

multiply by the number 10 a total of 11 times

$1.5 \times 10^{11} =$
$\quad 1.5 \times 10 \times 10 \times 10 \times 10 \times 10 \times 10 \times 10 \times 10 \times 10 \times 10 \times 10$

divide by the number 10 a total of 11 times

$5.0 \times 10^{-11} = \dfrac{5.0}{10 \times 10 \times 10 \times 10 \times 10 \times 10 \times 10 \times 10 \times 10 \times 10 \times 10}$

Significant Figures

Imagine you are given two pieces of string (**Figure 1**) and you wish to find the total length of string that you have. You measure the length of one piece of string with an accurate scale ruler and find that it is 12.72 cm long. You measure the length of the other piece of string with an old metre stick and find that piece of string to be 14 cm long. What is the total length of string that you have? If you claim the total length to be 26.72 cm, you have assumed that the second piece of string is 14 cm—was your old metre stick able to give you this level of precision? No. You can only come to the conclusion that you have a total rope length of 27 cm.

When we take measurements and manipulate numbers in science, we must look at the number of significant digits (or figures) in a number. The number of significant digits represents how carefully, and with what level of precision, the measurement was taken. Your calculator often registers a number with several significant digits regardless of where these numbers came from and what they mean.

Figure 1

Calculating Averages

There are many statistical methods that help us analyze the quantitative data that we collect. One of the most important of these is finding the average of a set of numbers. Calculating the average of a set of numbers allows you to reduce their findings to one representative value. For example, if we measure the heights of four students,

Suzanne	175 cm	Omar	145 cm
Jan	185 cm	Molly	180 cm

we could calculate that the average height of the group is

$$\text{average height} = \frac{175 + 185 + 145 + 180}{4} = 171 \text{ cm}$$

Slope of a Straight Line

When you plot data on a point-and-line graph (see Constructing Graphs, page 556), you may be able to connect the data points with a straight line. A straight line on a graph shows a direct relationship between the independent variable (usually on the horizontal axis) and the dependent axis (usually on the vertical axis). It is possible to represent this linear (or straight-line) behaviour with a mathematical equation that looks like this

$$y = mx + b$$

where y is the dependent variable, x is the independent variable, m is the slope (or steepness) of the line, and b is known as the y-intercept. Perhaps the most important of these constants is the slope of the line. To find the slope, choose two points on the line (x_1, y_1) and (x_2, y_2). The slope is equal to

$$m = \frac{\text{rise}}{\text{run}} = \frac{y_2 - y_1}{x_2 - x_1}$$

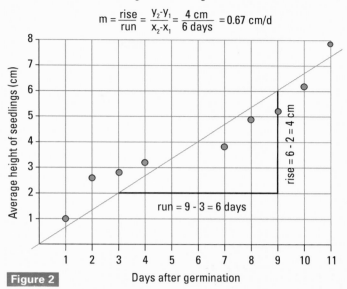

Height of Seedlings vs. Time

$$m = \frac{\text{rise}}{\text{run}} = \frac{y_2 \text{-} y_1}{x_2 \text{-} x_1} = \frac{4 \text{ cm}}{6 \text{ days}} = 0.67 \text{ cm/d}$$

run = 9 - 3 = 6 days

rise = 6 - 2 = 4 cm

Figure 2 Days after germination

Predicting Using Formulas

Using various mathematical or experimental techniques, scientists are often able to connect and explain scientific phenomena by a mathematical formula. For example, in this book, you use the following relationship:

$$\text{density} = \frac{\text{mass}}{\text{volume}}$$

The predictive power of a formula such as this is impressive. For example, if you know the density of an object and its mass, you can predict its volume.

Sample Problem: A bag of milk has a density of 0.95 kg/L and a mass of 1.26 kg. Will you be able to pour the entire bag of milk into a 1-L pitcher?

density = 0.95 kg/L

mass = 1.26 kg

$$\text{density} = \frac{\text{mass}}{\text{volume}} \text{, therefore}$$

$$\text{volume} = \frac{\text{mass}}{\text{density}}$$

$$= \frac{1.26 \text{ kg}}{0.95 \text{ kg/L}}$$

$$= 1.33 \text{ L}$$

So a 1-L pitcher will not hold all of the milk in the bag.

⑦D Reaching a Conclusion

Scientific investigations mean very little, if anything, unless the scientist states a conclusion about the results. A **conclusion** is a statement that explains the results of an investigation. This statement should refer back to your original hypothesis. State whether the results support, partially support, or reject your hypothesis. Don't worry if your hypothesis is incorrect—scientists usually need to revise and repeat experiments many times in order to obtain the solutions they are seeking. How many experiments do you think have been repeated in order to learn what we now know about cancer treatments?

Suppose you wanted to find out about the relationship between students' hearing ability and the volume of the music they listened to. Your hypothesis may be: "As students increase the volume of the music they listen to, their hearing ability will decrease." You design and conduct an experiment and obtain the data shown in **Figure 1**.

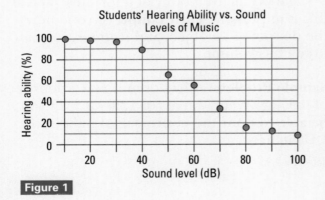

Students' Hearing Ability vs. Sound Levels of Music

Figure 1

One conclusion from these results may read, "My hypothesis was correct. The results show that as the volume increases, there is a decline in students' ability to hear because..." Complete this concluding statement by providing an explanation for the results obtained. Can you think of any other possible hypotheses and conclusions that would explain these results? For example, you might hypothesize, "Students increase the volume when their hearing ability is low, because they couldn't hear the music otherwise." What would your conclusion be?

Reaching conclusions in science allows you to critically analyze the results using a mix of logic, common sense, understanding, and patience.

Try This Concluding

Reach an educated guess (hypothesis) for each of the following problems. Think about the conclusions you made— remember to refer back to your original hypothesis.

1. How many cubic metres of soil are required to fill a hole 6 m long, 2 m wide, and 2 m deep?

2. If 3 cats catch 3 mice in 3 min, how many cats would be needed to catch 100 mice in 100 min?

⑦E Reflecting on Ideas

It is always important to reflect on events in order to learn from them. This is one aspect of scientific inquiry that is sometimes neglected, especially with beginning scientists. Once an investigation has been completed, it is always advantageous to step back and think about what you did. What went well? What were the challenges? What could be improved? What would you do differently if you were to complete the investigation again? It is through reflecting on your actions that you will improve your work and increase your learning.

When reflecting on any investigation you performed, be sure to identify the types of experimental error that may have emerged at any point in the experimental process. These sources of error should be identified in the conclusion or discussion of your results. For example, say you completed an experiment and were able to conclude that plants did grow toward sunlight. What sources of experimental error could have occurred while performing this experiment? These could include such things as variations in air temperature, inconsistent exposure to light, or differences in soil composition. Be as specific as possible when stating experimental errors—it is not good enough to simply state, "It was due to human error."

Experimental errors decrease the validity of an experiment. But, more importantly, they allow you to revise your procedure. That revision will increase the reliability of the results the next time you perform the same investigation. Remember, experiments are a repeating process. (Refer back to the flow chart on scientific inquiry on page 527.) By repeating the process, you are improving your ideas and investigative skills.

> **Try This** **Think About This**
>
> Reflect on your last science class. What worked well during that class? What could you have improved? What did you like about the class? What didn't you like about the class? What will you try to work on improving in your next science class?

8A Reporting Your Work

All investigators use a similar format to write reports, although the headings and order may vary slightly. Your report should reflect the process of scientific inquiry that you used in the investigation.

S. H. Secondary School
Grade 9 Science
Heating Ice
Prepared By: Dave Fisher
Partner(s): Keltie Morrison
Prepared For: Ms Feddery
Due Date: March. 14, 1999

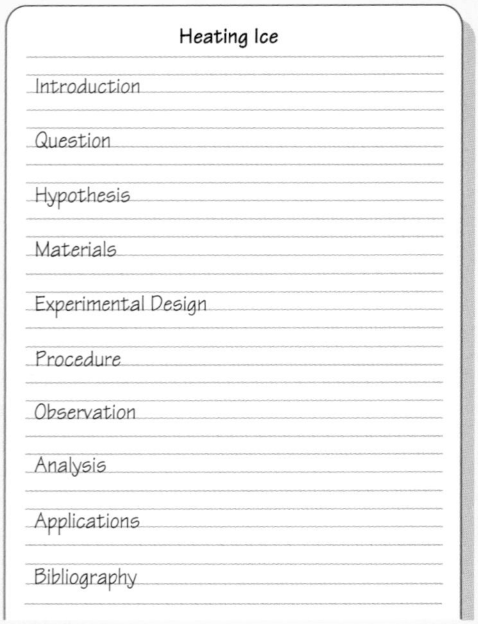

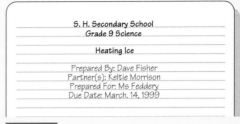

Figure 1
A cover page

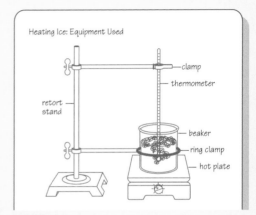

Cover Page: Make a cover page (**Figure 1**) that includes the following:
- the name of your school
- the course code
- the title of your investigation
- your name
- name of your partner(s) (if applicable)
- your instructor's name
- the due date

Title: At the beginning of your written report, write the title of your experiment.

Introduction: Always begin with a brief explanation of pertinent theory underlying the experiment. This includes the information you discovered by researching your topic. This section is also referred to as your "review of literature."

Figure 2
A diagram

Purpose, Question, or Problem: Make a brief statement about your investigation. This can also be written as a question.

Hypothesis: Write the hypothesis. Remember this is your "educated guess," based on your previous knowledge and the research you completed.

Materials: These include consumables (e.g., water, paper towels). Be specific about sizes and quantities. These also include nonconsumables (e.g., test tube, beaker). Make a detailed list of the materials you used.

Diagram: Make a full-page diagram of the materials and apparatus you used in the experiment (**Figure 2**). Remember to label and title the diagram. Your diagram can be placed at the end of your report, following the application.

Heating Ice
Introduction
Question
Hypothesis
Materials
Experimental Design
Procedure
Observation
Analysis
Applications
Bibliography

Figure 3
Your report should reflect the process of scientific inquiry

Experimental Design: List the independent, dependent, and controlled variables in your experiment and summarize the procedure.

Procedure: The most important part of an investigation, when others are trying to determine if it is "good" or "bad" science, is the procedure. Many researchers read only the procedure section in a report, to gain insight into a procedure they could use themselves. To be sure that your work is judged fairly, make sure you leave nothing out! Remember to write this

in numbered steps, past tense, and passive voice.

Observations/Results: Present your observations and results in a form that is easily understood. The data should be in tables, graphs, or illustrations, each with a title. Include any calculations that are used. The results of the calculations can be shown in a table.

Analysis/Conclusion: Summarize the investigation as you would if you were writing a book report. Refer back to your hypothesis. Was it correct, partially correct, or incorrect? Explain how you arrived at your conclusion(s). Justify your method and describe your results. Suggest a theory to support or interpret your results. If you were assigned questions with an investigation, you would answer them here. Discuss any sources of experimental error that may have affected your findings.

Applications: Describe how the new knowledge you gained from doing the investigation relates to your life and our society. It should answer the question, "Who cares?"

References/Bibliography: Give credit for the resources you used in your research. Always cite your source(s). Failing to do so is considered plagiarism (unacknowledged copying). It is unethical and illegal. Whenever you give credit to an author (including yourself from previous reports!) for the use of graphs, tables, diagrams, or ideas, use the following technique:

- Immediately after the information is used, give the last name of the source, the date of the publication, and the page reference. For example:

   The demand for electronic devices on board satellites has been a great stimulus to the development of microchips (Hawkes, 1992, p. 18).

   The bibliography, at the end of your report, should be in alphabetical order of authors' last name. Be sure to use "hanging" indents, that is, indent every line after the first one. The format for books, journals, web sites, newspapers, and CD-ROMs differs. To make

sure you cite your sources correctly, refer to the following examples:

- If the quotation or information comes from a book, use the following format, which includes the publisher's name and the city of publication:
   Hawkes, Nigel (1992), *Into Space.* New York: Gloucester Press.
- If the quotation or information comes from a journal or a magazine, use the following format. Include the volume number of the magazine (usually found on the Table of Contents page) and the page numbers of the article.
   Flamsteed, Sam (1997). "Impossible Planets." *Discover* 18 (9): 78–83.
- If the quotation or information comes from a newspaper, use the following format:
   Chandler, David. "New solar system is the first discovered outside our own." *Toronto Star.* April 16, 1999. Section A, p. 3.
- If the quotation or information comes from a web site, use the following format:
   Space Technologies Web Site, N.D. Available HTTP: http://www.space.gc.ca/ENG/About/SPATECH/welcome.html#Director [1999, April 6]
   If no date is given on the web site, use N.D.
- If the quotation or information comes from a CD-ROM, use the following format:
   Oxford English Dictionary Computer File: On Compact Disc. 2nd ed. CD-ROM. Oxford. Oxford UP, 1992.

Checklist for Writing a Report
Refer to this checklist whenever you are required to write a scientific report.
- ✔ Cover page
- ✔ Title of the investigation
- ✔ Introduction
- ✔ Purpose/Question
- ✔ Hypothesis
- ✔ Materials
- ✔ Diagram
- ✔ Experimental Design
- ✔ Procedure
- ✔ Observations/Results
- ✔ Analysis/Conclusion
- ✔ Applications
- ✔ References/Bibliography

(8B) Exploring an Issue

In *Nelson Science 9*, you will have opportunities to explore many issues. Advances in science and technology need to be evaluated from many different perspectives—particularly since these advances are being made at an ever increasing rate.

Figure 1 shows a helpful way of organizing the advantages and disadvantages of a given issue related to science or technology.

Should you use an electric toothbrush?

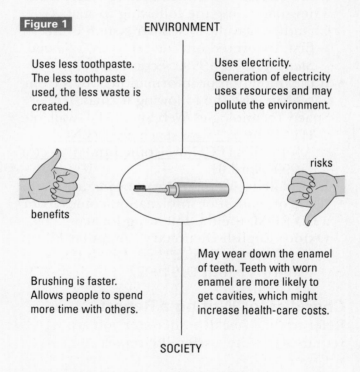

Figure 1

ENVIRONMENT

Uses less toothpaste. The less toothpaste used, the less waste is created.

Uses electricity. Generation of electricity uses resources and may pollute the environment.

benefits

risks

Brushing is faster. Allows people to spend more time with others.

May wear down the enamel of teeth. Teeth with worn enamel are more likely to get cavities, which might increase health-care costs.

SOCIETY

Toward an Educated Opinion

The following process will help you evaluate the pros and cons of an issue and provide supporting evidence and arguments to the position you ultimately take on that issue.

1. Initial Research:

Choose an issue that really interests you. Gather at least two initial sources of information about it. Consider all the sources of information available to you (page 530).

2. Formulation of a Question

Put together a question around the issue. The question should be answerable, important to our society, and debatable. You are encouraged to "try this question out" on your classmates, teacher, and others.

3. Further Research

Continue your research on this topic. Find at least two additional sources of information on the issue. Be sure to scrutinize the credibility of your information. Remember PERCS (page 533)!

4. Reaching Your Position

Answer your question. Explain your position thoroughly. Be sure to consider the following:
- Have you included information from at least three articles?
- Have you stated your position clearly?
- Have you shown why this issue is relevant and important to our society?
- Have you included at least two solid arguments (with solid evidence) supporting your position?
- Have you included at least two arguments against your position and shown their respective faults?
- Have you analyzed the strong points and weak points of each perspective?

5. Communicating Your Position

There are several ways of communicating your position, each with its own format:
- Write a position paper or brief essay.
- Participate in a formal debate.
- Participate in a role-play activity, taking on the role of a specific, affected party.
- Write a letter to the editor of a newspaper.

8C Science Projects

Presenting Your Project

After you have completed your project, you may want to display your results in a science fair or present them to your class (**Figure 1**).

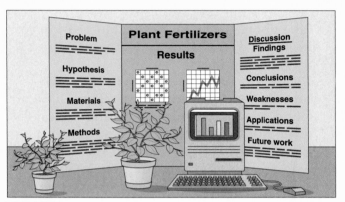

Figure 1
Your display

When constructing a display board, there are some things to keep in mind.
- Use a sketch to plan the layout.
- Collect a variety of display items (photos, sketches, charts, graphs, text).
- Use different sizes of letters for the text on your board. The most important ideas should have the largest letters.
- Place the title in a central location.
- Make all lettering neat and easy to read.
- Simple is best. Use the same kind of lettering throughout. If you are using colours or shapes to highlight important features, use only a few.
- Make sure your diagrams, graphs, and charts are neat.
- Place your results in a prominent position on the board.
- Try to place all your materials on the board before you start gluing or stapling. Have a classmate or a parent "critique" your display.
- Place only the most important information on the display board. If there is lots of space around wording and diagrams, the display will be more attractive.

- Make sure that nothing on your board is blocked by objects that will be in front of the board.
- Make sure things flow in a logical, easy-to-follow sequence.

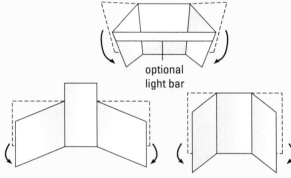

Three different display boards

You may also want to display your project on the Internet or put together a computer-based presentation. Have a look at on-line science fairs on the Internet for ideas on how to put together your electronic display.

Your Oral Presentation

Plan your oral presentation well in advance. Prepare a point-form "speech" on cue cards or a computer-based presentation and be sure to consider the following questions:
- What was the goal of your project?
- Why was the topic interesting?
- How did you carry out your project?
- What were the major results of your project?
- What conclusion did you reach?
- How might your findings help others?
- On the basis of your results, is there another project that could be done?

It is best not to read your speech when presenting. But don't memorize every word. Just practise it several times. If you get stuck, use your display board as a guide. Another trick is to memorize the seven questions above, or write them on cards, and answer them as you speak.

⑨Ⓐ Good Study Habits

Understanding anything—whether it is a life-saving technique in the swimming pool, a trumpet solo, or a science lesson—is an active process. Studying takes on many forms. It involves learning and understanding material. Developing the following study skills can help you in your learning. You can modify these tips to help you in other school courses and in recreational activities.

Your Study Environment

- **Organize your working area.** The place where you study at home and at school should be neat, tidy, and organized. Papers, books, magazines, or pictures that are strewn all over your working area will distract you from focusing on the work at hand.
- **Maintain a quiet study space.** Make sure that the place where you decide to study is removed from distractions such as the phone, stereo system, TV, friends, and annoying brothers and sisters. Popular study spaces include the school library, the public library, and your bedroom at home (if it has a working area). Any quiet place where you can be productive will work.
- **Make sure you feel comfortable in your study space.** You will be the most productive and study effectively if you are working in an area where you feel at ease—personalize your learning space.
- **Be prepared with all the materials you will need.** It is important that you have all your notebooks, textbooks, computer equipment, paper, pencil, pen, ruler, and anything else you use for your work in your study space. If you have to continually get up to find a book or eraser, you won't be able to accomplish as much as you had hoped.

Study Habits

- **Prepare for class by reading material ahead of time.** It is also helpful to read or view materials from other sources, such as science magazines, newspapers, the Internet, and television programs.
- **Take notes.** To make note taking easier, you may want to make up a shorthand method of recording ideas.
- **Review any notes you made in class the same day and add comments.** Then have a friend or relative quiz you on the material in the notes. Reinforce your understanding by answering the questions in the textbook—even if they are not assigned.
- **Use your notes and the textbook to prepare summaries.** Studying is most effective with a pen or pencil or your word processor, so you can write down the important ideas (**Figure 1**). You may want to write or type study cards to assist in making effective, point-form summaries of your notes. Look at the example in **Figure 2**. Notice there is a title at the top and all the information has been condensed into a point-form, easy-to-learn format. It's much more effective than learning material that is in paragraph form. Condense, condense, and condense!

Figure 1

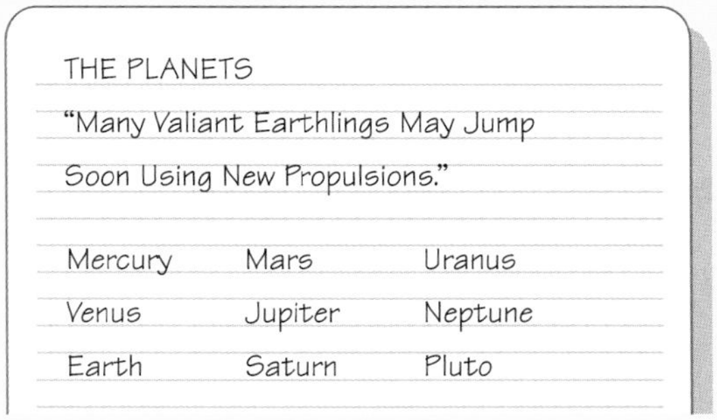

THE PLANETS

"Many Valiant Earthlings May Jump
Soon Using New Propulsions."

Mercury	Mars	Uranus
Venus	Jupiter	Neptune
Earth	Saturn	Pluto

Figure 2

- **Draw a concept map to help you summarize a chapter, unit, or lesson.** (See Concept Mapping on page 572.) You may want to use the textbook summary at the end of the chapter.
- **"Practice makes perfect."** This is as true for science as it is for playing piano and shooting baskets. If you practise your science skills until they become almost automatic, you will have more time to think about how you will use them.
- **Schedule your study time.** This will help you avoid that most ineffective of all study methods, "cramming" before assignments and tests. Use a daily planner and take it with you to every class. Write all homework assignments, tests, projects, and extra-curricular commitments in it. This will assist you in organizing a daily "To Do" list that will ensure maximum use of your time.
- **Know your strengths and your weaknesses.** Take advantage of all opportunities to get help with areas in which you may have trouble. Use your strengths to help yourself and others. Form a study group and have regular meetings. You may be able to help others in some parts of the course. In turn, you may receive help from them.
- **Teach the material you have learned to someone who has not yet learned it.** Their questions will help you see what areas of the subject you need to learn more about and what areas you don't completely understand (**Figure 3**).
- **Take study breaks.** It is important that you set study goals and take a short study break after meeting each of the goals you set. For example, you could decide that you will take a study break after crossing two items off

your "To Do" list. Taking study breaks will help you rejuvenate for the next tasks on your list and will assist you in completing all your work effectively and accurately.

Study Checklist
✔ Organize your study environment. (Is it quiet? comfortable? organized?)
✔ Read material ahead of time.
✔ Take notes.
✔ Review.
✔ Summarize. (Make study cards, if that technique works for you.)
✔ Draw a concept map.
✔ Practise.
✔ Schedule study time. (Use your daily planner and make a "To Do" list.)
✔ Know your strengths and weaknesses.
✔ Teach what you have learned.

Figure 3

 **Your Study Space**

Draw pictures of your study environment at home and at school. List five good points about your study spaces and five things that could be changed in each of your study spaces to give you an even more effective learning environment.

9B Using Your Computer Effectively

Computers are becoming more and more common in your learning. Most of you have access to this technology at some point every week. Like your notebook, it is important to keep your computer files organized. It is also necessary to use your computer effectively—there is a time for computer work and a time for computer games and recreational surfing of the Internet. The following hints can assist you in using your computer effectively in your learning. Remember, these hints can be modified to suit your needs.

Computer Hints

Create a "science" file for all your science work. It is important that any computer work you do is kept in one folder. Otherwise, your work could be saved in lots of different locations on your computer and this means that it could become easily misplaced or even lost.

Organize your science units in your science folder. Now that you have created a science folder, organize this folder into the various units that you will study this year. These include Electricity, Matter, Reproduction, and Space. Refer to the illustration on this page for a detailed glimpse of what your science folder might look like.

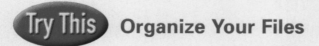 **Organize Your Files**

The next time you are using a computer to complete schoolwork, try organizing your science files in a way that makes sense to you. You may decide to organize all the computer work you do in each of your courses.

Think carefully about what to name each of your documents. Make sure that the names you give your documents allow you to easily identify what a particular document contains. Work that is called Science 1, Science 2, and Science 3 provides no indication of what the work actually is. Be as specific as possible.

Back up your work. Always make an extra copy of any work you do on the computer. Copy your work onto a disk or onto your school network. Too many times students lose their work because they do not take the time to do this. It really does save time in the long run. It's much better than having to complete a report or assignment all over again.

Maintain correct posture and form. It is important that you don't sit too close to the screen and that you sit in an upright position with your hands positioned correctly while typing. Incorrect hand positioning can eventually cause carpal tunnel syndrome. The study of how people interact effectively and safely with computers is part of what is termed "ergonomics."

9C Working Together

Scientific discoveries are almost always made by teams of people working together. Scientists share ideas, help each other design experiments and studies, and sharpen each other's conclusions. We have all worked in a group at one time or another. In the "real world," group work is necessary and usually more productive than working alone. It is therefore important for us to be able to work in teams.

While working with *Nelson Science 9*, you will spend much of your time doing science as scientists do it—in teams. In this activity, you will work as a team. After you finish, reflect on the advantages and disadvantages of teamwork.

Evaluate the effectiveness of the team—the strengths, weaknesses, opportunities, and challenges. Answer the following:

1. What were the *strengths* of your teamwork?

2. What were the *weaknesses* of your teamwork?

3. What were *opportunities* provided by working with your team?

4. What possible *challenges* did you see with respect to your teamwork?

Teamwork Tips

When you work together with other students in a team, follow these tips:
- Encourage all members to contribute to the work of the group.
- Respect everyone's contributions. There are many points of view and all perspectives should be considered. Keep an open mind. There should be no put-downs.
- Be prepared to compromise.
- Keep focused on the task at hand. Divide the various tasks among all group members.
- Support the team's final decision.

- When you are given the opportunity of picking your own team, be sure that the students you decide to work with complement your strengths and weaknesses.
- Can you think of any more tips that should be considered during group work?

Try This — Your Team Is Lost!

Your team is lost on a deserted island. You must survive on the island for at least five days. As a *team*, review the following list of items and choose the five things that your group would want on your island. The final decision should be reached by consensus! Be ready to share and support your decisions with the rest of the class.

flashlight with 4 batteries	12 chocolate bars
4 hats	groundsheet
box of matches	pen, pencil, paper
6 cans of beans	knife
map of island	rope
fruit (4 days' supply)	calculator
water	chewing gum
4 books	compass
Monopoly game	first-aid kit
can opener	comic books
laptop computer	

9D Concept Mapping

When you are trying to describe objects and events, it is sometimes helpful to record your ideas so that you can see them and compare them with those of other people. Instead of putting your ideas in sentences, they can be made into concept maps. A concept map is a collection of words (representing concepts) or pictures that you connect to each other with arrows and short descriptions. The map is a drawing of what is happening in your brain.

You may have seen concept maps similar to the one shown above. This concept map, called a food web, shows one way of thinking about animals and what they eat. The arrowhead points to the animal that eats the animal or plant at the other end of the arrow. The arrows describe the relationship between the organisms.

Concept maps can also be drawn of other topics and in other ways. For instance, the relationships between family members can be drawn using a concept map like the one to the right.

Concept maps can also be drawn to show a series of cause-and-effect relationships. You may find making this kind of map useful during a science investigation.

Sometimes concept maps can be used to show how your ideas change and become more complex as you work on a topic. You can see an example of this type of concept map on page 573.

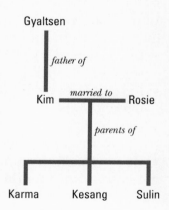

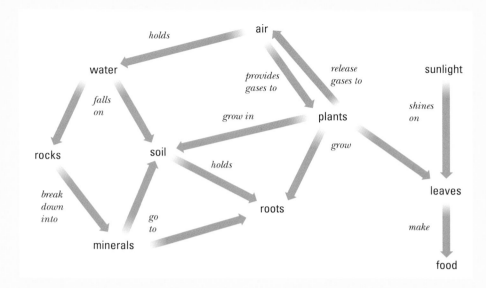

Making a Concept Map

Here are some steps you can take to help you make a concept map.

1. Choose the central idea of your concept map.

2. Write the central idea and all related ideas on small scraps of paper.

3. Move the scraps of paper around the central idea so that the ideas most related to each other are close to each other. Ask yourself how they are related, and then use that information in the next steps.

4. On a big sheet of paper, write down all your ideas in the same pattern that you have arranged the scraps. Draw arrows between the ideas that are related.

5. On each arrow, write a short description of how the terms are related to each other.

As you go, you may find other ideas or relationships. Add them to the map.

When you gain new ideas—whether from research, from your investigations, or from other people—go ahead and change your concept map. You may want to add new ideas in a different colour of ink, to indicate your new ways of thinking about the ideas.

Try This Mapping "Science"

Develop a concept map on the topic Science. Include at least 10 related ideas. Try to include some connections to your other subject areas. When you have finished your concept map, share it with a classmate. Other ideas or concepts may come to you while you discuss your map. Add these ideas in a different colour.

Glossary

A

absolute magnitude: the brightness of a star as it would be if it were a standard distance from Earth

aircraft: a vehicle that travels through the air, usually no higher than 20 km above ground

alchemist: an experimenter—part philosopher, part scientist, part magician; some alchemists believed that metal could be turned into gold

alkali metals: elements that occupy the far left column (first group) of the periodic table

alloy: a metal made by combining two or more different metals or metals and nonmetals

ampere: the SI unit used to measure electric current; symbol: A

anaphase: the phase of mitosis during which the chromosomes split and single strands of genetic information move to opposite ends of the cell

anode: a positive plate that produces electrons in a fuel cell

anther: part of the stamen, the male structure, in the flower of a plant; holds pollen grains

apparent magnitude: the brightness of a star as it appears to a person on Earth

asexual reproduction: any of several forms of reproduction that results in offspring that are genetically identical to the parent

asteroid: a small rocky object; from the Greek work *astron*, meaning "starlike"

asteroid belt: a ring around the Sun made up of thousands of asteroids

astronomical unit: a distance measure (a.u.) equal to the average distance between Earth and the Sun, used to compare large distances in the solar system

astronomy: the study of what is beyond Earth

atom: a particle in an element; from the Greek word *atomos*, meaning "indivisible"

atomic mass: the average mass of an atom of an element

atomic model: a theory proposed by John Dalton in 1808 to explain why elements differ from each other and from non-elements

atomic number: the number of protons in an atom

atomic radius: the average distance from the nucleus to the "outer edge" of a spherical atom

axis: an imaginary straight line joining the North Pole and the South Pole

B

benign: the term used to describe a cancerous tumour that remains in a confined area, causing little damage to the organism

Big Bang theory: a model of the beginning of the universe

binary fission: the form of asexual reproduction in which the organism splits directly into two equal-sized offspring, each with the parent's genetic material

black hole: the high-density core left when a star about 30 times the mass of the Sun dies

Bohr diagram: a diagram, representing the electronic structure of an element, that comprises electrons in a series of concentric circles (energy levels or orbits) drawn around the central nucleus, containing the symbol of the element

Bohr-Rutherford diagram: a diagram, summarizing the numbers and positions of all three subatomic particles in an atom, that comprises electrons in a series of concentric circles (energy levels or orbits) drawn around a central nucleus, containing the numbers of protons and neutrons in the element

bond: a concept used in models that represents the forces that hold atoms together

branch circuit: a separate circuit through which current passes to each load in a parallel circuit

brittle: the physical property of a substance that shatters easily

budding: the form of asexual reproduction in which the offspring begins as a small outgrowth of the parent and eventually breaks off, becoming an organism on its own

C

cancer: the term for a group of diseases associated with uncontrolled, unregulated cell division

carcinogen: a substance or energy that causes a mutation in the genes that regulate cell division

cathode: a negative plate in a fuel cell

cell cycle: the sequence of events in a cell from one division to another

cell membrane: a covering around a cell that controls the movement of materials into and out of the cell

cell wall: the non-living outermost covering of a plant cell

centriole: a small protein structure in animal cells, critical to cell division

ceramic: a material manufactured by heating minerals and rocks

CFCs: chlorofluorocarbons; compound made of carbon, chlorine, and fluorine atoms, developed by scientists in the 1930s

charge: a negative or positive quantity of electricity that builds up on an object

charging by contact: transferring an electric charge from one substance to another by touching

charging by friction: transferring an electric charge from one substance to another by a rubbing action

charging by induction: transferring an electric charge from one substance to another without direct contact

chemical change: the alteration of a substance into one or more different substances with different properties

chemical formula: the combination of symbols that represents a particular compound

chemical group: the set of elements that appears in the same column in the periodic table

chemical property: the characteristic behaviour that occurs when one substance interacts with another to become a new substance

chemical symbol: an abbreviation for the name of an element

chemical test: a test that produces a distinctive chemical reaction to identify an unknown substance

chloroplast: the organelle in a plant cell that contains chlorophyll

chromosome: a threadlike structure that contains genetic information

cilia: tiny small hairs on an animal cell

circuit breaker: a safety switch that controls the amount of current that flows in a circuit

circuit diagram: a drawing that uses a special set of symbols to represent the electrical components and wiring in an electric circuit

cloning: the technical process by which identical offspring are formed from a single cell or tissue

combining capacity: the ability of an element to combine with other elements

combustible: the chemical property of a substance that allows it to burn when exposed to flame and oxygen

combustion: the chemical reaction that occurs when a substance reacts rapidly with oxygen and releases energy

comet: a chunk of ice and dust that travels in a very long orbit around the Sun

composite: a material formed by combining two other materials

compound: a pure substance that contains two or more different elements in a fixed proportion

conception: the process of fertilization in which the head of the sperm cell penetrates the cell membrane of the egg

conjugation: the form of sexual reproduction in which two cells come together and exchange genetic information

connector: a conducting wire that provides a controlled path for electric current to flow to each part of the circuit

constellation: a group of stars that forms shapes or patterns

corpus luteum: a tissue inside the ovary that secretes hormones essential for pregnancy

corrosion: a slow chemical change that occurs when a metal reacts with oxygen to form oxide

cosmology: the study of the origin and changes of the universe

crossbreeding: the process of taking pollen from one plant and using it to fertilize the eggs of another

crystal: a solid mineral in which a regular pattern of three-dimensional shapes is visible

cytokinesis: the process in cell division in which the cytoplasm and its contents separate into equal parts

cytoplasm: the area of a cell where nutrients are absorbed, transported, and processed

D

decay: splitting apart of unstable nuclei to produce radioactive particles

density: the mass per unit volume of a substance, usually expressed in kilograms per cubic metre or grams per cubic centimetre

deoxyribonucleic acid: the genetic chemical found in all living things

discharge: to neutralize or remove all electric charges

discharge at a point: removing an electric charge by repelling electrons off the tip of a conductor that is pointed at the end

distribution panel: a metal box where all circuit breakers (or fuses) are connected to each of the separate circuits

DNA: *See* **deoxyribonucleic acid**

ductile: the physical property of a solid that allows it to be pulled into wires

E

Earth-centred universe: a belief that the stars were attached to a large ball that revolved around Earth once every day

efficiency: a comparison between the amount of useful energy produced (output energy) and the original amount of energy used (input energy)

egg: the female sex cell; a package designed to feed and protect a developing embryo

electric charge: a negative or positive quantity of electricity that builds up on an object

electric circuit: a controlled path through which electric current passes

electric current: a measure of the rate at which electric charges flow; symbol: I

electric potential: the electrical energy that an electron possesses

electrical conductor: a substance in which electrons can move freely from one atom to another

electrical energy: the energy released into an electrical load by moving electrons

electrical insulator: a substance in which electrons cannot move freely from one atom to another

electrical load: anything that converts electrical energy into the form of energy required

electrical power: the rate at which electrical energy is used

electrodes: metal plates, usually zinc and copper, that are placed in liquid or gel in a voltaic cell

electrolysis: the use of electricity to cause chemical changes in solutions

electrolyte: the liquid in a voltaic cell that conducts an electric current

electromagnetic spectrum: the broad band of energies that comprises radio waves, microwaves, infrared rays, visible light, ultraviolet rays, X rays, and gamma rays

electron: a negatively charged particle with a relative mass of about 1/2000 of the mass of a proton or a neutron

electrostatic series: a continuum of substances listed in order of increasing tendency to gain electrons

electrostatics: the study of static electric charge

element: a pure substance that cannot be broken down into simpler substances

embryo: the stage in development when a fertilized egg has divided to form a mass of at least 64 cells

endometrium: the thick lining of the uterus

endoplasmic reticulum: a series of "canals" that carry materials throughout a cell

energy: the ability to do work

energy level: the circular path or orbit around the nucleus associated with individual electrons

energy meter: a meter that measures the total amount of electrical energy used in a building

excited state: the condition of an electron when it is energized sufficiently to jump to a higher orbit

external fertilization: a process in which the male sperm cells are united with the female egg cells outside the female's body

F

fertilizer: a chemical substance added to the soil to increase plant growth

flagellum: a whiplike tail on the outside of the cell membrane that helps the cell move

flame test: an experimental technique using a flame to determine the identity of a metal

flammable: the chemical property of a substance that allows it to burn when exposed to flame and oxygen

follicle: a group of cells, including a reproductive cell, inside the ovary of the female

fossil fuel: coal, oil, natural gas, and gasoline formed from the long-buried remains of organisms

fragmentation: the form of asexual reproduction in which a new organism is formed from a part that breaks off from the parent

free fall: a continuous falling effect of a spacecraft created by the combination of the gravitational pull of Earth and the forward speed of the spacecraft

fuel cell: a cell that produces electricity without combustion, using hydrogen and oxygen

fuse: a piece of material that will melt (fuse) when heated to a high temperature by the current flowing through it

G

galaxy: a huge collection of gas, dust, and hundreds of millions of stars and planets

gas giant: planet with an atmosphere that consists mainly of the low-density gases hydrogen and helium

gene: a unit of the genetic information that determines a specific characteristic of an individual

geosynchronous orbit: the orbit of a satellite around Earth that takes 24 h, allowing it to remain in the same location above Earth's surface

GFCI: *See* **ground fault circuit interrupter**

Global Positioning System: the 24 satellites used for search and rescue operations that travel in 12-h orbits, 20 000 km above Earth's surface

Golgi apparatus: an organelle that stores proteins and packages them for release inside or outside the cell

gravity: the force that pulls objects toward each other

ground: connect to Earth

ground fault circuit interrupter: a special kind of combination outlet socket and circuit breaker that responds to very small changes in current; (GFCI)

ground state: the low-energy state that is the normal orbit of an electron

ground terminal: a third round hole underneath the other two holes in an outlet socket, which is connected to the ground terminal inside the distribution panel

grounding pin: the third pin, containing the grounding wire, in the three-pin plug of an appliance that grounds the appliance to the circuit breaker

group: a column of the periodic table

H

halogens: reactive nonmetals that occur in different states and occupy the seventeenth column of the periodic table

hardness: the physical property of a solid that is the measure of its resistance to being scratched or dented

heavy metal: an element that is shiny, malleable, and conducts electricity, is generally solid at room temperature, and has very high density

hermaphrodite: any organism that creates both male and female sex cells

heterogeneous mixture: a substance in which the different components are identifiable and can be separated by physical means

hormone: any chemical that acts as a messenger between cells

hydrocarbon: a compound containing hydrogen and carbon; found in fossil fuels

I

induced charge separation: a slight shift in position of electrons that produces opposite charges on the two sides of a particle

inner planet: one of four small planets close to the Sun, with a density roughly the same as the density of rock; also known as a terrestrial planet

internal fertilization: a process in which the male sperm is united with the female egg cell inside the female's body

International Space Station (ISS): a space station that involves the co-operation of space agencies from Brazil, Canada, Europe, Japan, Russia, and the United States

input energy: the amount of energy used to make electricity

interphase: the phase of mitosis during which the cell grows and prepares for cell division by duplicating its genetic information

ion: an atom that has become charged by gaining or losing one or more electrons

isotope: any of two or more forms of an element, each with the same number of protons but with different numbers of neutrons

J

joule: the SI unit for measuring energy; symbol: J

K

kilowatt hour: the unit for measuring energy; the number of kilowatts of electrical power used multiplied by the number of hours; symbol: kW·h

L

launcher: a device that carries a payload into space

lead-acid cell: a reusable energy source in which the chemical change is reversed by connecting the cell to a source of electrical energy until the cell is recharged and the electrodes return to their original state

light-year: the distance that light rays travel in one year $(9.46 \times 10^{12}$ km)

live wire: an insulated wire that carries electrical energy into a building

low Earth orbit: an orbit just above Earth's atmosphere

lysosome: a saclike organelle used to break down large compounds

M

main breaker switch: the central location in a building where the two live wires are each connected to the circuit breaker

malignant: the term used to describe a cancerous tumour that spawns cells that can break away and move to other areas of the body

malleable: the physical property that allows the shape of a solid to be changed

mass number: the sum of the protons and neutrons in the nucleus of an atom

matter: anything that takes up space and has mass

meiosis: the process that forms sex cells

Mendeleev's periodic law: a statement summarizing Dmitri Mendeleev's observations that elements arranged in order of increasing atomic mass reflect a pattern in which similar properties occur regularly

menstruation: the process in non-pregnant women during which the endometrium is shed from the uterus through the vagina

metal: a solid that is generally shiny and malleable and a good conductor of heat and electricity

metalloids: elements that possess both metallic and nonmetallic properties

metallurgy: the technology of separating metals from their ores

metaphase: the phase of mitosis in which the chromosomes line up in the middle of the cell

meteor: a bright streak of light across the sky caused by a meteoroid

meteorite: a meteoroid that reaches the ground

meteoroid: a lump of rock or metal trapped by Earth's gravity and pulled down through Earth's atmosphere

microgravity: *See* **free fall**

mineral: a naturally occurring compound, sometimes containing metal combined with oxygen, sulfur, or other elements

mitochondrion: an oval-shaped organelle that provides cells with energy

mitosis: the process by which nuclear material is divided during cell division

mixture: a substance that contains two different pure substances or types of particles

model: a way to represent a thing or process

modern periodic law: a law stating that elements arranged in order of increasing atomic number reflect a pattern in which similar properties occur regularly

molecule: the combination of two or more atoms

mutation: a change in the genetic code

N

nebula: a huge cloud of dust and gases in outer space; the beginning and ending of a star

negative charge: an excess of electrons

negative terminal: the plate in a voltaic cell where electrons collect

neutral wire: a wire that leads from the main electrical supply outside to the inside of a building where it is attached with a special wire to the plumbing system or to a metal stake driven into the ground

neutralize: to discharge or remove all electric charges

neutron: a neutral particle located in the nucleus, with a relative mass of 1

neutron star: an extremely dense star composed of neutrons; results when a star about 10 times the mass of the Sun dies

noble gases: inert gases, found on the far right column of the periodic table, that almost never form chemical compounds with other elements

nonluminous: not making or emitting its own light; reflecting light from other light sources

nonmetal: one of a class of elements that are not good conductors of heat or electricity

nonrenewable energy resources: sources of fuel that cannot be replaced in a reasonable amount of time

nuclear fusion: a process during which substances fuse to form new substances, releasing huge amounts of heat, light, and other forms of energy

nuclear model: Ernest Rutherford's model of an atom describing a dense, positive nucleus around which negative electrons appear to occupy a large amount of space

nucleolus: a spherical structure, within the nucleus of some cells, associated with the production of proteins

nucleus: 1. the central core of an atom, which contains two kinds of particles: the positively charged proton and the uncharged neutron; 2. the main organelle of the cell, which directs the cell's activities

nutrient: a chemical compound necessary for growth

O

observatory: a large building with an open dome through which a telescope provides a view of planets, stars, and other objects in the universe

ohm: the SI unit for electric resistance; symbol: Ω

ohmic resistor: a type of electrical load that does not change electrical resistance with temperature

open circuit: a circuit this is not operating and through which no current is flowing

orbit: 1. a circular path around the nucleus associated with individual electrons; 2. a circular or elliptical path followed by one object as it revolves around a much larger object

orbital period: the period of time required for an orbiting object to complete one revolution of the central object

ore: rock containing a valuable mineral

organelle: a specialized structure inside plant and animal cells

outer planet: a planet in the solar system beyond the four inner planets; one of the four gas giants or Pluto

outer space: everything outside of Earth's atmosphere

output energy: the amount of energy produced

ovary: an organ or structure designed to contain female sex cells; the primary reproductive organ of the female mammal

oviduct: the part of the female reproductive system where fertilization of the egg cell takes place

ovulation: the process during which the ovary wall bursts and the egg cell is released into the oviduct

parallel circuit: an electric circuit in which each electrical load is connected to the energy source by its own separate path or branch circuit

parturition: the process of human birth

payload: a satellite, piloted spacecraft, or cargo launched into space

period: a horizontal row of elements in the periodic table

periodic table: an organized arrangement of elements that explains and predicts physical and chemical properties

periodic trend: a gradual change in the properties of elements across a row in the periodic table

photoelectric cell: a cell that converts light energy directly into electrical energy

photosynthesis: a chemical process during which plants combine carbon dioxide from the air, water, and energy from the Sun to produce sugars and oxygen gas

physical change: a change in the state or form of a substance that does not change the original substance

physical property: a characteristic or description of a substance that can be used to identify it

pistil: the female reproductive structure in the flower of a plant, composed of the stigma, style, and ovary

placenta: an organ in a pregnant woman formed by the blood vessels from the mother and the embryo growing side-by-side

planet: a large piece of matter, generally spherical, that revolves around a star

planetary system: a group of objects that includes at least one planet in orbit around a star

polarized plug: a plug with two different-sized prongs, one narrow connected to the 120-V conductor through a terminal in a lamp or appliance and one wider connected to the neutral wire in a distribution panel

pollen: the male sex cells of a flower

pollination: the process by which pollen is moved from the anther to the egg cells to fertilize those cells

polymer: a material made of long molecules composed of repeating subunits

positive charge: the charge on a proton; a deficiency of electrons

positive terminal: the plate in a voltaic cell where positive charges collect

potential difference: the loss of electric potential produced by electrical resistance as a current flows through a conductor

precipitate: a solid, insoluble material that forms in a liquid solution

primary cell: a disposable energy source in which the chemical reactions use up the materials in the cell as electrons flow from it

products: the substances resulting from a chemical reaction

prophase: the phase of mitosis in which the individual chromosomes become visible

proton: a positively charged particle located in the nucleus, with a relative mass of 1

pulsar: a type of neutron star that emits pulses of very high-energy radio waves

pure substance: a substance that contains only one kind of particle

Q

quasar: an object that looks like a faint star but emits up to 100 times more energy than our entire galaxy; from the expression "quasi-stellar radio source," which means a starlike object that emits radio waves

R

radar: a device that emits bursts of radio waves and picks up their reflections to detect the location of objects and determine how far away they are

radio telescope: a device that receives radio waves from space

radioactive: the state of an unstable element in which nuclei may break apart, ejecting very high-energy particles

radioisotope: an atom with an unstable nucleus

reactants: the substances participating in a chemical reaction

red giant: a star, near the end of its life, that becomes larger and redder as it runs out of hydrogen fuel

red shift: a movement toward the red end of the spectrum

red supergiant: a star with a mass 10 times or more larger than the Sun's near the end of its life, that becomes larger and redder as it runs out of hydrogen fuel

reflecting telescope: an optical device that uses a concave mirror to gather light and make distant objects appear larger

refracting telescope: an optical device that uses lenses to gather and focus light and make distant objects appear larger

regeneration: the ability to regrow a tissue, an organ, or a part of the body

remote sensing: using imaging devices to make observations from a distance, such as from a satellite in low Earth orbit

renewable energy resources: sources of fuel that constantly replenish themselves

reproductive cells: cells that produce sex cells through the process of meiosis

resistance: the ability to impede the flow of electrons in conductors

resistor: an electrical device designed to impede the flow of electrons in conductors

revolution: the movement of one object travelling around another

ribosome: an organelle involved in building proteins essential for cell growth and reproduction

rotation: the spinning of an object around its axis

S

satellite: an object that travels in orbit around another object

secondary cell: a reusable energy source in which one chemical process discharges the cell and another recharges it to its original state

selective breeding: a method of reproduction that results in several generations of offspring all having the same desired characteristics

seminiferous tubules: tiny, twisting tubes inside the testis that produce sperm cells

series circuit: an electric circuit in which the electrical loads are wired to one another in a single path

sexual reproduction: reproduction in which two sex cells unite to form a zygote

solar system: the Sun and all the objects that travel around it, including the nine known planets and the moons of those planets

solubility: the ability of a substance to dissolve in a solvent

solution: a mixture made up of liquids, solids, or gases

somatic cells: cells that reproduce only by normal cell division

space probe: an unpiloted spacecraft that is sent to moons, planets, comets, the Sun, and other parts of the solar system to relay information back to Earth

spacecraft: a vehicle designed to travel in the near vacuum of space

spectroscope: a device that splits light energy into patterns of colour

spectrum: the band of colours produced when light is split into its component frequencies

sperm: the male sex cell

spinoff: an extra benefit from technology originally developed for another purpose

spore: a reproductive body encased within a protective shell

spore formation: the form of asexual reproduction in which the organism undergoes cell division to produce smaller, identical cells, called spores, that are usually housed within the parent cell

stamen: the male reproductive structure in the flower of a plant, composed of the anther and the filament

standard atomic notation: an internationally recognized system used to identify chemical substances

static electricity: a charge on a substance that stays in the same place

star: a large collection of matter that emits huge amounts of energy

star cluster: a group of stars that are relatively close and travel together

structural diagram: a drawing to explain molecules in which atoms are represented by chemical symbols and bonds are shown as straight lines connecting the symbols

subatomic particles: the protons, neutrons, and electrons that make up atoms

Sun: the star around which Earth and eight other planets revolve

Sun-centred solar system: a model reflecting the observation that Earth and other planets travel around the Sun

sunspot: a dark patch on the Sun's photosphere

superconductors: ceramics that conduct electricity with no resistance at low temperatures

supernova: a huge explosion that occurs at the end of a massive star's life

sustainability: the pursuit of economic prosperity, social justice, and protection of the natural environment, while simultaneously securing good health and enhancing well-being for all people and for future generations

synthetic: a material that is invented and produced by people

T

telophase: a phase of mitosis during which the two halves of the cell reorganize to form daughter cells

terrestrial planet: one of four small planets close to the Sun, with a density roughly the same as the density of rock; from the Latin *terra* for "earth"; also known as an inner planet

testis: an organ that produces sperm cells; the primary reproductive organ of the male mammal

thrust: the force that causes an object to move

triangulation: a method of measuring the distance to an object by measuring the angles between the baseline and the object and then drawing a scale diagram to calculate the distance to the object

trimester: one of three stages in human pregnancy

tumour: a mass of cancer cells formed by abnormal rapid cell division

U

umbilical cord: the cordlike structure that connects the embryo with the placenta

universe: everything that exits, including all matter and energy everywhere

uterus: the organ in the female reproductive system where the embryo is nourished as it grows

V

vacuole: an organelle filled with water, sugar, minerals, and proteins

vegetative reproduction: the form of asexual reproduction in which a section of a plant grows to form a new plant

viscosity: the physical property of a liquid that limits its ability to flow

visible spectrum: a small part of the electromagnetic spectrum that can be seen as a pattern of colours

volt: the SI unit used to measure electric potential; symbol V

voltage: electric potential

voltage drop: a measure of the energy each electron gives up as it moves through a circuit; commonly used for potential difference

voltaic cell: the primary wet cell, developed by Alessandro Volta, consisting of two plates made of different metals (electrodes) placed in a liquid (electrolyte) that conducts an electric current

W

watt: the SI unit for electrical power; symbol: W

watt hour: a unit for measuring energy; the number of watts of electrical power used multiplied by the number of hours; symbol: W·h

weight: the force of gravity acting on an object

white dwarf: a small star created by the remaining material when a red giant dies

word equation: a concise way to indicate a chemical reaction between substances

Z

zodiac constellation: a constellation named after an animal; from the Greek word *zodion* for "animal sign"

zygote: a fertilized egg cell; the product of sexual reproduction

Index

Credits

Photo Credits

Table of Contents: p. 4 CORBIS/Graham Neden/Ecoscene; p. 7 Visuals Unlimited/Science VU.

Matter: Unit Opener © Phil Norton; **Unit Overview:** p. 10 top Corel, bottom CORBIS/Paul A. Souders;

Chapter 1: p. 12 top CORBIS/Michael Neveux, bottom Canapress/Fabrice Coffrini; p. 13 top PhotoDisc, inset Jeremy Jones; p. 16 middle Image Network, bottom left J. A. Wilkinson/VALAN PHOTOS, bottom centre PhotoDisc, bottom right Al Handen/Image Bank; p. 17 top to bottom CORBIS/Wolfgang Kahler, G. Kahrner/VALAN PHOTOS, Visuals Unlimited/Bill Beatty, Image Network, Visuals Unlimited/Mark A. Schneider, CORBIS/Henry Diltz, Visuals Unlimited/Mark Skalney; p. 18 top PhotoDisc, middle CORBIS/Kit Kittle; p. 19 top Visuals Unlimited/Glenn Oliver, bottom PhotoDisc; p. 28 CORBIS/Michael Yamashita; p. 29 left p. A. Wilkinson/VALAN PHOTOS, right Kennon Cooke/VALAN PHOTOS; p. 31 top CORBIS/Jules T. Allen, bottom PhotoEdit/D. Young-Wolff; p. 34 top Peggy Rhodes, bottom inset CORBIS/Gianni Dagli Orti; p. 35 top Greg Kinch, bottom CORBIS/Galen Rowell; p. 36 Mazda; p. 38 top CORBIS, middle CORBIS/Doug Wilson, bottom Visuals Unlimited/p. Holden.

Chapter 2: p. 42 top Canapress/Denis Seguin, bottom Scott Camazine/Oxford Scientific Films; p. 43 Canapress; p. 45 CORBIS/Philadelphia Museum of Art; p. 47 Richard Megna/Fundamental Photographs; p. 50 top and middle Visuals Unlimited/Richard Treptow, bottom V. Wilkinson/VALAN PHOTOS; p. 51 Visuals Unlimited/L. S. Stepanowicz; p. 58 Jerry Mason/Science Photo Library; p. 62 top CORBIS/Digital Art, middle and bottom right Dave Starrett, bottom left Jean Bruneau/VALAN PHOTOS; p. 66 top Ariel Skelley/First Light, bottom Guy Lebel/VALAN PHOTOS; p. 67 A. B. Joyce/VALAN PHOTOS; p. 68 Visuals Unlimited/R. E. Muck; p. 69 top courtesy of Glenn Martin, bottom Visuals Unlimited/W. A. Banaszewski; p. 70 top to bottom John Cancalosi/VALAN PHOTOS, Visuals Unlimited/Bob Newman, Visuals Unlimited/A. J. Copely, CORBIS/José Manuel Sanchis Calvete, Visuals Unlimited/Ken Lucas, CORBIS/James L. Amos, Visuals Unlimited/Dane S. Johnson; p. 71 Shashinka Photo; p. 72 top courtesy of INCO Ltd., bottom CORBIS/Paul A. Souders; p. 73 Simon Fraser/Northumbria Circuits/Science Photo Library.

Chapter 3: p. 78 Peggy Rhodes; p. 79 David M. Glass/Image Network; p. 80 Peggy Rhodes; p. 82 CORBIS/Ted Spiegel; p. 83 CORBIS/Bettmann; p. 84 NASA; p. 86 top courtesy of Shree Mulay, bottom Photo Edit/Amy Etra; p. 90 top Scott Camazine/Oxford Scientific Films, bottom CORBIS/Lester V. Bergman; p. 94 Imperial College, Dept. of Physics/Science Photo Library; p. 95 Visuals Unlimited/Rich Treptow; p. 97 top Visuals Unlimited/SIU, middle James King-Holmes/Science Photo Library, bottom Visuals Unlimited/Deneve Feigh Bunde; p. 98 Sun Syndicate;

Chapter 4: p. 102 CORBIS/Lynn Goldsmith; p. 103 Julia Lee; p. 104 top Visuals Unlimited/Rich Treptow, bottom Burndy Library; p. 110 left Visuals Unlimited/Garry Carter, centre CORBIS/Michael Neveux, right CORBIS/Gunter Marx; p. 111 top Richard Megna/Fundamental Photos, bottom Janet Mortimer/VALAN PHOTOS; p. 112 top Canapress/Bill Sandford, bottom CORBIS/Graham Neden/Ecoscene, bottom inset CORBIS/Bettmann; p. 117 V. Whelan/J. A. Wilkinson/VALAN PHOTOS; p. 118 Corel; p. 119 top Science Photo Library/© David Parker/IMI/University of Birmingham High Technology Consortium, bottom Michael Keller/First Light; p. 125 courtesy of Colin Haskin; p. 129 PhotoDisc.

Reproduction: Unit Opener Dr. Jeremy Burgess/Science Photo Library; **Unit Overview:** p. 136 top CORBIS/Lester V. Bergman, bottom Scott Camazine/Oxford Scientific Films; p. 137 top PhotoDisc, bottom CORBIS/Peter Johnson;

Chapter 5: p. 138 top left CORBIS/David Spears, top right CORBIS/Lester V. Bergman, middle left Carolina Biological Supply Co., middle right CORBIS/Lester V. Bergman, bottom left David E. Meyers/Tony Stone Images, bottom right CORBIS/Galen Rowell; p. 139 Alex Bartel/Science Photo Library; p. 140 Sinclair Stammers/Science Photo Library, inset D. P. Wilson/Photo Researchers; p. 141 top right Geoff Thompkinson/Science Photo Library, inset Dr Ann Smith/Science Photo Library/Photo Researchers, bottom left Takeshi Takahara/Photo Researchers, inset Biophoto Assoc./Photo Researchers; p. 148 top CORBIS/Guy Stubbs, bottom CORBIS/Lester V. Bergman; p. 149 CORBIS/Phil Schermeister; p. 150 top CORBIS/David Spers, middle Visuals Unlimited/Science VU; pp. 154–5 Andrew Bajer, University of Oregon; p. 158 top left courtesy of Harriet Simard, bottom right Michael Gadowski/Science Photo Library; p. 160 top left A. B. Dowsett/Science Photo Library, middle CORBIS/Jeffrey L. Rotman, bottom John Cooke/Oxford Scientific Films; p. 161 top CNRI/Science Photo Library, bottom Harry Taylor/Oxford Scientific Films; p. 164 CORBIS/Gary Braasch; p. 165 CORBIS/ Bettmann; p. 167 Visuals Unlimited/Gerald and Buff Corsi; p. 170 CORBIS/Vince Streano.

Al Harvey/The Slide Farm; middle J.A.Wilkinson/VALAN PHOTOS, bottom Jeremy Jones; p. 371 courtesy of Natural Resources Canada; p. 373 table data from *Staying Connected* Winter '98 © Ontario Hydro; p. 377 Dave Starrett; p. 379 courtesy of Robert Williams; p. 380 top J.A.Wilkinson/VALAN PHOTOS, bottom V. Wilkinson/VALAN PHOTOS; p. 381 CORBIS/Richard Hamilton Smith; p. 382 left Visuals Unlimited/John Sohlden, right Corel; p. 384 © Lee Foster/ FPG International L.L.C.; p. 385 Kennon Cooke/VALAN PHOTOS.

Space: **Unit Opener** Pekka Parviainen/Science Photo Library; **Unit overview:** p. 396 top VALAN PHOTOS, bottom Visuals Unlimited/Science VU; p. 397 top Visuals Unlimited/Science VU, bottom NASA;

Chapter 13: p. 398 top and bottom Jack Zehrt/FPG International; p. 398–399 Visuals Unlimited/George East; p. 399 Terence Dickinson; p. 400 Ian Davis-Young/VALAN PHOTOS; p. 405 CORBIS/Roger Ressmeyer; p. 406 T.C. Middleton/Oxford Scientific Films; p. 410 John Cancalosi/VALAN PHOTOS; p. 411 Visuals Unlimited/Cheryl Hogue; p. 417 Corel; p. 422 NASA/JPL/Cal Tech; p. 424 left NASA/Hawaiian Astronomical Society, right Visuals Unlimited/Ishtar Terra; p. 425 left Oxford Scientific Films/NASA, right NASA; p. 426 Visuals Unlimited/Science VU; p. 427 top left NASA/JPL/Cal Tech, top right NASA, bottom Visuals Unlimited/Science VU; p. 428 NASA; p. 429 top courtesy Mary Lou Whitehorne, bottom Terence Dickinson; p. 430 Oxford Scientific Films/NASA; p. 431 top NASA, middle left Visuals Unlimited/Science VU, middle right NASA; p. 432 top NASA/D. Roddy LPI, bottom NASA/JPL; p. 433 NASA/JPL/Cal Tech.

Chapter 14: p. 436 top Visuals Unlimited/George East, bottom Pekka Parviainen/Science Photo Library; p. 437 Visuals Unlimited/Science VU; p. 440 J. Baum/Science Source/Photo Researchers; p. 445 Images BC/Image Network; p. 448 left David Nunuk/First Light, right Visuals Unlimited/Science VU; p. 449 top National Optical Observatory, bottom CORBIS/Stephanie Maze; p. 450 top Stan Osolinski/Oxford Scientific Films, bottom NASA/JPL/Cal Tech; p. 451 top Dave Starrett, bottom Jack Zehrt/FPG International; p. 452 CORBIS/John Wilkinson/Ecoscene; p. 454 Visuals Unlimited/Science VU; p. 456 CORBIS/Roger Ressmeyer; p. 458 Al Harvey/The Slide Farm; p. 459 Dept. of Physics, Imperial College/Science Photo Library; p. 461 top Dennis Di Cicco/Science Photo Library, bottom left NASA, bottom middle Dr. Jeanne Lorre/Science Photo Library, bottom right Pekka Parviainen/Science Photo Library; p. 462 top left Visuals Unlimited/Science VU, top right Karl Henize, bottom NASA; p. 463 Visuals Unlimited/Science VU.

Chapter 15: p. 466 top CORBIS, bottom NASA; p. 467 top CORBIS/Roger Ressmeyer; p. 468 NASA; p. 470 top and bottom, NASA; p. 474 top courtesy Ron Thorpe, bottom painting by Paul Fjeld courtesy Ron Thorpe; p. 479 Brad Whitmore (STSCI)/NASA; p. 480 NASA; p. 482 John Morse/University of Colorado/NASA; p. 483 NASA.

Chapter 16 pp. 486–487 NASA; p. 489 bottom NASA; p. 491 Dave Starrett; p. 493 top NASA; p. 496 courtesy Canada Centre for Remote Sensing; p. 499 Visuals Unlimited/Science VU; p. 500 NASA; p. 501 bottom NASA; p. 503 NASA; p. 506 bottom left CORBIS/Lindsay Hebberd, bottom right courtesy Fyrepel, A Division of Lakeland Industries, Inc.; p. 507 top courtesy Amyn Samji, bottom Elliott Coleshill; p. 508 Dave Starrett; p. 511 NASA; p. 514 top courtesy Learning Technologies Inc., bottom Ames Research Centre/NASA; p. 515 Dave Starrett.

Skills Handbook: opener and p. 523 CORBIS/Phil Schermeister; pp. 523–525, 534, 546, 560, 569 Dave Starrett, p. 547 Comstock.

Career Profile Writing Credits

Alison Armstrong (*Madeline Boscoe, p. 247; Colin Haskin, p. 125; Dr. Shree Mulay, p. 86; Harriet Simand, p. 158; Helder Sousa, p. 31; Dr. Albert Yuzpe, p. 219*); Jim Dawson (*Gary Masse, p. 286; Shelley Harding-Smith, p. 306; Robert Williams, p. 379*); Julia Lee (*Karen Cheung, p. 355; Anya Martin, p. 69; Lap-Chi Tsui, p. 179*); Ron Thorpe (*Paul Fjeld, p. 474; Amyn Sanji, p. 507; Ivan Semeniuk, p. 451; Mary Lou Whitehorne, p. 429*).